INTRODUCTION TO GOOGLE SKETCHUP

INTRODUCTION TO GOOGLE SKETCHUP

SECOND EDITION

AIDAN CHOPRA

with
LAURA TOWN AND CHRIS PICHEREAU

John Wiley & Sons, Inc.
New York • Chichester • Weinheim • Brisbane • Toronto • Singapore

PUBLISHER	Don Fowley
EXECUTIVE EDITOR	John Kane
MARKETING MANAGER	Christopher Ruel
SENIOR EDITORIAL ASSISTANT	Tiara Kelly
SENIOR PRODUCTION MANAGER	Janis Soo
ASSOCIATE PRODUCTION MANAGER	Joel Balbin
CREATIVE DIRECTOR	Harry Nolan
COVER DESIGNER	Ngieng Seng Ping

Cover Image: Proposed student housing facility courtesy of Niles Bolton Associates, Architects, and The Clark Construction Group. Modeling by Jeremy Fretts and Vicki Kuan. This image was exported directly from Google Sketchup. Foreground trees are 2D "face-me" trees created from original images. Balcony railings and canopy struts modeled using "follow-me". Roof structure composed of groups and components. Windows were created using the "Windowizer" plug-in from Rick Wilson. In this image, style settings are set to "display edges," with no profiles, and shadows are turned on.

This book was set in Times New Roman by Aptara, and printed and bound by Courier Westford. The cover was printed by Courier Westford.

To order books or for customer service, please call 1-800-CALL WILEY (225-5945).

Library of Congress Cataloging-in-Publication Data

Chopra, Aidan.
Introduction to Google Sketchup / Aidan Chopra with Laura Town and Chris Pichereau. — 2nd ed.
p. cm.
Includes index.
ISBN **978-1-118-07782-5** (pbk.)
1. Computer graphics. 2. SketchUp. 3. Three-dimensional display systems. 4. Engineering graphics. I. Town, Laura. II. Pichereau, Chris. III. Title. IV. Title: Introduction to Google Sketch Up.
T385.C52175 2012
006.6'93—dc23

2011051142

Printed in the United States of America

10 9 8 7 6 5 4 3 2 1

PREFACE

Today's students have different goals, different life experiences, and different academic backgrounds, but they are all on the same path to success in the real world. This diversity, coupled with the reality that these learners often have jobs, families, and other commitments, requires a flexibility that our nation's higher education system is addressing. Distance learning, shorter course terms, new disciplines, evening courses, and certification programs are some of the approaches that colleges employ to reach as many students as possible and to help them clarify and achieve their goals.

The second edition of *Introduction to Google SketchUp,* offers specially designed suite of services and content, which helps you address this diversity and the need for flexibility. *Introduction to Google SketchUp, Second Edition's* content puts a focus on the fundamentals to help students grasp the subject, bringing them all to the same basic understanding. Content from *Introduction to Google SketchUp, Second Edition,* has an emphasis on teaching job-related skills and practical applications of concepts with clear and professional language. The core competencies and skills help students succeed in the classroom and beyond, whether in another course or in a professional setting. A variety of built-in learning resources allow the students to practice what they need to perform and help instructors gauge students' understanding of the content. These resources enable students to think critically about their new knowledge and apply their skills in any situation.

Our goal with *Introduction to Google SketchUp, Second Edition,* is to celebrate the many students in your courses, respect their needs, and help you guide them on their way.

LEARNING SYSTEMS

To meet the needs of working college students, the second edition of *Introduction to Google SketchUp* uses a learning system based on Bloom's Taxonomy. Key topics are presented in easy-to-follow chapters. The text then prompts analysis, evaluation, and creation with a variety of learning aids and assessment tools. Students move efficiently from reviewing what they have learned, to acquiring new information and skills, to applying their new knowledge and skills to real-life scenarios.

Using this learning system, students not only achieve academic mastery of *concepts* in the Google SketchUp software, but they also master real-world *skills* related to that content. The learning system also helps students become independent learners, giving them a distinct advantage in the field, whether they are just starting out or seeking to advance in their careers.

ORGANIZATION, DEPTH, AND BREADTH OF THE TEXT

Modular Format

Research on college students shows that they access information from textbooks in a non-linear way. Instructors also often wish to reorder textbook content to suit the needs of a particular class. Therefore, although *Introduction to Google SketchUp, Second Edition* proceeds logically from the basics to increasingly more challenging material, chapters are further organized into sections that are self-contained for maximum teaching and learning flexibility.

Numeric System of Headings

Introduction to Google SketchUp, Second Edition, uses a numeric system for headings (for example., 2.3.4 identifies the fourth subsection of Section 3 of Chapter 2). With this system, students and teachers can quickly and easily pinpoint topics in the table of contents and the text, keeping class time and study sessions focused.

Core Content

The topics in *Introduction to Google SketchUp, Second Edition,* are organized into 16 chapters with an online bonus chapter.

Part I: Introduction to Google SketchUp

Chapter 1, "Meeting Google SketchUp," provides a basic introduction to the SketchUp software. Students evaluate the program's limitations and capabilities, as well as learn how SketchUp compares to other available 3D modeling software. The last part of the chapter provides a quick tour of the program, which helps prepare users to create their own models.

Chapter 2, "Establishing the Modeling Mindset," explores the basic concepts related to modeling using Google SketchUp. Edges and faces, which are the central components of all SketchUp models, are explained. The chapter then examines the major differences between modeling in 2D and 3D. Finally, it closes with a look at some of the tools within SketchUp that allow users to carry out essential tasks such as navigating around a model, drawing lines, selecting objects, and working with accurate measurements.

Chapter 3, "Building Simple Models," walks students through the entire process of creating a basic 3D model—in this case, one of a doghouse. Here, students not only learn how to create and view their model, but they also discover how to alter that model by changing its color, texture, and style and by adding shadows. Students are guided through the workflow of creating and sharing a model.

Part II: Creating Models in Google SketchUp

In Chapter 4, "Modeling Buildings," readers take these fundamentals even further by discovering how to draft a floorplan of a simple, rectilinear building and convert this 2D plan to a 3D model. The chapter then explains how to add elements such as stairs, doors, windows, and a roof to the resulting model.

Chapter 5, "Keeping Your Model's Appearance," is about creating and using SketchUp components. The chapter starts by addressing groups which are a lot like small components. After that, component features like finding them, managing them, making your own and finally Dynamic Components are discussed.

Chapter 6, "Creating Everyday Objects," explores SketchUp's capabilities for creating forms other than buildings. Learn how students, interior designers and even archeologists use SketchUp tools and different techniques for modeling terrain, characters, and other objects. Readers also learn how to create symmetrical models and extrude various 2D shapes into 3D objects.

Chapter 7, "Keeping Your Model Organized," describes SketchUp's two main tools that it provides for organizing your models: outliner and layers. The chapter ends with a detailed example of how these tools can be used together to make modeling easier through the modeling of a house.

Chapter 8, "Modeling with Photographs," describes several methods for incorporating digital photos into SketchUp models. Specifically, the chapter explains how to paint the faces in a model with photographs, how to model on top of photo-textured faces, and how to build a model from scratch using a tool called Photo Match.

Part III: Designing and Viewing Your Model in SketchUp

Chapter 9, "Changing Your Model's Appearance," provides a complete rundown of styles, which are groups of settings that allow users to easily adjust the appearance of their faces and edges. Readers learn how to prepare a model that incorporates SketchUp's built-in styles, as well as how to make changes to various elements of these pre-made styles. Readers will learn how to create new styles, save styles, construct styles libraries, and share styles with other SketchUp users. This chapter also provides a brief description of how the light and shadow controls work. It then moves on to show how users can make their shadows more realistic, and it closes with a look at how to animate shadows to see how they change over time.

Chapter 10, "Presenting Your Model Inside SketchUp," explains three methods for displaying a model without ever leaving SketchUp. First, it shows how to "walk" around and through models of 3D buildings, just like in a video game. Next, it describes how to create animated slide shows by setting up scenes with different camera views, times of day, and even visual styles. Last, it explores how to show what's inside a model by cutting sections through it without taking it apart.

Part IV: Sharing Your SketchUp Design

Chapter 11, "Working with Google Earth and the 3D Warehouse," focuses on making SketchUp models that anyone can see on Google Earth. Students learn how to navigate in Google Earth and how to build a model in SketchUp for Google Earth. Readers also discover how they can contribute to the 3D Warehouse, a large online repository of free 3D models that anyone can add to or borrow from.

Chapter 12, "Printing Your Work," explains how to print views of a SketchUp model; methods for printing using both the Windows and Mac versions of the software are also described. The final portion of the chapter is devoted to the topic of scaled printing, which is somewhat difficult in SketchUp although it is still much easier than drawing things by hand.

Chapter 13, "Exporting Images and Animations," focuses on the export file formats that are common to both the Windows and Mac versions of Google SketchUp. Various 2D formats, such as TIFS, JPEGS, and PNGs, are explored, as is the process of exporting animations as movie files that anyone can open and view.

Chapter 14, "Exporting to CAD," Illustration, and Other Modeling Software, applies to features that are only available in the Pro (for purchase) version of SketchUp. The first half of the chapter shows to use SketchUp Pro to generate 2D files for CAD and illustration software. The second half describes how to export a model to a number of different 3D modeling programs.

Chapter 15, "Creating Presentation Documents with LayOut," also pertains to a feature that is only available in SketchUp Pro: a separate piece of software called Google SketchUp LayOut. This is a program that lets users create documents for presenting 3D models both on paper and on-screen. In this chapter, readers learn about the different tasks LayOut can accomplish, how to navigate the LayOut user interface, and how to create a simple presentation drawing set from a SketchUp model.

Finally, Chapter 16, "Troubleshooting and Using Additional Resources," closes out the book by describing some actions that may be helpful when SketchUp is slow or crashes, as well as when faces, colors, and edges are not working the way they should. The chapter also discusses a number of plugins that can enhance the SketchUp experience, and it concludes with a list of resources that can help users take their SketchUp skills to the next level.

The Online Bonus Chapter, "Building Your Own Dynamic Components," found in SketchUp Pro, covers authoring your own Dynamic Components that scale intelligently when you use the Scale tool to resize them, automatically add or subtract elements as you scale them, and animate when you click them with the Interact tool.

This second edition has been updated and modified in response to user suggestions:

All techniques and features have been updated to reflect the changes made in Google SketchUp 8.

- Chapter 2 now includes a section on mastering the rotate tool.
- Chapter 6 has been revised to include sections on creating new and editing existing terrain. This chapter also explains solids and demonstrates the proper use of solid tools to students.
- Chapter 8 now includes techniques to optimize photo textures, demonstrates how to wrap an image around a cylinder, and how to match more than one photo.
- Chapter 10 has a brief section on using scenes to manage your model.
- Chapter 11 was updated to include information on organizing your collections in the Google 3D warehouse and controlling access to your online models.
- Chapter 15 now provides an in-depth look at LayOut from working with dimensions, managing styles, and working with templates and scrapbooks.

PRE-READING LEARNING AIDS

Each chapter in the second edition of *Introduction to Google SketchUp* features a number of new learning and study aids, described in the following sections, to activate students' prior knowledge of the topics and orient them to the material.

Do You Already Know?

This bulleted list focuses on *subject matter* that will be taught. It tells students what they will be learning in this chapter and why it is significant for their careers. It also helps students understand why the chapter is important and how it relates to other chapters in the text.

What You Will Find Out and What You Will Be Able To Do

This bulleted list emphasizes *capabilities* and *skills* students will learn as a result of reading the chapter and notes the sections in which they will be found. It prepares students to synthesize and evaluate the chapter material and relate it to the real world.

WITHIN-TEXT LEARNING AIDS

The following learning aids are designed to encourage analysis and synthesis of the material, support the learning process, and ensure success during the evaluation phase.

Introduction

This section orients the student by introducing the chapter and explaining its practical value and relevance to the book as a whole. Short summaries of chapter sections preview the topics to follow.

In the Real World

These boxes tie section content to real-world organizations, scenarios, and applications. Engaging stories of professionals and institutions—challenges they faced, successes they had, and their ultimate outcome.

Google SketchUp in Action

These margin boxes point out places in the text where professional applications of a concept are demonstrated. An arrow in the box points to the section of the text and a description of the application is given in the box.

For Example

These margin boxes highlight documents and Web sites from real companies that further help students understand key concepts. The boxes can reference a figure or the Tool that can be found at the end of each chapter.

Career Connection

Case studies of people in the field depicting the skills that helped them succeed in the professional world.

Tips from the Professionals

A list of tips that provide relevant advice and helpful tools for students.

Pathways to . . .

This boxed section provides how-to or step-by-step lists to help students perform specific tasks.

Summary

Each chapter concludes with a summary paragraph that reviews the major concepts in the chapter and links back to the "Do You Already Know" list.

Key Terms and Glossary

To help students develop a professional vocabulary, key terms are bolded when they first appear in the chapter and are also shown in the margin of page with their definitions. A complete list of key terms with brief definitions appears at the end of each chapter and again in a glossary at the end of the book. Knowledge of key terms is assessed by all assessment tools (see below).

EVALUATION AND ASSESSMENT TOOLS

The evaluation phase of *Introduction to Google SketchUp, Second Edition*'s learning system consists of a variety of within-chapter and end-of-chapter assessment tools that test how well students have learned the material and their ability to apply it in the real world. These tools also encourage students to extend their learning into different scenarios and higher levels of understanding and thinking. The following assessment tools appear in every chapter.

Self-Check

Related to the "Do You Already Know" bullets, and found at the end of each section, this battery of short-answer questions emphasizes student understanding of concepts and mastery of section content. Though the questions may be either discussed in class or studied by students outside of class, students should not go on before they can answer all questions correctly.

Summary Questions

These exercises help students summarize the chapter's main points by asking a series of multiple-choice and true/false questions that emphasize student understanding of concepts and mastery of chapter content. Students should be able to answer all of the Summary Questions correctly before moving on.

Apply: What Would You Do?

These questions drive home key ideas by asking students to synthesize and apply chapter concepts to new, real-life situations and scenarios.

Be a Designer, Architect, and Developer...

Found at the end of each chapter, "Be a . . ." questions are designed to extend students' thinking and are thus ideal for discussion or writing assignments. Using an open-ended format and sometimes based on Web sources, they encourage students to draw conclusions using chapter material applied to real-world situations, which fosters both mastery and independent learning.

Online Resources

Introduction to Google SketchUp, Second Edition, is accompanied by a variety of online elements, including pre-tests, post-tests, and other supplemental materials. These resources are available to instructors and students through the text's Book Companion Website, www.wiley.com/go/chopra/googlesketchup2e.

In addition, Google offers SketchUp users numerous other computer-based tools, including the following:

- **Video tutorials:** Google has created several video tutorials to facilitate learning of SketchUp. These tutorials are available for free and can be accessed by anyone with an Internet connection.
- **Self-paced tutorials:** In addition to the video tutorials, a number of self-paced online tutorials are also available for free through the Google SketchUp website.
- **Online help center:** SketchUp's extensive online help center offers a list of frequently asked questions (FAQs), as well as a knowledge base of technical support issues and solutions.
- **Quick reference card:** Through both the SketchUp website and the program's Help menu, users can access a quick reference card that lists all of SketchUp's toolbars, tools, and modifier keys.
- **Component libraries:** SketchUp features extensive component libraries that allow users to easily add detail to their models by inserting any number

of pre-drawn objects. Specific libraries exist for a variety of fields, including architecture, construction, film and stage design, landscape architecture, mechanical design, and transportation.

Instructor and Student Package

Introduction to Google SketchUp, Second Edition, is also available with the following teaching and learning supplements: All supplements are available online at the text's Book Companion Website, located at www.wiley.com/go/chopra/googlesketchup2e.

Instructor's Resource Guide

The Instructor's Resource Guide provides the following aids and supplements for teaching an introduction to SketchUp course:

- **Sample syllabus:** This syllabus serves as a convenient template that instructors may use for creating their own course syllabi.
- **Text summary aids:** For each chapter, these include a chapter summary, learning objectives, definitions of key terms, and answers to in-text question sets.
- **Teaching suggestions:** For each chapter, these include a chapter summary, learning objectives, definitions of key terms, lecture notes, answers to select text question sets, and at least three suggestions for classroom activities, such as ideas for speakers to invite, videos to show, and other projects.

PowerPoints

Key information is summarized in 10 to 15 PowerPoint slides per chapter. Instructors may use these in class or choose to share them with students for class presentations or to provide additional study support.

ACKNOWLEDGMENTS

Taken together, the content, pedagogy, and assessment elements of *Introduction to Google SketchUp, Second Edition*, offers the career-oriented student exposure to the most important aspects of SketchUp, as well as ways to develop the skills and capabilities that current and future employers seek in the individuals they hire and promote. Instructors will appreciate the book's practical focus, conciseness, and real-world emphasis.

Special thanks are extended to Jorge Paricio Garcia of The Art Institute of Pittsburgh for acting as an academic advisor to the text. His careful review of the manuscript, significant contributions to the content, and assurance that the book reflects the most recent techniques and features of Google SketchUp, were invaluable assets in our development of the manuscript. We also thank Chris Pichereau for all of her valuable contributions to the revision and hard work in preparing the manuscript for production.

We would like to thank the following reviewers for their feedback and suggestions during the text's development. Their advice on how to shape *Introduction to Google SketchUp, Second Edition,* into a solid learning tool that meets both their needs and those of their busy students is deeply appreciated:

- Greg Stier, The Art Institute of Indianapolis
- Michael R. Aehle, St. Louis Community College
- Nancy Bredemeyer, Indian River State College
- Patricia Combes, Brighton, East Los Angeles
- Michael Lorenz, St. Louis Community College—Meramec
- Stan G. Guidera, Architect, Bowling Green State University
- Frances Rampey, Wilson Community College
- Chris Tornow, Chapman University
- James Dozier, University of Minnesota

BRIEF CONTENTS

CONTENTS

CHAPTER 1

MEETING GOOGLE SKETCHUP

A Brief Overview

Do You Already Know?

- What is Google SketchUp and why you would use it?
- The types of modeling does SketchUp do?
- How does SketchUp work with other programs?

WWW

For the answers to these questions, go to **www.wiley.com/go/chopra/googlesketchup2e**

What You Will Find Out	What You Will Be Able To Do
1.1 SketchUp and Google's relationship.	Install SketchUp and assess basic functions.
1.2 How SketchUp is different from other 3D software.	Compare SketchUp with other 3D software.
1.3 The capabilities of SketchUp.	Evaluate and understand SketchUp's capabilities and limitations.
1.4 How to navigate the SketchUp software.	Assess the role of the five main parts of SketchUp in constructing models.

INTRODUCTION

Years ago, software for building three-dimensional (3D) models of things like buildings, cars, and other objects was difficult—people went to school for years to learn it! In addition, 3D modeling software was expensive. It was so expensive that the only people who could use it were professionals and software pirates—in other words, people who used the program without buying it. Then SketchUp entered the marketplace.

After seeing that lots of people wanted and needed to make 3D models, the people who invented SketchUp decided to design a more intuitive program. Instead of making users think about 3D models as complex mathematical constructs (the way computers think about them), the software engineers created an interface that allows users to build models using elements they are already familiar with: lines and shapes. So, do you need to know how to draw to use SketchUp? If you use the latest version of the software, then the answer is no. Traditional drawing is about *translating* what you see onto a flat piece of paper—or going from 3D to 2D, which is hard for most people. In SketchUp, however, you are always in 3D, therefore, no translation is involved—you just *build*, and SketchUp takes care of features like perspective and shading for you.

This first chapter is about putting SketchUp in context. In this chapter, you will learn how to do the following:

- Assess the basics of SketchUp.
- Evaluate its limitations and capabilities.
- Understand why Google offers SketchUp for free.
- Compare SketchUp to other 3D software.

In the last part of the chapter, you'll receive a quick tour of SketchUp, which prepares you to create models with this exciting program.

1.1 SKETCHUP BASICS

Before beginning a more detailed discussion of SketchUp, it's helpful to have some basic background information on the software. For starters, there are two different versions of SketchUp:

- You can get the free Google SketchUp version by downloading it from the Internet. Just type http://sketchup.google.com into your web browser, and read through the first page of the Google SketchUp website. Click the links to download the application to your computer, and then follow the installation instructions.
- SketchUp works with Windows and Mac OS X. Google SketchUp is available for both operating systems, and it looks and operates nearly the same way on both platforms.
- **A professional version is available.** Google also offers a professional version of SketchUp (called Google SketchUp Pro) that can be purchased. It includes a few terrific features not included in the free version that

certain users—such as architects, production designers, and other design professionals—need for exchanging files with other software. SketchUp Pro also includes a whole new sub-application for creating presentation documents that works natively with SketchUp models. It's called LayOut, and it's discussed later in this book. If you think you might need SketchUp Pro, you can download a free trial version from http://sketchup.google.com. Furthermore, if you are a student and you meet certain requirements, you can obtain the professional version of SketchUp at a reduced price at http://sketchup.google.com/industries/edu/students.html.

GOOGLE SKETCHUP IN ACTION

Install SketchUp and assess basic functions.

Many years ago, when photography was first invented, there was suddenly a new way to make pictures of things that didn't involve drawing, engraving, or painting. In this day and age, you can't throw a rock without hitting a photograph of something. *Everything* (it seems) can take pictures, including cellular phones. Consequently, more than a century after its creation, photography remains the primary way that visual information is communicated.

But what comes after photography? Google believes it's 3D, and here's why: You live in 3D. The furniture you buy (or build) is 3D, and so is the route you take to work. Because so many of the decisions you make (e.g., buying a couch or finding your way) involve 3D information, doesn't it make sense to be able to experience that information in 3D?

Software like SketchUp lets you see 3D information on a 2D screen, which is good. However, affordable 3D printers and holography that produce holograms are just over the horizon. All that's left is to build a model of every single thing in the world—and guess who's going to do it? You are! By making SketchUp free and easily accessible for all users, Google is leading the 3D charge. Rather than relying on a small number of extensively trained users to get around to modeling everything in the universe, Google has made SketchUp available to anyone who wants to participate. After all, the idea behind Google is to organize the world's information, not to

SELF-CHECK

1. SketchUp allows you to work first in 2D and then translate your work into 3D. True or false?
2. Although there is a professional version available for purchase, you can download a free version of SketchUp from the Internet. True or false?
3. SketchUp is only available for PC users. True or false?

Apply Your Knowledge You are beginning to create a model of a new deck for your house and want to use SketchUp. Download the software to use on your computer.

recreate it. By giving SketchUp away, the company has created an entirely new kind of information to organize. You'll learn more about Google's methods for storing and organizing this information later in the book, in the discussion of Google Earth and Google 3D Warehouse.

1.2 COMPARING SKETCHUP TO OTHER 3D MODELING PROGRAMS

If you're reading this book, you're most likely interested in two things: building 3D models and doing so using SketchUp. The following sections explain how SketchUp compares to other 3D modeling programs—specifically, how long it takes to learn to use the software, and what kind of models the software produces.

1.2.1 Traveling the SketchUp Learning Curve

When it comes to widely available 3D modeling software, SketchUp is the easiest to learn. This software has been successful for one reason: within a few hours of initially launching the program, most people are able to learn SketchUp well enough to build a simple model. SketchUp doesn't require reading thick manuals or understanding special geometric concepts; instead, it simply requires initiating the program and diving right in. Part of the reason for this is that the software is easy to use; another part is that SketchUp comes with a number of introductory tutorials that assist you in getting up and running as quickly as possible.

So, how long will it take you to discover how SketchUp works? It depends on your background and experience, but generally, you can expect to be able to make a recognizable object in less than four hours. That's not to say you'll be an expert—it just means that SketchUp's learning curve is extremely low. You don't need to know much to get started, and you'll continue to learn new skills years from now.

But is SketchUp *easy*? Lots of people say so, but it's all relative. SketchUp is without a doubt easier than most other modeling programs currently in the marketplace, but 3D modeling can be tricky. Some people catch on right away, and others take longer. One thing is certain though: if you want to build 3D models and you have an afternoon to spare, there is no better tool to start with than SketchUp.

GOOGLE SKETCHUP IN ACTION

Compare SketchUp with other 3D software.

1.2.2 Understanding the Types of Models SketchUp Produces

Three-dimensional modeling software creates two basic types of models: **solid models** and **surface models.** Figure 1-1 illustrates the basic difference between the two.

Solid models
Models that are not hollow but are dense throughout.

Surface models
Models that are hollow.

Figure 1-1

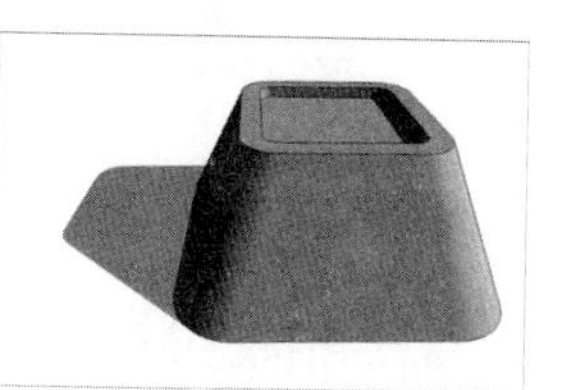

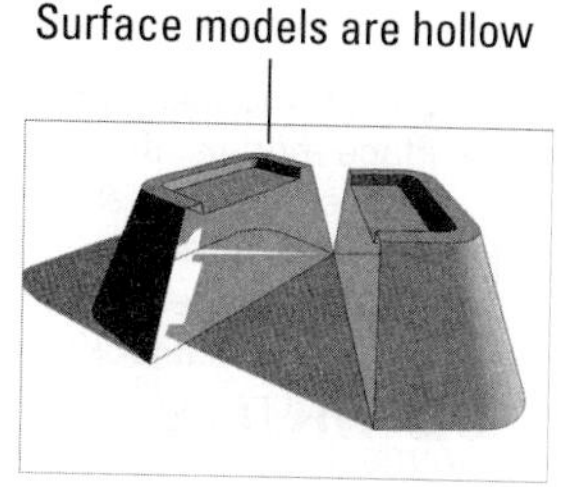

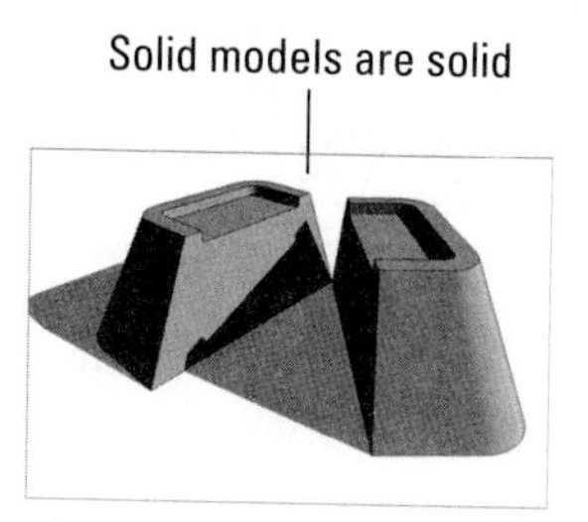

Surface models are hollow, whereas solid models are not.

Faces
In SketchUp, infinitely thin surfaces.

Edges
In SketchUp, straight lines.

Google SketchUp is a "surfaces" modeler. In other words, everything in SketchUp is basically made of the following **Faces** and **Edges.**

That's it. Even things that look thick, such as cinder-block walls, are in fact hollow shells. Making models in SketchUp is a lot like building things out of paper—albeit *extremely* thin paper. Surface modelers like SketchUp are great for quickly making models, because all you really need to concern yourself with is modeling what things *look* like. That's not to say that these modelers are less capable; it's just that they're primarily intended for visualization.

In contrast, using a "solids" modeler is more like working with clay. When you cut a solid model in half, you create new surfaces where you cut; that's because objects are, of course, solid. Programs like SolidWorks, Form•Z, and Inventor create solid models. People who manufacture parts, like mechanical engineers and industrial designers, tend to work with solid models because they can perform precise calculations. For example, being able to calculate the volume of an object means that you can figure out how much it will weigh. Also, special machines can produce real-life prototypes directly from solid-model files. This is helpful for modeling and visualizing how lots of small components will fit together.

An important point to reinforce here is that there is no "best" type of modeling software. It all depends on three things:

- how you like to work,
- what you are modeling, and
- what you plan to do with your model when it is done.

GOOGLE SKETCHUP IN ACTION

Evaluate and understand SketchUp capabilities and limitations.

In SketchUp Pro 8 (the non-free version of the software), there is a set of tools that allows you to manipulate special solid objects in your models. The Solid Tools feature offers a new way to work in SketchUp and is covered in Chapter 6.

Polygonal modelers
Modeling programs that use straight lines and flat surfaces to define everything; within these modelers, even things that look curvy aren't actually curvy.

Curves-based modelers
Modeling programs that use curves to define lines and surfaces.

FOR EXAMPLE

POLYGONAL VS. CURVES-BASED MODELERS

While 3D modeling programs are classified based upon whether they create solid or surface models, they are also categorized based on the type of math they use to produce their models. More specifically, programs can be thought of as either **polygonal modelers** (of which SketchUp is an example) or **curves-based modelers.** Polygonal modelers use straight lines and flat surfaces to define everything; here, even things that *look* curvy aren't actually curvy. Curves-based modelers, in contrast, use true curves to define lines and surfaces. This yields organic, flowing forms that are much more realistic than those produced by polygonal modelers; however, curvy modelers put a lot more strain on the computers that have to run them (as well as on the people who have to figure out how to use them). Ultimately, the choice between the two types of modelers involves a trade-off between simplicity and realism.

SELF-CHECK

1. Explain the difference between a surface model and a solid model.
2. What is a surface model made of?
3. A surface model created in SketchUp is a solid object. True or false?
4. SketchUp is especially designed to help in the production of real-life prototypes. True or false?

Apply Your Knowledge You're beginning to work with SketchUp. Think of three surface models you would want to create in this computer program, according to your own interests and design preferences.

1.3 WHAT YOU SHOULD (AND SHOULDN'T) EXPECT SKETCHUP TO DO

It's wise to be skeptical of tools that claim to be able to do everything. Typically, it's better to rely on specialists, or on tools that are designed to do one thing *really* well. SketchUp is an example of a specialist—one that was created to build 3D models. In fact, here's a list of the various model building tasks that you can do with SketchUp:

- **Start a model in lots of different ways:** With SketchUp, you can begin a model in whatever way makes sense for what you are building:
  - *From scratch:* When you first launch SketchUp, you see a small person standing in the middle of your screen. To create a completely blank screen, delete the person. You can then model anything you want.

- *From a photograph:* You can use SketchUp to build a model based on a photo of the item you want to build. This is not a beginner-level feature, but it's there.
- *With another computer file:* SketchUp can import images and CAD (computer-aided drawings produced with other software) files creating a starting point for what you want to make.
- *In Google Earth:* Google Earth has aerial imagery and 3D terrain data for the entire world. SketchUp 8 makes it easy to grab a geo-location snapshot—any size from the planet—and use it as a site for your model.
- *From Building Maker:* Google has a very specialized tool, called Building Maker, for modeling real-world structures. You can begin modeling using Building Maker and then modify it in SketchUp 8. This capability is particularly useful and saves time if you are modeling existing buildings.

- **Work "loose" or "tight":** One of the best things about SketchUp is that you can model without worrying about exactly *how big* something is. You can make models that are super-sketchy, but if you want, you can also make models that are absolutely precise. In this way, SketchUp is just like paper—the amount of detail you add is entirely up to you.
- **Build something real or make something up:** What you build with SketchUp really isn't the issue. You only work with lines and shapes, and how you arrange them is entirely up to you. SketchUp isn't intended for making buildings any more than it is for creating other things. It's just a tool for drawing in three dimensions.
- **Share your models:** After you've made something you want to show off, you can do a number of things:
  - *Print:* Yes, you can print from SketchUp.
  - *Export images:* If you want to generate an image file of a particular view, you can export one in any of several popular formats.
  - *Export movies:* Animations are a great way to present 3D information, and SketchUp can create them easily.
  - *Export other 3D model formats:* The Pro version of SketchUp allows you to share your model with other pieces of software to create CAD drawings, generate photorealistic renderings, and more.
  - *Upload to the 3D Warehouse:* This is a giant, online repository of SketchUp models that you can add to (and take from) all you want.
  - *Contribute to Google Earth:* Models you create of actual buildings that are efficient, accurate, and *photo-textured* (painted with photographs of the actual building) are welcome on Google Earth's default 3D Buildings layer. You simply submit your work for consideration and if it is accepted, it goes live in a place where millions of people can view it.

So, what *can't* SketchUp do? A few things, actually—but that's okay. SketchUp was designed from the outset to be the friendliest, fastest, and most useful modeler available. Fantastic programs are available that do the things in the following list, and SketchUp can exchange files with most of them:

- **Photorealistic rendering:** Most 3D modelers have their own built-in photo renderers, but creating model views that look like photographs is a rather specialized undertaking. SketchUp has always focused on something called **nonphotorealistic rendering (NPR)** instead. NPR is essentially technology that makes things look hand-drawn. If you want to make realistic views of your models, consider using one of the many third-party renderers that work well with SketchUp (discussed in Chapter 16).
- **Animation:** Although SketchUp can export animations, the movies you can make with this software only involve moving your "camera" around your model. True animation software also lets you move around the things *inside* your model. SketchUp doesn't do this; however, the professional version does allow you to export your models to a number of different programs that offer this more advanced type of animation.

Nonphotorealistic rendering (NPR)
Technology that makes objects look hand-drawn or otherwise not like a photograph. SketchUp offers nonphotorealistic rendering.

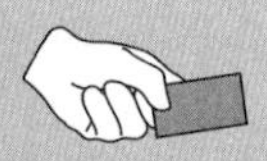

CAREER CONNECTION

Anthony DiFeo is an industrial designer who uses Google SketchUp constantly in his job because it is a program that allows him to prepare quickly presentations that look professional. He likes to combine different viewing styles, such as making a model appear as if it were part of a hand-sketch or a watercolor, without having to make it too finished, because "clients get nervous about time spent and changes to be made." He often starts his models by importing AutoCAD plans and elevations, but he admits to using SketchUp later during the phases of concept and design and even on the final creation of construction documents; he says the details are "easier to explain in 3D." He often combines his models with Google Earth to make presentations that include site images, too; he often models the terrain as well, as he concedes that it is rare that he would get a flat site.

Tips from the Professionals

SketchUp Benefits

1. **Intuitive:** Easy for staff to learn.
2. **Cost Effective:** Because it's free, the only cost is of time to learn the software. High-end products are cost-prohibitive.
3. **Adaptable:** Client changes and modifications are easily put into the models and quickly communicated back to the client.

That's it. When you use SketchUp to draw a bunch of edges and faces in the shape of a staircase, all SketchUp knows is how many edges and faces it has to keep track of, and where they all need to go. There is nothing called a *stair* in SketchUp, just the edges and the faces.

IN THE REAL WORLD

Fredericton, New Brunswick, Canada in 3D Program

"[I] discovered the Google Earth 3D building models and SketchUp utility at the GeoWeb Conference in 2008. City employees were already using Google Earth for work or personal purposes.

"I knew there was a program to upload data to Google and getting data to Google was a big initiative for us. We wanted to find avenues to release available data and information to be public at large and the Cities in 3D Program seemed like a perfect solution."

"We decided to go through the Cities in 3D Program because of time and having Google folks involved helped speed it up. We want our city to be discoverable through Google as well as through our local website. But if there are people going straight to Google, we want them to discover it there too."

—Rob Lunn, GIS Coordinator for the City

SELF-CHECK

1. Every computer program has a specialty and SketchUp has always focused on photorealistic rendering. True or false?
2. With SketchUp you can do which of the following?
   a. Share your models
   b. Export movies
   c. Export images
   d. All of the above
3. Can you enumerate the different ways we can start a project in SketchUp?
4. The recent addition of Dynamic Components has expanded the way the program understands how to treat and modify objects. True or false?
5. It is not possible to create animations in SketchUp file. True or false?

Apply Your Knowledge Investigate on the Internet at least two different photorealistic third-party rendering programs you can use in SketchUp as plugins. Compare the results you might be able to obtain. You can visit the official blog of SketchUp at http://sketchupdate.blogspot.com/ or http://sketchup.google.com/intl/en/download/plugins.html for more information.

Having said that, the previous version of SketchUp introduced an exciting new development called Dynamic Components. These are preprogrammed objects that *know* what they are. A dynamic staircase, for example, is smart enough to know that it should add or subtract steps when you make it larger or smaller. Dynamic Components are a big improvement for SketchUp. Suddenly, there is a class of components in the program that has what is known as *intelligence*. What this means for you is that SketchUp is easier to learn than ever before.

With the exception of Dynamic Components, components in SketchUp do not have any idea what they are supposed to represent. This can be disconcerting to some people. If you want a model of something you are required to make it out of edges and faces. Keep in mind, SketchUp was designed to allow you to model *anything*, not just buildings, so its tools are designed to manipulate geometry. This is good news in the long run because you are not restricted in any way; you can model anything you can imagine.

1.4 TAKING THE TEN-MINUTE SKETCHUP TOUR

Before learning how to use SketchUp, let's take a quick look at where you can find most of the important windows, menus, and related components within the program.

As with most programs you already use, the SketchUp user interface has five main parts. Figure 1-2 shows all of them, in both the Windows and Mac versions of the program.

Figure 1-2

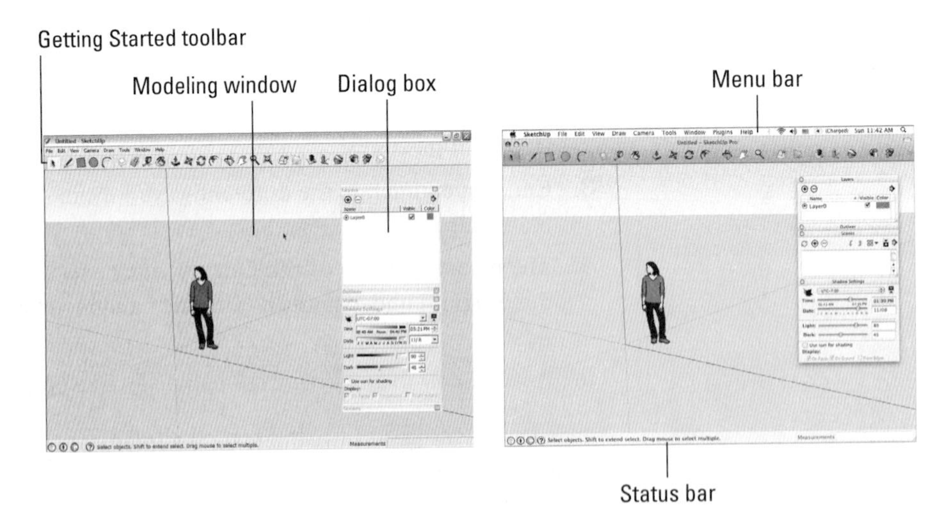

The main parts of SketchUp, in both the Windows (left) and Mac (right) versions.

These five parts, plus an additional feature, are described as follows:

- **Modeling window:** Notice the big area in the middle of the computer screen? That's the modeling window, and it's where you spend 99 percent of your time in SketchUp. You build your model there; in fact, it's sort of a frame into a 3D world inside your computer. What you see in your modeling window is *always* a 3D view of your model, even if you happen to be looking at it from the top or side.
- **Menu bar:** For anyone who is familiar with computers, the menu bar is nothing new. Each menu contains a long list of options, commands, tools, settings, and other features that pertain to just about everything you can do in SketchUp.
- **Toolbars:** These contain buttons that you click to activate tools and commands; they are faster than using the menu bar. SketchUp has a few different toolbars, but only one is visible when you launch the program for the first time: the Getting Started toolbar. (Note that if your modeling window is too narrow to show all the tools on the Getting Started toolbar, you can click the arrow on the right to see the rest of them.)
- **Dialog boxes:** Some programs call these boxes *palettes*, and some call them *inspectors*. In this book we will call them *dialog boxes*, as they are referred to as such in the SketchUp Help weblink at http://sketchup.google.com/support/?hl=en.
- **Status bar:** Consider this your SketchUp dashboard. It contains contextual information you use while you're modeling.
- **Context menus:** Right-clicking with your mouse items in your modeling window usually causes a context menu of commands and options to open. These are always relevant to whatever you happen to right-click (and whatever you're doing at the time), so the contents of each context menu vary.

GOOGLE SKETCHUP IN ACTION

Assess the role of the five main parts of SketchUp in constructing models.

In addition, although the following aren't part of the SketchUp user interface (or the items described in the previous list), they're a critical part of modeling in SketchUp:

- **A mouse with a scroll wheel:** On most mice, you usually find a left button (the one you use the majority of the time), a right button (the one that opens the context menus), and a center scroll wheel that you can both roll back and forth and click down like a button. If your mouse doesn't have a scroll wheel, you should consider getting one that does—it will improve your SketchUp experience more than any single other thing you could buy.
- **A keyboard:** This might amaze you, but some people have tried to use SketchUp without a keyboard; nevertheless, it's just not possible. So many of the things you do regularly (like making copies) involve your keyboard, so you better have one handy if you are planning to use SketchUp.

1.4.1 Using the Menu Bar

SketchUp's menus are generally straightforward; you won't find anything surprising in any of them. All the same, here's what they contain:

- **File:** This menu includes options for creating, opening, and saving SketchUp files. It's also where to go if you want to import or export a file, or to make a printout of your model view.
- **Edit:** The Edit menu has all the commands that affect the bits of your model that are selected.
- **View:** This menu is less obvious. You would assume it would contain all the options for flying around in 3D space, but it doesn't—those are found on the Camera menu. Instead, the View menu includes all the controls you use to affect the *appearance* of your model itself: what's visible, how faces look, and so on. The View menu also contains settings for turning on and off certain elements of SketchUp's user interface.
- **Camera:** This menu contains controls for viewing your model from different angles. In SketchUp, your "camera" is literally your point of view.
- **Draw:** The Draw menu includes tools for drawing edges and faces in your modeling window.
- **Tools:** Most of SketchUp's tools are contained in this menu, except of course, the ones you use for drawing.
- **Window:** If you are ever searching for a dialog box that you want to use, this is the place to look; they're all right here.
- **Plugins:** You can get extra tools for SketchUp—in other words, smaller programs that "plug in" to it and add functionality. Some of these programs show up on this menu after they are installed.
- **Help:** The Help menu contains some incredibly useful resources for understanding SketchUp while you are using the tool. Give particular attention to the video tutorials and the special SketchUp community link.

1.4.2 Checking the Status Bar

Just below the modeling window lies an often overlooked feature of SketchUp: the status bar. This part of the screen contains helpful information, including the following:

- **Context-specific instructions:** Often, you can check the status bar to see what options are available for whatever you are doing in SketchUp. Modifier keys (keyboard strokes that you use in combination with certain tools to perform additional functions), step-by-step instructions, and general information about what you are doing all appear here in one place.

- **The Measurements Box:** Simply put, the Measurements Box is where numbers appear. Chapter 2 goes into more detail about the Measurements Box, but its basic purpose is to allow you to be precise while you are modeling.
- **Status indicator icons:** These three small icons appear in the lower-left corner of your screen. They change to inform you about your model. You click them to find out what they do. The most important one to note at this time in your learning is the one that looks like a question mark. When you click this icon, it opens the Instructor dialog box, which contains information about the tool you are currently using.

1.4.3 Using the Toolbars

The Getting Started toolbar contains a small subset of the tools that you can use in SketchUp. The idea behind this toolbar is that seeing all the tools right away tends to overwhelm new users, so having a limited selection is more helpful for most people.

To gain access to more tools (through toolbars, that is—you can always access everything through the menus), do the following, depending on which operating system you're using:

- **Windows:** Choose View ⇨ Toolbars to see a list of all the toolbars that are available. You may want to start with the Large Tool Set to begin, and then add toolbars as you need them (and as you figure out what they do).
- **Mac:** Choose View ⇨ Tool Palettes ⇨ Large Tool Set. To add even more tools, right-click the Getting Started toolbar (the one right above your modeling window) and choose Customize Toolbar. Now drag whatever tools you want onto your toolbar, and click the Done button.

1.4.4 Understanding the Dialog Boxes

Most graphics programs have multiple small controller boxes that float around your screen, and SketchUp is no exception. After the dialog boxes are open, you can "dock" them together by moving them close to one another (although many users end up with these boxes all over their screen). Dialog boxes in SketchUp contain controls for all kinds of things; here are a few that deserve special attention:

- **Preferences:** While the Model Info dialog box (see the next bullet point) contains settings for the SketchUp file that is open, the Preferences dialog box has controls for how SketchUp behaves *no matter what* file is open. Pay particular attention to the Shortcuts panel, where you can set up keyboard shortcuts for any tool or command in the program. (Note that on a Mac, the Preferences dialog box is on the SketchUp menu, which does not exist in the Windows version of SketchUp.) Some Preference settings changes do not take effect until you open another SketchUp file, so don't be concerned if you can't see a difference right away.

- **Model Info:** This dialog box is the most important of all dialog boxes. It has controls for everything imaginable; you should definitely open it and take your time going through it. Chances are, the next time you can't find a particular setting you're looking for, it's in Model Info. It is important that you choose your settings before you start a new model because in most cases, it will not replace the older settings (such as new dimension style) with the new ones (an old dimension style might not be refreshed; it just might need to be redrawn with the newer style instead).
- **Entity Info:** This box is small, but it shows information about entities—edges, faces, groups, components, and lots of other things—in your model. It is highly recommended that you keep this open, because it helps you see what you have selected. This is especially useful when your model has many parts, some of which may be difficult to see!
- **Instructor:** The Instructor box only does one thing: it shows you (often with animations) how to use whatever tool happens to be activated. While you're discovering SketchUp, you should keep the Instructor dialog box open off to the side. It's a great way to learn about all the things the tool you're "holding" can do.

IN THE REAL WORLD

Denice Murphy from Funktional Design Inc. discovered the potential of Google SketchUp while completing her Bachelor's Degree in Interior Design and fell in love with it. She enjoys working with it and often starts with a plan view that was imported from AutoCAD. She uses SketchUp to quickly visualize her ideas in three dimensions during the construction phase. Her expertise in the program has allowed her to use it extensively to create either exploratory design concepts, to create space planning concepts, or to prepare finished renderings.

She uses her animations in Google SketchUp to provide the client with a much more accurate and realistic view of the project while it's still in the design phase. She states, "As an interior designer, I personally do not have a problem envisioning my designs in a two dimensional plan. But I learned early on in my career that most of my clients do! By using Google SketchUp, my clients are never confused on what their final design in going to look like. It makes my job so much easier, and the clients so much more relaxed during the entire design process."

SELF-CHECK

1. What box shows you how to use an activated tool in SketchUp?
2. What menu allows you to control the options for changing the appearance of your model?
3. Name the five main parts that we can use in SketchUp's user interface.
4. You can get extra plugins for SketchUp. True or false?

Apply Your Knowledge You have installed SketchUp. Go through some of the toolbars such as Large Tool Set, Styles, Views, or Shadows and make them active. Then stack or minimize them in your modeling window or make them be a part of the Getting Started tool bar.

SUMMARY

Section 1.1

- SketchUp is free and you can download the latest version at http://sketchup.google.com/.
- SketchUp operates in both the Windows and Mac OS X environments.
- There is also a professional version of SketchUp available for a fee. Students may purchase it at a reduced rate.

Section 1.2

- Three-dimensional modeling software creates two basic types of models: **solid models** and **surface models**. SketchUp is a surface modeler.
- Everything is made up of **faces** and **edges**.

Section 1.3

- You can build models:
  - from Scratch
  - from a photograph
  - with another computer file
  - in Google Earth
  - from Building Maker
- Your models can be real or imaginary.

Section 1.4

- Most tools and commands in SketchUp can be found in multiple places, Menus, toolbars, and dialog boxes often contain the same things. There is built-in redundancy in the user interface to allow for multiple ways to work with the program.

ASSESS YOUR UNDERSTANDING

SUMMARY QUESTIONS

1. SketchUp is a ________ program.
2. One of the features of SketchUp is that it can produce images that look like they are hand-drawn. True or false?
3. Which one of these two statements is correct? SketchUp is a surface model program. SketchUp is a solid surface model program.
4. SketchUp has animation capabilities, which include the option of creating a walk-through around or inside your model. True or false?
5. Modelers that use straight lines and flat surfaces to define everything are known as:
   a. Curves-based modelers
   b. Polygonal modelers
   c. Solid modelers
   d. Surface modelers
6. Straight lines are also known in SketchUp as:
   a. Surfaces
   b. Faces
   c. Edges
   d. Solids
7. The Instructor dialog box can be can activated to show us what a particular tool can do. True or false?
8. Explain in a brief sentence what the toolbar portion of SketchUp contains. Explain why they are faster than using the menu bars?
9. Dialog boxes in SketchUp can be moved around and docked together. True or false?

APPLY: WHAT WOULD YOU DO?

1. You want to build a deck on the back of your house, but you would like to draw a model of it first in SketchUp so that you can experiment with different lengths and widths and determine what size would work best. Explain the first steps you would take to start the model.
2. Would you advise a mechanical engineering student who needs to make a model of a car part that shows the precise weight of the part to use SketchUp? Please explain the reasons for your answer.

3. What are the five main parts of SketchUp's user interface and what is the purpose of each?
4. What are the main advantages of using SketchUp? Can you describe one specific situation in which you would choose SketchUp instead of other modeling programs?

BE A DESIGNER

Experimenting with SketchUp

Download SketchUp and spend an hour navigating through the software. Write down what you find easy to do and what you find difficult. Keep this list and write down directions for how to complete the difficult tasks as you work through this text.

KEY TERMS

Curves-based modelers
Edges
Faces
Nonphotorealistic rendering (NPR)
Polygonal modelers
Solid models
Surface models

CHAPTER 2

ESTABLISHING THE MODELING MINDSET

Model Basics

Do You Already Know?

- How SketchUp represents three-dimensional space on a two-dimensional screen?
- What inferences are and how to use them?
- How to use the Orbit, Pan, and Zoom tools?
- How to add color and texture to your work?

For the answers to these questions, go to **www.wiley.com/go/chopra/googlesketchup2e**

What You Will Find Out	What You Will Be Able To Do
2.1 How to use edges and faces.	Assess when and how to use edges and faces.
2.2 What inferences are and how to use them.	Work with inferences.
2.3 How to use SketchUp tools for precision drawing.	Use SketchUp tools such as Orbit, Zoom, Pan, and Line.
	Apply precision in your model using the Measurement Box.
	Select parts and move, copy, rotate, and apply color to your model.

INTRODUCTION

SketchUp has many great tools, and it is tempting to start modeling right away. However, that is not the best approach. Think about when you decided to learn how to drive a car. You probably didn't just get behind the wheel, step on the gas, and figure it out as you went along. You should approach SketchUp in much the same way you approached learning to drive; in other words, you should really understand several things about the software before you get started. This chapter is dedicated to introducing those concepts that will make your first few hours with SketchUp a lot more productive and fun.

After reading this chapter, you will be able to do the following:

- Assess when to use edges and faces—the basic components that SketchUp models are made of.
- Evaluate how SketchUp lets you work in 3D (three dimensions) on a 2D (flat) surface (namely, your computer screen).
- Understand how SketchUp represents that depth is everything when it comes to making models. If you've never used 3D modeling software before, pay close attention to the middle part of this chapter.
- Assess SketchUp tools and understand when to use them to accomplish a variety of tasks—things like navigating around your model, drawing lines, selecting objects, and working with accurate measurements.

2.1 ALL ABOUT EDGES AND FACES

In SketchUp, everything is made up of one of two kinds of components: edges and faces. They are the basic building blocks of every model you will ever make.

Collectively, the edges and faces in your model are called **geometry.** Thus, when someone refers to geometry when discussing SketchUp, they are talking about edges and faces. Other modeling programs have other kinds of geometry, but SketchUp keeps it simple. That's a good thing—it means there's less to keep track of.

Geometry
Edges and faces in SketchUp models.

GOOGLE SKETCHUP IN ACTION

Assess when and how to use edges and faces.

The drawing on the left in Figure 2-1 is a basic cube drawn in SketchUp. It's composed of 12 edges and 6 faces. The model on the right is a lot more complex, but the geometry is the same: it's all just edges and faces.

2.1.1 Understanding Edges

Edges are lines. You can use lots of different tools to draw them, erase them, move them, hide them, and even stretch them. Here are some basic things you need to know about SketchUp edges:

- **Edges are always straight.** Not only is everything in your SketchUp model made up of edges, but all of these edges are also perfectly straight.

Figure 2-1

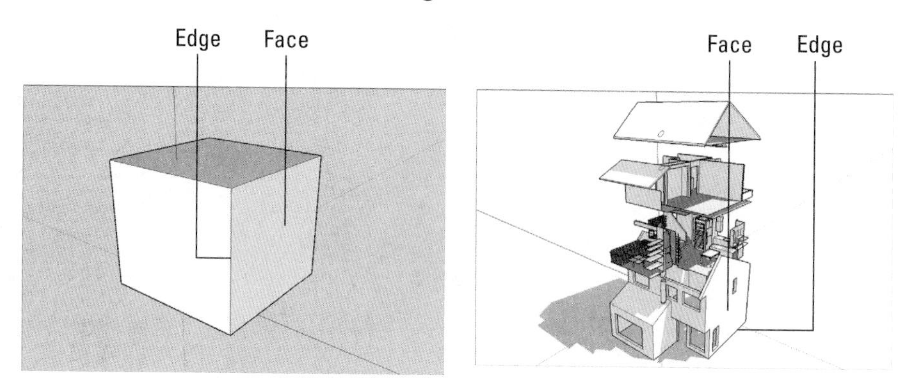

SketchUp models are made from edges and faces.

Even arcs and circles are made of small straight-line segments, as shown in Figure 2-2.

- **Edges don't have a thickness.** This idea can be a bit difficult to grasp. In SketchUp, you never have to worry about how thick the edges in your model are because that's just not how the software works. Depending on how you choose to *display* your model, your edges may look like they have different thicknesses. In reality, however, your edges themselves don't have a built-in thickness. You can read more about making your edges look thick in Chapter 8.
- **Just because you can't see the edges doesn't mean they aren't there.** Edges can be hidden so that you can't see them; doing so is a popular way to make certain forms. Take a look at Figure 2-3. On the left is a model that looks rounded. On the right, the hidden edges are represented as dashed lines. See how even surfaces that appear smoothly curved are made of straight edges?

GOOGLE SKETCHUP IN ACTION

Assess when and how to use edges and faces.

Figure 2-2

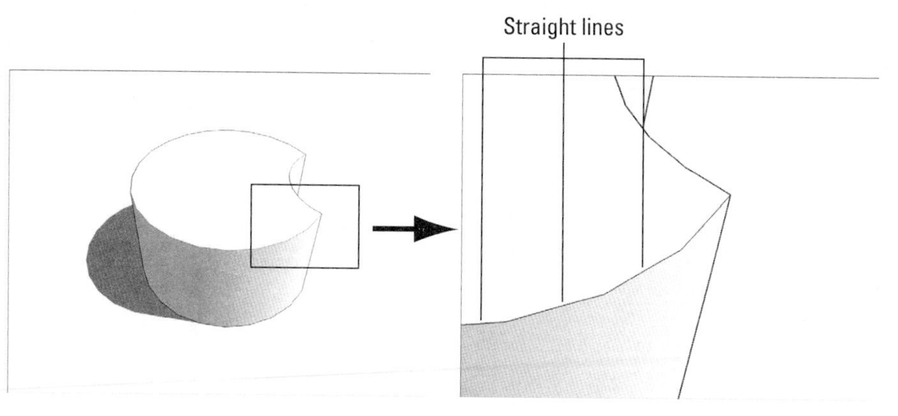

In SketchUp, even curved lines consist of straight edges.

Figure 2-3

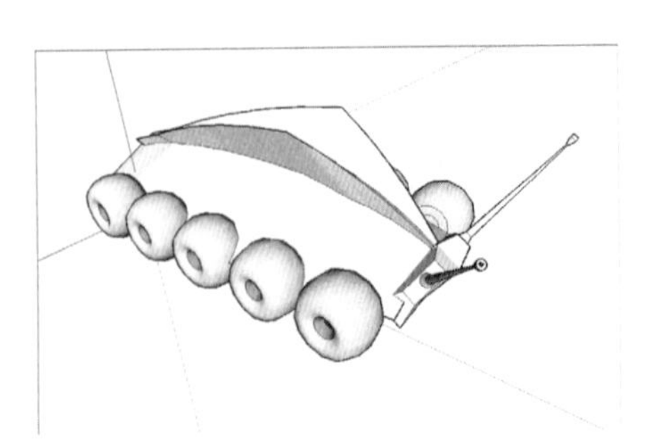

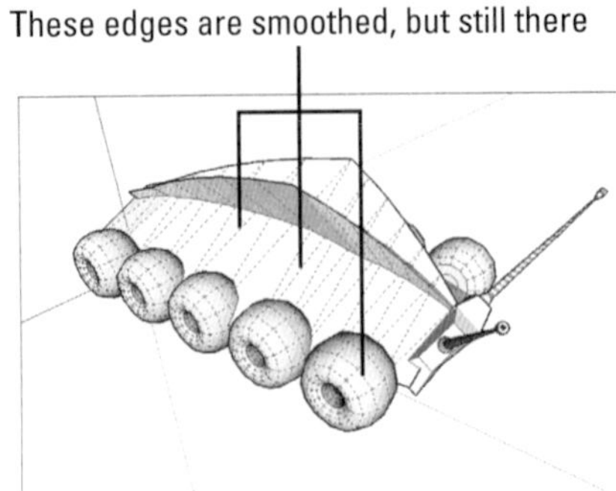

Even organic shapes and curvy forms are made of straight edges.

2.1.2 Understanding Faces

Faces are surfaces. If you think of SketchUp models as being made of toothpicks and paper, faces are basically the paper. Here's what you need to know about them:

Coplanar
On the same plane.

- **You cannot have faces without edges.** To have a face, you need to have at least three **coplanar** edges that form a loop. In other words, a face is defined by the edges that surround it, and those edges all have to be on the same, flat plane. Because you need at least three straight lines to make a closed shape, faces must have at least three sides. There's no limit to the number of sides a SketchUp face can have, though, as long as they are coplanar. Figure 2-4 illustrates what happens when you get rid of an edge that defines one or more faces.
- **Faces are always flat.** In SketchUp, even surfaces that look curved are made up of multiple, flat faces. In the model shown in Figure 2-5, you can see that what looks like an organically shaped surface (on the left) is really made up of many smaller faces (on the right). To make multiple flat faces look like one big, curvy surface, the edges between them are smoothed. You can find out more about smoothing edges in Chapter 6.

Figure 2-4

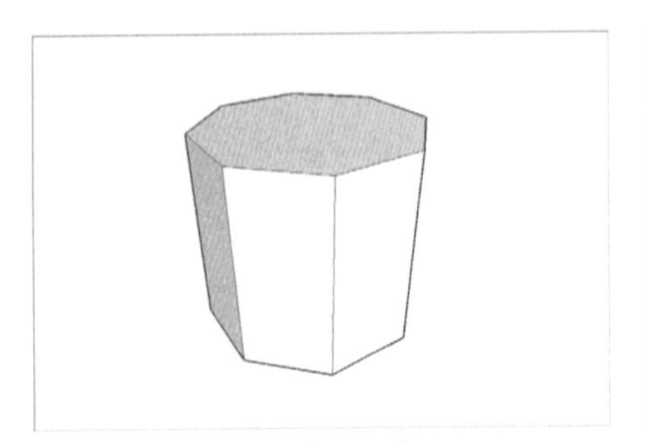

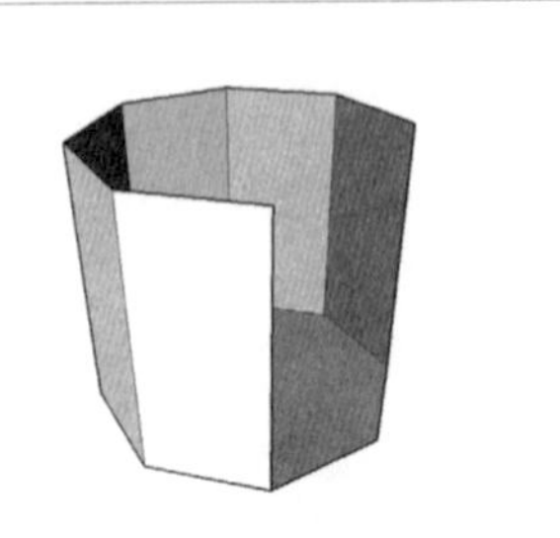

You need at least three edges to make a face.

Figure 2-5

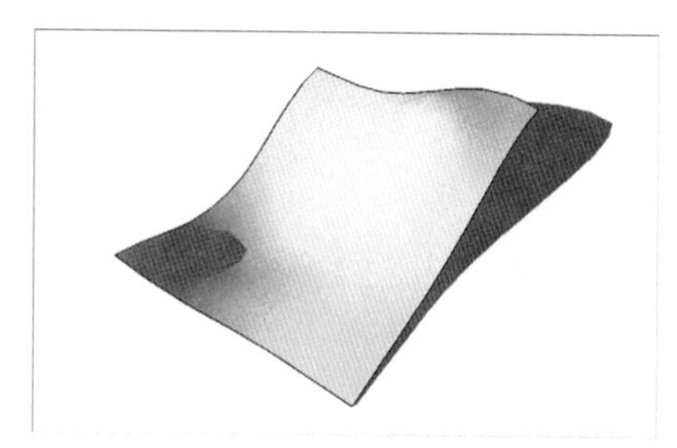

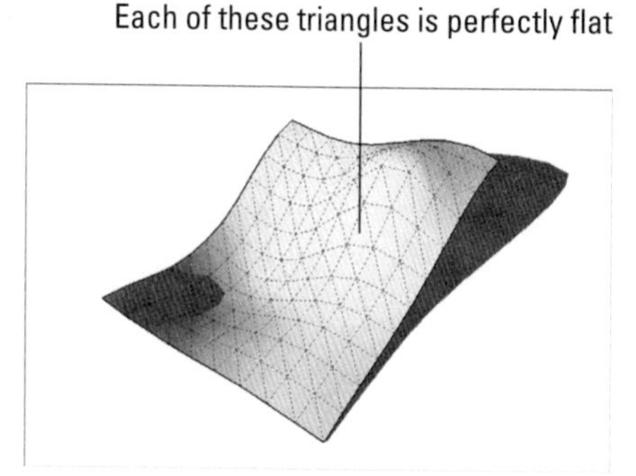

All faces are flat, even the ones that make up larger, curvy surfaces.

- **Just like edges, faces don't have any thickness.** If faces are a lot like pieces of paper, they are *infinitely thin* pieces of paper—in other words, they don't have any thickness. To make a thick surface (for example, a 6-inch-thick wall), you need to use two faces side by side.

2.1.3 Understanding the Relationship Between Edges and Faces

Now that you know that models are made from edges and faces, you have learned most of how SketchUp works. Here is some additional information that will help fill in the gaps:

- **Every time SketchUp can make a face, it will.** There is no such thing as a "Face tool" in this software. SketchUp automatically makes a face every time you finish drawing a closed shape out of three or more coplanar edges. Figure 2-6 shows this in action: as soon as the last edge that is drawn is connected to the first one to close the "loop," SketchUp creates a face.
- **You can't stop SketchUp from creating faces, but you can erase them if you want.** If a face you don't want ends up getting created, just right-click

Figure 2-6

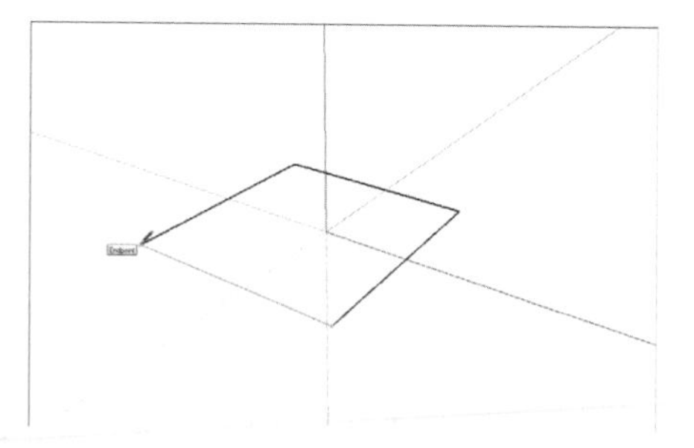

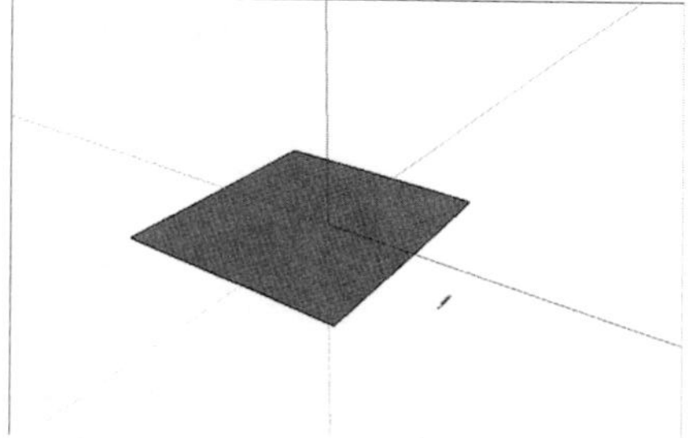

SketchUp automatically makes a face whenever you create a closed loop of coplanar edges.

Figure 2-7

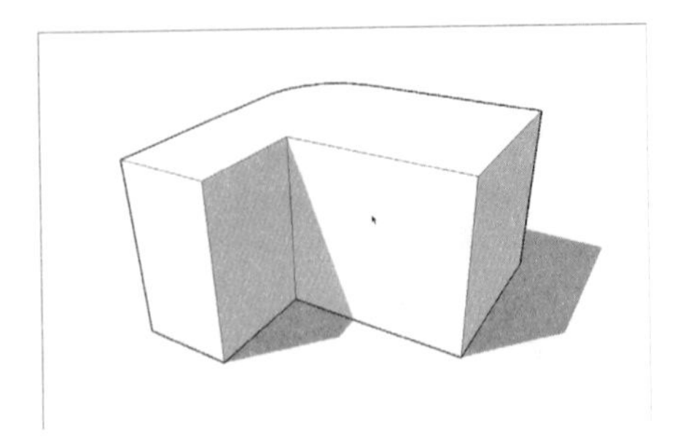

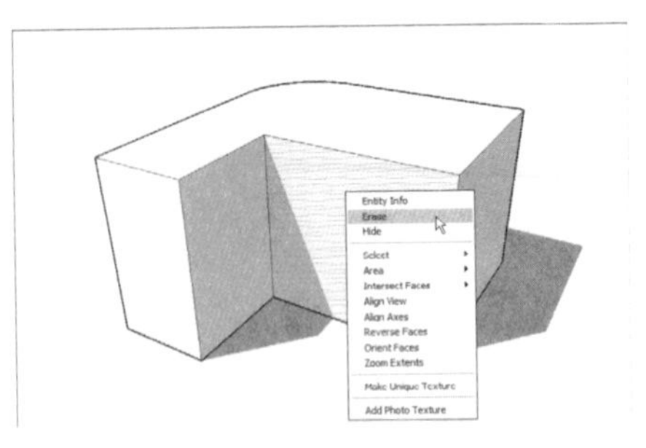

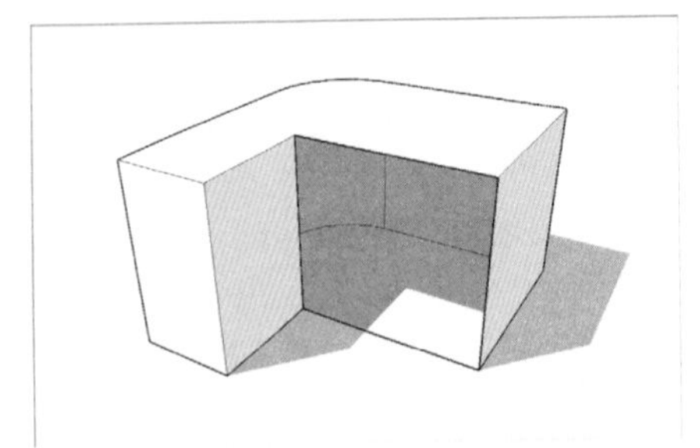

You can delete a face without deleting the edges that define it.

it and choose Erase from the context menu. That face will be deleted, but the edges that defined it will remain (see Figure 2-7).

- **If you delete one of the edges that defines a face, that face will be deleted, too.** For example, when one of the edges in the cube is erased (with the Eraser tool, in this case), *both* of the faces that were defined by that edge disappear. This happens because it's impossible to have a face without having all its edges.
- **Retracing an edge re-creates a missing face.** If you already have a closed loop of coplanar edges but no face (because you erased it, perhaps), you can redraw one of the edges to make a new face. Just use the Line tool to trace over one of the edge segments and a face will reappear (see Figure 2-8).

Figure 2-8

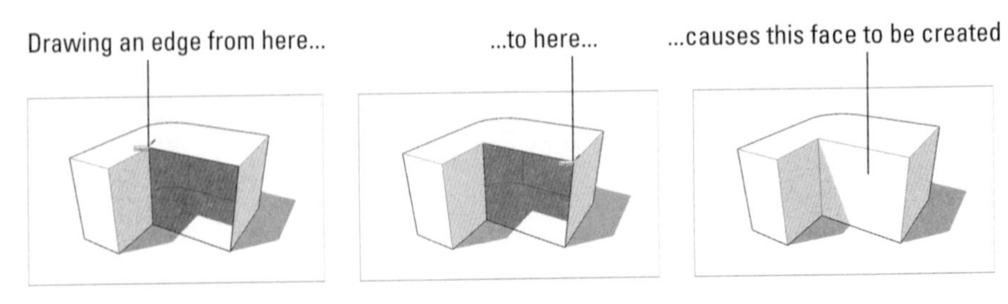

Just retrace any edge on a closed loop to tell SketchUp to create a new face.

Figure 2-9

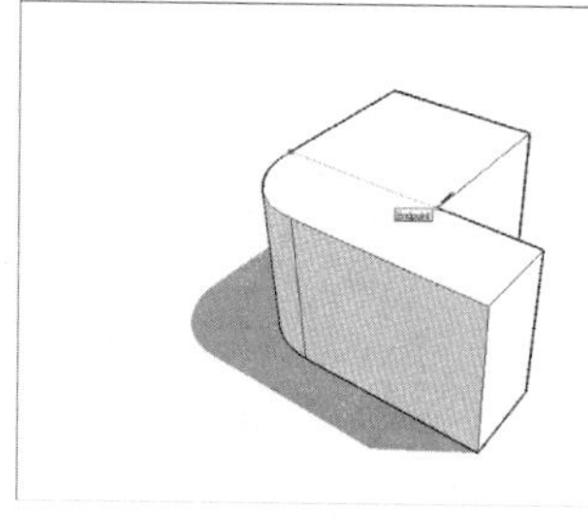
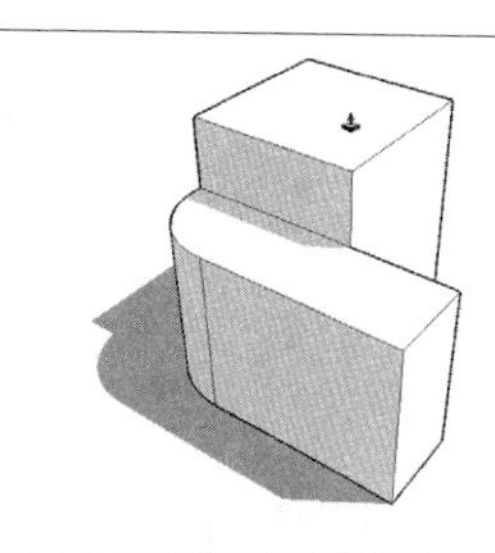

Splitting a face with an edge, and then extruding one of the new faces.

- **Drawing an edge all the way across a face splits the face in two.** When you draw an edge (such as by using the Line tool) from one side of a face to another, you cut that face in two. The same thing happens when you draw a closed loop of edges (like a rectangle) on a face—you end up with two faces, one "inside" the other. In Figure 2-9, the face is split in two with the Line tool, and then one of the faces is extruded a bit with the Push/Pull tool.
- **Drawing an edge that crosses another edge automatically splits both edges where they touch.** You can split simple edges you draw with the Line tool, as well as edges that are created when you draw shapes like rectangles and circles. Most of the time, this auto-slicing is desirable, but if it's not, you can always use groups and components to keep things separate. Reference the first part of Chapter 5 for more information.

SELF-CHECK

1. Individual edges are always straight, even when they appear to form a curve in SketchUp. True or false?
2. If we draw three coplanar lines that form a triangle, SketchUp will automatically create a plane. True or false?
3. Drawing a line from edge to edge across a closed shape in SketchUp will automatically split it. True or false?
4. Faces are thick, whereas edges are thin. True or false?

Apply Your Knowledge Practice building a box that has a long dimension (think of a box made to ship a coat rack, for example). Practice adding a few lines on one of the longer faces and use Push/Pull to change the original shape. Also, practice removing a face or an edge.

2.2 DRAWING IN 3D ON A 2D SCREEN

For computer programmers, letting you draw 3D objects on your screen is a difficult problem. You wouldn't think it would be so challenging; after all, people have been drawing in perspective for a very long time. If someone could figure it out 500 years ago, why should it give your computer any problems?

Drawing axes
Three colored lines visible in the SketchUp modeling window that enable users to work in three-dimensional space.

It's because, human perception of depth on paper is a trick of the eye. And, of course, your computer doesn't have eyes that enable it to interpret depth without thinking about it. You need to give your computer explicit instructions. In SketchUp, this means using **drawing axes** and inferences, as explained in the sections that follow.

IN THE REAL WORLD

Robert Pearce, AIA, explains that his choice of using Google SketchUp was clear, as "it is easy to use and it is reasonably priced even in the professional version." He also states that the free version would allow you to learn the majority of the software, prior to making a financial commitment toward the professional version.

At Robert Pearce Design Studio LLC, they usually prepare high-resolution images that they send to their clients, but occasionally they include in the presentations animations that allow them to walk their clients through the model. He thinks that the option of "animating the scene transition is a plus." Also, he likes the ability to change colors quickly and that Google SketchUp allows him to represent shadows at different times of day or the year, which can be a very important factor to communicate when presenting the designs to his clients.

In his day-to-day practice, he typically imports a CAD floor plan and works toward creating a full exterior 3D model in SketchUp. He later captures 2D CAD roof plans, elevations, and preliminary building sections for further analysis. For his final presentation, he creates photorealistic renditions by using SU Podium plugin (that is a third party plugin that adds functionality to SketchUp). Finally, the image output might be later modified in Adobe Photoshop.

Drawing in Perspective?

Contrary to popular belief, modeling in SketchUp does not involve drawing in perspective and letting the software "figure out" what you mean. There are two reasons for this, which both turn out to be very good things:

- **Computers are not very good at figuring out what you are trying to do.** This has probably happened to you: You're working away at your computer, and the software you are using tries to "help" by guessing what you are doing. Sometimes it works, but most of the time it doesn't, and eventually, it can become quite annoying. Thus, even if SketchUp were *able* to interpret your perspective drawings, you would probably spend more time correcting its mistakes than actually building something.

- **Most people can't draw in perspective anyway.** Unfortunately, creating drawings with proper perspective just isn't one of the things most of us are taught. So even if SketchUp did work by turning your 2D perspective drawings into 3D models (which it most certainly doesn't), the vast majority of those people who "can't draw" wouldn't be able to use it.

TIPS FROM THE PROFESSIONALS

Working with Colored Axes

When you're working with the colored axes, you need to keep three important things in mind:

1. The red, green, and blue drawing axes define 3D space in your model. If you were standing at the spot where all three axes meet (the axis origin), the blue axis would run vertically, passing through your head and feet. The red and green axes define the ground plane in SketchUp; you would be standing on top of them. The axes are all at right angles to one another and extend to infinity from the origin.
2. When you draw, move, or copy something parallel to one of the colored axes, you're working in that "color's direction." Take a look at Color Plate 2. In the first image, we're drawing a line parallel to the red axis, so we would say we're drawing "in the red direction." We know that the line we're drawing is parallel to the red axis because it turns red to let us know. In the second image, we're moving a box parallel to the blue axis, so we're "moving in the blue direction." We know we're parallel to the blue axis because a dotted, blue line appears to tell us so. (Note that the red and green axes define the ground plane in SketchUp, and the blue axis is vertical.)
3. The point of using the red, green, and blue axes is to let SketchUp know what you mean. Remember that one problem with modeling in 3D on a computer is the fact that you are working on a 2D screen. Consider the example shown in Color Plate 3. If we click the cylinder with the Move tool and move the cursor up, how is SketchUp supposed to know whether we mean to move it up in space (above the ground) or back in space? That's where the colored axes come into play. If we want to move it up, we go in the blue direction. If we want to move it back, we follow the green direction (because the green axis happens to run from the front to the back of the screen).

2.2.1 Giving Instructions with the Drawing Axes

Color Plate 1 is a shot of the SketchUp modeling window, right after you create a new file. See the three colored lines that cross in the lower-left corner of the screen? These are the **drawing axes,** and they are the key to understanding how SketchUp works. Simply put, you use SketchUp's drawing axes to figure out where you are (and where you want to go) in 3D space.

When you are working in SketchUp, use the colored drawing axes *all the time.* They're not just handy; they're what make SketchUp work. Having colored axes (instead of axes labeled x, y, and z) lets you draw in 3D space without having to type commands to tell your computer where you want to draw. They make modeling in SketchUp quick, accurate, and relatively intuitive. All you have to do is make sure that you're working in your intended color direction as you model by lining up your objects with the axes and watching the screen tips that tell you what direction you're working in. After your first couple of hours with the software, paying attention to the colors becomes second nature.

2.2.2 Watching for Inferences

If you've spent any time experimenting with SketchUp, you've surely noticed all the colored squares, dotted lines, yellow tags, and other similar objects that appear as you move your cursor around your modeling window. All these elements are collectively referred to as SketchUp's **inference engine.** The sole purpose of this engine is to help you while you're building models. Without inferences, SketchUp wouldn't be very useful.

Inference engine
Objects such as colored squares, dotted lines, yellow tags, and other similar objects that appear as the user moves his or her cursor around the SketchUp modeling window.

Point Inferences

Generally, SketchUp's inferences help you be more precise. **Point inferences** (see Color Plate 4) appear when you move your cursor over specific parts of your model. They look like small colored shapes, and if you pause for a second, they're accompanied by a yellow tag that says what they are. For example, watching for the small green Endpoint inference (which appears whenever your cursor is over one of the ends of an edge) helps you accurately connect an edge you are drawing to the end of another edge in your model.

Point inferences
Small colored shapes that appear when a SketchUp user moves the cursor over specific parts of their model.

Here is a list of SketchUp's various point inferences:

- Endpoint (green)
- Midpoint (cyan or light blue)
- Intersection (black)
- On Edge (red)
- Center (of a circle, green)
- On Face (dark blue)

FOR EXAMPLE

See Figure 2-1 for an example of edges and faces.

Keep in mind that in SketchUp lines are called *edges* and surfaces are called *faces*. Everything in your model is made up of edges and faces.

Linear Inferences

As you've probably already noticed, color plays a big part in SketchUp's **user interface** (the way it looks). Perhaps the best example of this is in the software's **linear inferences,** or the "helper lines" that show up to help you work more precisely. Color Plate 5 is an illustration of all of the linear inferences in action, and here's a description of what they do:

User interface
The visual elements in a software program that the user uses to interact with the software, such as dialog boxes and buttons.

Linear inferences
Helper lines that allow a user to work more accurately.

- **On Axis:** When an edge that you are drawing is parallel to one of the colored drawing axes, the edge turns the same color as that axis.
- **From Point:** This inference is a little harder to describe. When you are moving your cursor, sometimes you will see a colored, dotted line appear. This means that you're "lined up" with the point at the end of the dotted line. Naturally, the color of the From Point inference corresponds to whichever axis you are lined up "on." Sometimes, From Point inferences show up on their own, and sometimes you have to *encourage* them; see the "Encouraging Inferences" section later in this chapter for additional details.
- **Perpendicular:** When you are drawing an edge that's perpendicular to another edge, the edge that you are drawing turns magenta (reddish purple).
- **Parallel:** When the edge you are drawing is parallel to another edge in your model, the edge that you are drawing turns magenta to notify you. You tell SketchUp which edge you are interested in "being parallel to" by encouraging an inference.
- **Tangent at Vertex:** This inference only applies when you are drawing an arc (using the Arc tool) that starts at the endpoint of another arc. When the arc you are drawing is tangent to the other one, the one that you're drawing turns cyan. Tangent, in this case, means that the transition between the two arcs is smooth.

In addition, one of the most important inferences in SketchUp is one that you probably didn't even realize was an inference. It's the fact that, unless you specifically start on an edge or a face in your model, you will always be drawing on the ground plane by default. In other words, if you just start creating content in the middle of nowhere, SketchUp assumes that you want to be drawing on the ground.

2.2.3 Using Inferences to Help You Model

A big part of using SketchUp's inference engine involves locking and encouraging inferences—sometimes even simultaneously. At first, this seems difficult, but with practice, it gets easier.

> **FOR EXAMPLE**
>
> Color Plate 6 shows a situation in which it would be useful to lock a blue On Axis inference while using the Line tool.

Locking Inferences

If you hold down Shift when you see any of the first four types of linear inferences described previously, that inference gets locked, and it stays locked until you release Shift. When you lock an inference, you constrain whatever tool you're using to only work in the direction of the inference you locked.

PATHWAYS TO... DRAWING A VERTICAL LINE

1. Select the pencil tool and click anywhere to start drawing an edge.
2. Move the cursor up until we see the edge we're drawing turn blue. This is the blue On Axis inference that lets us know that we're exactly parallel to the blue drawing axis.
3. Hold down Shift to lock the inference and raw a vertical line. Here, our edge gets thicker to let us know it's locked, and now we can only draw upward in the blue direction (no matter where we move the cursor).
4. Release Shift to unlock the inference.

Encouraging Inferences

Sometimes, an inference you need doesn't appear; when this happens, you have to encourage it. To encourage an inference, just hover your cursor over the part of your model you would like to "infer" from, and then slowly go back to whatever you were doing when you decided you should use an inference. The following example demonstrates how to encourage an inference.

Color Plate 7 shows a model of a cylinder.

PATHWAYS TO... ENCOURAGING INFERENCES

1. Hover (don't click) over the edge of the circle for about two seconds.
2. Move slowly toward the middle of the circle until the Center Point inference appears.
3. Hover (still don't click) over the center point for a couple of seconds.
4. Move the cursor slowly in the direction of where we want to start drawing our edge. A dotted From Point inference should appear.
5. Click to start drawing the edge.

SELF-CHECK

1. What appears when you move your cursor over certain parts of your model?
   a. Linear inferences
   b. Point inferences
   c. Yellow tags
   d. All of the above
2. When you draw, move, or copy something parallel to one of the colored axes, you are not locked on that particular color axis. True or false?
3. To lock an inference, you hold down Shift. True or false?
4. Explain the reason for inferences of different colors.

Apply Your Knowledge Build a simple shed with a hip roof using our different inferences. For the On Axis inference, draw a simple rectangle that is parallel to the axis that we have in the model. To practice with From Point, draw the outline of a fence that you would build from one corner of your structure. You can use the Perpendicular and Parallel inferences to help you draw the position of a diamond-shaped window. Finally, you can practice with the Tangent at Vertex inference to build an arc door.

2.3 WARMING UP FOR SKETCHUP

2.3.1 Getting the Best View of What You Are Doing

Using SketchUp without learning how to orbit, zoom, and pan is like trying to build a ship in a bottle . . . in the dark, with your hands tied behind your back, while using chopsticks. In other words, it's virtually impossible!

GOOGLE SKETCHUP IN ACTION

How to use SketchUp tools such as Orbit, Zoom, Pan, and Line.

At least half of modeling in SketchUp involves using the aforementioned navigation tools, which allow you to change your view so you can see what you are doing. Most people who try to learn SketchUp on their own take too long to understand this. Thus, the following sections will help you avoid this problem.

The Orbit Tool

Orbit
Ability to look at a SketchUp model from every angle; this is accomplished with the Orbit tool.

Hold a glass of water in your hand. Now twist and turn your wrist around in every direction so that the water's all over you and the rest of the room. Stop when the glass is completely empty. This is basically how the **Orbit** tool works.

Figure 2-10

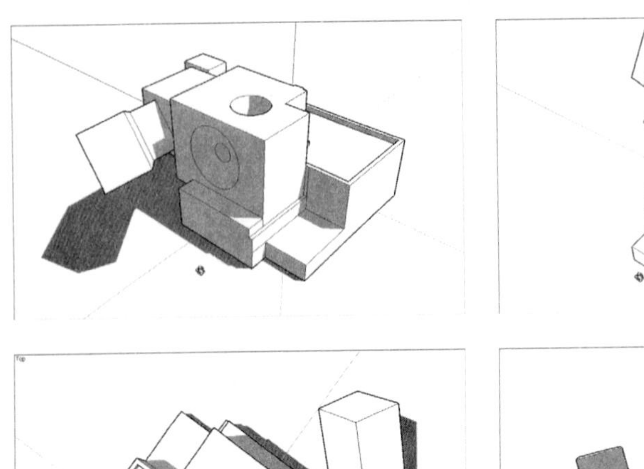
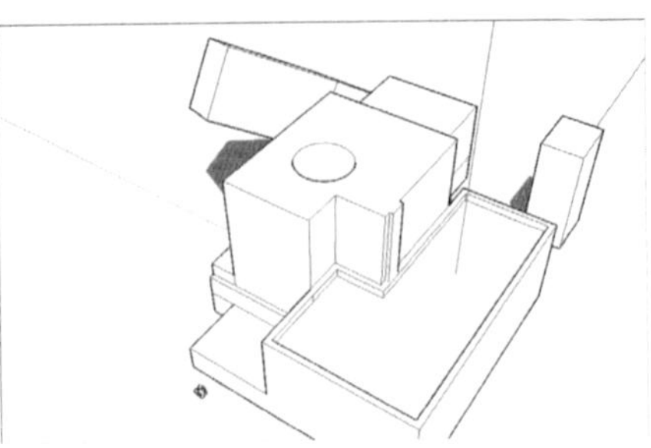
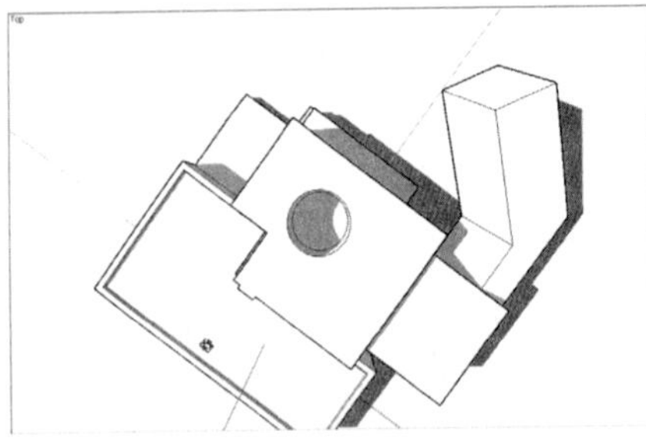
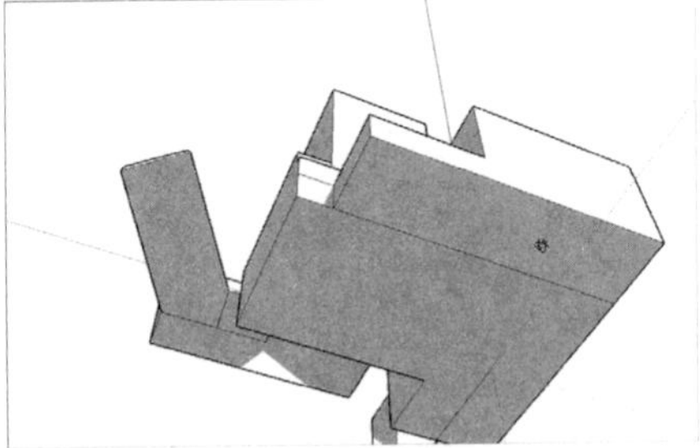

The Orbit tool lets you see your model from any angle.

Just as your wrist helps you twist and turn the glass to see it from every angle, think of using Orbit as a way to "fly around" your work. Figure 2-10 shows the Orbit tool at work.

Here are some things you should know about using Orbit:

- **The Orbit tool is on the Camera menu.** However, by far the least productive way to use Orbit is to choose it from the Camera menu.
- **Orbit is also on the toolbar.** The second-least productive way to activate Orbit is to click its button on the toolbar. This button looks like two blue arrows trying to form a ball.
- **You can orbit using your mouse.** Here is how you should *always* orbit: Click the scroll wheel of your mouse and hold it down. Now move your mouse around. See your model swiveling around? Release the scroll wheel when you're done. Using your mouse to orbit means that you don't have to switch tools every time you want a better view, which saves you a significant amount of time.

The Zoom Tool

Zoom
Getting closer or further away from the model; this is accomplished by using the Zoom tool.

Hold your empty glass at arm's length. Close your eyes, and then bring the glass rushing toward you, stopping right before it hits you in the nose. Now throw the glass across the room, noticing how it shrinks as it gets farther away. That, in a nutshell, describes the **Zoom** tool.

Figure 2-11

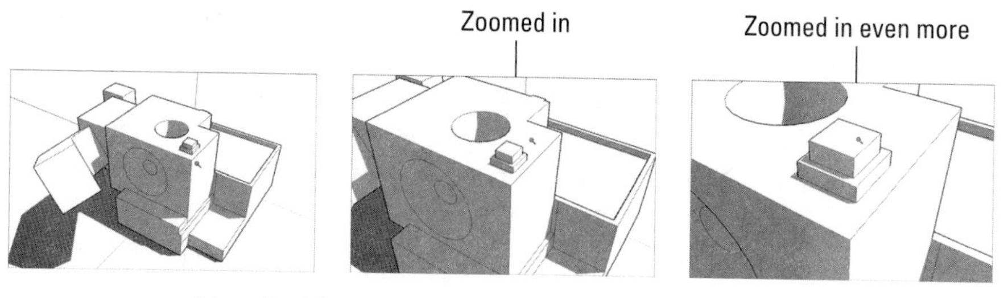

Use the Zoom tool to get closer to your model.

You use Zoom to get closer to (and farther from) your model. If you're working on something small, you zoom in until it fills your modeling window. To see everything at once, zoom out. Figure 2-11 is a demonstration of the Zoom tool at work.

TIPS FROM THE PROFESSIONALS

Getting the Best Use Out of Zoom

There are a few things you need to know about Zoom:

- **Just like Orbit, you can activate the Zoom tool in several ways.** The worst way to activate Zoom is from the Camera menu; the second-worst way is to click the Zoom button in the toolbar. If you use Zoom in either of these two ways, you actually zoom in and out by clicking and dragging up and down on your screen. The best way to zoom is to roll your finger on the scroll wheel of your mouse to zoom in and out. Instead of clicking the scroll wheel to orbit, just roll your scroll wheel back and forth to zoom. Using your mouse to zoom means that you don't have to switch tools—as soon as you stop zooming, you revert to whatever tool you were using before.
- **Use Zoom Extents to see everything.** Technically, Zoom Extents is a separate tool, but it's related closely enough to be mentioned here. If you want your model to fill your modeling window (which is especially useful when you "get lost" with the navigation tools), just choose Camera ⇨ Zoom Extents.

When you use the Zoom tool, SketchUp zooms in on your cursor; just position it over whatever part of your model you want to zoom in on (or zoom out from). If your cursor is not over any of your model's geometry (faces and edges), the Zoom tool won't work very well—you'll end up zooming either too slow or too quick.

Pan
Sliding the model view around the modeling window; this is accomplished by using the Pan tool.

The Pan Tool

Using the **Pan** tool is a lot like washing windows; in other words, you move the paper towel back and forth, but it stays flat and it never gets any closer or farther away from you. The Pan tool is basically for sliding your model view around in your modeling window. To see something that's to the right, you use Pan to slide your model to the left. It's as simple as that.

There are three things you should know about Pan:

- **Pan is on the Camera menu:** Once again, however, that's not where you should go to activate it.
- **Pan is also on the toolbar:** Again, you *could* access the Pan tool by clicking its button on the toolbar (it looks like a severed hand), but there's a better way…
- **Hold down your mouse's scroll wheel button and press Shift.** When you do both at the same time—basically, Orbit+Shift—your cursor temporarily turns into the Pan tool. When it does so, simply move your mouse to pan.

GOOGLE SKETCHUP IN ACTION

Use the Orbit, Pan, and Zoom tools and apply color and texture

2.3.2 Drawing Edges with Ease

Being able to use the Line tool without having to think too much about it is *the* secret to being able to model anything you want in SketchUp. You use the Line tool to draw individual edges, and because SketchUp models are really just fancy collections of edges (albeit carefully arranged), anything you can make in SketchUp, you can make with the Line tool.

Earlier in this chapter was discussed that all SketchUp models are made up of edges and faces. Anytime you have three or more edges that are *connected* and *on the same plane,* SketchUp creates a face. If you erase one of the edges that defines (or borders) a face, the face disappears too. Reference section 2.1.3 Understanding the Relationship Between Edges and Faces for more information.

SketchUp lets you draw lines in two ways: You can either use the click-drag-release method or the click-move-click one. They both work, of course, however, you should train yourself to do the latter. You will have more control with this method, and your hand won't get as tired. When you draw edges by clicking and *dragging* your mouse (click-drag-release), you're a lot more likely to "drop" your line accidentally. Because the Line tool only draws straight lines, think about using it less like a pencil (even though it looks like one) and more like a spool of sticky thread.

Finally, note that the Eraser tool is specifically designed for erasing edges; use it by clicking the edges you don't like to delete them. You can also drag over edges with the Eraser, but that's a little harder to get used to.

PATHWAYS TO... DRAWING EDGES

1. Select the Line tool (some people call it the Pencil tool).
2. Click where you want your line to begin.
3. Move your cursor to the desired endpoint for your line, and click again to end. Figure 2-12 demonstrates the basic idea. When you draw a line segment with the Line tool, notice how SketchUp automatically tries to draw another line? This is called **rubber banding**—the Line tool lets you "continue" to draw edge segments, automatically starting each new one at the end of the previous one you drew.
4. When you want the Line tool to stop drawing lines, press Esc to "snip" the line at the last spot you clicked.

Rubber banding
Drawing edge segments, automatically starting each new one at the end of the previous one.

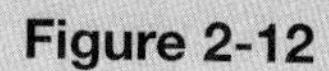

Figure 2-12

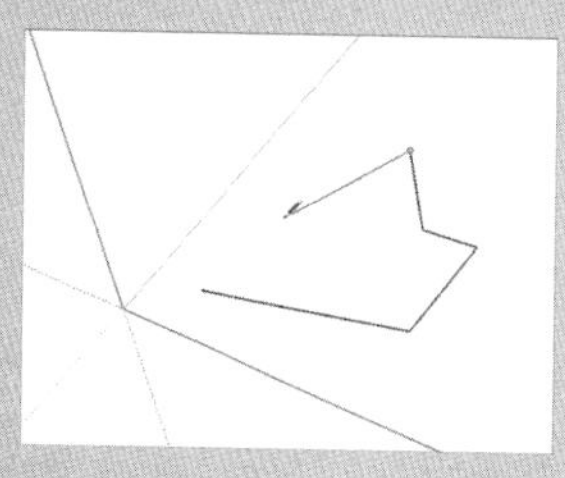

Use the Line tool to draw edges.

Turning Off "Rubber-Banding" Lines

Depending on what you're making and how you work, you might want to turn the Line tool's rubber-banding behavior off. To do so, follow these steps:

1. Choose Window ⇨ Preferences (SketchUp ⇨ Preferences on a Mac).
2. Choose the Drawing panel from the list on the left in the Preferences dialog box.
3. Deselect the Continue Line Drawing check box.

2.3.3 Injecting Accuracy into Your Model

GOOGLE SKETCHUP IN ACTION
Apply precision in your model using the Measurement Box.

It's all well and fine to make a model, but most of the time you need to make sure that your model is accurate. Without a certain level of accuracy, a model is not as useful for figuring things out. The key to accuracy in SketchUp is the small text box that lives in the lower-right corner of your SketchUp window (and is pointed out in Figure 2-13). This text box is

Figure 2-13

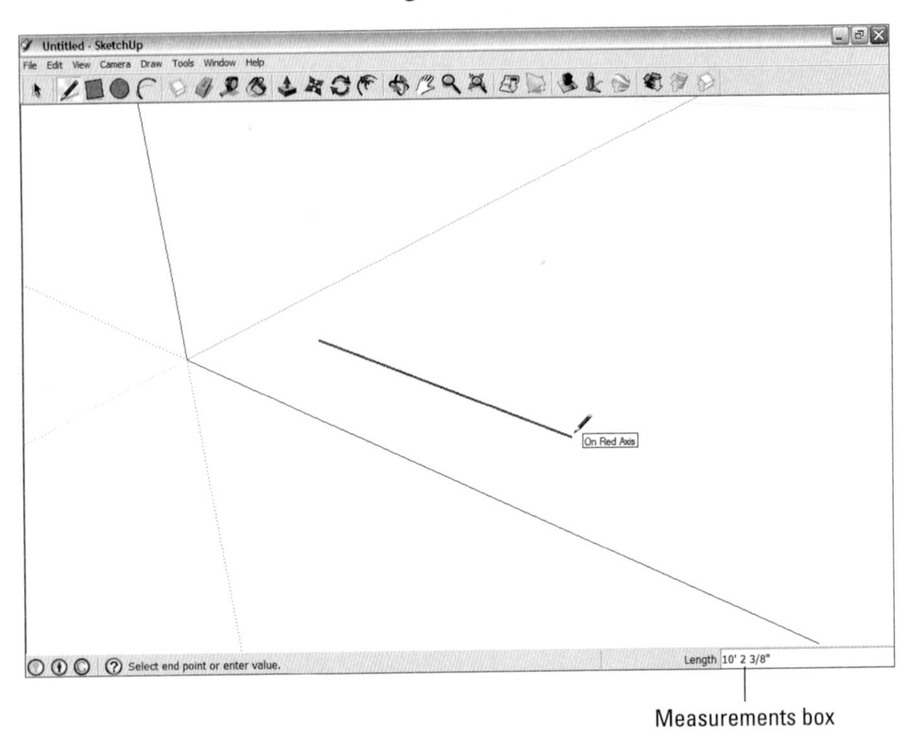

SketchUp's Measurements box is the key to working precisely.

called the Measurements box, and here are some of the things you can use it to do:

- Make a line a certain length.
- Draw a rectangle a certain size.
- Push/pull a face a certain distance.
- Change the number of sides in a polygon.
- Move something a given distance.
- Rotate something by a certain number of degrees.
- Make a certain number of copies.
- Divide a line into a certain number of segments.
- Change your field of view (how much you can see).

Here are some things you should know about the Measurements box:

- **You don't have to click in the Measurements box to enter a number.** When they're first starting out with SketchUp, many people assume that they need to click in the Measurements box (to select it, presumably) before they can start typing. This isn't the case—just start typing, and whatever you type shows up in the Measurements box automatically. When it comes to being precise, SketchUp is always "listening" for you to type something in this box.

- **The Measurements box is context-sensitive.** This means that what it controls depends on what you happen to be doing at the time. If you're drawing an edge with the Line tool, the Measurements box knows that whatever you type is a length; if you're rotating something, it knows to "listen" for an angle.
- **You the set the default units for the Measurements box in the Model Info dialog box.** Perhaps you want a line you're drawing to be 14 inches long. If you're set up to use inches as your default unit of measurement, just type **14** into the Measurements box and press Enter—SketchUp assumes that you mean 14 inches. If you want to draw something 14 *feet* long, you would type in **14'**, just to let SketchUp know that you mean feet instead of inches. You can override the default unit of measurement for the Measurements box by typing in any unit you want. If you want to move something a distance of 25 meters, type in **25m** and press Enter. You set the default units for the Measurements box in the Units panel of the Model Info dialog box (which is on the Window menu).
- **Sometimes, the Measurements box does more than one thing.** In certain circumstances, you can change the Measurement boxes mode (what it's "listening for") by typing in a unit type after a number. For example, when you're drawing a circle, the default "value" in the VCB is the radius—thus, if you type **6** and press Enter, you'll end up with a circle with a radius of 6 inches. But if you type in **6*s***, you're telling SketchUp that you want 6 *sides* (and not inches), so you'll end up with a circle with 6 sides. If you type in **6** and press Enter, and then type in **6s** and press Enter again, SketchUp will draw a hexagon (a 6-sided circle) with a radius of 6 inches.
- **The Measurements box lets you change your mind.** As long as you don't do anything after you press Enter, you can always type a new value into the Measurements box and press Enter again; there's no limit to the number of times you can change your mind.
- **You can use the Measurements box *during* an operation.** In most cases, you can use the Measurements box to be precise while you're using a tool. Here's how that works:

  1. Click once to start your operation (such as drawing a line or using the Move tool).
  2. Move your mouse so that you're going in the correct color direction. If you're using the Line tool and you want to draw parallel to the green axis, make sure that the edge you're drawing is green. Be sure not to click again.
  3. Without clicking the Measurements box, just type in the dimension you want; you should see it appear in the Measurements box.
  4. Press Enter to complete the operation.

- **You can also use the Measurements box *after* an operation.** Doing this revises what you've just done. These steps should give you an idea of what this means:

  1. Complete your operation. This might be drawing a line, moving something, rotating something, or any of the other things mentioned at the beginning of this section.
  2. Before you do anything else, type in whatever dimension you intended, and then press Enter. Whatever you did should be redone according to what you typed in.

TIPS FROM THE PROFESSIONALS

To give you a more concrete example of using the Measurements box after an operation, say you want to move a box (shown in Figure 2-14) a total of 5 meters in the red direction (parallel to the red axis). Here's what you should do:

1. Using the Move tool, click the box once to "pick it up."
2. Move your mouse until you see the linear inference that tells you that you're moving in the red direction.
3. Type in 10m, and then press Enter. Your box will be positioned exactly 10 meters from where you picked it up.
4. Now, say you aren't happy with the 10 meters, so you decide to change it. Simply type in 15m, and then press Enter again, and the box will move another 5 meters in the blue direction. You can keep doing this until you're happy with the results.

Figure 2-14

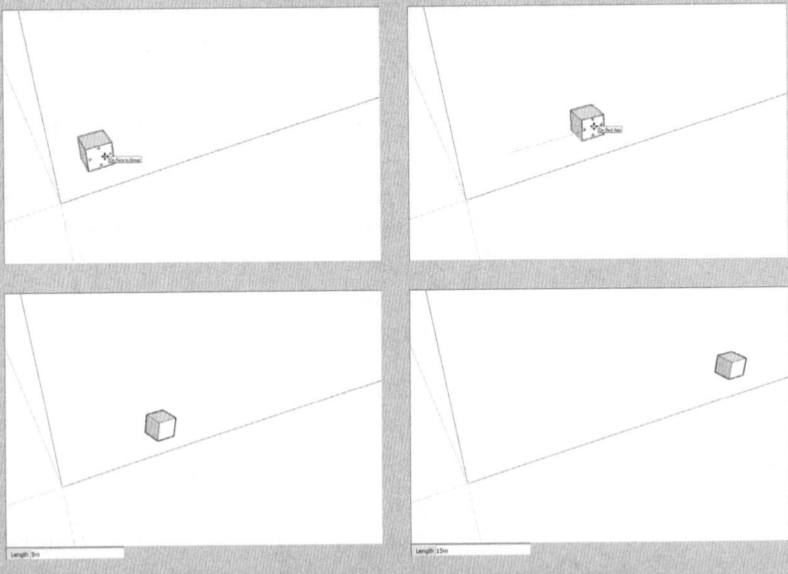

Here, the box is first moved 10 meters, and then it's moved 15 meters.

IN THE REAL WORLD

Imagine that you've been working away in SketchUp, not paying particular attention to how big anything in your model is when you suddenly decide that you need what you've made to be a specific size. SketchUp has a terrific trick for taking care of this exact situation: You can use the Tape Measure tool to resize your whole model based on a single measurement.

Here's how it works: Say, you've started to model a simple staircase, and you now want to make sure that it's the right size. You know you want the riser height (the vertical distance between the steps) to be 7 inches, so this is what you do:

1. Select the Tape Measure tool (choose Tools ⇨ Tape Measure).
2. Make sure that the Tape Measure is in Measure mode by pressing Ctrl (Option on the Mac) until you don't see a plus sign (+) next to the Tape Measure cursor.
3. To measure the current distance (in this case, the riser height), click once to start measuring, and click again to stop.

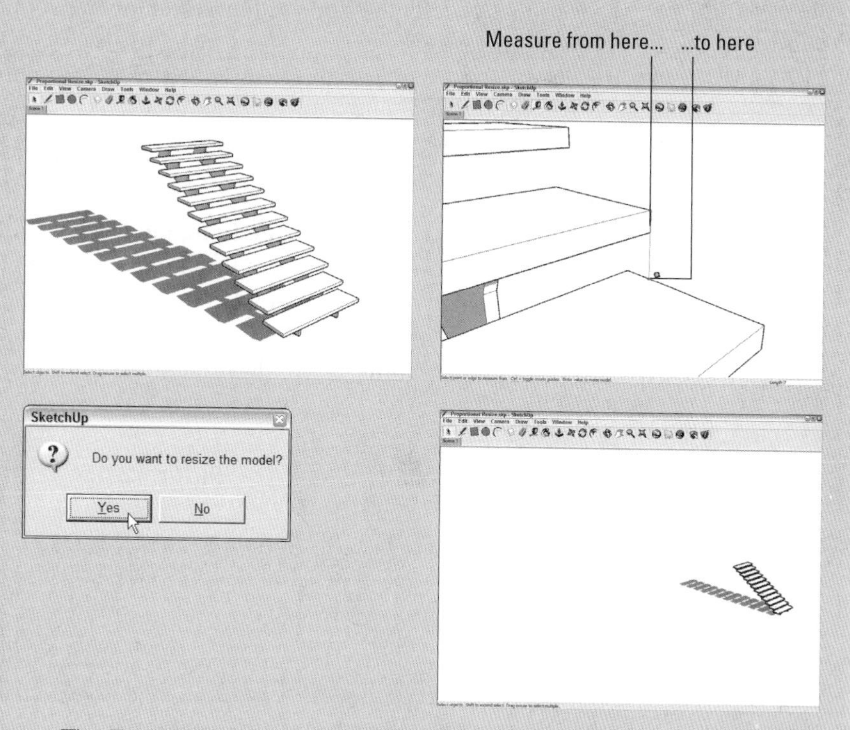

The Tape Measure tool allows you to resize your entire model at one time.

(continued)

IN THE REAL WORLD (continued)

4. Type in the dimension you want what you just measured to be: 7 (for 7 inches).
5. In the dialog box that appears, that asks whether you want to resize the whole model, click the Yes button.

When you click the Yes button, the whole model is resized proportionately to the dimension you entered.

The Tape Measure tool allows you to resize your entire model at one time.

2.3.4 Selecting Parts of a Model

GOOGLE SKETCHUP IN ACTION

"Learn to select parts, move, copy, rotate and apply color to your model"

If you want to move something in your model, or rotate it, or copy it, or do any number of other things to it, you need to select it first. When you select an element, you are telling SketchUp that *this* is the part of the model you want to work with. To select things, use the Select tool, which looks exactly the same as the Select tool most other graphics programs—it's an arrow.

Here's everything you need to know about selecting things in SketchUp:

- Simply click anything in your model to select it. Do this while you are using the Select tool.
- **To select more that one item, hold down Shift while clicking all the things you want to select (see Figure 2-15).** Shift works both ways when it comes to the Select tool. You can use it to *add* to your set of selected objects, and you can also use it to *subtract* an object from your selection. In other words, if you have multiple objects selected and you want to

Figure 2-15

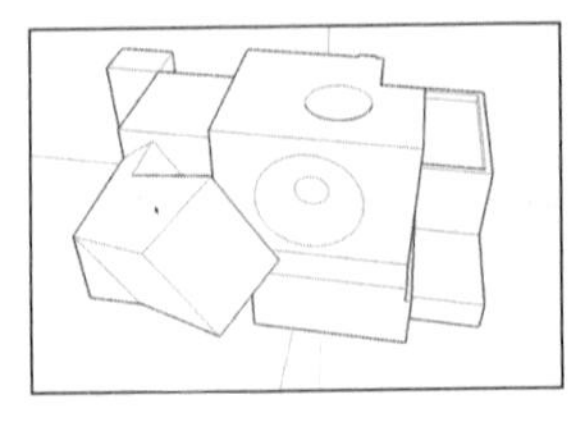

Click to select a face

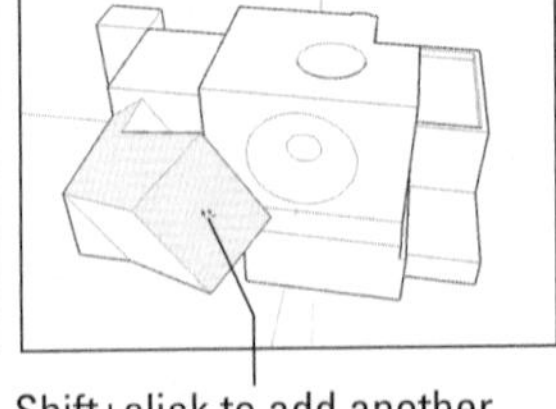

Shift+click to add another face to your selection

Click things with the Select tool to select them. Hold down Shift to select more than one thing.

Figure 2-16

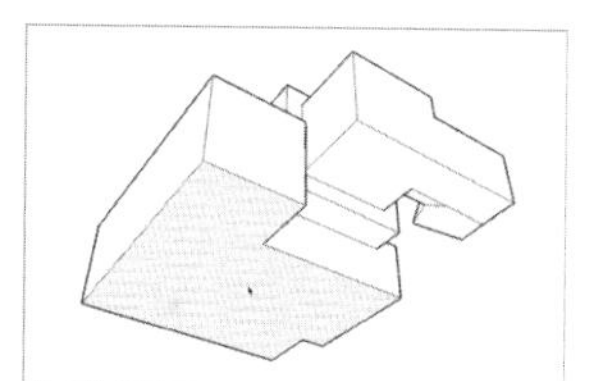

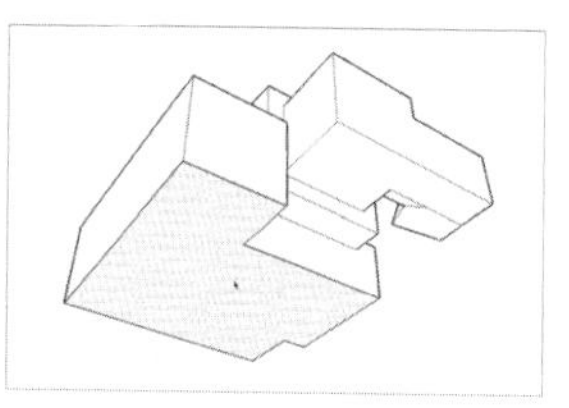

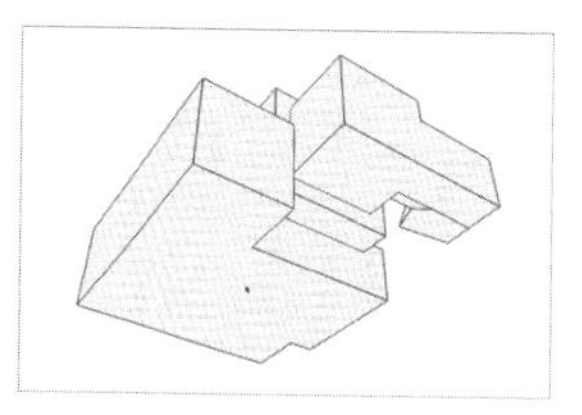

Try single-, double-, and triple-clicking edges and faces in your model to make different kinds of selections.

deselect something in particular, just hold down Shift while you click that object, and it won't be selected anymore.

- Selected objects in SketchUp look different depending on what kind of objects they are:
  - Selected edges and guides turn blue.
  - Selected faces are covered in tiny blue dots.
  - Selected groups and components are surrounded by a blue box. (Reference Chapter 5 for more information)
  - Selected section planes turn blue. (See Chapter 10 for more information.)

- **A much fancier way to select things in your model is to double- and triple-click them.** When you double-click a face, you select that face and all the edges that define it. Double-clicking an edge gives you that edge plus all the faces that are connected to it. When you triple-click an edge or a face, you select the whole object that it's a part of. Figure 2-16 illustrates this.
- **You can also select several things at once by dragging a box around them.** You have two kinds of selection boxes; which one you use depends on what you're trying to select (see Figure 2-17):
  - *Window selection:* If you click and drag from *left to right* to make a selection box, you create a window selection. In this case, only things that are entirely inside your selection box are selected.
  - *Crossing selection:* If you click and drag from *right to left* to make a selection box, you create a crossing selection. With one of these, anything your selection box touches (including what's inside) ends up getting selected.

Note that just because you can't see something doesn't mean it isn't selected. Whenever you make a selection, it's a very good idea to orbit around to make sure you've got only what you intended to select. Accidentally selecting too much is an easy mistake to make.

Figure 2-17

Only this is selected

All this is selected too

Dragging left to right selects everything completely inside your selection box. Dragging right to left selects everything that your selection box touches.

IN THE REAL WORLD

Patrick Reynolds is an artist and designer who utilizes SketchUp for his 3D models. In his own words, "my interest in 3D visualization focuses on utilizing SketchUp for historic preservation, planning, design and education."

He has done 3D modeling for museums and historic properties and recently has worked on creating some renderings of proposed modifications to a restaurant inside a museum. He rendered the space in SU and all the additional elements (exhibits and interactive items) were modeled to scale. While his model was proportionally correct, he did not feel that he needed to add a lot of detail, because the focus was on selling the concept.

He also has worked on other renderings, such as an entrance of a temporary exhibit with all the necessary scenic elements in proper scale, but focusing on the scale more than in the technical construction.

To create photorealistic renderings, he uses IRender Nxt and touches the work up in Adobe Photoshop. For his movies he uses Movie Maker to stitch the pieces together. He states that "SketchUp is lightning fast and the open source nature of ruby scripts has made it grow to be an amazing tool."

Does It Have to Be Blue?

Previous portions of this chapter say that selected items turn blue in SketchUp, but you can make them turn any color you want. Blue is just the default color for new documents you create. The "selected things" color is one of the settings you can adjust in the Styles dialog box. If you are interested, you can read all about styles in Chapter 9.

2.3.5 Moving and Copying

To move things in SketchUp, use the Move tool. To make a copy of something, use the Move tool in combination with a button on your keyboard: Ctrl in Windows, and Option on a Mac. It's really that simple.

Moving Things Around

The Move tool is the tool that looks like crossed red arrows. Using it involves clicking the entity you want to move, moving it to where you want it to be, and clicking again to drop it. It's not a complicated maneuver, but getting the hang of it takes a little time. Here are some tips for using Move successfully:

- **Click, move, and click. Don't drag your mouse.** As with the Line tool, try to avoid the temptation to use the Move tool by clicking and dragging with your mouse; doing so makes things a lot more difficult. Instead, practice clicking once to pick things up, moving your mouse without any buttons held down, and clicking again to put down whatever you're moving.
- Click a point that will let you position an object *exactly* where you want when you drop it (instead of just clicking anywhere on the thing you're trying to move to pick it up). Figure 2-18 shows two boxes that you want to precisely stack on top of each other. If you just click anywhere on the first box and move it over the other one, you can't place it where you want; SketchUp doesn't work that way. To stack the boxes precisely, you have to click the *bottom corner* of the soon-to-be top box to grab it there, and then move your cursor over the *top corner* of the bottom box to drop it. Now the boxes are lined up perfectly. In addition, now that these two boxes are stacked, they are also "glued together." This means that they cannot be pulled apart without distorting their shapes.
- **Press Esc to cancel a move operation.** Here's something beginners do all the time: They start to move something (or start moving something accidentally), and then they change their minds. Instead of pressing Esc, they try to use Move to put things back the way they were. Inevitably, this doesn't work. If you change your mind in the middle of moving something, just press Esc, and everything will go back to the way it was.
- **Don't forget about inferences.** To move something in one of the colored directions, just wait until you see the dotted On Axis linear inference appear, and then hold down Shift to lock yourself in that direction.

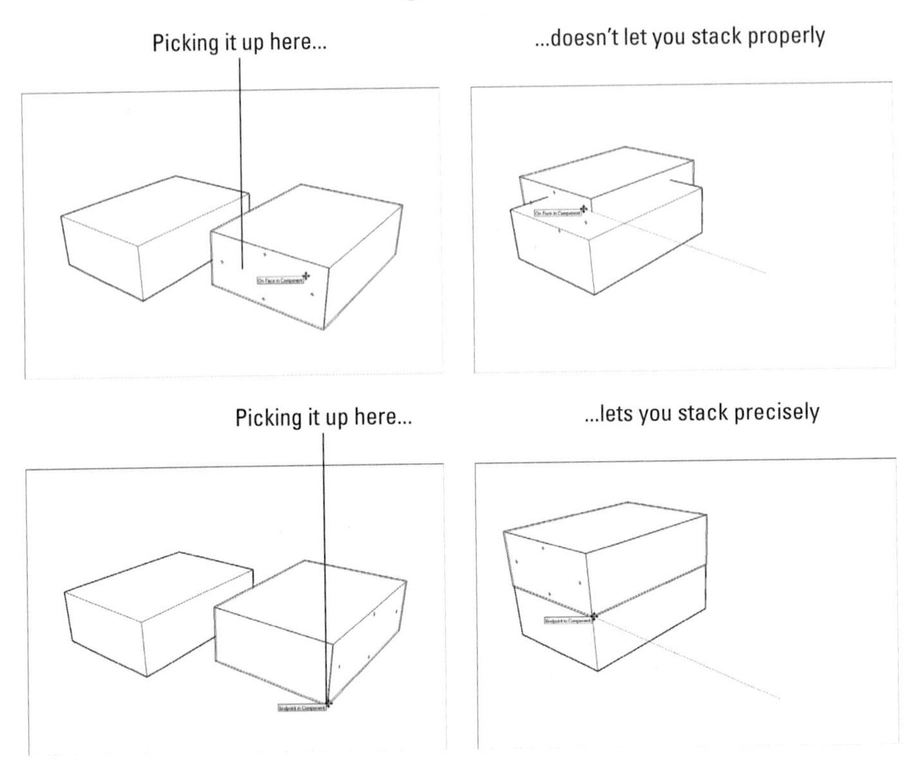

To move objects precisely, choose precise points at which to grab the objects and put them down.

- **Don't forget about the Measurements box.** Always remember, you can move things precise distances using the Measurements box.

Modeling with the Move Tool

Vertices
The endpoints of edges.

In SketchUp, the Move tool is very important for modeling; it's not just for moving entire objects around. You can also use it to move just about anything, including **vertices** (edges' endpoints), edges, faces, and combinations of any of these. By only moving certain entities (all the things just mentioned), you can change the shape of your geometry pretty drastically. Figure 2-19 illustrates this.

Using the Move tool to create forms (instead of just moving them around) is an incredibly powerful way to work, but isn't particularly intuitive. After all, nothing in the physical world behaves like the Move tool—you can't just grab the edge of a hardwood floor and move it up to turn it into a ramp in real life. In SketchUp, however, you can—and you should.

Figure 2-19

Moving a vertex

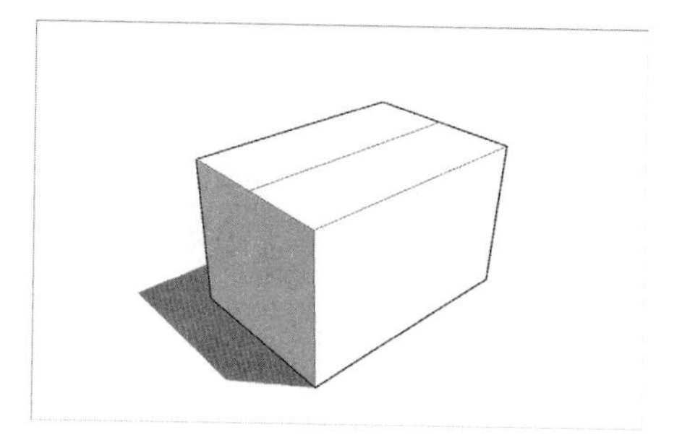

Moving an edge

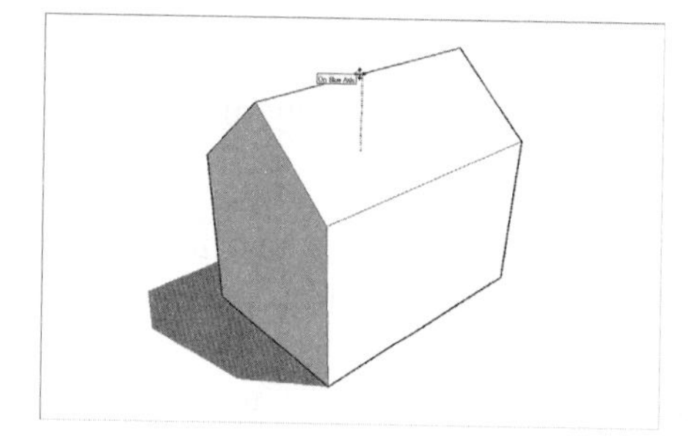

Moving a face

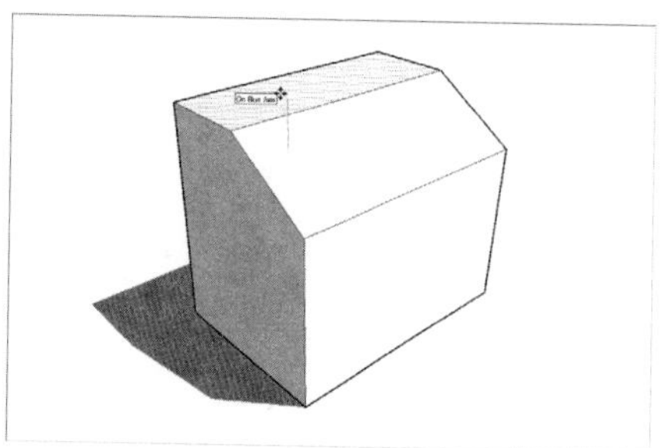

You can use the Move tool on vertices, edges, and faces to model different forms.

Deciding When to Preselect

The Move tool works in two different ways. Eventually, you'll need to use both of them, depending on what you're trying to move:

- **Moving a selection:** When you have a selection of one or more entities, the Move tool only moves the things you've selected. This comes in handy every time you need to move more than one object. Figure 2-20 shows a selection being moved with the Move tool.
- **Moving without a selection:** If you don't have anything selected, you can click anything in your model with the Move tool to move it around. Only the thing you click is moved. Figure 2-21 illustrates this.

Making Copies with the Move Tool

Lots of people spend time hunting around in SketchUp, trying to figure out how to make copies. This process is actually very simple: you just press a **modifier key** while you are using the Move tool. Consequently, instead of

Modifier key
A button on the keyboard that users can push to take a different action than what they are currently doing.

Figure 2-20

Using the Move tool when you have a selection only moves the things in that selection.

moving something, you move a copy of it. Here are a couple of things to keep in mind when doing this:

- **Press Ctrl to copy in Windows, and press Option to copy on a Mac.** This tells SketchUp to switch from Move to Copy while you're moving something with the Move tool. Your cursor will have a small "+" appear next to it, and you'll see your copy moving when you move your mouse.

Figure 2-21

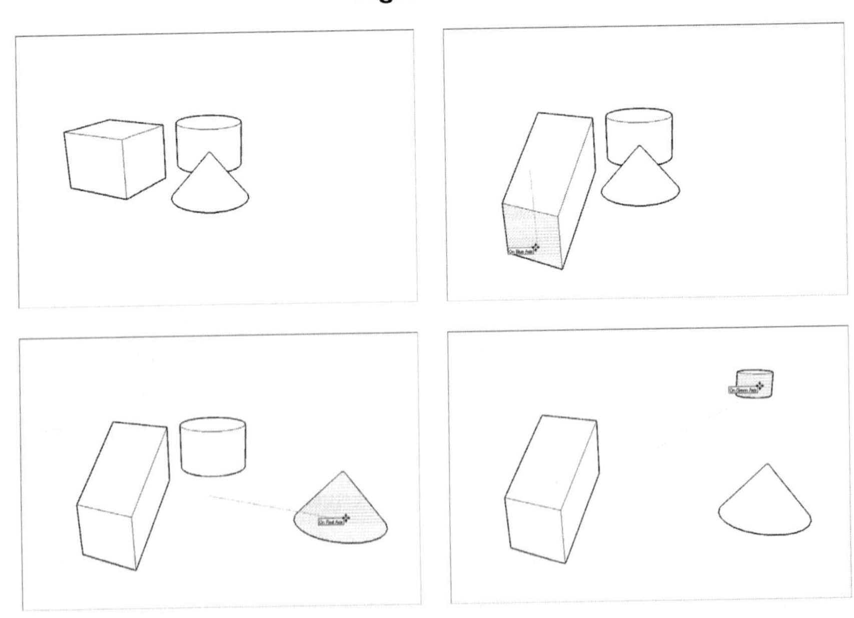

Without anything selected, you can click anything in your model with the Move tool to start moving it around.

Figure 2-22

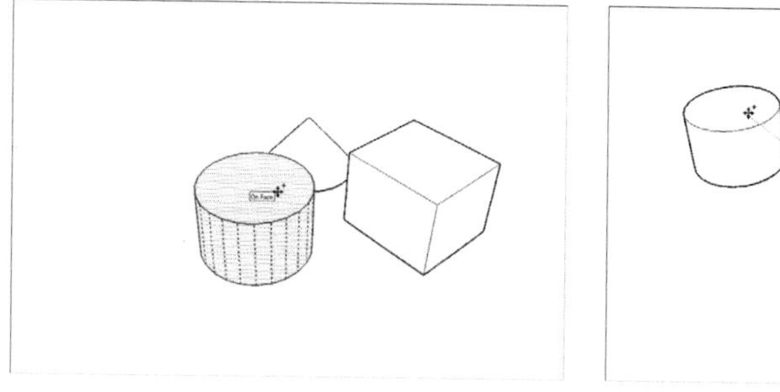

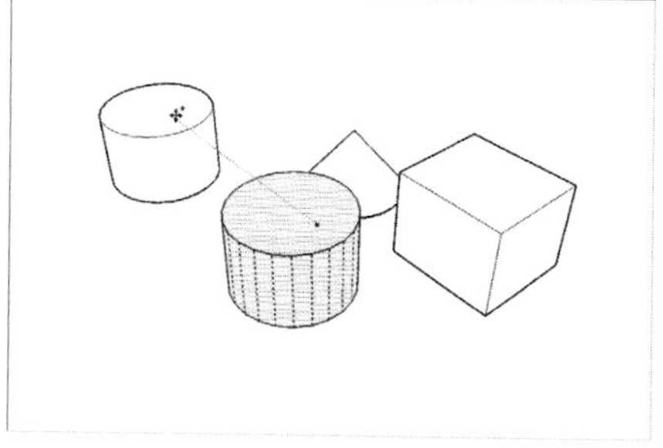

Press Ctrl (Option on a Mac) to tell SketchUp to make a copy while you're moving something.

Figure 2-22 shows this in action. If you decide you don't want to make a copy, just press Ctrl (Option) again to toggle back to Move; the "+" sign will disappear.

- **Copying is just like moving, except you're moving a copy.** This means that all the same rules that apply to using the Move tool apply to making copies, too.
- **You can make more than one copy at a time.** Perhaps you want to make five equally spaced copies of a column, as shown in Figure 2-23. All you

Figure 2-23

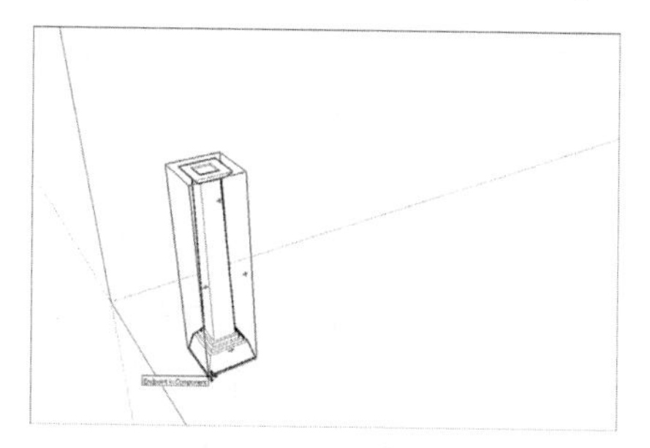

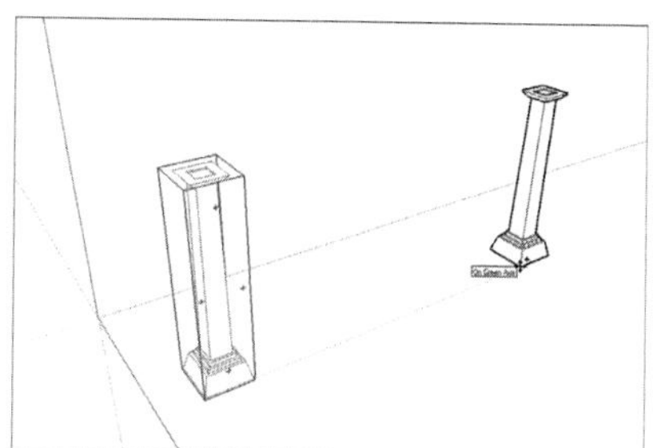

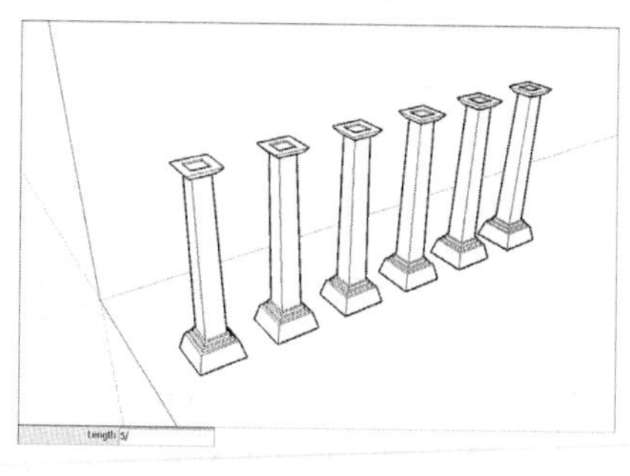

To make evenly spaced copies, type in the number of copies you want followed by a slash (/), and then press Enter.

Figure 2-24

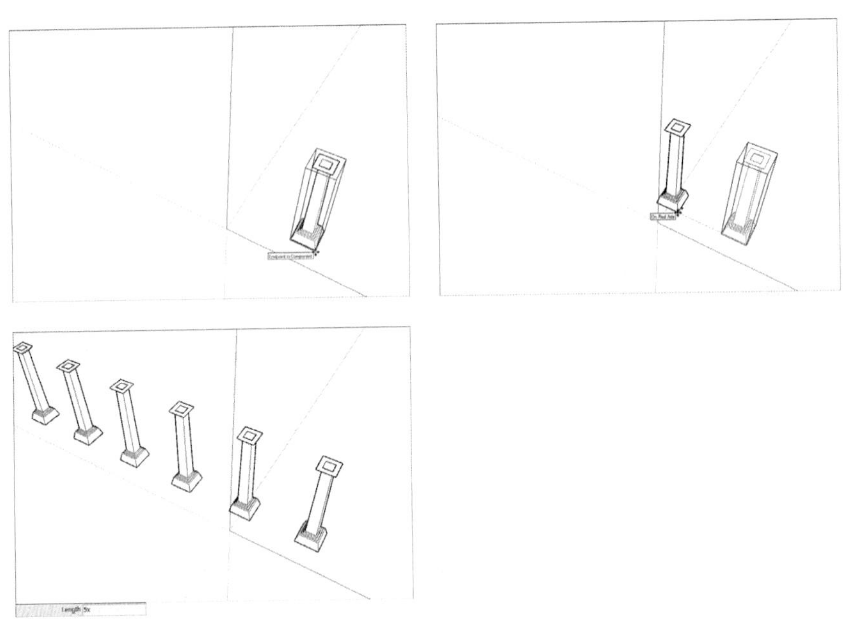

To make multiple copies in a row, type in the number of copies you want, followed by an x, and press Enter.

have to do is move a copy to where you want the last column to be, then type in **5/** and press Enter. This makes five copies of the column and spaces them evenly between the first and last column in the row. Alternatively, if you know how far apart you want your copies to be, you can move a copy that distance, type in **5x**, and press Enter. Your five copies will appear equally spaced in a row (see Figure 2-24).

2.3.6 Mastering the Rotate Tool

Using SketchUp's Rotate tool is a lot like using the Move tool. Despite the fact that rotation is pretty straightforward, included is a section about it in this chapter for one specific reason: The Rotate tool has a trick up its sleeve that most new modelers don't discover until hours after they could've used it.

Keep the following in mind when working with the Rotate tool:

- **It's better to preselect.** As with the Move tool, rotating something you have already selected is usually easier.
- **The Rotate tool can also make copies.** Press the Ctrl key (Option on a Mac) to switch between rotating your original or rotating a copy. You can also make several copies at once just like was presented earlier in the chapter under Making Copies with the Move Tool.

- **You can be precise**. Use your keyboard and the Measurements box to type exact angles while you're rotating. Reference Injecting Accuracy in Your Model earlier in this chapter to find out more.

Using Auto-Fold

Folds
Edges and faces that are created in place of a single face.

This will happen to you sooner or later when using SketchUp: You'll be trying to move a vertex, an edge, or a face, and you won't be able to go in the direction you want. SketchUp doesn't like to let you create **folds** (when extra faces and edges are created in place of a single face) with the Move tool, so it constrains your movement to directions that won't end up adding them. If SketchUp won't let you move whatever you're trying to move, *force* it to do so by doing one very important thing: pressing and holding down Alt (Command on a Mac) while you're moving. When you do this, you're telling SketchUp that it's okay to proceed—in other words, to create folds if it has to. This is called Auto-Fold, and Figure 2-25 illustrates how it works.

Figure 2-25

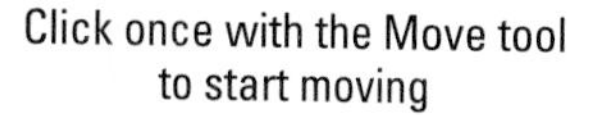

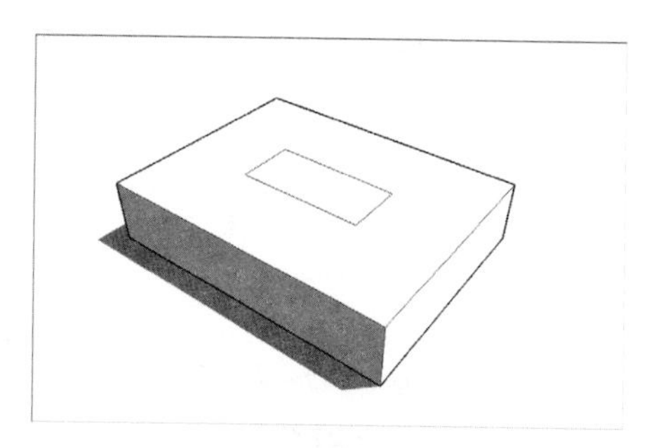

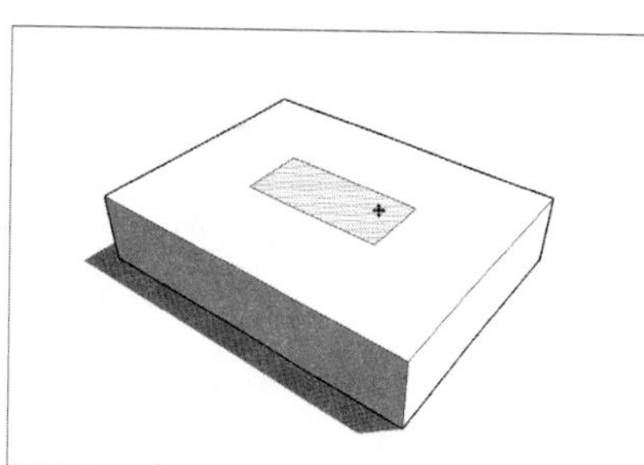

Hold down Alt (⌘ on a Mac)
and move your mouse

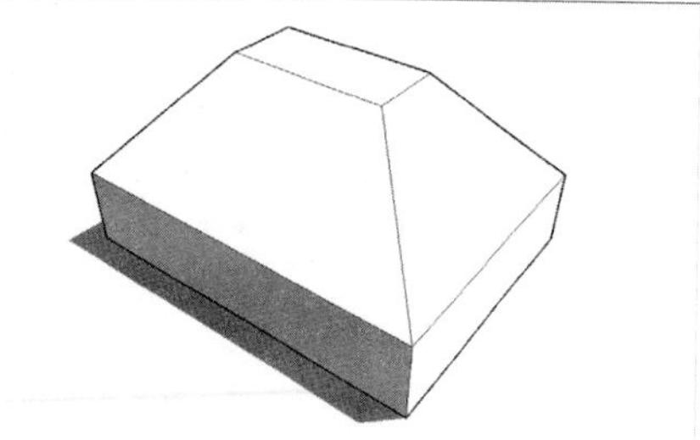

Use Auto-Fold to force SketchUp to create folds in a model.

PATHWAYS TO... USING THE ROTATE TOOL: THE BASIC METHOD

1. Select everything you want to rotate.
2. Activate the Rotate tool.

The default keyboard shortcut for Rotate is Q, just in case you're wondering.

3. Click once to establish an axis of rotation.

Your axis of rotation is the theoretical line around which the selected objects will rotate; picture the axle of a wheel. Although it would be nice if SketchUp drew the axis of rotation in your model, you'll just have to imagine it.

As you move the Rotate tool's big protractor cursor around your screen, notice that the cursor sometimes changes orientation and color. When you hover over a face, the cursor re-aligns itself to create an axis of rotation that's perpendicular to that face. When the cursor is red, green, or blue, its axis of rotation is currently parallel to that colored axis.

You can (and should) use inference locking when you are using the Rotate tool. Just hover over any face in your model that's perpendicular to the axis of rotation you want, hold down the Shift key to lock in that orientation, and click where you want your axis to be.

4. Click again to start rotating.

Clicking part of the thing you're rotating is helpful, especially if you're rotating visually instead of numerically (by typing an angle).

5. Move your mouse; then click again to finish rotating. Now is a good time to type a rotation angle and press Enter. As with everything else in SketchUp, you can be as precise as you want—or need—to be.

The basic method of using Rotate is fine when you need to rotate something on the ground plane, but this method isn't as useful when your axis of rotation isn't vertical. Finding a face to use to orient your cursor can be tricky or impossible, and that's where a lot of SketchUp modelers get hung up. In version 6 of SketchUp, the software's designers introduced a feature that pretty much everybody realizes is great: You can establish a precise axis of rotation (the invisible line around which you're rotating) without having any preexisting faces to use for orientation. This makes rotating things about a million times easier, and those of us who use SketchUp a lot danced little jigs (albeit awkwardly) when we heard the news.

In this case, using Rotate goes from being a five-step operation to a seven-step one (check out Figure 2-26 for a visual explanation):

1. Select everything you want to rotate.
2. Activate the Rotate tool.
3. Click once to establish your axis of rotation, but don't let go—keep your finger on your mouse button.
4. Drag your cursor around (still holding down the mouse button) until your axis of rotation is where you want it.

Figure 2-26

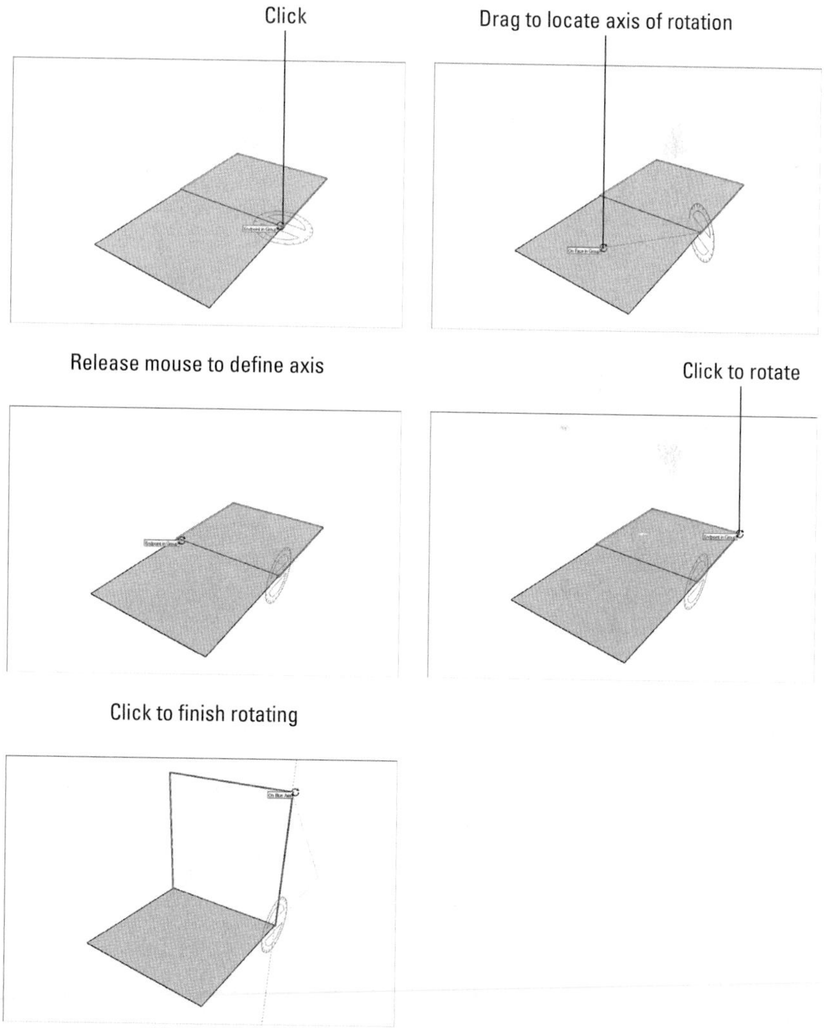

Click

Drag to locate axis of rotation

Release mouse to define axis

Click to rotate

Click to finish rotating

Define a custom axis of rotation by click dragging your mouse.

As you drag, notice your Rotate protractor changes orientation; the line from where you clicked to your cursor is the axis of rotation.

5. Release your mouse button to set your axis of rotation.
6. Click (but don't drag) the point at which you want to "pick up" whatever you're rotating.
7. Click again to drop the thing you're rotating where you want it.

2.3.7 Making and Using Guides

Guides
Temporary lines that users can create to work more accurately in SketchUp.

Sometimes you need to draw temporary lines while you're modeling. These temporary lines, called **guides,** are useful for lining things up, making things the right size, and generally adding precision and accuracy to what you are building.

In previous versions of SketchUp, guides were called construction geometry, because that's basically what they are: a special kind of entity that you create when and where you need them. They aren't part of your model, because they're not edges or faces. This means that you can choose to hide them or delete them, and they don't affect the rest of your geometry.

Figure 2-27 shows an example of guides in action. Here, we have positioned guides 12 inches from the wall and 36 inches apart to draw the sides

Figure 2-27

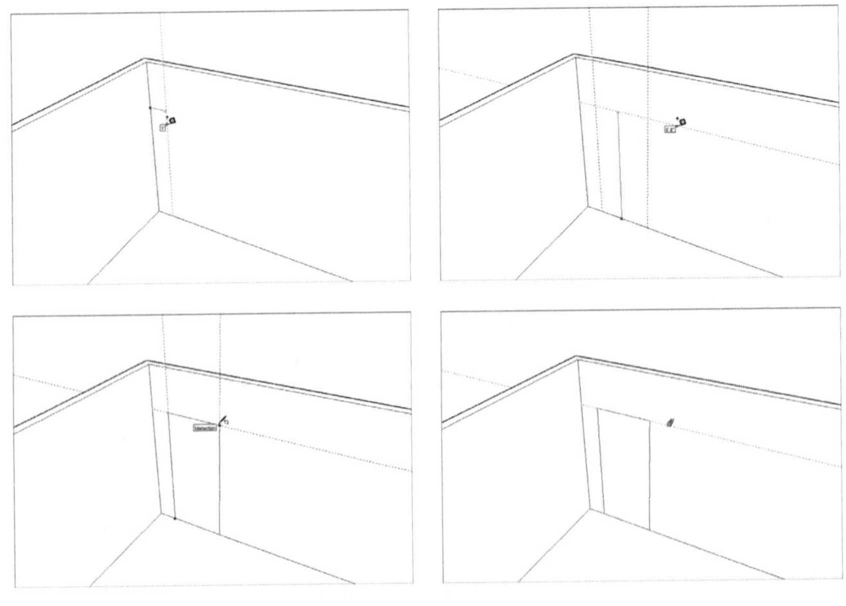

Use guides to measure things before you draw.

of a doorway. We then use another guide 6 feet, 8 inches from the floor to indicate the top. Next, we draw a rectangle bounded by the guides, which we know is exactly the right size. When we're done, we can erase the guides with the Eraser tool, as will be explained momentarily.

Creating Guides with the Tape Measure Tool

You can create three different kinds of guides, and the Tape Measure tool is used to make all of them (see Figure 2-28).

- **Parallel guide lines:** Clicking anywhere (except the endpoints or midpoint) along an edge with the Tape Measure tells SketchUp that you want to create a guide parallel to that edge (see Figure 2-28). Just move your

Figure 2-28

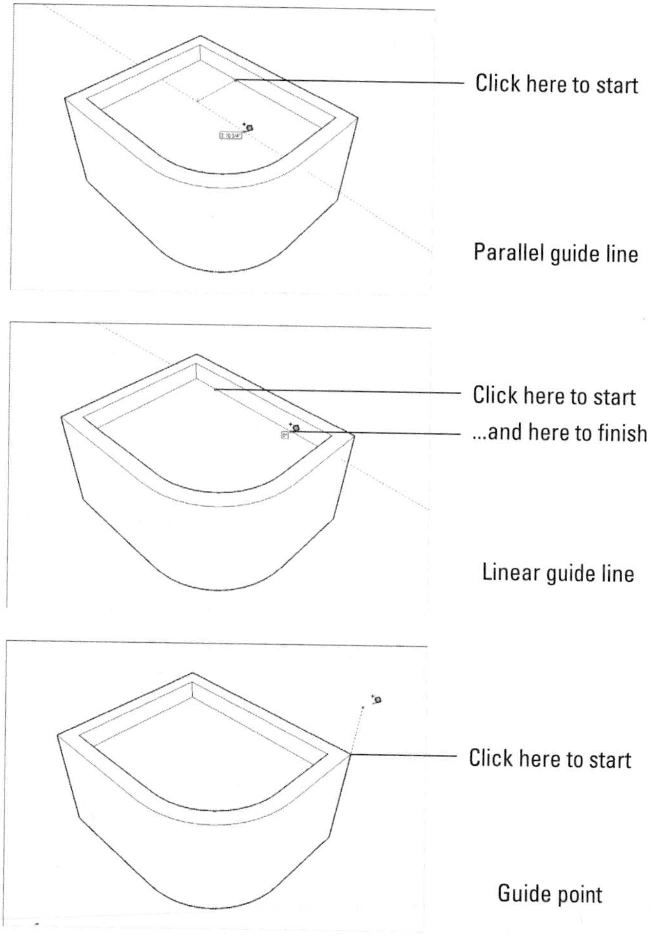

Click here to start

Parallel guide line

Click here to start

...and here to finish

Linear guide line

Click here to start

Guide point

Use the Tape Measure tool to create guide lines and points.

mouse and you'll see a parallel, dashed line; click again to place it wherever you want.

- **Linear guide lines:** To create a guide along an edge in your model, click one of the endpoints or the midpoint once, and then click again somewhere else along the edge.
- **Guide points:** You might want to place a point somewhere in space; you can do exactly that with guide points. With the Tape Measure tool, click an edge's midpoint or endpoint, and then click again somewhere else in space. A little *x* appears at the end of a dashed line—that's your new guide point.

Here's an important point about the Tape Measure tool: It has two modes, and it only creates guides in one of them. Pressing Ctrl (Option on a Mac) toggles between the modes. When you see a + next to your cursor, your Tape Measure can make guides; when there's no +, it can't.

Using Guides to Make Life Easier

As you're working along in SketchUp, you'll find yourself using guides all the time; they're an indispensable part of the way modeling in this software works. Here are some things you should know about using guides:

- **Position guides precisely using the Measurements box.** Check Section 2.3.3 ("Injecting Accuracy into Your Model") earlier in this chapter to find out how.
- **Erase guides one at a time.** Just click or drag over them with the Eraser tool to delete guides individually. You can also right-click them and choose Erase from the context menu.
- **Erase all your guides at once.** Choosing Edit ⇨ Delete Guides does just that.
- **Hide guides individually or all at once.** Right-click a single guide and choose Hide to hide it, or deselect View ⇨ Guides to hide all of them. It's a good idea to hide your guides instead of erasing them, especially while you're still modeling.
- Select, move, copy, and rotate guides just like any other entity in your model. Guides aren't edges, but much of the time, you can treat them as such.

2.3.8 Painting Faces with Color and Texture

Materials
The colors and textures in SketchUp models.

When it comes to adding colors and textures—collectively referred to in SketchUp as **materials**—to your model, there's really only one place you need to look, and one tool you need to use: the Materials dialog box and the Paint Bucket tool, respectively.

The Materials Dialog Box

To open the Materials dialog box (or Colors dialog box on the Mac), choose Window ⇨ Materials. Figure 2-29 illustrates what you see when you do this. The Materials dialog box is radically different in the Windows and Mac versions of SketchUp, but that's okay, because they basically do the same thing.

In SketchUp, you can choose from two different kinds of materials to apply to the faces in your models:

- **Colors:** These are simple—colors are always solid colors. You can't have gradients (where one color fades into another), but you can make almost any color you want.
- **Textures:** Basically, a SketchUp texture is a tiny image (a photograph, really) that gets tiled over and over to cover the face you apply it to. If you paint a face with, say, a brick texture, what you are really doing is telling SketchUp to cover the surface with however many brick photo tiles it takes to do the job. The preview image you see in the Materials dialog box is actually a picture of a single texture image tile. (Note that on a Mac, you have to click the little brick icon in the Materials dialog

Figure 2-29

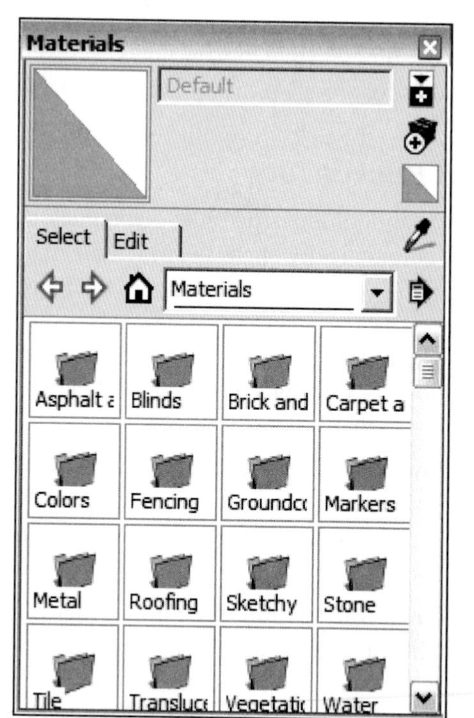

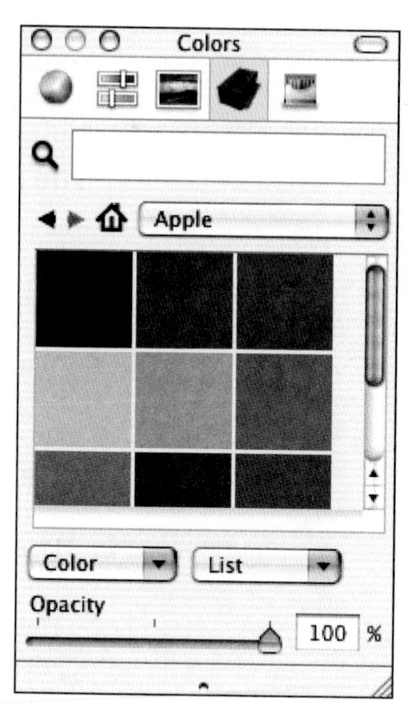

The Materials dialog box in Windows (left) and on the Mac (right).

box to see the textures libraries that ship with SketchUp; it's the drop-down list next to the little house icon.) SketchUp comes with many textures, but you can get even more from the Google SketchUp website (http://sketchup.google.com/bonuspacks.html). If that's still not enough, you can go online and choose from thousands more available that are for sale. And if that's still not enough, you can make your own, although the process of doing so is well beyond the scope of this book.

Here's some more interesting information about SketchUp materials:

- **Materials can be translucent.** Adjusting the Opacity slider makes the material you've selected more or less translucent, which makes seeing through windows in your model a lot easier.
- **Textures can have transparent areas.** If you take a look at the materials in the Fencing library, you'll notice that a lot of them look kind of strange; they have areas of black that don't seem right. These black areas are areas of transparency. When you paint a face with one of these textures, you'll be able to see through the areas that look black.
- **You can edit materials, and even make your own.** This can be considered a relatively advanced use of SketchUp, so it won't be covered in this book.

There is a third thing (besides colors and textures) you can apply to the faces in your models: photos. In fact, photo-texturing is an incredibly important part of some SketchUp workflows—especially those that relate to building models for Google Earth. Chapter 8 covers this subject in detail.

The Paint Bucket Tool

The Paint Bucket tool looks just like—you guessed it—a bucket of paint. Activating it automatically opens the Materials dialog box, which is handy. Here's everything you need to know about the Paint Bucket:

- **You "fill" the bucket by clicking in the Materials dialog box.** Just click a material to load your bucket, and then click the face you want to paint. It's as simple as that.
- **Holding down Alt (Command on a Mac) switches to the Sample tool.** With the Sample tool, you can click any face in your model to load your Paint Bucket with that face's material. Release the Alt key to revert to the Paint Bucket tool.
- **Holding down Shift paints all similar faces.** If you hold down Shift when you click to paint a face, all faces in your model that match the one you click will be painted, too. If things don't turn out the way you want, just choose Undo from the Edit menu to go back a step.

CAREER CONNECTION

SketchUp is not just for architects. An interior design firm in Vancouver, BC, uses Google SketchUp to support their interior planning and design, renovations, and project management service.

They use SketchUp from importing site dimensions through schematics and design, and onto working drawings. They love the instant feedback they get when working with SketchUp. Building and designing to scale blurs the line between schematic and working drawings allowing for using the same drawing. They have even created a market niche by building 3D models and rendering perspectives for other designers and homeowners—not to mention the great visuals that they use for their firm's presentations.

Tips from the Professionals

1. SketchUp can be used as an effective interior design tool.
2. You can leverage one tool for schematic through design.
3. The possibility of creating new markets is possible, so keep an open mind when using the tool.

SELF-CHECK

1. Three SketchUp navigation tools are:
   a. Fly, Jump, and Pan
   b. Orbit, Jump, and Pan
   c. Fly, Zoom, and Pan
   d. Orbit, Zoom, and Pan
2. You should always use Orbit by using your keyboard. True or false?
3. What key do we press to snip off a line when we are done drawing?
4. The Measurement box is context-sensitive. True or false?
5. You can only make one copy at a time. True or false?

Apply Your Knowledge Using the tool shed that you had built previously as your starting point, practice with the different tools that we have covered in this chapter. For example, try moving or rotating the shed, increase the angle of the pitched roof, or Zoom in to work on a small detail on the front door. Try to apply different materials to the roof, walls, and door and use the Guides and the Measurement box to add a window outline at a specific height and with specific dimensions.

SUMMARY

Section 2.1

- Edges and faces are the basic building blocks of every model you will ever make.
- Collectively, edges and faces in your model are called geometry.
- Edges are lines, are always straight, and don't have thickness.
- Faces are surfaces, are always flat, and they need to be bound by at least three edges.
- You cannot have faces without edges.

Section 2.2

- You use the drawing axes to figure out where you are (and where you want to go) in 3D space.
- Point Inferences include:
  - Endpoint (green)
  - Midpoint (cyan or light blue)
  - Intersection (black)
  - On edge (red)
  - Center (of a circle, green)
  - On face (dark blue)
- Linear Inferences are:
  - On Axis
  - From Point
  - Perpendicular
  - Parallel
  - Target of Vertex
- You can lock and encourage inferences.

Section 2.3

- There are several SketchUp tools available to aid in precision modeling:
- Orbit tool
- Zoom tool
- Pan tool
- Line tool
- Measurements box
- Move tool
- Rotate tool
- Materials dialog box
- Paint Bucket tool

ASSESS YOUR UNDERSTANDING

SUMMARY QUESTIONS

1. What tool do we use to slide the model view around the modeling window?
2. There is only one way to activate the Zoom tool. True or false?
3. You can make copies by using a modifier key. True or false?
4. The key to accuracy in SketchUp is:
   a. The Measurements box
   b. The user interface
   c. The inference engine
   d. Rubber banding
5. Choose the correct work for the following statement: SketchUp automatically makes a ______ every time you finish drawing a closed shape out of three or more coplanar edges.
   a. Face
   b. Backup copy
   c. Cube
6. If you want to get closer to your model, you can use:
   a. The Pan tool
   b. The Orbit tool
   c. Guides
   d. The Zoom tool
7. What tool would we use to create guides?
8. Describe the rubber banding effect when we draw a line segment with the Line tool.
9. When selecting colors for your model, you cannot produce a gradient but you can almost make any color you want. True or false?
10. If you want your model to fill your modeling window, you would use:
    a. Pan
    b. Orbit
    c. Zoom extents
    d. Zoom

APPLY: WHAT WOULD YOU DO?

1. You are introducing a friend of yours how to use SketchUp to draw a model of a doghouse. What are the first three things you would teach him or her and why?

2. Describe where we find the Measurements box in our screen and four ways of using it.
3. You've started to model a staircase and you want to make sure the riser height is 6 inches. How do you do this?
4. You are modeling a house with five equal-sized columns. After you have drawn the first column, what is the easiest way to add the other four columns?

BE A DESIGNER

Zoom, Pan, and Orbit

Open the Components dialog box (from the Window menu) and select a pre-created model. Zoom, pan, and orbit around this model using the appropriate tools.

Draw a Box

In SketchUp, draw a box. Use the Tape Measure tool to create linear and point guides. Save the box for future exercises.

KEY TERMS

Coplanar
Drawing axes
Folds
Geometry
Guides
Inference engine
Linear inferences
User interface
Vertices
Zoom
Materials
Modifier key
Orbit
Pan
Point inferences
Rubber banding

CHAPTER

3 BUILDING SIMPLE MODELS

Step-by-Step Instructions for Creating a Model of a Doghouse

Do You Already Know?

- How to set up SketchUp and change the default settings?
- The first steps to take in building simple models?
- The different ways you can change the appearance of a model?
- How to add shadows to a model?

For the answers to these questions, go to **www.wiley.com/go/chopra/googlesketchup2e**

What You Will Find Out	What You Will Be Able To Do
3.1 How to initialize SketchUp.	Establish initial settings through the Welcome menu.
3.2 The process for creating a simple model.	Learn to create a simple building.
3.3 How to use paint tools.	Be able to apply colors to your model.
3.4 How to work with different styles.	Change styles on your model.
3.5 How to work with shadows and light.	Apply shadows and light for specific locations and time of day.
3.6 How to share files.	Create JPEG files for sharing through e-mail.

INTRODUCTION

If you can't wait to start using SketchUp, you've come to the right chapter! In this chapter, you'll learn how to make a simple model step-by-step, spin it around, paint it, and even apply styles and shadows. These pages are about *doing* and about the basics of putting together the various SketchUp features to produce a great model in no time.

More specifically, in this chapter, we'll walk through the steps necessary to create a model of a doghouse. The nice thing about doghouses is that they're a lot like human houses in several important ways: they have doors and roofs, and just about everybody has seen one.

In this chapter, you will build a doghouse. Don't get nervous! The chapter presents the steps in chronological order with very simple and easy-to-understand instructions, beginning with how to set up the program. In addition, you'll learn to alter how your model looks by changing its color, texture, and style and by adding shadows. Finally, you'll discover how to share your model with others.

Keep in mind that while other chapters can be referenced out of order, this chapter needs to be followed step-by-step for the lesson to be effective.

3.1 SETTING UP SKETCHUP

Because setup can be boring, this discussion is as brief as possible. To make sure that you're starting at the right place, simply follow these steps when setting up SketchUp on your computer:

1. **Launch Google SketchUp.** In other words, open the program on your computer.
2. **Choose your default settings.** If you've never launched SketchUp on your computer before, you'll see the *Welcome to SketchUp dialog box* appear upon startup (see Figure 3-1).

   Here's what to do if the Choose Default Settings box appears on your screen:

   - Click the Choose Template button.
   - Choose one of the Architectural Design templates—it doesn't matter if you prefer Feet and Inches or Millimeters.
   - Click the Start using SketchUp button to close the dialog box and open a new SketchUp file.

     If the Welcome to SketchUp dialog box doesn't appear, you (or someone else) has told the dialog box not to show up automatically on startup. Don't worry—just follow these steps to set things straight:

   - Choose Help ⇨ Welcome to SketchUp from the menu bar.

Figure 3-1

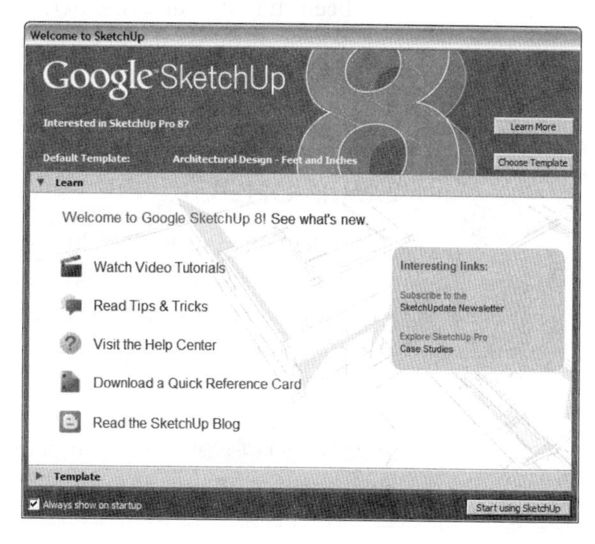

The Welcome to SketchUp dialog box appears the first time you launch SketchUp.

- Follow the first two steps in the preceding steps list.
- Open a new file by choosing File ⇨ New.

  If you are using the Pro version of SketchUp 8, the Welcome to SketchUp dialog box looks a little different—it includes information about your software license. The best place to go for help when you are having trouble with your license is the SketchUp Help Center. Choose Help ⇨ Help Center to go there directly from SketchUp.

3. **Make sure that you can see the Getting Started toolbar.** Figure 3-2 shows the Getting Started toolbar. If it's not visible in your modeling window, choose View ⇨ Toolbars ⇨ Getting Started to make it appear. If you're on a Mac, choose View ⇨ Show Toolbar.

Figure 3-2

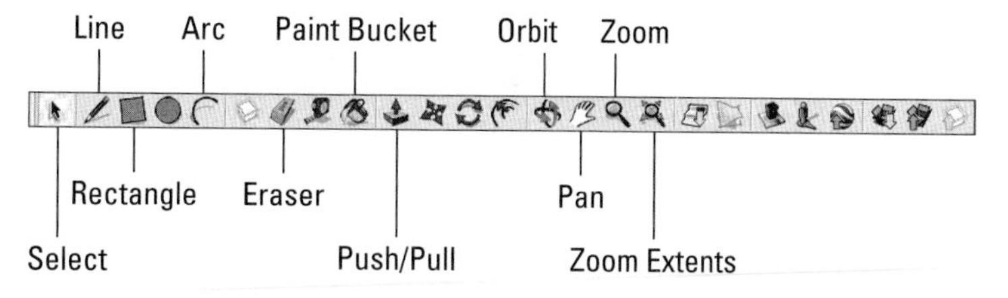

The Getting Started Toolbar is located at the top of your modeling window.

4. **Clear your modeling window.** If this isn't the first time SketchUp has been run on your computer, you might see dialog boxes all over your screen. If this is the case, open the Window menu and make sure that everything is deselected to get rid of these boxes.

SELF-CHECK

1. If the Welcome to SketchUp dialog box doesn't appear, then this is because someone has told the dialog box not to show up automatically on startup. True or false?
2. At startup, you will be able to choose your preferred measuring method, from the Architectural Design templates. True or false?

Apply Your Knowledge You are beginning to model a dog house. Practice launching and setting up your SketchUp environment.

3.2 MAKING A QUICK MODEL

GOOGLE SKETCHUP IN ACTION

Learn to create a simple building.

Figure 3-3 shows what your computer screen should look like at this point. You should see a row of tools across the top of your modeling window, a small figure of a woman, and three colored drawing axes (red, green, and blue lines).

Now, you're ready to start creating your model of a doghouse! Simply follow these steps to build your model.

1. **Delete the little person on your screen.** Using the Select tool (the arrow on the far left of your toolbar), click on the woman to

Figure 3-3

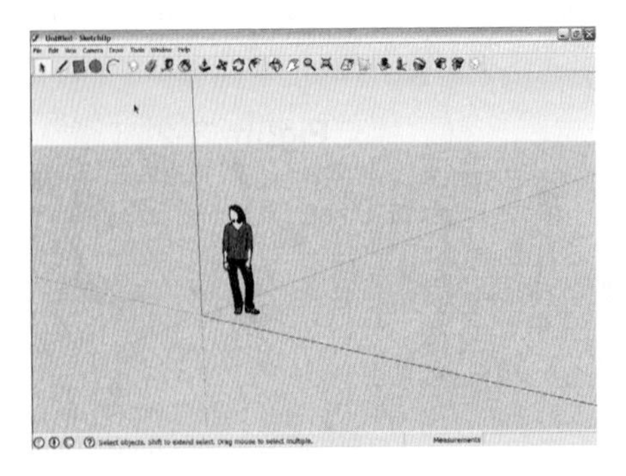

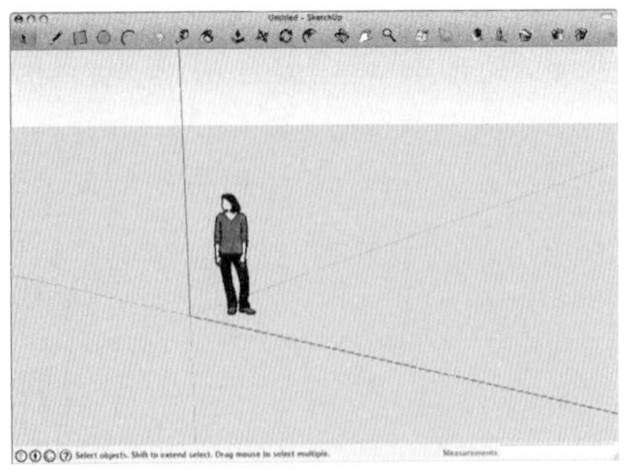

This is what your screen should look like in Windows (left) and on a Mac (right).

select her (her name is Susan, by the way), and then choose Edit ⇨ Delete.

2. **Choose Camera ⇨ Standard Views ⇨ Iso.** This switches you to an **isometric view** (3D view) of your model, which allows you to build an object without having to "move around."

Isometric view
A three-dimensional view of a model.

3. **Draw a rectangle on the ground.** Use the Rectangle tool (between the pencil and the circle on your toolbar) to draw a rectangle by doing the following:
   - Click once to place one corner on the left side of your screen.
   - Click again to place the opposite corner on the right side of your screen.

     Remember that you're in a 3D, *perspective*, view of the world, so your rectangle will look more like a diamond—90-degree angles don't look like 90-degree angles in perspective. Figure 3-4 shows what you should be aiming for in this step.

     It's important to draw the right kind of rectangle for this example (or for any model you're trying to create in Perspective view), so try it a few times until it looks like the rectangle in Figure 3-4. To go back a step, choose Edit ⇨ Undo Rectangle, and the last thing you did will be undone. You can use Undo to go back as many steps as you like, so feel free to use it anytime.

Figure 3-4

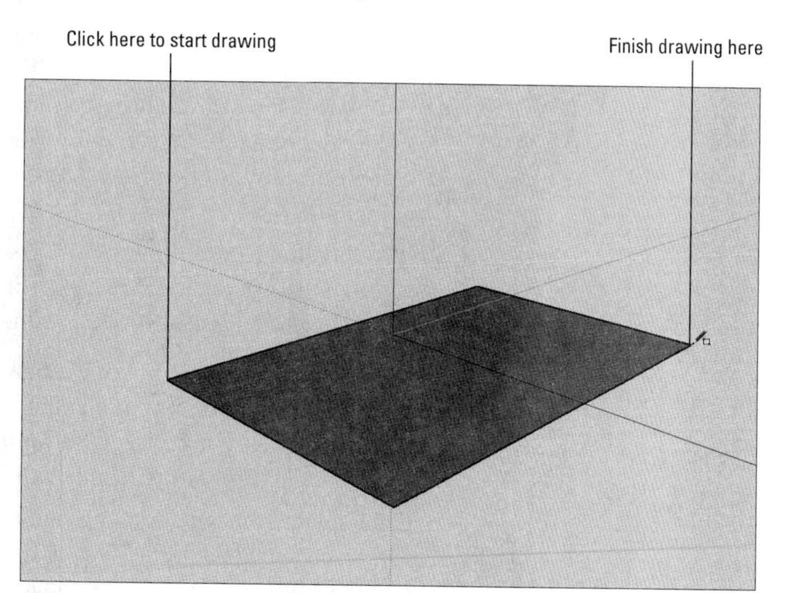

Draw a 3D rectangle on the ground.

Figure 3-5

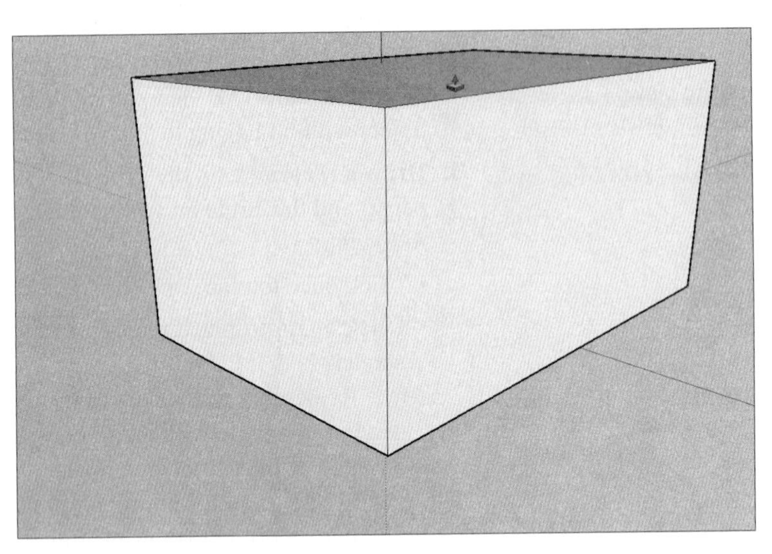

Use the Push/Pull tool to extrude your rectangle into a box.

4. **Use the Push/Pull tool to extrude your rectangle into a box.** Use this tool (it looks like a brown box with a red arrow coming of out the top) to "pull" your rectangle into a box by following these steps:

   - Click the rectangle once to start the push/pull operation.
   - Click again, somewhere above your rectangle, to stop pushing/pulling.

     At this point, you should have something that looks like Figure 3-5. If you don't, use Push/Pull again to make your box look about the right height.

     Note that if you are pushing/pulling on your box and everything suddenly disappears, it's because you pushed/pulled the top of your box all the way to the ground. Just choose Edit ⇨ Undo and keep going.

5. **Draw diagonal lines for your roof.** Use the Line tool (it's shaped like a pencil) to draw two diagonal edges (lines) that will form your peaked roof, as shown in Figure 3-6. Follow these steps:

   - Click once at the midpoint of the top of your box's front face to start your line. You'll know you're at the midpoint when you see a small, light-blue square and the word *Midpoint* appear. Move slowly to make sure that you see this indicator.
   - Click again somewhere along one of the side edges of your box's front face to end your line. Wait until you see a red *On Edge* cue (just like the Midpoint indicator in the last step) before you click; if you don't, your new line won't end on the edge like it's supposed to.

Figure 3-6

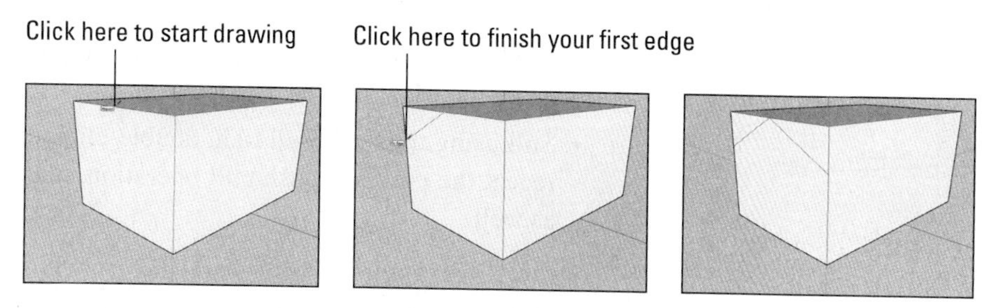

Draw two diagonal lines that will become your peaked roof.

- Repeat the previous two steps to draw a similar (but opposite) line from the midpoint to the edge on the other side of the face. Don't worry about making your diagonal lines symmetrical; for the purposes of this exercise, it's not important that they are.

6. **Push/pull the triangles away to leave a sloped roof.** Use the Push/Pull tool (the same one you used in step 4) to get rid of the triangular parts of your box, leaving you with a sloped roof. Have a look at Figure 3-7 to see this in action, and follow these steps:

   - Select the Push/Pull tool, and then click the right triangular face once to start the push/pull operation.

Figure 3-7

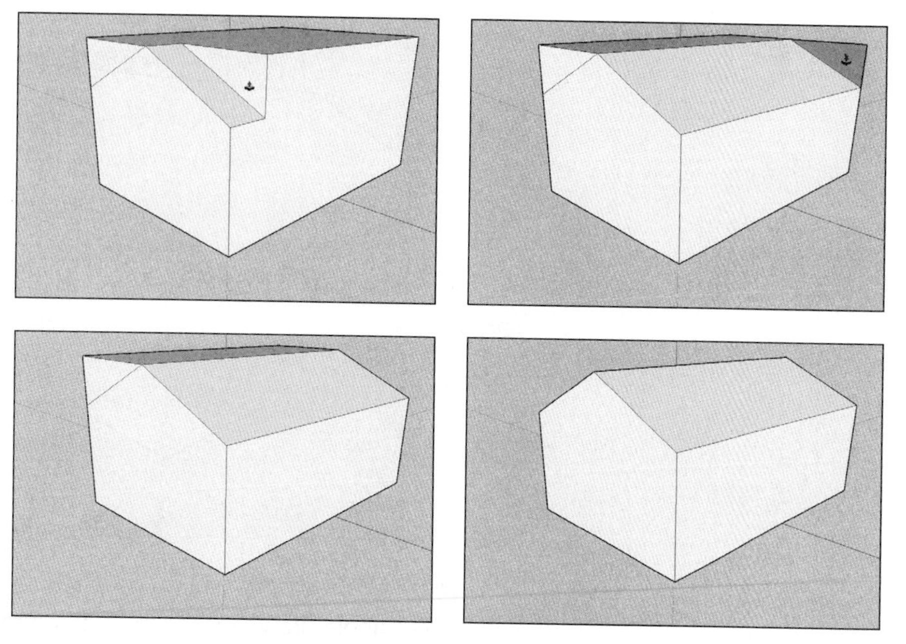

Use the Push/Pull tool to form a peaked roof on your box.

- Move your cursor to the right to "push" the triangle as far as it will go (even with the end of your box).
- Click again (on the triangle) to end the push/pull operation and to make the triangular face disappear.
- Still using the Push/Pull tool, double-click the left triangular face to repeat the previous push/pull operation, making that face disappear as well.

7. **Draw a rectangle on your front face.** Switch back to the Rectangle tool (which you used in step 3) and draw a rectangle on the front face of your pointy box. Make sure that the bottom of your rectangle is flush with the bottom of your box by watching for the red On Edge cue to appear before you click. Check out Figure 3-8 to see what it should look like when you're done.

   Remember, using the rectangle tool is a two-step process: You click once to place one corner and again to place the opposite corner. Try not to draw lines and shapes in SketchUp by dragging your cursor; doing so makes matters more difficult. Practice clicking once to start an operation (like drawing a rectangle) and clicking again to stop.

Figure 3-8

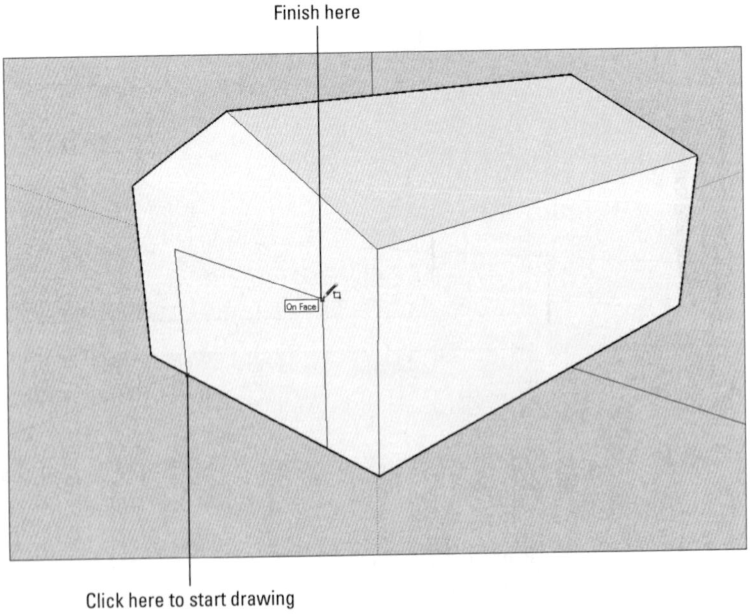

Finish here

On Face

Click here to start drawing

A rectangle drawn on the front of your pointy box.

Figure 3-9

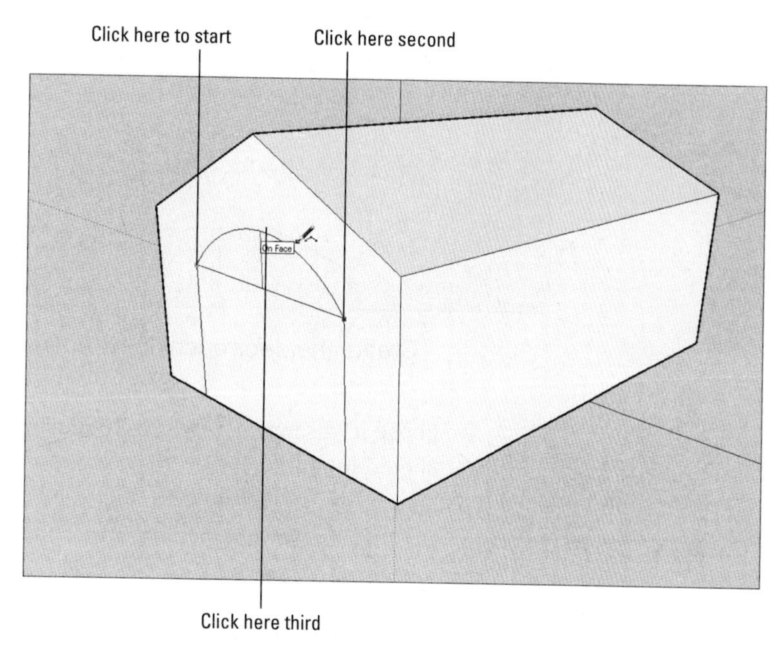

Draw an arc on top of your rectangle.

8. **Draw an arc on top of the rectangle you just drew.** Use the Arc tool (to the right of the Circle tool) to draw an arc on top of your rectangle (see Figure 3-9). Follow these steps to draw an arc:

   - Click the upper-left corner of the rectangle to place one endpoint of your arc. Make sure that you see the green Endpoint hint before you click.
   - Click the upper-right corner of the rectangle to place the other endpoint of your arc.
   - Move your cursor up to "bow out" the line you're drawing into an arc, and click when you're happy with how it looks.

9. **Select the Eraser tool and then click the horizontal line between the rectangle and the arc to erase that line.**

10. **Push/pull the doorway inward.** Use the Push/Pull tool to push the "doorway" face you created in Steps 7 through 9 in just a bit. Remember, you use Push/Pull by clicking a face once to start, moving your cursor to "push/pull" it in or out, and then clicking again to stop.

11. **Erase the horizontal line at the bottom of the doorway by clicking it with the Eraser tool.** This makes the line (and the whole face

Figure 3-10

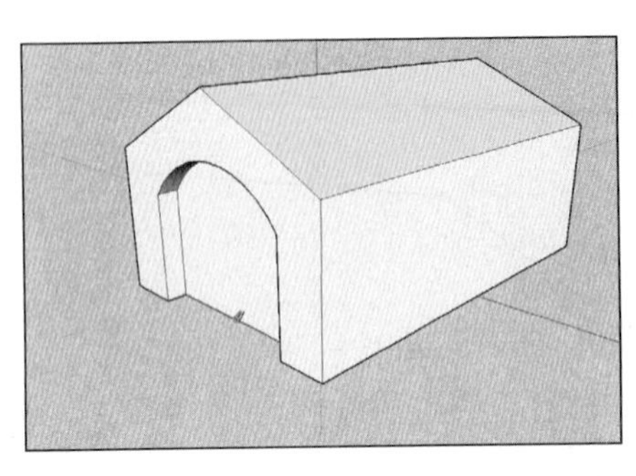
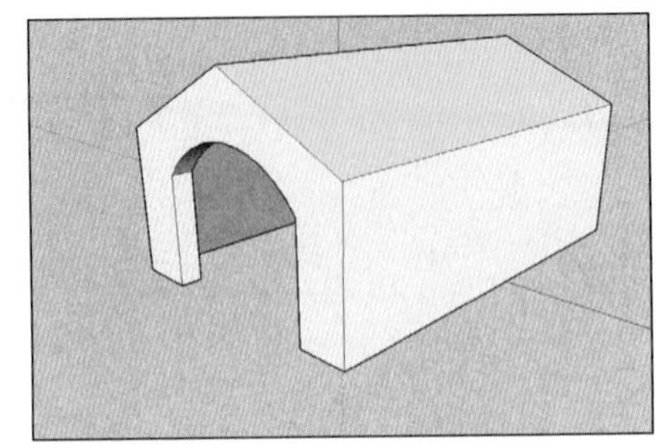

Create the door opening by erasing its bottom edge.

above it) disappear. Figure 3-10 illustrates what your finished doghouse should look like.

SELF-CHECK

1. In SketchUp, when you are working in a perspective view, any rectangle that you draw will fit to that perspective view and the right angles will look like a diamond. True or false?
2. Explain in a brief sentence how you can use the Push/Pull tool to transform a rectangle into a box.
3. Explain briefly how you would raise a peaked roof using the line tool, using a 3D box as a starting point.

Apply Your Knowledge Practice changing the roof of your doghouse using the steps in this section.

3.3 PAINTING YOUR MODEL

To be able to paint your model, you must first understand how to spin it around. Moving your model around is *the most important* skill to develop when you are first learning SketchUp. Practice these steps to apply colors (and textures) to the faces in your model—and to learn more about moving your model while you're doing it:

1. **Choose Window ⇨ Materials to open the Materials dialog box.** See Figure 3-11 for a picture of this dialog box. Click a color or texture that you like. When you do, you automatically "pick up" the Paint Bucket tool and fill it with your chosen material.

Figure 3-11

Click here to see your materials libraries

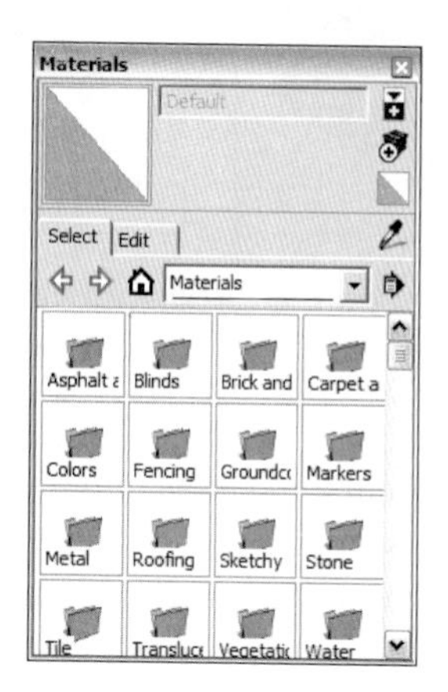

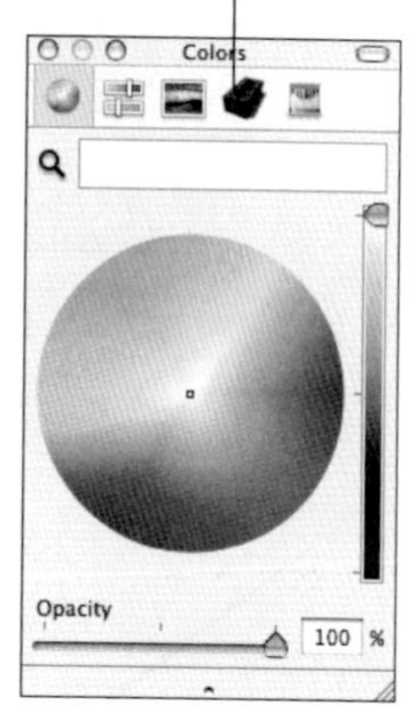

The Materials dialog box in Windows (left) and on a Mac (right).

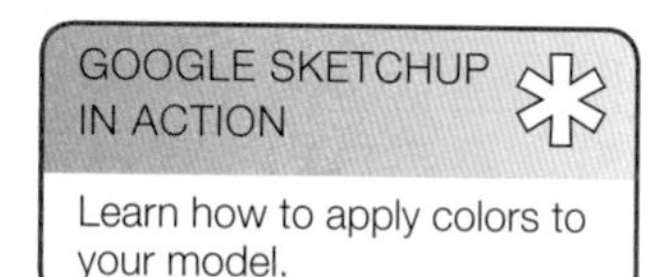

2. **Paint some of the faces in your model.** Do this by clicking any face you want to paint with the Paint Bucket tool.
3. **Switch materials.** Choose another material from the Materials dialog box by clicking it.
4. **Paint the rest of the faces you can see.** Refer to Figure 3-12 to see how this is done. Loop through Steps 2 to 4 for as long as you like.
5. **Choose the Orbit tool.** This is the tool that's just to the left of the white hand on the toolbar. By selecting this tool, you are preparing to move your model around.
6. **Click somewhere on the left side of your screen and *drag* your cursor to the right (see Figure 3-13). Release your mouse button when you are done.** Your model spins! This is called orbiting. Orbit around some more, just to get the hang of it. If you're orbiting, and you've dragged

Figure 3-12

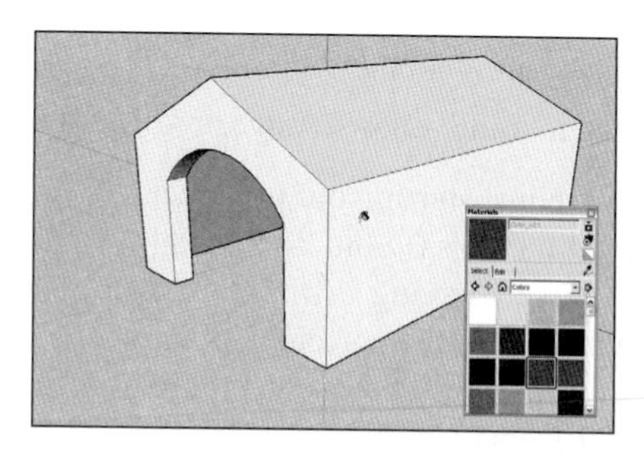

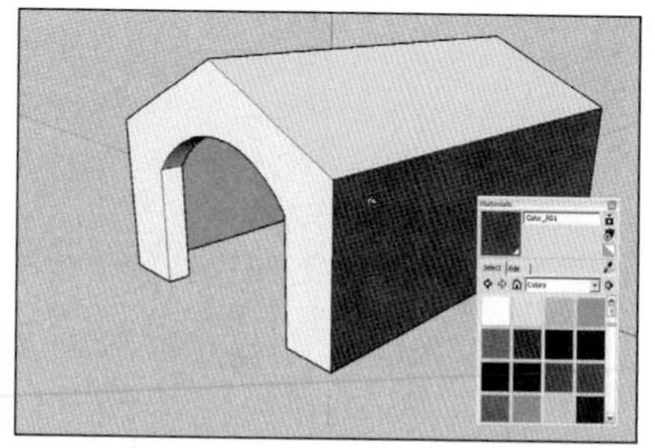

Use the Paint Bucket tool to paint everything you can see.

Figure 3-13

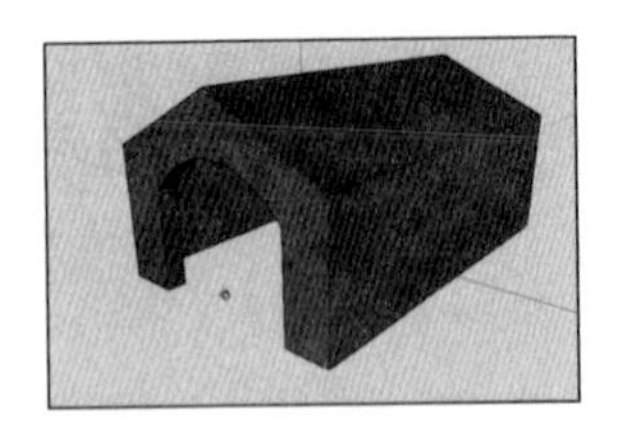

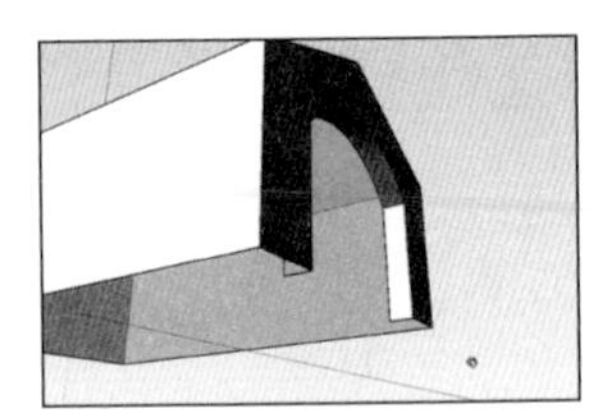

Choose the Orbit tool and drag your cursor to spin your model around.

your cursor over as far as it will go but you haven't orbited as much as you want to, don't fret. Just release the mouse button, move your cursor to where it was when you started orbiting, and orbit some more by clicking and dragging. You usually can't see what you want to see with a single orbit; typically, you'll need a bunch of separate "drags" to get things looking the way you want them to.

7. **Zoom in and out if you need to by selecting the Zoom tool and dragging your cursor up and down in your modeling window.** The Zoom tool looks like a magnifying glass, and it's on the other side of the white hand icon. Dragging up zooms in, while dragging down zooms out.
8. **If needed, move around in two dimensions with the Pan tool by selecting it and then clicking and dragging the Pan cursor inside your modeling window.** The Pan tool is the white hand found between Orbit and Zoom. You use Pan to "slide" your model inside your modeling window without spinning it or making it look bigger or smaller. You can pan in any direction.
9. **Use the Orbit, Zoom, Pan, and Paint Bucket tools to finish painting your doghouse.** Now that you know how to move around your model, paint it as follows (Color Plate 8 shows what it should look like):
   - Paint the exterior walls reddish brown.
   - Paint the roof light blue.
   - Paint the interior yellow-orange.

When you're just starting out in SketchUp, it's easy to get a little lost with the navigation tools (Orbit, Zoom, and Pan); it happens to everybody. If you find yourself having difficulties, just choose Camera ⇨ Zoom Extents. When you do this, SketchUp automatically places your model directly in front of you; check out Figure 3-14 to see Zoom Extents in action. Zoom Extents is also a button on the toolbar; it's located next to the Zoom tool.

Figure 3-14

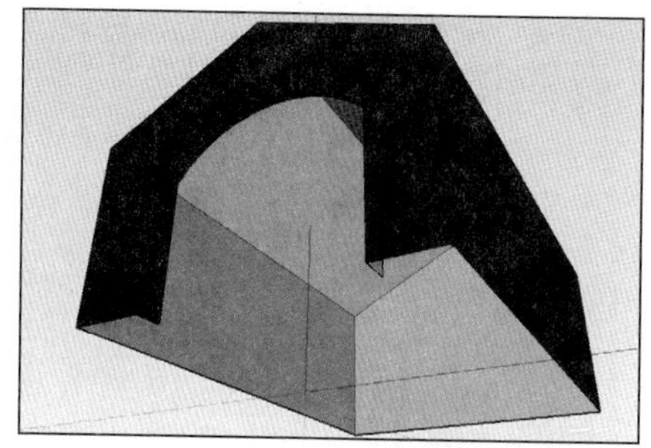

Use Zoom Extents anytime you can't determine where your model went.

SELF-CHECK

1. To paint your model, what tool do we use?
   a. Color tool
   b. Material tool
   c. Appearances tool
2. The Materials dialog box includes colors and _______.
3. Orbit, Zoom, and Pan are navigation tools. True or false?
4. Which tool would you use to move your model around?
   a. Orbit tool
   b. Zoom tool
   c. Pan tool
   d. Zoom extents

Apply Your Knowledge Change your doghouse's color scheme.

3.4 GIVING YOUR MODEL SOME STYLE

SketchUp Styles allow you to change your model's appearance—basically, the way it's drawn—with just a few clicks of your mouse. You can create your own styles, of course, but SketchUp also comes with a library of premade ones that you can use without knowing anything about how they work.

Follow these steps to try several different styles on your doghouse:

1. **Choose Window ⇨ Styles.** This opens the Styles dialog box.
2. **Click the Select tab.** This lets you see the Select pane.
3. **In the Libraries drop-down menu, choose the Assorted Styles library.** This is shown in Figure 3-15.

Figure 3-15

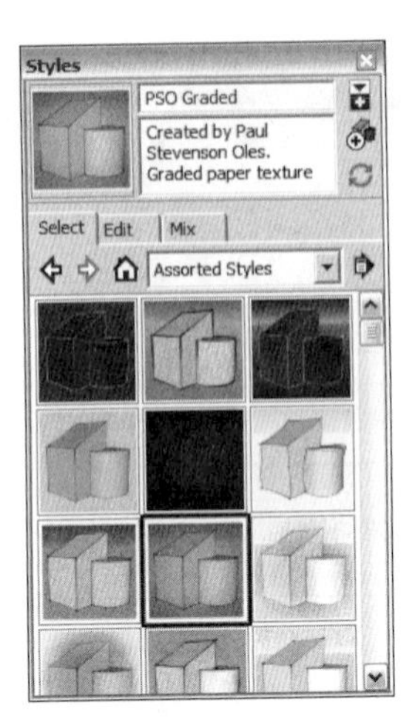

The Assorted Styles library is a sampler of ready-mixed SketchUp styles.

Figure 3-16

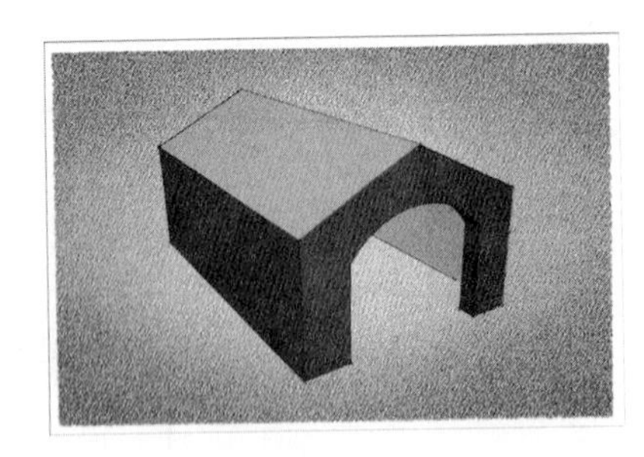

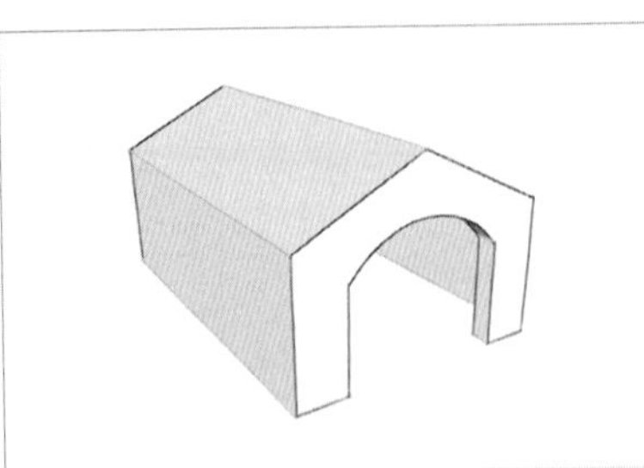

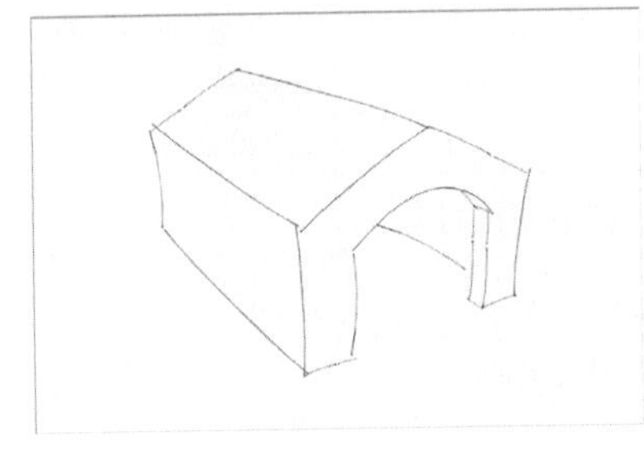

The same doghouse with four very different styles applied to it.

4. **Click through the different styles to see what they're about.** When you click a style in the Styles dialog box, that style is applied to your model. Figure 3-16 shows a model of a doghouse with a few different styles applied. Can you figure out which styles have been used?
5. **Go back to the original style.** Click the little house icon in the Styles dialog box to see a list of all of the Styles you have applied to your model. Find the Architectural Design Style (it should be first in the list) and click to choose it. For more information on styles, see Chapter 9.

SELF-CHECK

1. How do you open the Styles dialog box?
2. SketchUp does not come with a library of premade styles but you can create your styles and save them for future use. True or false?

Apply Your Knowledge Apply three different styles to your doghouse. Then return to the original style.

3.5 ADDING SHADOWS

You are about to use what some people consider to be the one of SketchUp's best features: shadows. When you turn on shadows, you are activating Sketch-Up's built-in sun. The shadows you see in your modeling window are *accurate*

Figure 3-17

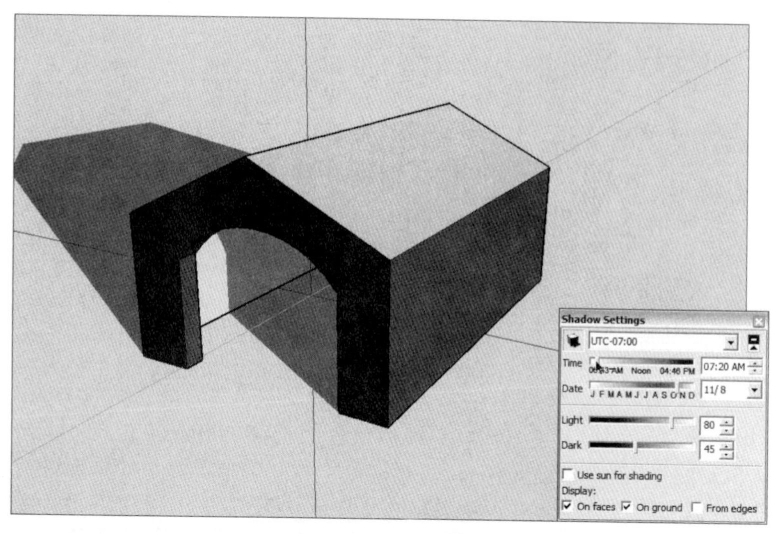

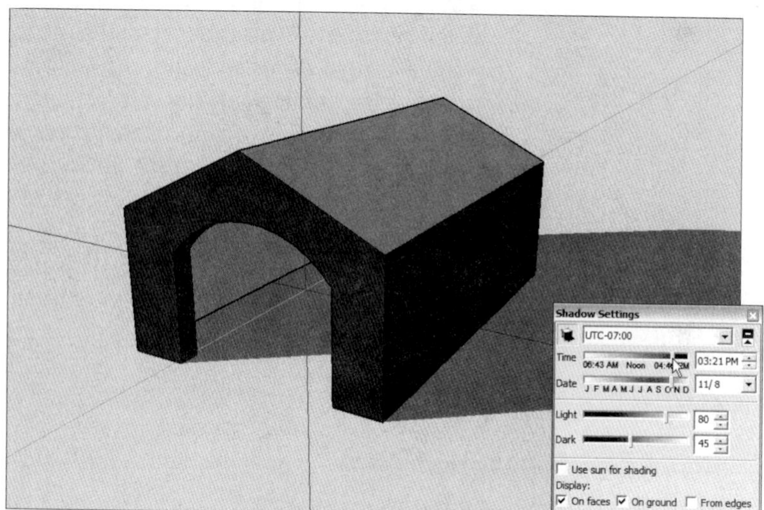

Use Orbit, Zoom, and Pan to navigate until your model looks something like this.

Figure 3-18

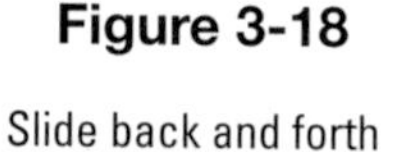

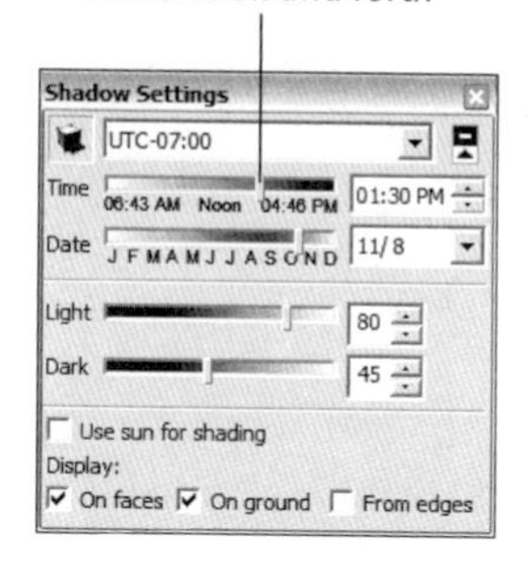

The Shadow Settings dialog box controls the position of SketchUp's built-in sun.

Apply shadows and light for specific locations and time of day.

for whatever time and location you set. For the purposes of this tutorial, though, don't worry about accuracy. Go through these steps to let the light shine in:

1. **Use Orbit, Zoom, and Pan to get an aerial, three-quarter view of your doghouse.** Your view should look much like the one shown in Figure 3-17.
2. **Choose Window ⇨ Shadows.** This opens the Shadow Settings dialog box, as shown in Figure 3-18.
3. **Select the Display Shadows check box to turn on the sun.** To do that, you need to click on the top left corner, on the icon of a cube casting a shadow on the ground. Your doghouse should now be casting a shadow.

4. **In the Shadow Settings dialog box, move the Time slider back and forth.** Changing the time of day means that you're moving SketchUp's sun around in the sky. For more about fine-tuning light and shadows, refer to Chapter 9.

IN THE REAL WORLD

Lucas Quijada, an industrial designer, practices in Denver, Colorado. He enjoys importing his Google SketchUp screen captures to Photoshop for fine tuning and also enjoys the process of importing 2D CAD floor plans to extrude in three dimensions the structure from these plans. He has also experimented with exporting SketchUp models into AGi32 (a specific daylight analysis software) to determine precise light values at desk height to retrofit lighting fixtures.

He likes to combine different viewing styles, shadow settings, perspective settings, and components such as people and objects to reach a satisfactory result and then saves his work then saves the images as 2D Graphics for use in his presentation boards. He also likes to combine the Podium plugin quite extensively to do more realistic rendering, and the response he gets from his clients is generally quite positive, mostly because of the speed in which ideas can be generated and communicated.

For Lucas, the combination of Google SketchUp and SUPodium is good mix, although he uses SketchUp to render most of his color views. If later on his ideas need to be polished up in more detail, he exports his views into AutoCAD or Solidworks for further refinement. To compose his boards, he prefers to import his screen captures done in SketchUp into Adobe Illustrator and Photoshop, to create visually rich boards with images and text.

SELF-CHECK

1. When you turn on shadows, you're activating SketchUp's built-in:
   - **a.** Sun and Moon
   - **b.** Artificial light sources
   - **c.** Sun
2. The shadows in SketchUp are accurate for the time and location you set. True or false?
3. How do you open the Shadow Settings dialog box?

Apply Your Knowledge Practice setting different times and locations to see the effects of shadowing on your dog house.

3.6 SHARING YOUR MODEL

JPEG
A compression technique for digital images.

Now that your model looks the way you want it to, you probably want to show it to someone. The easiest way to do this is by exporting your model to a **JPEG** Joint Photographic Experts Group, image that you can attach to an email. JPEG compression reduces file sizes to a small percentage of their normal size. Users can often choose the degree of compression they desire when saving files as JPEGs. The more compressed a file is, the lower the quality of the image. Follow these steps, and you'll be on your way:

PATHWAYS TO... SHARING YOUR MODEL

1. Navigate around (using Orbit, Zoom, and Pan) until you like the view of your model that you see in your modeling window.
2. Choose File ⇨ Export ⇨ 2D Graphic.
3. In the Export dialog box that opens, choose JPEG from the Export Type drop-down menu.
4. Pick a location on your computer system, and give your exported image a name.
5. Click the Export button to create a JPEG image of what's visible in your modeling window.

Exporting a JPEG file is just one way to share models. You can share your model on Google Earth, as a printout, as an image or animation, or as a slick presentation that will (hopefully) impress all your friends.

SELF-CHECK

1. What is a JPEG and when would you use it in SketchUp?
2. Name two ways you can share your models.

Apply Your Knowledge Using your newly created dog house file, create a JPEG file and e-mail it to yourself.

SUMMARY

Section 3.1

- When you initially launch SketchUp you work through the *Welcome to SketchUp dialog box* to establish your initial architectural settings.

Section 3.2

- Switching to an **isometric view** (3D view) of your model allows you to build an object without having to "move around."
- Using the rectangle tool is a two-step process: You click once to place one corner and again to place the opposite corner.
- Try not to draw lines and shapes in SketchUp by dragging your cursor; doing so makes matters more difficult.

Section 3.3

- Moving your model around is the most important skill to develop when you are first learning SketchUp.
- Choose the Orbit tool and drag your cursor to spin your model around.

Section 3.4

- SketchUp comes with a library of premade styles that you can use without knowing anything about how they work.

Section 3.5

- When you turn on shadows, you are activating SketchUp's built-in sun.
- The shadows you see in your modeling *window* are *accurate* for whatever time and location you set.

Section 3.6

- The easiest way to share a model is by exporting it to a **JPEG** image that you can attach to an email.
- JPEG stands for Joint Photographic Experts Group and is a compression technique for digital images.
- JPEG compression reduces file sizes to a small percentage of their normal size.
- Users can often choose the degree of compression they desire when saving files as JPEGs.
- The more compressed a file is, the lower the quality of the image.

ASSESS YOUR UNDERSTANDING

SUMMARY QUESTIONS

1. Describe the steps you would take to change your model into an isomeric view.
2. What might a rectangle look like in 3D?
   a. A rectangle
   b. A diamond
   c. A box
   d. A circle
3. If you are turning a rectangle into a box using the Push/Pull tool and everything disappears, it is because you have pushed/pulled the top of your box to the ground. True or false?
4. The line tool is shaped like a diagonal line. True or false?
5. If you suddenly "lost" your model, name the feature you would choose under Camera, to find it.
6. If you make a mistake, you can always go back one or more steps by choosing _______ from the Edit menu.
7. Styles allow you to change your model's appearance. True or false?
8. Write one way you can share your model.
9. You can share your model on Google Earth. True or false?
10. The Push/Pull tool looks like:
    a. A pencil
    b. A rope
    c. A small man
    d. A brown box with a red arrow coming out of it
11. You can use Undo for only the last two steps you completed. True or false?
12. Describe where to find textures in Google SketchUp.

APPLY: WHAT WOULD YOU DO?

1. SketchUp gives you the ability to add shadows to your models. Why would you want to add shadows?
2. Imagine you would want to expand your two-car garage and build an in-law apartment on top of it, and want to share it with your friends. Name three ways you can share the model of your design.

3. Continuing with the previous question, how would you add color and textures to the in-law apartment? What specific colors and materials would you choose?
4. What tools would you use to navigate around your model in three dimensions, and why is this important during the design phase?

BE A DOGHOUSE DESIGNER

Altering Your Model

Follow the directions in this chapter and build a doghouse. Once you are done, alter your model by:

a. Changing the color

b. Adding shadows

c. Changing the style

d. Adding a door

Sharing Your Model

Follow the directions in this chapter and build a doghouse. After you are done, share your model by doing the following:

a. Exporting it as a JPEG file

b. E-mailing it to a friend

KEY TERMS

Isometric view

JPEG

CHAPTER

4 MODELING BUILDINGS

From Drafting a Floor Plan to Adding a Roof

Do You Already Know?

- Components of a floor plan?
- SketchUp's different drafting tools?
- How to model interior and exterior walls?
- Two different methods for creating stairs?
- The different types of roofs?

For the answers to these questions, go to **www.wiley.com/go/chopra/googlesketchup2e**

What You Will Find Out	What You Will Be Able To Do
4.1 How to draft a simple floor plan.	Learn to go from 2D to 3D. Understand how to add floors to buildings. Learn how to create and insert doors and windows. Work with SketchUp's components.
4.2 How to create stairs.	Learn different methods to create stairs.
4.3 How to create a roof.	Understand how to build flat roofs with parapets. Learn about gabled and hip roofs. Understand and practice putting your roof together.

INTRODUCTION

Even though SketchUp lets you make just about anything you can think of, certain forms are easier to make than others. Fortunately, these kinds of shapes are exactly the ones that people want to make with SketchUp most of the time. That's no accident! SketchUp was designed with architecture in mind, so the whole paradigm—the fact that SketchUp models are made of faces and edges, as well as the fact that the software offers certain tools—is perfect for making things like buildings.

But what about those curvy, swoopy buildings? You can use SketchUp to make those, too, but they're a little harder, so they're not really a good place to start. Because most of us live in boxy places with right-angled rooms and flat ceilings, that kind of architecture is relatively easy to understand, and it is also easier to model with SketchUp.

In this chapter, you'll learn some of the fundamentals of SketchUp modeling in terms of making simple, rectilinear buildings. You will draft a simple floor plan. You will convert a 2D plan to a 3D model, and you will add stairs, doors, windows, and a roof to your model. By reading about how to build certain kinds of things, instead of just reading about what the individual tools do, you should find it easier to get started. Even if you're not planning to use SketchUp to model any of the things described in this chapter, you should still be able to apply these concepts to your own creations.

Most of the concepts in this chapter reference those that were introduced in Chapter 2, so you might find a need to refer back to that chapter while working in this one.

4.1 DRAWING FLOORS AND WALLS

Most floors and walls are flat surfaces, so it's easy to model them with straight edges and flat faces in SketchUp. In fact, chances are good that the first thing you ever modeled in SketchUp looked a lot like the floor and walls of a building.

There are two different kinds of architectural models that most people want to create in SketchUp. Exactly how you approach modeling floors and walls depends entirely on the type of model you're making:

Exterior model
A model of the outside of a building, which does not include interior walls, rooms, or furniture.

Interior model
A model of the inside of a building, which must take into account interior wall thicknesses, floor heights, ceilings, and furnishings.

- **Exterior models:** An **exterior model** of a building is basically just an empty shell; you don't have interior walls, rooms, or furniture to worry about. This makes for a slightly simpler proposition for users who are just starting out.
- **Interior models:** An **interior model** of a building is significantly more complicated than an exterior-only one; dealing with interior wall thicknesses, floor heights, ceilings, and furnishings involves a lot more modeling prowess.

Keep in mind, everything in SketchUp is made up of super-flat faces (they have no thickness), the only way to model a wall that's, say, 8 inches thick is to use two faces side-by-side and 8 inches apart. For models where you need to show wall thicknesses—namely, interior models—this is what you'll have to do. Exterior models are easier to make because you can use single faces to represent walls. Figure 4-1 shows the difference between single- and double-face walls.

Figure 4-1

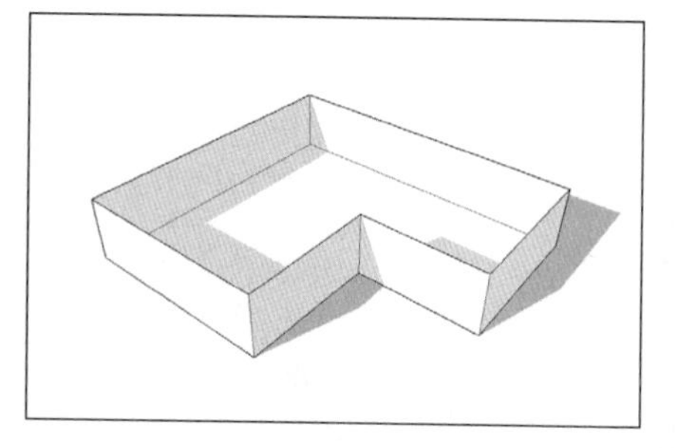

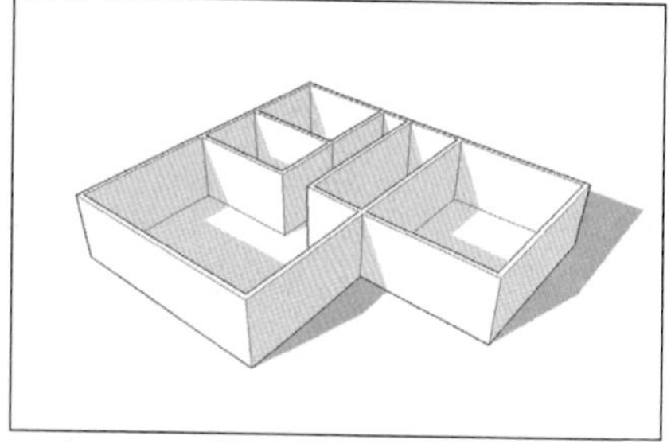

Use single faces for exterior models and double faces for interior ones.

One of the biggest mistakes new SketchUp users make is to attempt an "inside-outside" model right away. Making a model that shows both the interior and the exterior of a building at the same time is, quite frankly, extremely difficult when you are just getting started. Instead, at this point, you should build two separate models if you need both interior and exterior views. If you require a combination model later on, you will be able to build it in a quarter of the time it took you to build either of the first two!

4.1.1 Beginning in 2D

GOOGLE SKETCHUP IN ACTION

Learn to go from 2D to 3D.

Of course, you can make a 3D model of a building's interior in many ways, but we're going to begin with a process that involves drawing a 2D floor plan that includes all your interior and exterior walls, and then extruding it upward to be the right height. In this method, don't worry about doors, windows, or stairs until after your model is extruded. You will put these elements in afterward.

FOR EXAMPLE

SKETCHUP VS. DRAFTING SOFTWARE

If you're importing a floor plan from another piece of software like AutoCAD or VectorWorks, you'll most likely appreciate this approach in that it lets you take 2D information and make it 3D, regardless of where it comes from.

Although SketchUp is a 3D modeling program through and through, it's not a bad tool for drawing simple 2D plans. The toolset is adequate and easy to use, and doing a couple of things before you get started will help you a great deal. One important idea to keep in mind, however, is that SketchUp isn't a full-fledged drafting program, and it probably never will be. If you're an architect-type who needs to do heavy-duty CAD (computer-aided drawing) work, you should probably draft in another piece of software and import your work into SketchUp whenever you need 3D models. If you're just drawing your house or the place where you work, look no further—SketchUp should meet your needs.

PATHWAYS TO... SWITCHING TO 2D VIEW

1. **Create a new SketchUp file.** Depending on the template you have set to open when you create a new SketchUp file, you may already be in a 2D view. If all you see are the red and green axes on a white background, you can skip Step 2. Remember that you can always switch templates by choosing Help Welcome to SketchUp and clicking the Templates section of the dialog box that pops up.
2. **Choose Camera ⇨ Standard Views ⇨ Top.** This changes your viewpoint so that you're looking directly down at the ground.
3. **Choose Camera ⇨ Parallel Projection.** Switching from Perspective to Parallel Projection makes it easier to draw plans in 2D. At this point, your modeling window should look like the one shown in Figure 4-2.

Figure 4-2

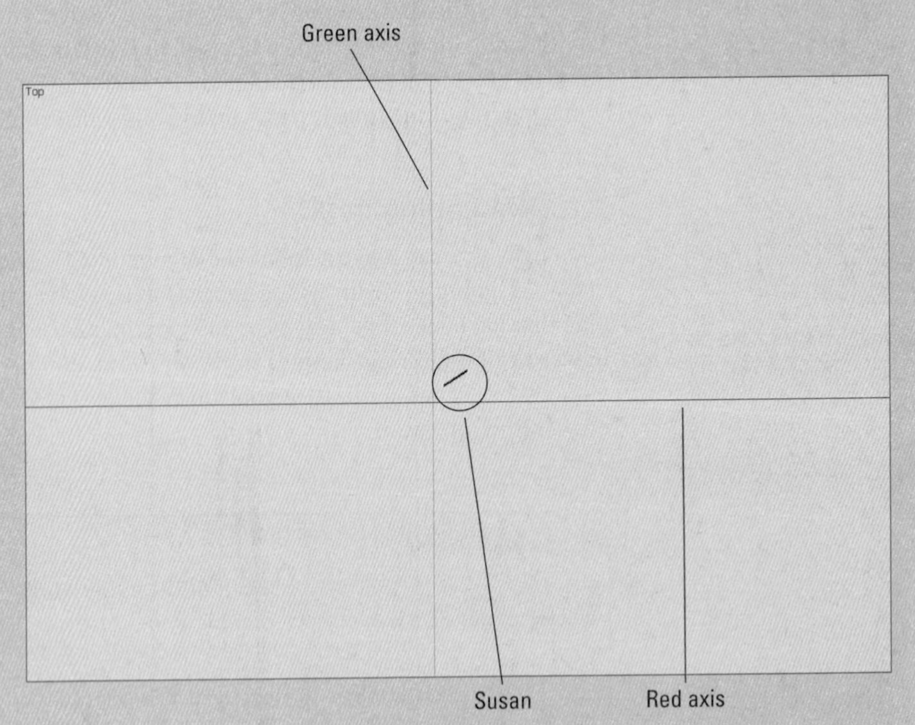

This is what your modeling window should look like before you start drawing in 2D.

Switching to 2D View

If you're going to use SketchUp to draw a 2D plan, the first thing you need to do is orient your point of view. It's easiest to draw in 2D when you're directly above your work, looking down at the ground plane. You also want to make sure that you're not seeing things in perspective, which distorts your view of what you have.

As always, feel free to delete Susan (the human figure that appears on your screen when you create a new SketchUp file) at this point. When working in 2D, Susan appears as a small diagonal line that's visible in your modeling window when you're in Top view. To get rid of Susan, just right-click her and choose Erase from the context menu.

SketchUp's Drafting Tools

Here's some good news: You don't need many tools to draft a 2D plan in SketchUp. Figure 4-3 shows the basic toolbar; everything you need is located right there.

Let's take a closer look at the options on this toolbar:

- **Line tool:** You use the Line tool (which looks like a pencil) to draw edges, which are one of the two basic building blocks of SketchUp models. Fundamentally, you click to start drawing an edge and click again to finish it. (You can find more information about drawing lines in Chapter 2.)
- **Eraser tool:** You use the Eraser to erase edges (see Figure 4-4). Keep in mind that you can't use the Eraser to delete faces, though erasing one of the edges that defines a face automatically erases that face, too. Take a look at the section about edges and faces at the beginning of Chapter 2 for more detail on this. You can use the Eraser tool in two different ways:
- **Clicking:** Click on edges to erase them one at a time.
- **Dragging:** Drag over edges to erase them; this is faster if you have lots of edges you want to erase.

Figure 4-3

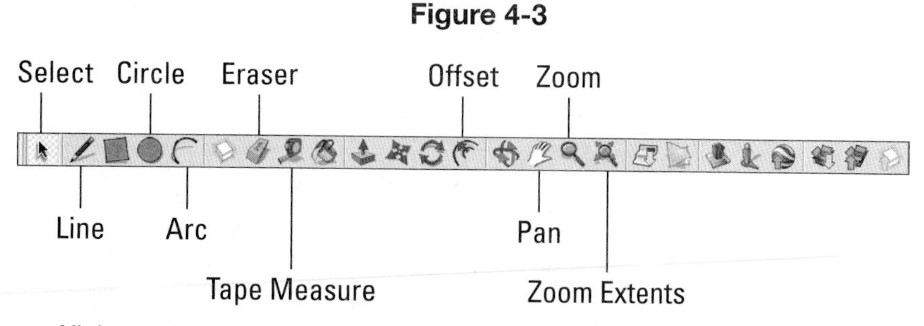

All the tools you need to draft in 2D in SketchUp are on the basic toolbar.

Figure 4-4

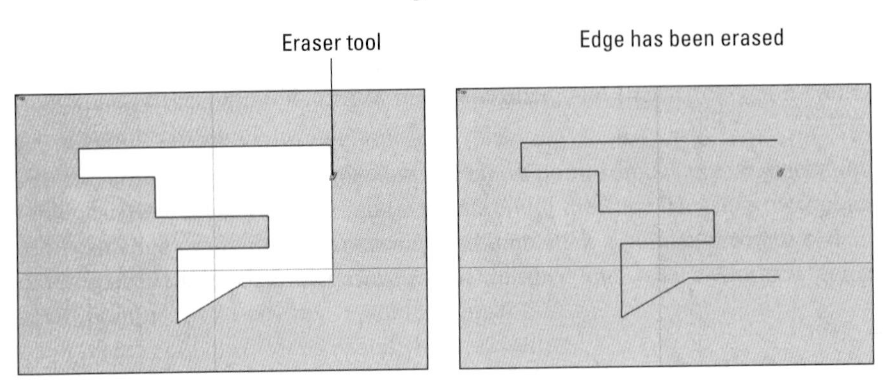

Use the Eraser tool to erase edges. Erasing an edge that defines a face erases that face, too.

- **Circle tool:** Drawing circles in SketchUp is quite easy. You simply click once to define the center and again to define a point on the circle (which also defines the radius). To enter in a precise radius, just draw a circle, type in a radius, and press Enter (see Figure 4-5). For more information on typing while you draw, review the section on model accuracy in Chapter 2.
- **Arc tool:** To draw an arc, click once to define one end of the arc, again to define the other end and a third time to define the bulge (how much the arc sticks out). If you'd like, you can type in a radius after you draw your arc by entering the radius, the units, and the letter *r*. If you want an arc with a radius of 4 feet, for instance, you first would draw your arc however big, then type in **4'r**, and then press Enter. This is shown in Figure 4-6.
- **Offset tool:** The Offset tool helps you draw edges that are a constant distance apart from edges that already exist in your model. Take a look at Figure 4-7. Using Offset on the shape in this figure lets us create another shape that's exactly 6 inches bigger all the way around (middle image) or exactly 6 inches smaller all the way around (right image). Offsetting edges is a useful way to create things like doorways and window trim.

Figure 4-5

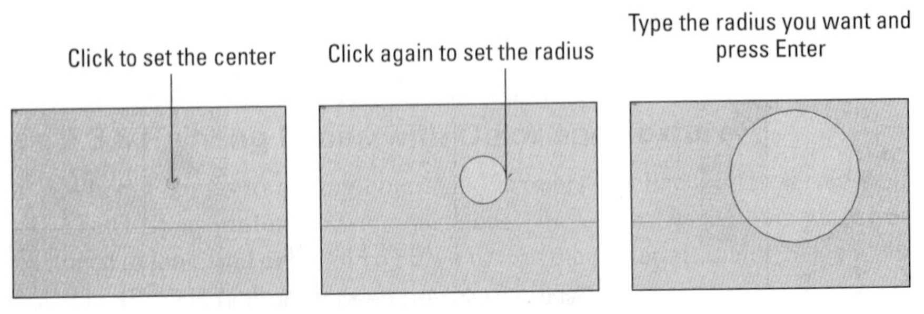

Drawing circles is easy with the Circle tool.

Figure 4-6

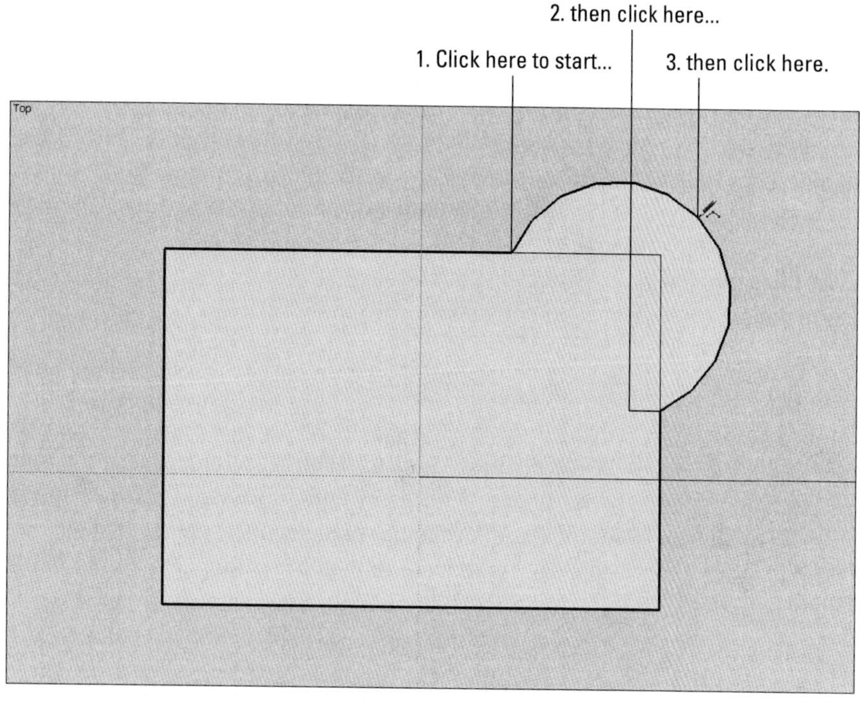

Using the Arc tool is a three-step operation.

You can use Offset in two ways; with both, you click once to start offsetting and again to stop:

- **Click a face to offset all its edges.** If nothing is selected, clicking a face with the Offset tool lets you offset all that face's edges by a constant amount, as shown in Figure 4-7.
- **Preselect one or more edges, and then use Offset.** If you have some edges selected, you can use Offset on just those edges. This comes in handy for drawing things like doorframes and balconies, as shown in Figure 4-8.

Figure 4-7

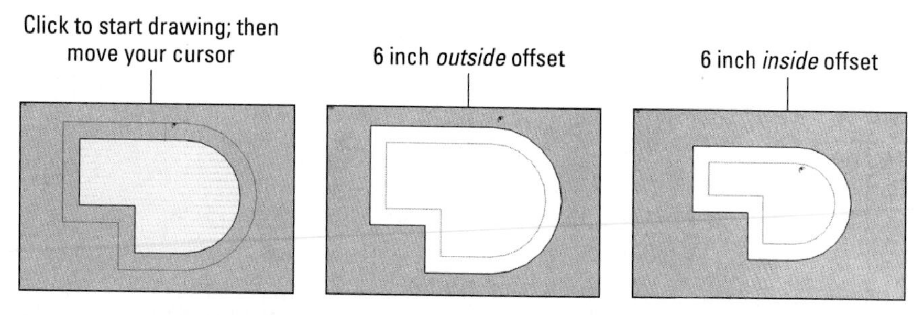

The Offset tool lets you create edges based on other edges.

Figure 4-8

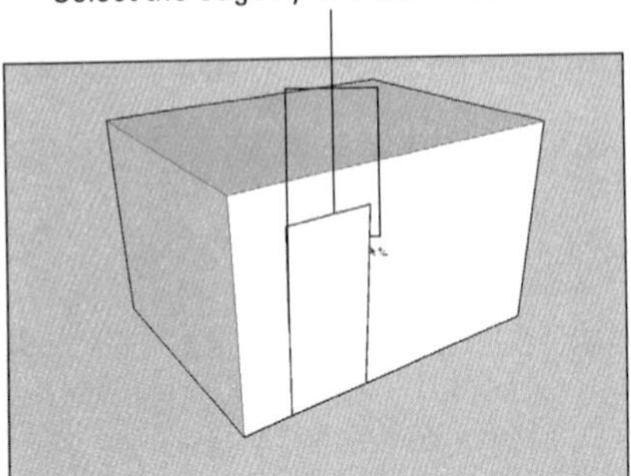

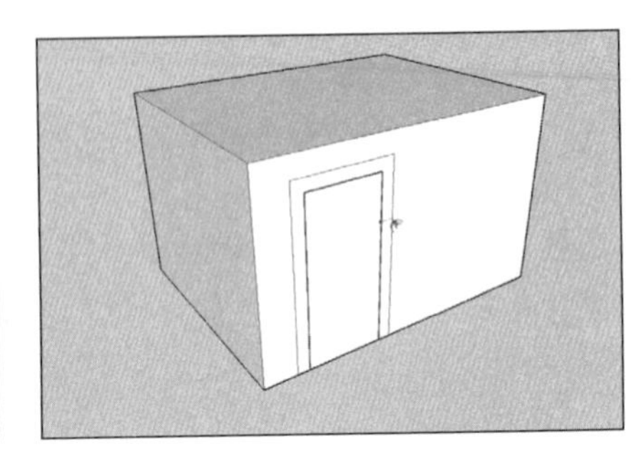

Using Offset on a set of preselected edges is handy for drawing things like doorframes.

- **Tape Measure tool:** The Tape Measure is one of those tools that does many different things. To use it for measuring distances, click any two points in your model to measure the distance between them. The distance "readout" is in the Measurements box in the lower-right corner of your modeling window. You can also use this tool for sizing a model and for creating guides, as explained in Chapter 2.

4.1.2 Creating a Simple Plan

If all you're trying to do is model an exterior view of a building, simply measure around the perimeter, draw the outline of the building in SketchUp, and proceed from there (see Figure 4-9). Your walls will be only a single face thick (meaning paper-thin), but that's okay—you're only interested in the outside, anyway.

If, on the other hand, you want to create an *interior* view, matters are somewhat more complicated. The business of measuring an existing building so that you can model it on the computer is easier said than done—even experienced architects and builders often get confused when trying to create

Figure 4-9

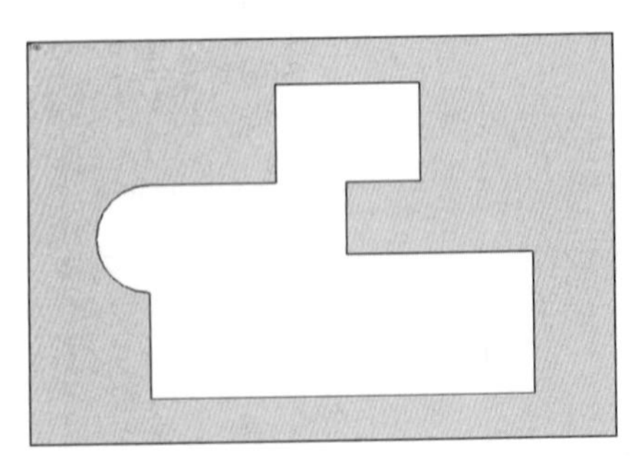

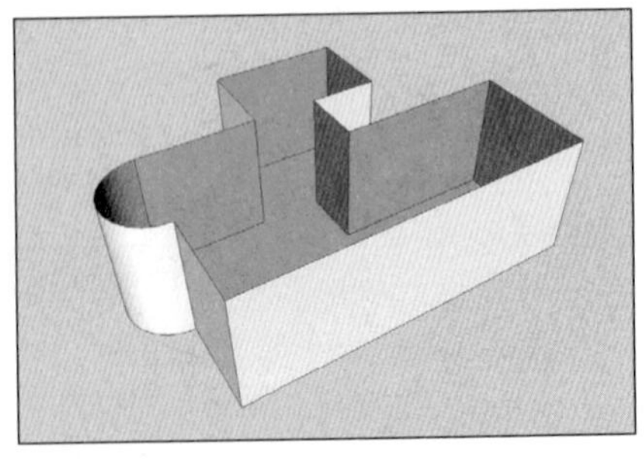

To make an exterior model, just measure the outside of your building to draw an outline in SketchUp.

As-built
A drawing of an existing building.

as-builts, as drawings of existing buildings are called. Closets, ventilation spaces, interior walls, and all kinds of other obstructions inevitably get in the way of obtaining good measurements. Thus, most of the time, you just have to give it your best shot and then tweak things a bit to make them right.

Drawing an Interior Outline

Because the main goal of creating an interior model of a building is to end up with accurate interior spaces, you will need to work from the inside out. If your tape measure is long enough, try to figure out a way to get the major dimensions first—this means the total interior width and length of the inside of your building. You might not be able to, but do your best. After that, it's really just a matter of working your way around, using basic arithmetic and logic to figure things out.

Often, it is extremely helpful to make a paper drawing before you start creating an interior outline with SketchUp. Figure 4-10 shows an example of a paper sketch that was used to model an existing house.

Figure 4-10

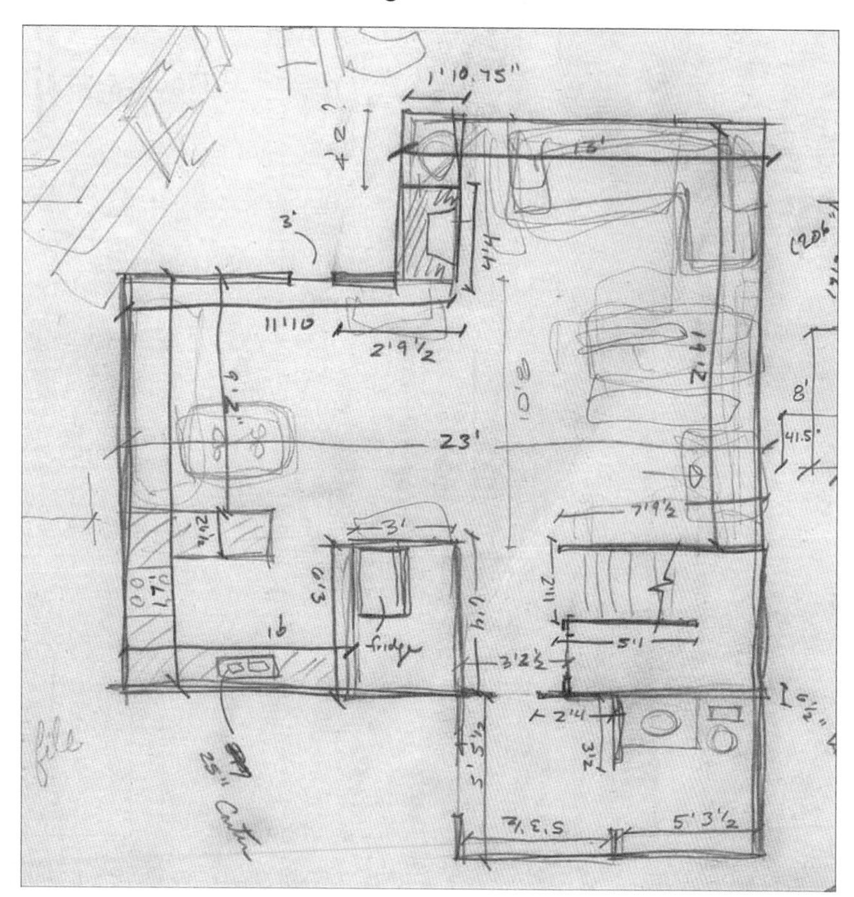

This paper sketch was used to model a house in SketchUp.

From this paper drawing, here's how you would draw a basic interior outline of the house:

1. **First, switch into 2D, overhead view.** The section "Switching to 2D View," found earlier in this chapter, explains how to do this.
2. **Using the Line tool, draw an edge 17 feet long (see the top two images in Figure 4-11), representing the eastern wall of the house.** To draw this, click once to start the edge, move your cursor up until you see the green linear inference (indicating that you are drawing parallel to the green axis), and click again to end your line. To make the edge 17 feet long, type in 17' (you also have the option of typing 204", which is its equivalent in inches) and then press Enter—the line will automatically resize itself to be exactly 17 feet in length. If you'd like, you can use the Tape Measure to double-check what you've done.
3. **Draw an edge 11 feet 10 inches long, starting at the end of the first edge, heading to the right in the red direction. (See the bottom two images in Figure 4-11.)** To do this, do exactly what you did to draw the

Figure 4-11

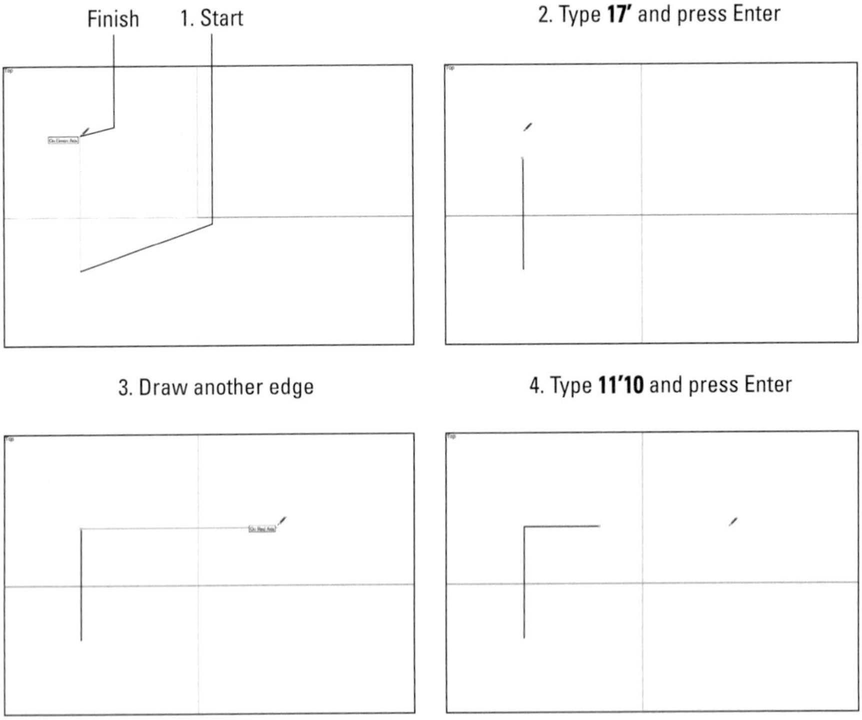

Finish
1. Start
2. Type **17'** and press Enter
3. Draw another edge
4. Type **11'10** and press Enter

Start by drawing an edge 17 feet long, and then draw a perpendicular edge 11 feet 10 inches long.

Figure 4-12

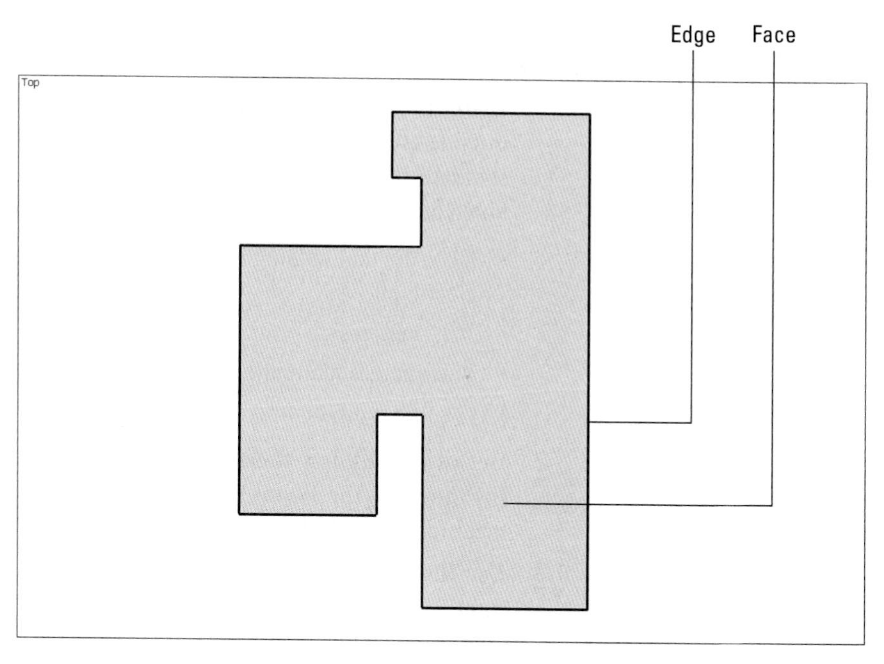

The completed interior perimeter of the house.

first edge, except move parallel to the red axis this time, type in **11'10** (you also have the option of typing 142", which is its equivalent in inches), and then press Enter.

4. **Keep going all the way around the house, until you get back to where you started (see Figure 4-12).** If you make a mistake, either use the Eraser to get rid of edges you're unhappy with or choose Edit ⇨ Undo to go back a step or two.
5. **If all your measurements don't add up, adjust things so that they do, and complete the outline.** When you complete your outline (forming a closed loop of edges that are all on the same plane), a face automatically appears. In this example, we have a total of 11 edges and 1 face.

When you are drafting in 2D, whatever you do, *do not* use the Orbit tool. Because you're working in 2D, you only need to use Zoom and Pan to navigate around your drawing (see Chapter 2 for more information on this). If you accidentally end up orbiting your model into a 3D view, follow the steps in the section "Switching to 2D View," found earlier in this chapter, to get things back in order. Finally, remember, if you ever get lost and no amount of zooming and panning gets you back to a view of your floor plan, choose Camera ⇨ Zoom Extents. Think of this as an emergency lever you can pull to fill your modeling window with your geometry.

Offsetting an Exterior Wall

You can offset an exterior wall thickness, just to make it easier to visualize your spaces. Here's how to do this:

1. **Using the Offset tool, offset your closed shape by 8 inches to the outside (see Figure 4-13, left panel).** An offset of 8 inches is a pretty standard thickness for an exterior wall, especially for houses. This is how you can use the Offset tool to create 8-inch thick walls:
   - Make sure that nothing in your drawing is selected by choosing Edit ⇨ Select None.
   - Click once inside your shape.
   - Click again outside the shape to make a second, bigger shape.
   - Type in **8,** and then press Enter.
2. **Because you know there are no alcoves on the outside of the house (as shown in the sketch), use the Line tool to draw across them.** See the middle panel of Figure 4-13 for an illustration of how this is done.
3. **Use the Eraser tool to get rid of the extra edges** (see Figure 4-13, right). By deleting the extra edges, you go back to having only two faces: one representing the floor, and one representing the exterior wall thickness. It doesn't matter that the wall is thicker in several places than in others; you can always go back and remedy this later on.
4. **With the Line tool, draw edges that define the thickness of your exterior wall**. (See Figure 4-13, upper-right.) In the case of this house, this means separating the bulges (which actually represent a fireplace and a mechanical closet) from the part of the wall that goes all the way up to the roof, two stories up. When you are done, you end up with several faces: one for the floor, one for the exterior wall (whose thickness should be more or less uniform), and a few for the bulges.
5. **Select the face that defines the exterior wall.** (See Figure 4-13, middle-right.) The easiest way to do this is to click once on the face with the Select tool.
6. **Make the face you just selected into a group.** (See Figure 4-13, lower right.) Chapter 5 explains these groups and their components, but this is all you need to know for now. Making groups allows you to separate different parts of your model. Turning your exterior wall into a separate group makes it easier to edit, hide, and move. Groups also simplify the process of adding more levels to your building, if that becomes necessary. To turn the face you selected in Step 5 into a group, choose Edit ⇨ Make Group. You see a perimeter of blue lines around your face; that's the group you just created. Congratulations! You have now just moved from beginner to intermediate SketchUp user!

Figure 4-13

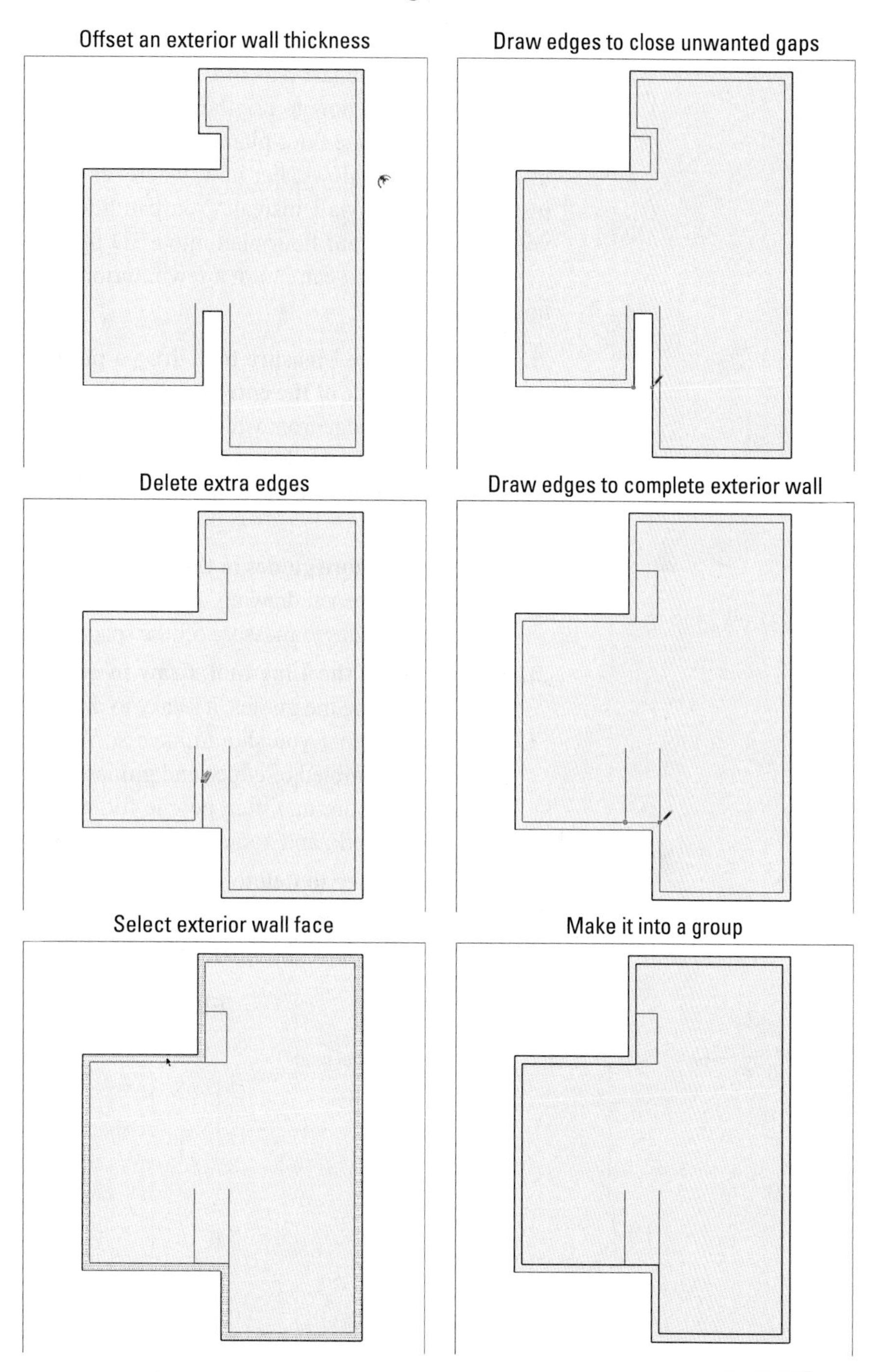

Use Offset to create an exterior wall thickness, and then clean up the image using the Line and Eraser tools.

Putting in Interior Walls

For this part of the process, guides are primarily used. If you have not done so already, read the last part of Chapter 2, where you'll find a full description of guides and how to use them.

When drafting a floor plan in SketchUp, it often helps to ignore things like doors and windows. For now, wherever a doorway should be in a wall, just draw a solid wall instead. You can add in doors and windows after you've extruded your floor plan into a 3D figure.

Here's how you can put in a few interior walls on the first floor of your house:

1. **With the Tape Measure tool, drag a parallel guide 5 feet, 3 ½ inches from the inside of the entryway (see Figure 4-14, left panel).** To do this, just click the edge from which you want to draw the guide, move your cursor to the right (to tell SketchUp which way to go), type in 5'3.5, and press Enter (you also have the option of typing 65.5", which is its equivalent in inches).
2. **Draw a few more guides in the same way you drew the first one.** Working from the pencil drawing, figure out the location of each interior wall, and create guides to measure off the space (see Figure 4-14, right panel).
3. **Switching to the Line tool, draw in edges to represent the interior walls.** By using the guides, it's easy to draw your edges correctly. Figure 4-15 shows what you should have so far. Remember to zoom! When you have a jumbled of edges and guides and you can't see what you are doing, just zoom in. Often people forget to change their point of view while they work, and zooming makes all the difference.
4. **Use the Eraser to delete your guides.**
5. **Use the Eraser to get rid of all extra edge segments (see Figure 4-16).** The goal is to have the smallest-possible number of 2D faces to extrude

Figure 4-14

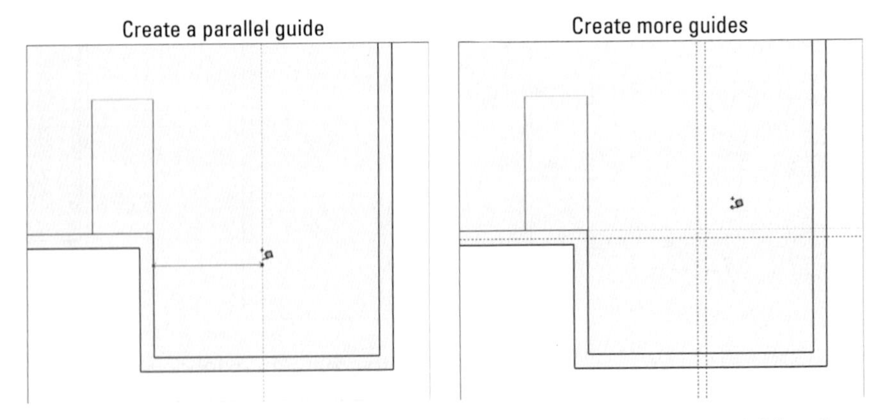

Draw a guide to help locate your first interior wall, create additional guides, and then draw edges using these guides.

Figure 4-15

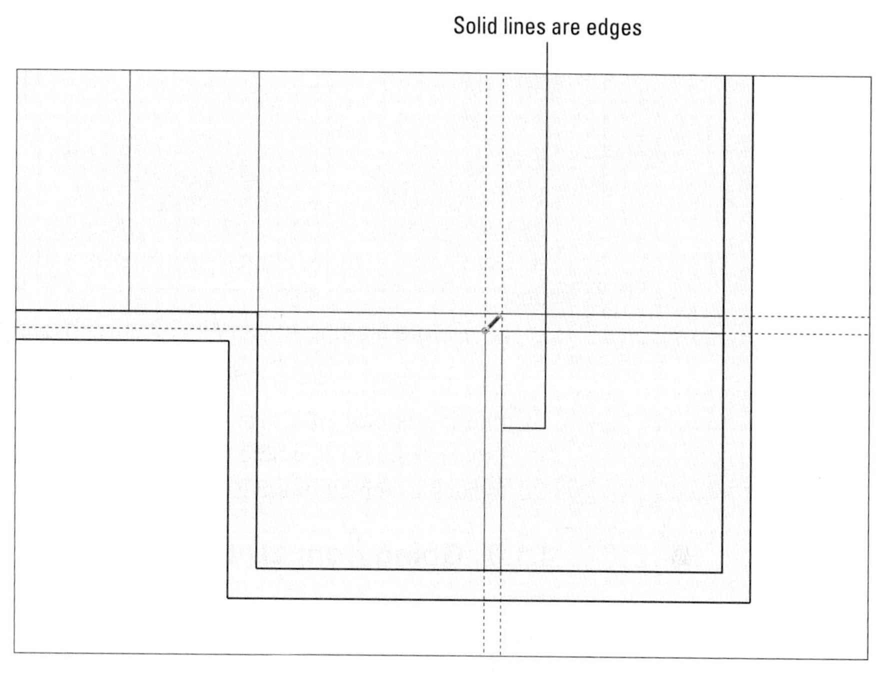

Use the Line tool to create edges where your guides come together.

Figure 4-16

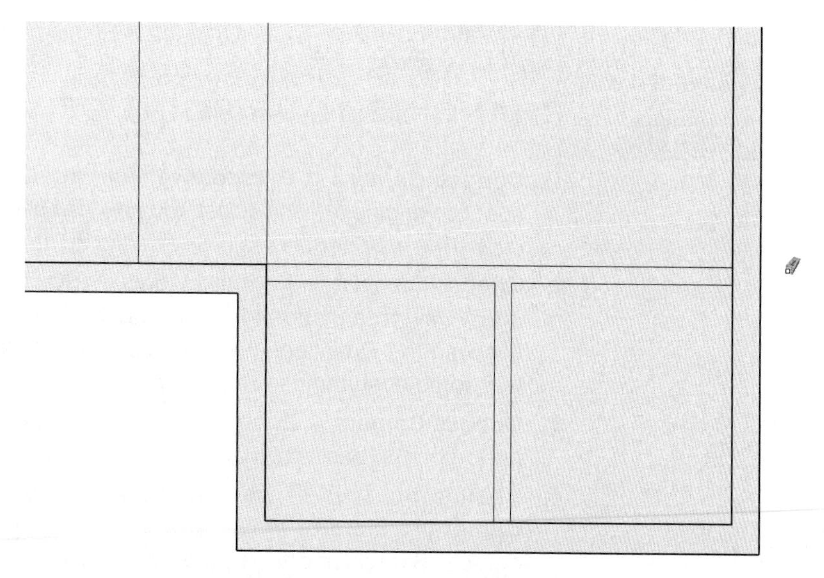

Using the Eraser, delete your guides and all the small edge segments left over from drawing the interior walls.

Figure 4-17

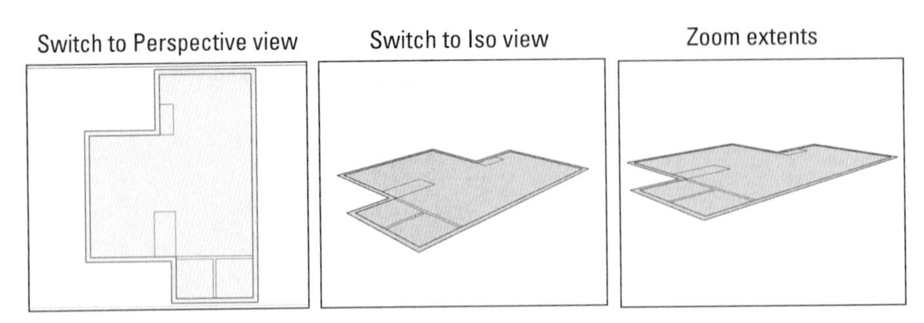

Before starting to work in 3D, switch over to a 3D view.

into 3D walls. Because the exterior-wall face—and the edges that define it—is part of a separate group, accidentally nicking it with the Eraser deletes the whole thing at once. If this happens, just use Undo to back a step, zoom in a little and try again.

4.1.3 Going from 2D to 3D

Once you have a 2D plan in hand, the next step is to extrude it into a 3D model. This is basically a one-step process, and it involves one important tool: Push/Pull. In the following sections, you'll take the simple floor plan you drew earlier in the chapter and turn it into 3D walls.

Changing Your Point of View

Before you pop up your plan into the third dimension, you need to change your point of view to get a better view of what you're doing. (See Figure 4-17.)

PATHWAYS TO...
CHANGING YOUR POINT OF VIEW

1. **Choose Camera ⇨ Perspective.** This "turns on" SketchUp's perspective engine, meaning that you now can see things more realistically—the way people really see things in 3D.
2. **Choose Camera ⇨ Standard Views ⇨ ISO.** This switches you from a top view to an isometric (three-quarter) view. You could also do this with the Orbit tool. (Remember, there is always more than one way to do everything in SketchUp.)
3. **Choose Camera ⇨ Zoom Extents.** Zoom Extents also has its own button on the basic toolbar.
4. **Change the field of view from 35 to 45 degrees by choosing Camera ⇨ Field of View, typing in 45, and pressing Enter.** By default, SketchUp's field of view is set to 35 degrees. (For more information on what this means, check out Chapter 10.)

Pushing and Pulling

The Push/Pull tool is a simple device; you use it to extrude flat faces into 3D shapes. It works (like everything else in SketchUp) by clicking—you click a face once to start pushing/pulling it, move your cursor until you like what you see, and then click again to stop push/pulling. That's it. (For more detail on Push/Pull, see the nearby sidebar, "More Fun with Push/Pull.")

Note that Push/Pull only works on flat faces; if you need to do something to a curved face, you'll have to use another tool (such as the Intersect with Model feature, which is discussed later in this chapter).

To use Push/Pull to extrude your house's first-floor plan into a 3D model (as shown in Figure 4-18), this is what you should do:

1. **Select the Push/Pull tool from the toolbar.** This tool looks like a little box with a red arrow coming out of the top.
2. **Click the "wall's" face once to start extruding it.** If you click the "floor" face, you would end up extruding that instead. If you choose the wrong face by accident, press Esc to cancel the operation, and try again.

Figure 4-18

1. Push/pull one interior wall

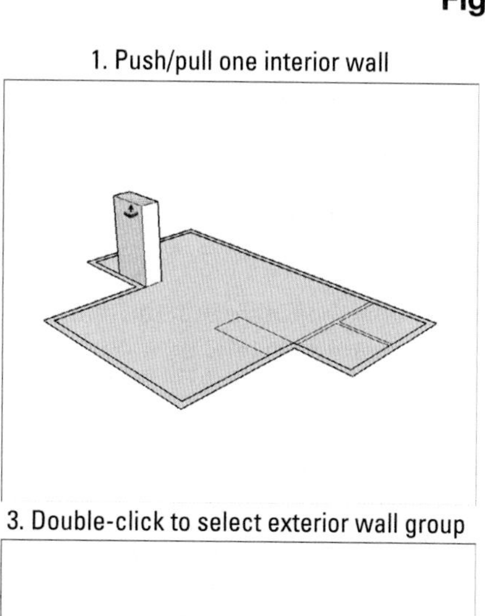

2. Push/pull remaining interior walls

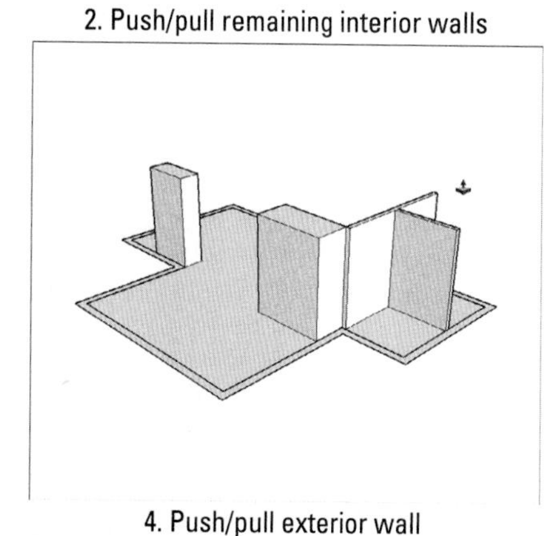

3. Double-click to select exterior wall group

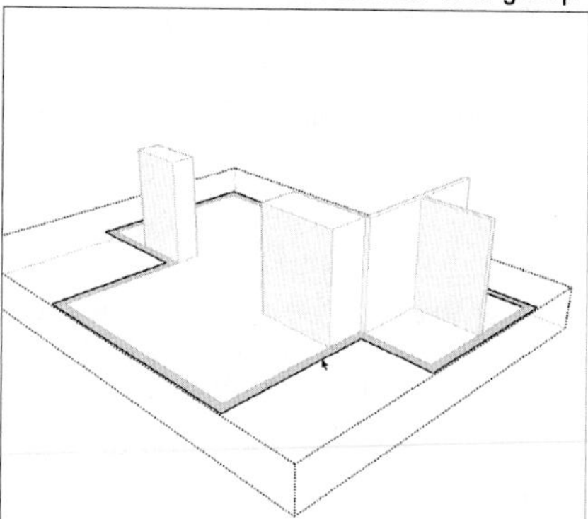

4. Push/pull exterior wall

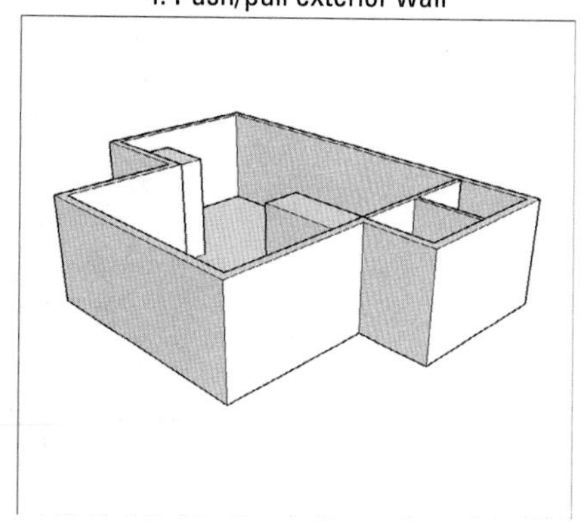

Use Push/Pull to extrude one of your faces into all the walls in your house.

3. **Move your cursor up to pull up the walls, and click to stop extruding.** It doesn't matter how much you extrude your face, because you're going to add precision in the next step.
4. **Type in 8' and press Enter.** When you do this, your push/pull distance is revised to exactly 8 feet (you also have the option of typing 96", which is the equivalent in inches).
5. **Repeat Steps 2–4 for all the interior walls in the house.** Using Orbit helps you view what you are doing as you work around the model.
6. **Push/pull the exterior wall to match the height of these interior walls.** Because the exterior wall face is part of a group, you need to edit the group before you can do anything to it. Right-click the exterior wall face and choose Edit Group from the context menu to get "inside," where you can follow Steps 2–4 in the preceding steps, to make the exterior wall group 3D. Clicking anywhere in space exits the group when you are done.

TIPS FROM THE PROFESSIONALS

More Fun with Push/Pull

Because Push/Pull is the tool that most people think of when they think of SketchUp, it might be helpful to know more about what it can do. The people who invented SketchUp actually *started* with the idea for Push/Pull—that's how closely SketchUp and Push/Pull are linked! Here are four things about Push/Pull that aren't immediately obvious when you first start using it:

- **Double-click with the Push/Pull tool to extrude a face by the last distance you pushed/pulled.** When you double-click a face, it automatically gets pushed/pulled by the same amount as the last face you used Push/Pull on.
- **Press Ctrl (Option on a Mac) to push/pull a *copy* of your face.** The first graphic in the following figure shows what this means. Here, instead of using Push/Pull the regular way, you can use a modifier key to extrude a copy of the face you're pushing/pulling. This comes in handy for quickly modeling things like multi-story buildings.
- **While pushing/pulling, hover over other parts of your geometry to tell SketchUp how far to extrude.** Take a look at the second graphic. Perhaps you want to use Push/Pull to extrude a cylinder that is exactly the same height as a box. Before you click the second time to stop pushing/pulling, hover over a point on the top of the box; now the cylinder is exactly that tall. To complete the operation, click while you're still hovering over the box. It's pretty simple, and it'll save you hours of time after you're used to doing it.

- **Pushing/pulling a face into another, coplanar face automatically cuts a hole.** In fact, this is how you make openings (like doors and windows) in double-face walls. The last graphic shows this in action.
- **If you happen to have trouble push/pulling anything, chances are that the face that you are trying to select is grouped.** Right-click on the grouped element and choose Edit and try again. If nothing is happening, you might have a group nested inside another group.

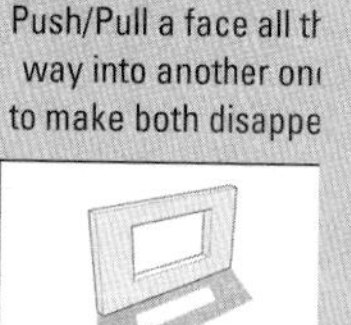

GOOGLE SKETCHUP IN ACTION

Understand how to add floors to buildings.

4.1.4 Adding Floors to Your Building

Adding a second (and third, and fourth) floor to your model is not as hard as it may seem. The key is to think of each level as a separate "tray" consisting of interior walls, a floor surface, and the ceiling of the level below. Each floor is modeled as an individual group, making it easier to hide, edit, and move. For the same reasons, exterior walls are also made as a separate group; they act kind of like a "box" into which your floor levels stack, see Figure 4-19.

Making Groups to Keep Things Separate

If you've been following along since the beginning of this chapter, the edges and faces that make up your exterior walls are already enclosed in a group by themselves. If they're not, seriously consider going back a few steps—taking the time to set things up now will save you hours of headache later. Trust this. Otherwise well-meaning people who have worked with other CAD or 3D modeling programs often take this opportunity to bring layers into the discussion. It's true, SketchUp has a Layers feature and the floor "trays" are a lot like layers conceptually. But, Layers would not be used when modeling multiple levels of the same building. Layers in SketchUp simply don't work the way you might think they do. Reference Chapter 7 to learn more.

Provided your exterior walls are already a group, the next step is to turn the rest of your first floor's geometry into another group. Follow these steps:

1. **Select the floor and interior walls of the first level.** You can accomplish this efficiently (with the Select tool): Just triple-click a face on any interior wall to select everything that's attached to it. Take a look at Chapter 2 for plenty of tips on selecting things.
2. **Make a group by choosing Edit ⇨ Make Group from the menu bar.** Chapter 5 is all about groups and components; peruse the first few pages if you need a quick tutorial.

Figure 4-19

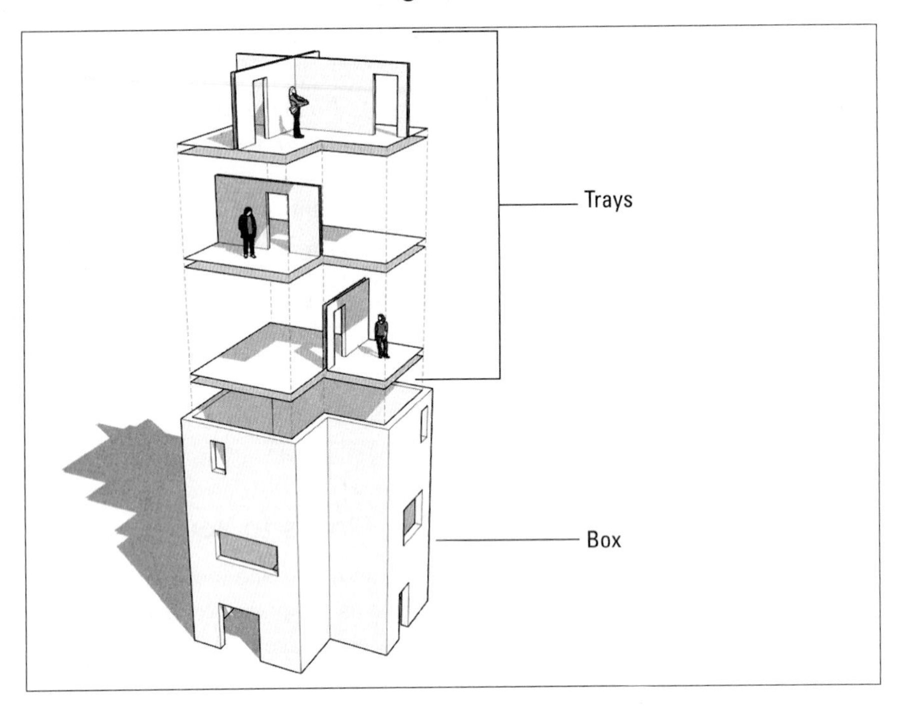

Floor levels are like trays stacked inside a box consisting of your exterior walls.

Drawing the Next Floor

Modeling each new floor, directly on top of the one underneath, guarantees that everything in your building lines up. Some folks advocate for working "off to the side" and putting things together later, but I think that's a recipe for trouble. Here's how you add a second floor to the house model (check out Figure 4-20 to see the steps as pictures):

1. **Trace the inside perimeter of the exterior wall to create a new face.** Use the Line tool to do this. Keep in mind that this works only if everything you touch is already part of another group; if it isn't, your new edges stick to your existing ones, and your model becomes messy.
2. **Push/Pull your new face into a thick slab.** Thickness depends on your building, but a reasonable ceiling-to-floor distance between levels for houses is about a foot. You can figure this out with a tape measure and a calculator. The underside of the new slab is the ceiling of the first floor. It is best to model buildings this way because it improves visibility when hiding a floor group to see the one below it.
3. **Draw the interior walls of the new floor.** This is just like drawing the first floor. Switch to the Top view (Camera ⇨ Standard Views ⇨ Top) and then use the Tape Measure, Eraser, and Line tools to draft your floorplan.

Figure 4-20

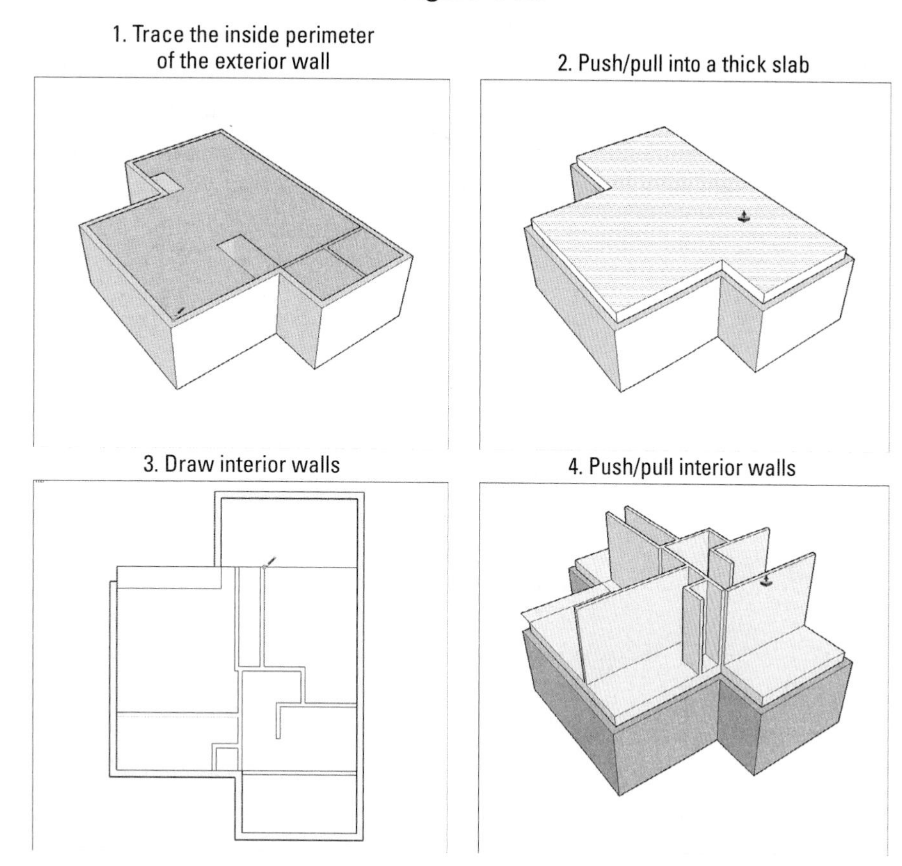

Draw right on top of the lower floor, then push/pull the interior walls to ceiling height.

Just start at the very beginning of this chapter for a refresher. If the floor you are drawing is bigger than the one below it, its outline overlaps the exterior walls. That's okay—just pay special attention to where your edges and faces end up as you draw. Use the Orbit tool every once in a while to check that everything's in order. In the event that your new floor is smaller than the one underneath, represent the inside boundary of the new exterior walls with a single edge. The next section deals with what to do when your first and second floorplans don't exactly match up.

4. **Push/Pull your interior walls to the correct height.** In this example it's 8 feet.
5. **Group together your interior walls, your floor, and the ceiling of the level below.** If you are unsure of how to do this, take a look at the steps in "Making groups to keep things separate," a few pages back.
6. **If your upper floor isn't bigger or smaller than your lower floor, pull up your exterior walls to match your interior ones**. In this example,

you are extending the box that holds your floor trays up another level. Follow these steps:

a. Edit your exterior wall group by double-clicking it with the Select tool.

b. Use Push/Pull to extrude it up.

c. Exit (stop editing) the group by clicking somewhere off to the side of your model.

Chances are your newest floor doesn't line up exactly with the one below it. Keep reading to find out what to do.

Creating Additional Exterior Walls

Most buildings are not simple extrusions; they bump in and out as they rise. Second-floor decks sit atop first-floor garages; bedrooms cantilever over gardens; intermediate roofs shelter new room additions. Buildings—especially multilevel houses—are complicated assemblies. Figuring out where walls, floors, and ceilings come together takes time, trial and error, and a good dose of spatial reasoning. It's best not to attempt the steps in this section when you are tired or distracted. If your building does happen to be one of the few with perfectly aligned floorplans, you can skip this section entirely; however, we know that's not likely the case. In this house, the second floor both overhangs and under-hangs the first floor. Wherever this happens, add a new section of exterior wall. The photograph in Figure 4-21 illustrates this.

Figure 4-21

Second floor bumps out

First floor bumps out

The outline of the second floor does not exactly match that of the first.

Follow these steps to solve this tricky problem:

1. **Draw faces to define any new exterior walls (see Figure 4-22):**
   a. Use the Line tool to trace the inside perimeter of your new exterior walls.
   b. Hide the group that includes your second-floor interior walls by right-clicking it and choosing Hide.

Figure 4-22

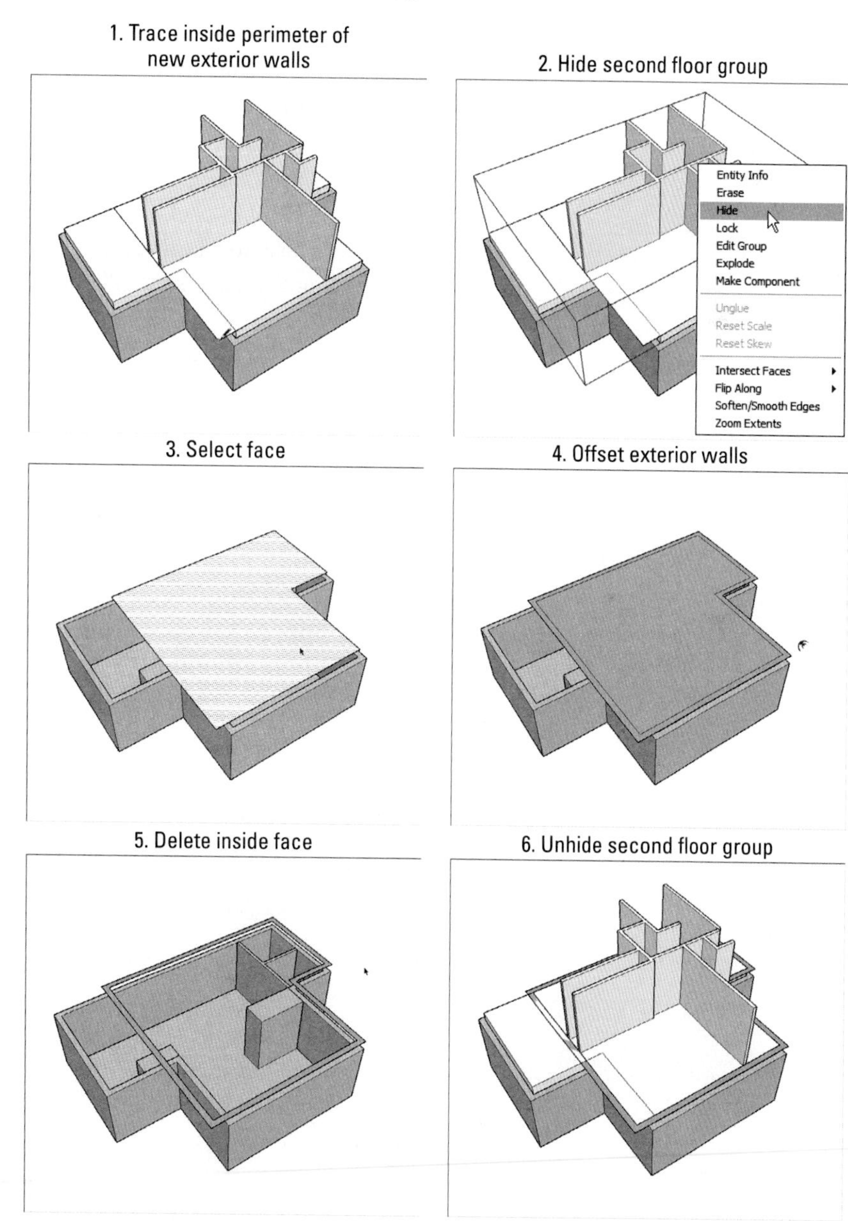

Use Offset to draw faces that represent new exterior walls.

c. Select the face that you created when you traced the inside perimeter in Step a. If you do not see a face, maybe you forgot to draw an edge somewhere.

d. Use the Offset tool to offset the edges of your selected face by the thickness of your exterior walls. *In this case, 8 inches.*

e. Delete the face in the center, leaving only a face that represents your new exterior wall thickness.

f. Unhide the group you hid in Step b by choosing Edit ⇨ Unhide ⇨ Last from the menu bar.

2. **Make a group out of your new exterior wall face by selecting it and then choosing Edit ⇨ Make Group.**

3. **Delete any floor geometry that doesn't belong. (See Figure 4-23, top.)**

   In this example, part of the floor extends past the exterior wall on the left side of the figure. Double-click the group with the Select tool to edit it, and then use the Eraser to take away only what you need to, being careful to leave the ceiling that covers the first floor, below.

Figure 4-23

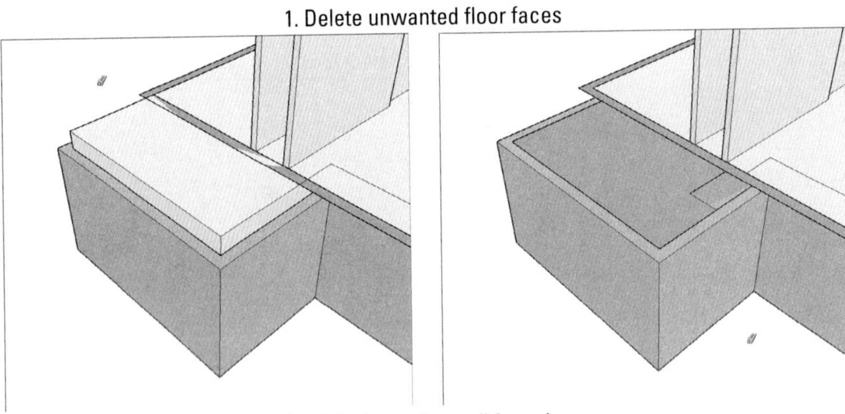

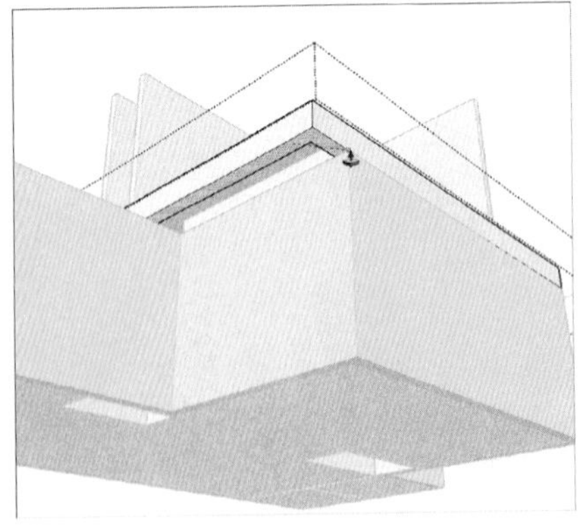

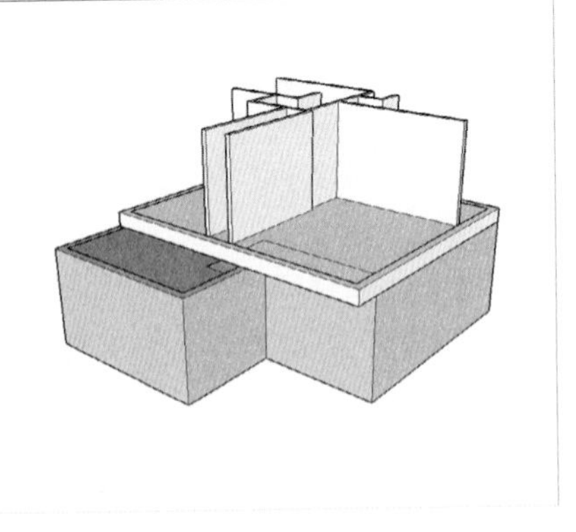

Delete extra floor faces; then push/pull down the walls.

4. **Push/pull down any wall faces to meet the top of the lower floor's exterior wall. (See Figure 4-23, bottom.)** Double-click a group with the Select tool to edit it; you need to do this before you can push/pull any of the faces you created in Step 1.
5. **Make all your exterior walls part of the same group**:
   a. Select the group that contains your new exterior walls and then choose Edit ⇨ Cut.
   b. With the Select tool, double-click the group containing your lower exterior walls. *You're "inside" that group.*
   c. Choose Edit ⇨ Paste in Place.
   d. Choose Edit ⇨ Group ⇨ Explode to ungroup the edges and faces in the selected group, sticking them to those in the lower group.
6. **Use Line, Erase, and Push/Pull to extrude your exterior walls up to the height of your interior walls (see Figure 4-24).** After everything

Figure 4-24

1. Make all exterior walls part of the same group

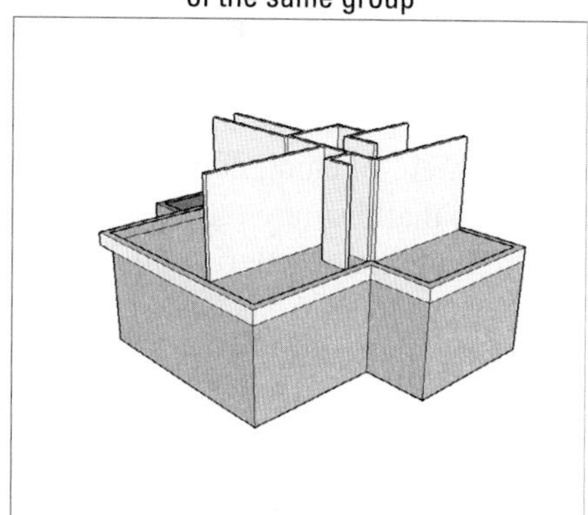

2. Delete edges between faces

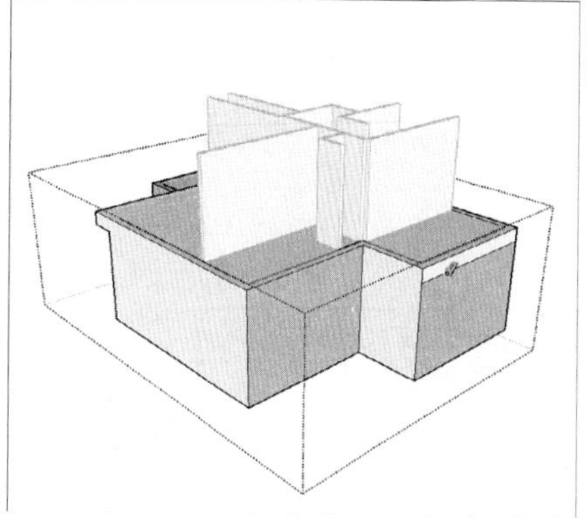

3. Push/pull exterior walls

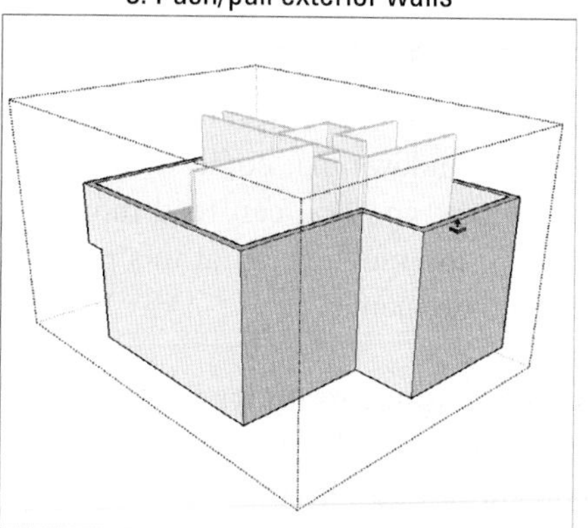

4. Orbit to make sure everything looks right

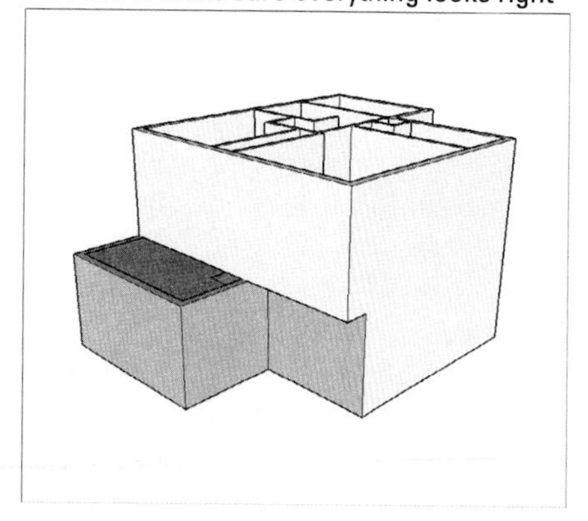

Make your exterior walls look accurate.

is in the same group, adding necessary edges, deleting extraneous ones, and pushing/pulling faces all at once is easier. Using your SketchUp knowledge, watch the colors as you draw, use the Shift key to lock inferences, and remember to zoom in on what you are doing. Reviewing Chapter 2 provides useful pointers on these actions.

Moving Forward

Now that you have learned the technique for modeling multilevel buildings, you can build up as high as you like. As you proceed, the following may be helpful:

4.1.5 Adding Doors and Windows

You can make openings in your walls in a couple of different ways. What you choose to do depends on what kind of building you are modeling, whether you are using single-face or double-face walls, and how much detail you plan to include in your model. You have two options:

- **Use SketchUp components that cut openings themselves.** The Google 3D Warehouse (read more in Chapter 5) contains scores of doors and windows that you can download and use in your models. Some of them cut their own openings when you insert them in a face. There is a catch, though: SketchUp's cut-opening components work only on single-face walls, which means that they are only really useful for exterior building models. If you're building an interior model, you have to cut your own openings.

Learn how to create and insert doors and windows.

- **Cut openings yourself.** For double-face walls, this is your only option; luckily, it's easy to do. Basically, you draw an outline for the opening you want to create, and then you use Push/Pull to create the opening—it works the same way for both doors and windows.

Using SketchUp's Components

As mentioned earlier, as long as you're making an exterior model, you can use the doors and windows that come with SketchUp. These are components of the program, and you can read more about them in Chapter 5. Without going into too much detail at this point, however, here are a few basics related to components:

- **They're in the Components dialog box.** Choose Window ⇨ Components to open the dialog box and then choose the Architecture collection from the Navigation drop-down list (it looks like an upside-down triangle). The Doors, Windows, and DC Doors and Windows collections are contained in there. Components that can cut their own openings generally contain

Figure 4-25

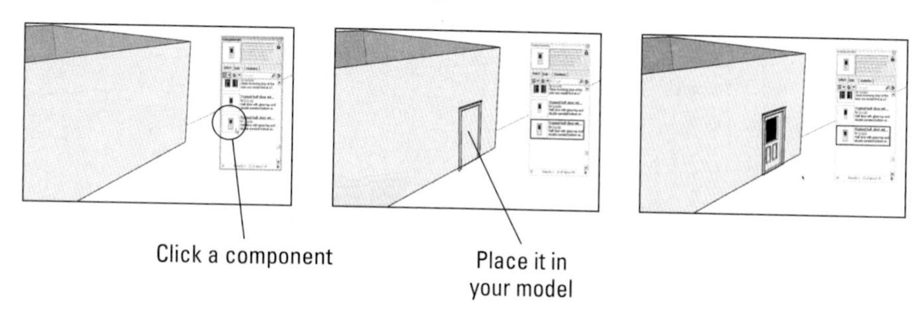

Placing window and door components in a model is simple.

gluing or cutting in their descriptions. Keep in mind that you need to be online to access the Google 3D Warehouse.

GOOGLE SKETCHUP IN ACTION

Work with SketchUp's components.

- **You can find hundreds more online.** If you're connected to the Internet, you can type any search query (such as revolving door) into the little search area at the top of the Components dialog box. This scours the 3D Warehouse for whatever you're looking for and shows the results below. Some advice: The 3D Warehouse holds so much information, making your query specific helps you sort through the results.
- **They are editable**. There's much more detail about this in Chapter 5, but for now, here's the gist: If you don't like something about one of SketchUp's built-in doors or windows, you can change it.
- **Some components are dynamic.** Dynamic Components have special abilities that make them easier to resize and otherwise reconfigure. Read more about Dynamic Components in Chapter 5.
- **They cut their own openings, but the openings are not permanent.** When you move or delete a door or window component you've placed, its opening is deleted with it. Adding a hole-cutting component to your model is easy, as you can see from Figure 4-25.

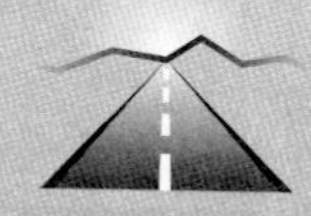

PATHWAYS TO... WORKING WITH COMPONENTS

1. In the Components dialog box, click the component that you want to place in your model.
2. Place the component where you want it to be.
3. If you don't like where it is, use the Move tool (explained in Chapter 2) to reposition your component.

Figure 4-26

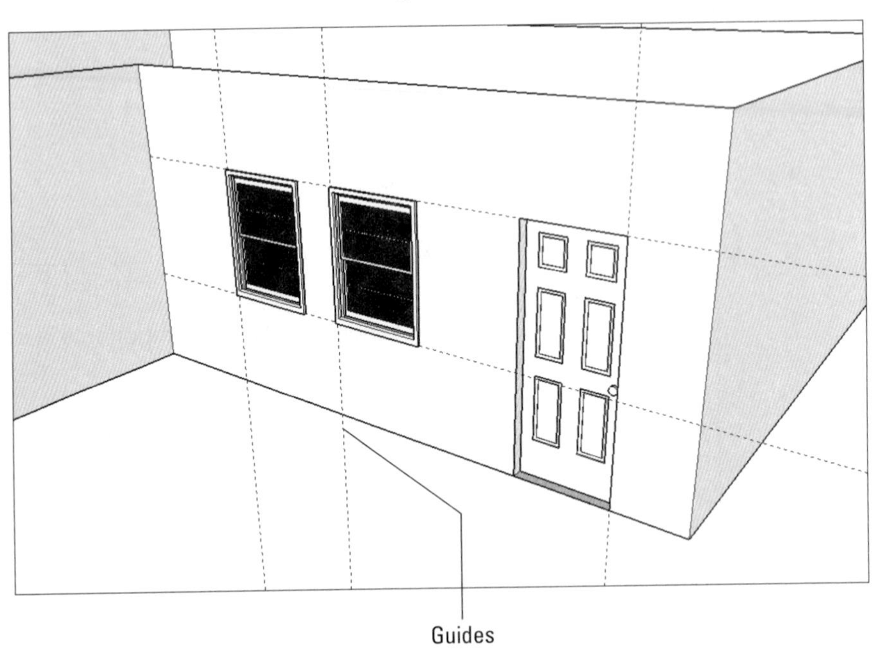

Use guides to help line up the components you add to your model.

Figure 4-26 shows a simple building to which a door and a couple of window components have been added. Notice how guides have been used to line things up—doing this is the best way to make sure everything's in the right spot.

Making Your Own Openings

Most of the time, you won't be able to get away with using SketchUp's built-in door and window components—the fact that they can't cut through two-faced walls means that they're limited to external use only. That's okay though, because cutting your own holes in walls is quick and easy, and you'll always end up with exactly what you want.

Figure 4-27 illustrates the basic steps in cutting a precise opening in a double-face wall.

Here's a more detailed discussion of how to carry out these steps:

1. **Mark where you want your opening to be with guides.** For a refresher on using guides, refer to Chapter 2. If you are drawing on a

Figure 4-27

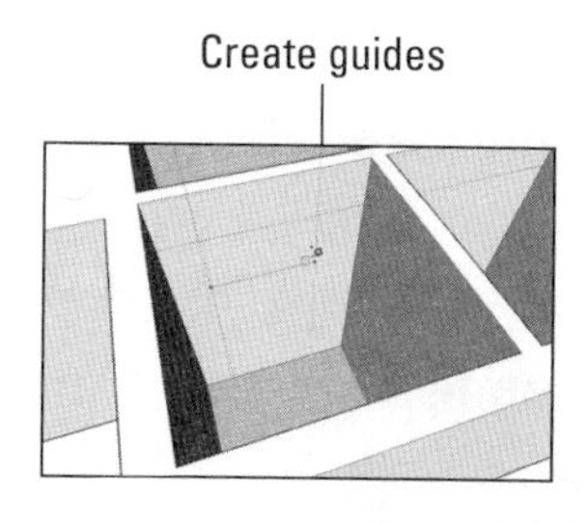

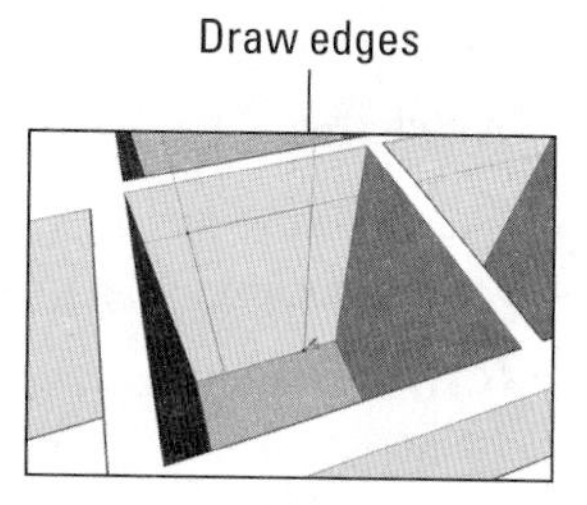

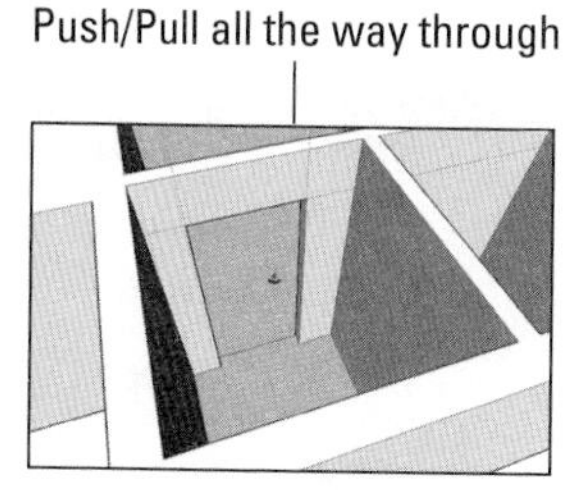

Use guides to plan where you want an opening, and then push/pull all the way through both faces.

wall that is part of a group, you need to edit that group in order to punch holes in the wall. To edit a group, double-click it with the Select tool. To stop editing, click somewhere off to the side of your model. Reference the first part of Chapter 5 for more info about working with groups.

2. **Draw the outline of the opening you want to create, making sure to create a new face in the process.** You can use any of the drawing tools to do this, though it's usually best to stick with the Line tool when you're starting out. Remember to pay close attention to what you're drawing; keep an eye out for the colored inferences, which let you know where you are.
3. **Use Push/Pull to extrude your new face back into the thickness of the wall until it touches the face behind it.** If everything goes well, your face should disappear, taking with it the corresponding area of the face behind it. Now you have an opening in your wall. If your face doesn't disappear and no opening is created, it's probably for one of the following reasons:

   - **Your faces aren't parallel to each other.** This technique only works if both faces are parallel. Keep in mind that just because two faces *look* parallel doesn't mean that they are.
   - **You hit an edge.** If you push/pull your face into a face with an edge crossing it, SketchUp gets confused and doesn't cut an opening. Use Undo, get rid of the pesky edge (if you can), and try again.

Throughout this process, remember to orbit! If you can't quite push/pull what you mean to push/pull, orbit around until you can see what you're doing.

IN THE REAL WORLD

Working Through Thick and Thin

Pay attention to which edges look thick and which ones look thin. When you're drawing in 2D, you can tell a lot from an edge's appearance:

- **Thin edges cut *through* faces.** Edges that are thin are ones that have "sunk in"; you can think of them like cuts from a razor-sharp knife. When you successfully split a face with an edge that you draw with the Line tool, it appears thin. The first image in the following figure shows what this looks like.
- **Thick edges sit *on top* of faces.** If the edge you just drew looks thicker than some of the other edges in your model, it isn't actually cutting through the face it's on—it's only sitting on top. An edge can sit on top of a face for a couple of reasons:
- *It has one end free*. So-called "free" edges are ones that aren't connected to other edges at both ends (as in the second image).
- *It crosses another edge*. Because edges don't automatically cut other edges where they cross, you have to manually split them. To make a thick edge thin (in other words, to make it "sink in"), use the Line tool to trace over each segment, as shown in the third image.

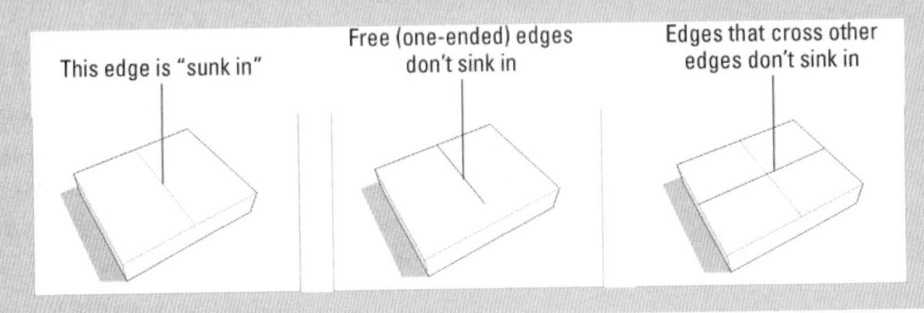

SELF-CHECK

1. To switch to 2D view, we would have to both select Camera ⇨ Standard Views ⇨ Top, and also Camera ⇨ Top. True or false?
2. Describe briefly what would be the process of showing wall thicknesses in SketchUp.
3. What is the most important tool that we can use to extrude a 2D plan into a 3D model?
4. Door and window components can cut through two-faced walls and are therefore ideal for interior models. True or false?

Apply Your Knowledge Using your own home or a building you know, model the exterior in 2D and migrate to 3D. Use the various techniques to create the interior part of building—apply floors, add doors and windows and practice push/pull.

4.2 CREATING STAIRS

There are probably a million different ways to make stairs in SketchUp, but everyone has their own preferred method. In the following sections, you'll find two different methods that work equally well. Take a look at all of them, and then decide which works best for your situation. Chapter 5 explains a third, slightly trickier, but more powerful, way of making stairs using components.

SketchUp 7 introduced Dynamic Components, which has some nice implications for people who need stairs to use in their models. So-called dynamic stair components automatically add or subtract individual steps as you make them bigger or smaller with the Scale tool. Depending on what you want to accomplish, a pre-made dynamic stair component may save you a lot of time. Find out more about them in Chapter 5.

GOOGLE SKETCHUP IN ACTION

Learn different methods to create stairs.

Before we explore these methods, here's some simple stairway vocabulary, just in case you need it. Each of these elements is also shown in Figure 4-28.

- **Rise and run:** The rise is the total distance your staircase needs to climb. If the vertical distance from your first floor to your second floor (i.e., your floor-to-floor distance) is 10 feet, that's your rise. The run is the total horizontal distance your staircase takes up. Thus, a set of stairs with a big rise and a small run would be very steep.
- **Tread:** A **tread** is an individual step—in other words, it's the part of the staircase you step on. When people refer to the size of a tread, they're talking about the tread's depth, or the distance from the front to the back of the tread. Typically, this is anywhere from 9 to 24 inches, but treads of 10 to 12 inches are most comfortable to walk on.
- **Riser:** The **riser** is the part of the step that connects each tread in the vertical direction. Risers are usually about 5 to 7 inches high, but that depends on the building. Not all staircases have actual risers (think of steps with gaps between treads), but they all have a riser height. Sometimes the riser does not connect to the leading edge of the thread abruptly at 90 degrees but rather with a nosing, or a slight rounded bump.

Nosing
A bump at the leading edge of a tread on a stair.

- **Landing:** A **landing** is a platform somewhere around the middle of a set of stairs. Landings are necessary in real life, but modeling them can be a pain; figuring out staircases with landings is more definitely more complicated. It's sometime easier if you think of your landings as really big steps.

Figure 4-28

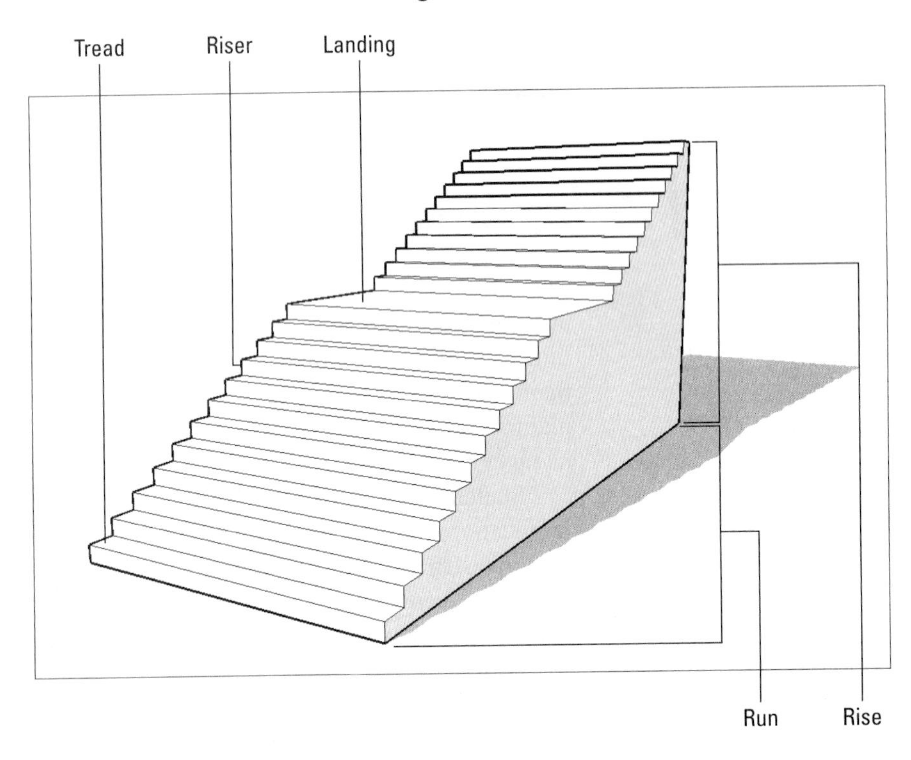

The anatomy of a staircase.

4.2.1 The Subdivided Rectangles Method

This is the method most people use to draw their first set of stairs. It's intuitive and simple, but it's also a bit more time-consuming than the other methods described in this chapter.

The key to the subdivided rectangles method is to use a special trick you can do with edges: Called *Divide*, it lets you pick any edge and divide it into as many segments as you want. If you know how many steps you need to draw, but not how deep each individual tread needs to be, this comes in handy.

Figure 4-29 illustrates the subdivided rectangles method at work.

Here's a detailed look at how to use this method of creating stairs:

1. **Start by drawing a rectangle the size of the staircase you want to build.** It almost always works best to model steps as a group, separate from the rest of your building, and then move them into position when they're done. (You can read all about groups in Chapter 5.)
2. **With the Select tool, right-click one of the long edges of your rectangle and choose Divide from the context menu.** If your staircase is wider than it is long, right-click one of the short edges instead.

Figure 4-29

Divide edge into smaller edges, marking off treads

Connect new endpoints

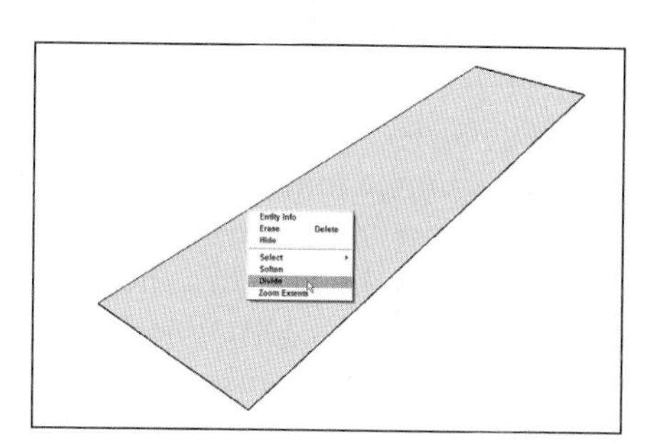

Divide vertical edge marking off vertical risers

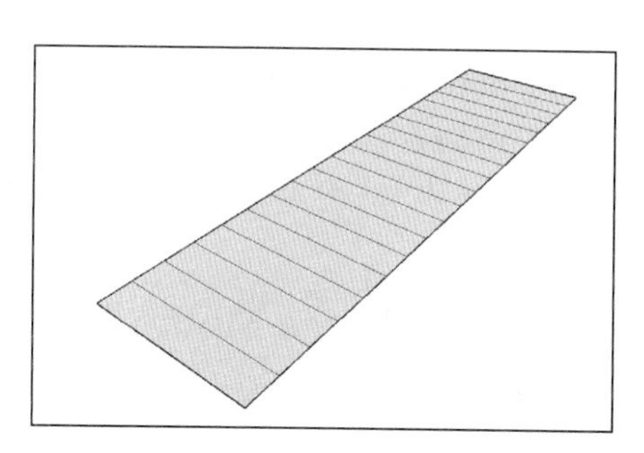

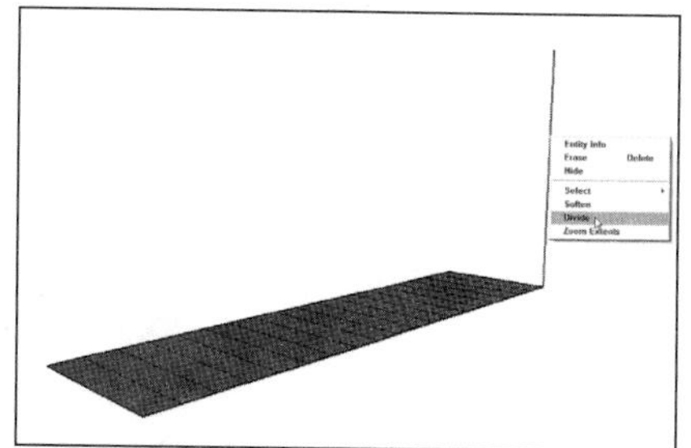

Infer to the endpoints on this divided edge

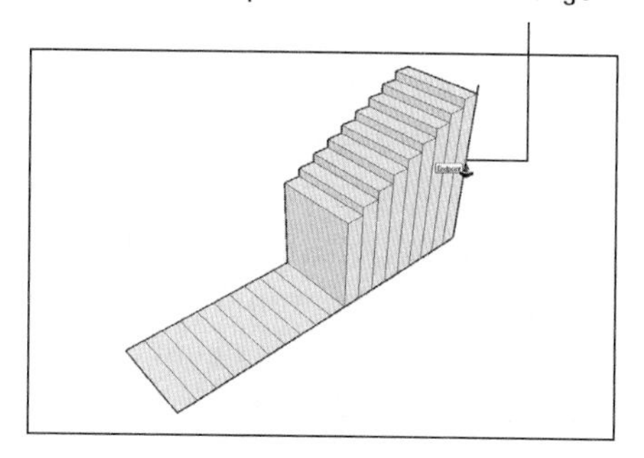

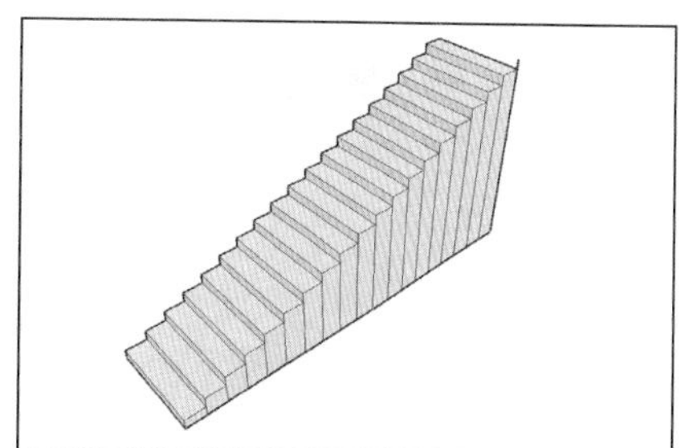

The subdivided rectangles method of building stairs.

3. **Before you do anything else, type in the number of treads you want to create and press Enter.** This command automatically divides your edge into many more edges, eliminating the need to calculate how deep each of your treads needs to be. Essentially, each of your new edges will become the side of one of your treads.
4. **Draw a line from the endpoint of each of your new edges, dividing your original rectangle into many smaller rectangles.** You can use the Line or the Rectangle tool to do this; pick whichever one you're most comfortable with.

5. **From one of the corners of your original rectangle, draw a vertical edge.** This edge should be the height of your staircase's total rise.
6. **Use the Divide command to split your new edge into however many risers you need in your staircase (this is generally your number of treads, plus one).** Repeat steps 2 and 3 to do this. The endpoints of your new, smaller edges will tell you how high to make each of your steps.
7. **Push/Pull the rectangle that represents your last step to the correct height.** Here's where you need to use the "hover-click" technique described earlier in this chapter. Just click once to start pushing/pulling, hover over the endpoint that corresponds to the height of that tread, and click again. Your step will automatically be extruded to the right height. It's a good idea to start extruding your highest step first, but keep in mind that it doesn't go all the way to the top; you always have a riser between your last step and your upper floor.
8. **Repeat step 7 for each of your remaining steps.**
9. **Use the Eraser to get rid of any extra edges you don't need.** Be careful not to accidentally erase geometry on the part of your staircase you can't see.

4.2.2 The Copied Profile Method

This method for modeling a staircase relies, like the last one, on using Push/Pull to create a three-dimensional form from a 2D face, but it's a lot more elegant. In a nutshell, you draw the *profile*—or the side view—of a single step, and then you copy as many steps as you need, create a single face, and extrude the whole thing into shape (see Figure 4-30).

Here's a more detailed explanation of how to use the copied profile method:

1. **Start with a large, vertical face, making sure that it's big enough for the flight of stairs you want to build.** You're going to end up pushing/pulling the whole staircase out of the side of this face.
2. **In the bottom corner of the face, draw the profile (side outline) of a single step.** You can use the Line tool to do this, although you might want to use an arc or two, depending on the level of detail you need. (For a refresher on drawing accurately, refer to Chapter 2.)
3. **Select all the edges that make up your step profile.** Remember that you can hold down Shift while clicking with the Select tool to add multiple objects to your selection.
4. **Make a copy of your step profile and place it above your first profile.** If you're unfamiliar with how to make copies using the Move tool, refer to the section on moving and copying found near the end of Chapter 2.
5. **Type in the number of steps you'd like to make, type the letter x, and then press Enter.** For example, if you wanted ten steps, you would

Figure 4-30

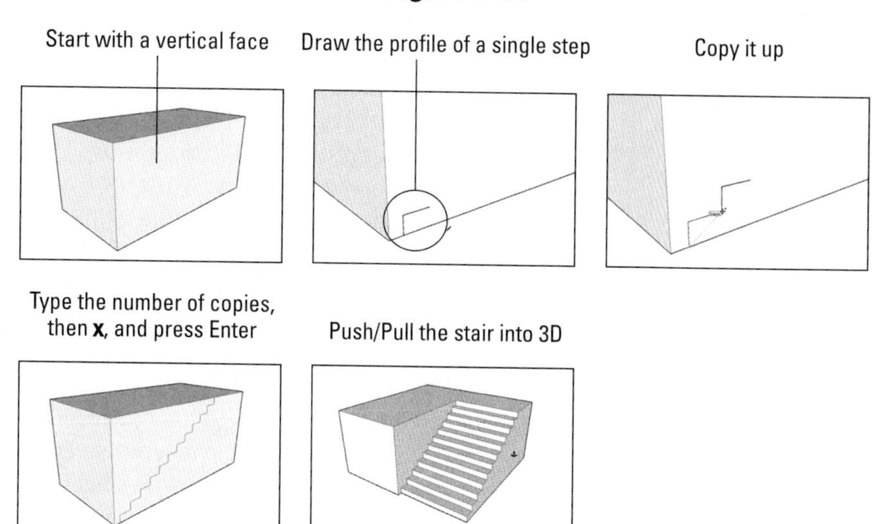

The copied profile method of drawing a staircase.

type in 10x. This technique repeats the copy operation you just did by however many times you tell it to; adding an x at the end of the number tells SketchUp you want to make copies.

6. **Draw an edge to make sure that all your step profiles are part of a single face.**
7. **Push/Pull the staircase face out to be the width you need it to be.** This is the part that seems like magic to most people!

This method of stairway building also works great in combination with the Follow Me tool, which is described in Chapter 6. For now, Figure 4-31 provides a brief preview of using Follow Me when creating stairs.

Figure 4-31

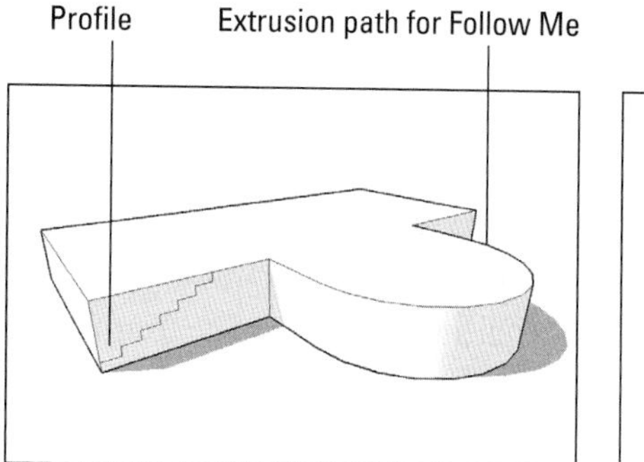

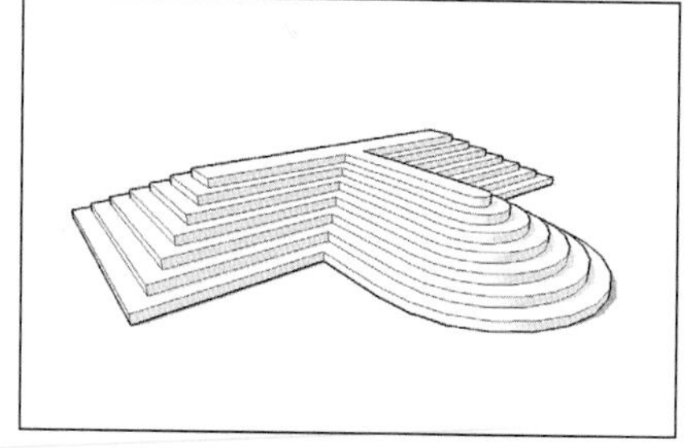

Using Follow Me with the copied profile method produces some impressive geometry.

IN THE REAL WORLD

Lateo Soletic founder of www.sketchupartists.org and owner of www.concepto-illustrations.com shares his technique of building stairs in SketchUp, "different types of stairs require different techniques. My most common approach is to draw one or two on a face and copy the rest; then, extrude. Another technique with circle stair for example is to draw one stair then use the 'Copy along Path' plugin (http://www.smustard.com/script/PathCopy)."

About the process of building and raising walls in three dimensions, Lateo likes to create perfectly clean models. In the past he used to import AutoCAD files and use the plugin "Make Faces" to create faces, but that is something that belongs to the past. Currently, he imports a floor plan as a JPEG file and save it in a separate layer (floorplan 01) so that he can always turn it off. Then he draws his walls on top of it, making them transparent so he can see the floor plan bellow.

Lateo explains why he abandoned the AutoCAD import: "The reason I abandoned the AutoCAD import process is the fact that you always get many stray lines around and what seems at first a quicker process, but in the end one ends up wasting time getting rid of the odd lines here and there. Definitely a model created from scratch in SketchUp is the best and the cleanest way to create a perfect model and it is also easiest to manipulate later."

SELF-CHECK

1. When drawing stairs, most beginners use the ______ method, which is simple and intuitive but also the most time-consuming?
   - **a.** Subdivided rectangles
   - **b.** Copied profile
   - **c.** "Treads are components"
   - **d.** Half rectangles
2. In the Subdivided Rectangles method, what trick do we use to break one long segment into smaller ones?
   - **a.** Right-click and choose Divide from the context menu.
   - **b.** Left-click and choose Segment from the context menu.
   - **c.** Right-click and choose Break from the context menu.
   - **d.** Right-click and choose Separate from the context menu.
3. Name the four basic parts of a staircase.

4. If we want to draw eight steps in the copied profile method, what would we have to type exactly when we are ready to copy the first step?

Apply Your Knowledge Using what you have learned about building stairs, create a staircase using both methods mentioned in this chapter.

4.3 CREATING A ROOF

If you are lucky, the roof you want to model is fairly simple. Unfortunately, home builders sometimes get creative, constructing roofs with dozens of different features that make modeling them difficult. For this reason, we are going to keep things fairly simple: The following sections are dedicated to showing you how to identify and model some of the basic roof forms. After that, you will learn about a great tool, Intersect Face, you can use to assemble complicated roofs from less-complicated pieces.

The tricky thing about roofs is that they are hard to see. If you want to make a model of something that already exists, it helps to get a good look at it—that's not always possible with roofs. One nice way to get a better view of a roof you're trying to build is to find it in Google Earth. For more information, reference Chapter 11.

Remember to always make a group out of your whole building before you work on your roof.

Before learning how to create roofs, let's review some general terminology related to roof types and components. Figure 4-32 provides an illustration of each of these roof types and components:

- **Flat roof:** Flat roofs are just that, except they aren't—if a roof were really completely flat, it would collect water and leak. That's why even roofs that look flat are sloped very slightly.
- **Pitched roof:** Any roof that isn't flat is technically a pitched roof.
- **Shed roof:** A shed roof is one that slopes from one side to the other.
- **Gabled roof:** Gabled roofs have two planes that slope away from a central ridge.
- **Hip roof:** A hip roof is one where the sides and ends all slope together.
- **Pitch:** The angle of a roof surface is referred to as its pitch.
- **Gable:** A gable is the pointed section of wall that sits under the peak of a pitched roof.
- **Eaves:** Eaves are the parts of a roof that overhang the building.
- **Fascia:** Fascia is the trim around the edge of a roof's eaves where gutters are sometimes attached.
- **Soffit:** A soffit is the underside of an overhanging eave.

Figure 4-32

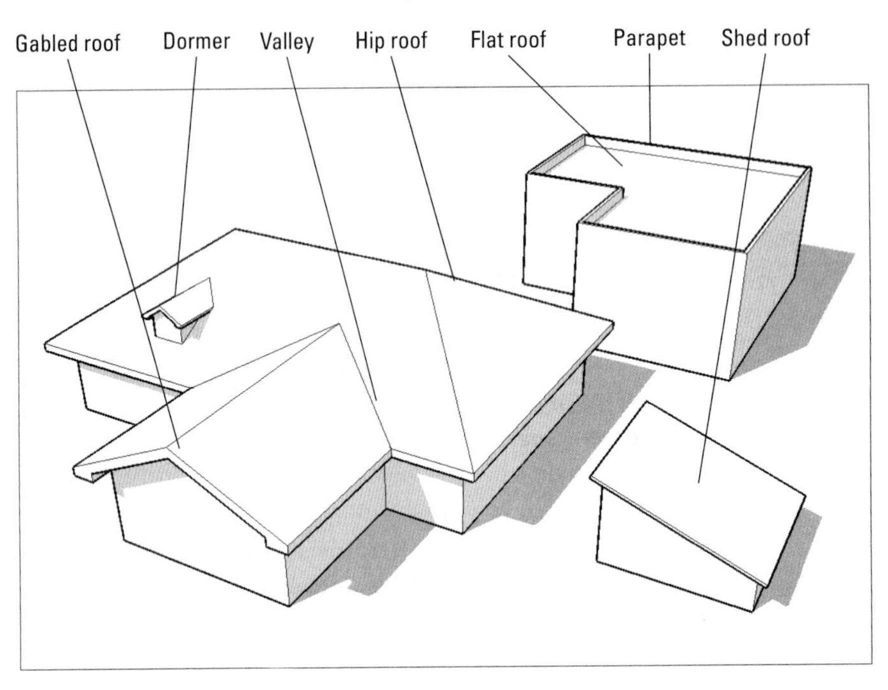

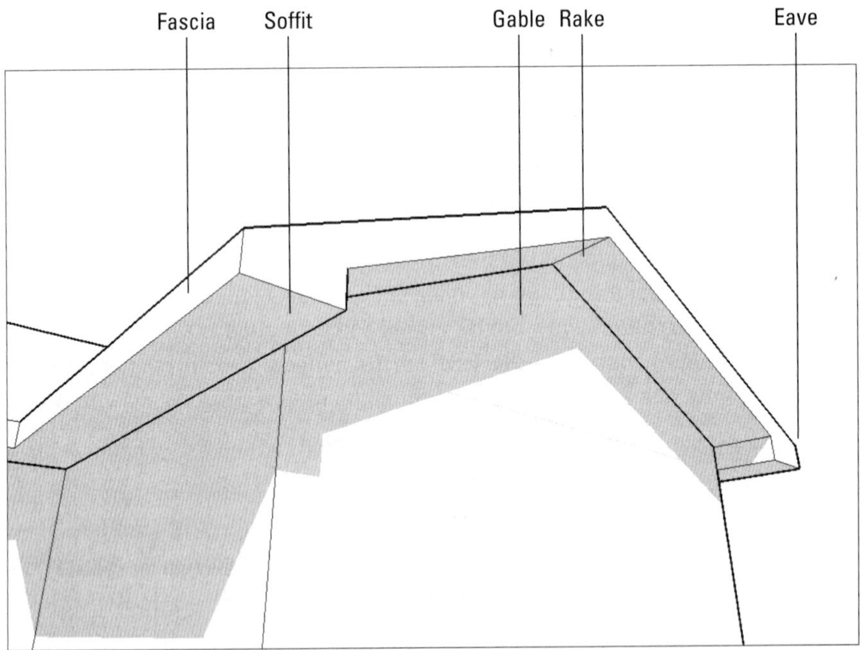

Some different kinds of roofs and their various parts.

Figure 4-33

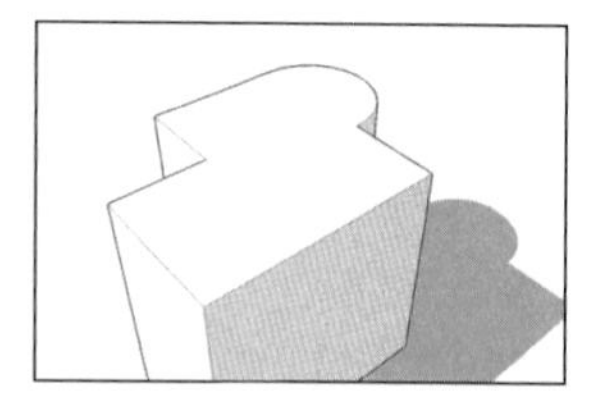

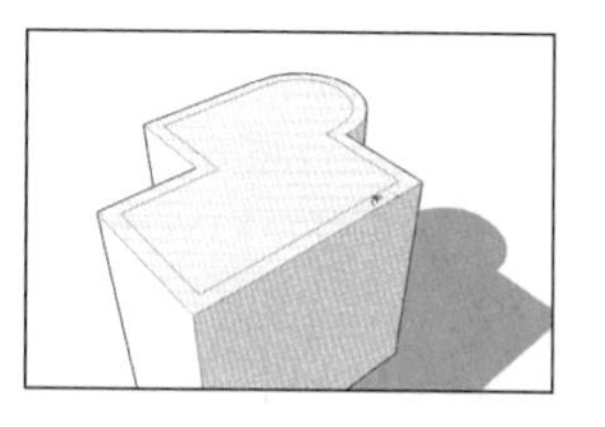

Offset to the inside

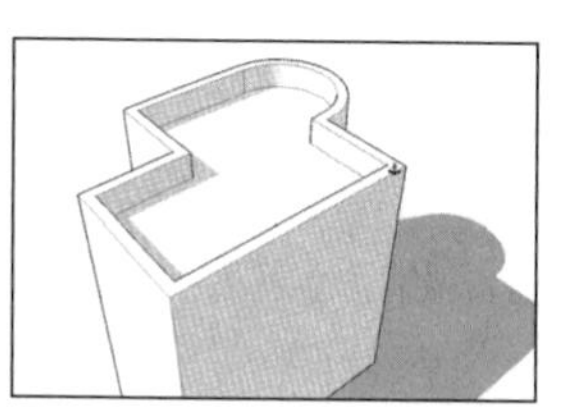

Push/pull your parapet up

Modeling parapets on flat-roofed buildings is easy.

- **Rake:** The rake is the part of a gabled roof that overhangs the gable.
- **Valley:** A valley is formed when two roof slopes come together; this is where water flows when it rains.
- **Dormer:** Dormers are the little things that pop up above roof surfaces. They often have windows, and they serve to make attic spaces more usable.
- **Parapet:** Flat roofs that don't have eaves have parapets. These are extensions of the building's walls that go up a few feet past the roof itself.

4.3.1 Building Flat Roofs with Parapets

GOOGLE SKETCHUP IN ACTION

Understand how to build flat roofs with parapets.

SketchUp is ideal for modeling flat roofs that feature parapets. In fact, by using a combination of the Offset tool and Push/Pull, you should be able to make a parapet in under a minute. Figure 4-33 provides a quick visual representation of this process.

PATHWAYS TO...
MODELING FLAT ROOFS THAT FEATURE PARAPETS

1. With the Offset tool, click the top face of your building.
2. Click again somewhere inside the same face to create another face.
3. Type in the thickness of your parapet, and then press Enter. This redraws your offset edges to be a precise distance from the edges of your original face. How thick should your parapet be? It all depends on your building, but most parapets are between 6 and 12 inches thick.
4. Push/pull your outside face (the one around the perimeter of your roof) into a parapet.
2. Type in the height of your parapet, and then press Enter.

4.3.2 Building Pitched Roofs

Modeling pitched roofs can be a complicated and often frustrating process. Thus, before building such roofs, it is helpful to keep the following tips in mind:

- **Start by making the rest of your building a group.** Always, *always* make a group out of your whole building before you start working on your roof. If you don't, your geometry will start sticking together and you'll end up erasing walls by accident. Beyond that, it's very handy to be able to separate your roof from the rest of your building whenever you'd like. You can also group your roof, if that makes sense for what you're doing. Consult Chapter 5 for a full rundown on making and using groups.
- **Draw a top view of your roof on paper first.** Drawing a top view can help you get a clearer idea of the roof's shape. Adding measurements and angles is even better, as it makes it easier for you to know what you need to do when you get around to using SketchUp.
- **Learn to use the Protractor tool.** This tool (which is on the Tools menu) is for measuring angles and, more importantly, creating angled guides. Because sloped roofs are all about angles, you will probably need to use the Protractor sooner or later. The best way to find out how it works is to open the Instructor dialog box by choosing Window ⇨ Instructor and then activating the Protractor tool.

Creating Eaves for Buildings with Pitched Roofs

One good way to create eaves (overhangs) on pitched roofs is to use the Offset tool, as shown in Figure 4-34.

Figure 4-34

Offset an overhang

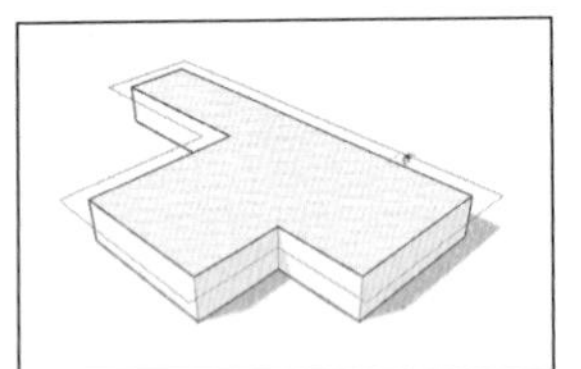

Delete the inside face

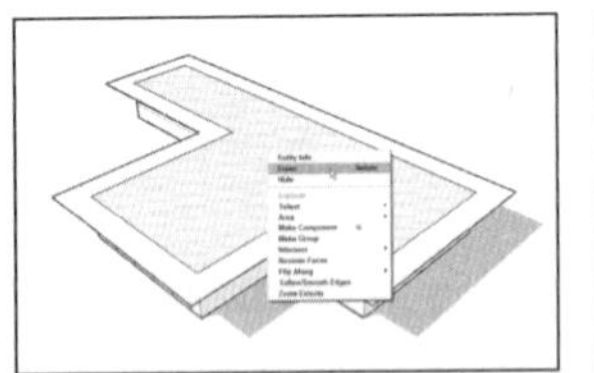

Push/Pull a fascia thickness

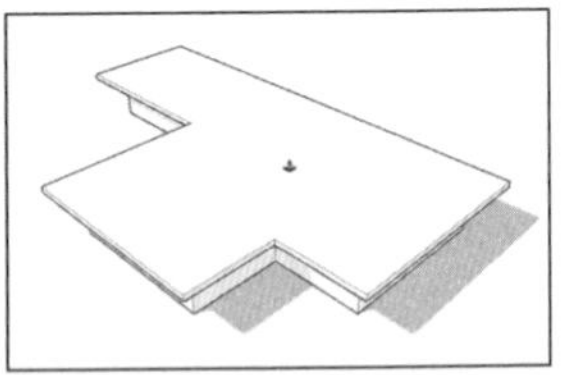

Eaves are the parts of the roof that overhangs a building's walls.

PATHWAYS TO...
CREATING EAVES ON PITCHED ROOFS

1. **Make a group out of your whole building before you start modeling the roof.** This makes it easier to keep your roof separate, which in turn makes your model easier to work with.

2. **Use the Line tool to create an outline of the parts of your roof that will have eaves of the same height.** The goal here is to end up with a single face to offset. A lot of buildings have complex roofs with eaves of all different heights. For the sake of this step, just create a face which, when offset, will create roof overhangs in the right places.
3. **Use the Offset tool to create an overhanging face.** For instructions on how to use Offset, see the section earlier in this chapter.
4. **Erase the edges of your original face.** A quick way to do this (with the Select tool) is as follows:
   - Double-click inside your first face. This selects both it and the edges that define it.
   - Press Delete to erase everything that's selected.
5. **Push/Pull your overhanging roof face to create a thick fascia.** Different roofs have fasciae of different thicknesses. If you don't know yours, just take your best guess.

4.3.3 Building Gabled Roofs

GOOGLE SKETCHUP IN ACTION

Learn about gabled and hip roofs.

You can approach the construction of a gabled roof in many different ways, but the method described below and depicted in Figure 4-35 works well on a consistent basis.

Figure 4-35

Create an angled guide with the Protractor

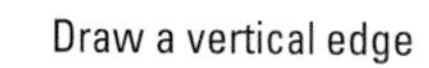

Draw a vertical edge

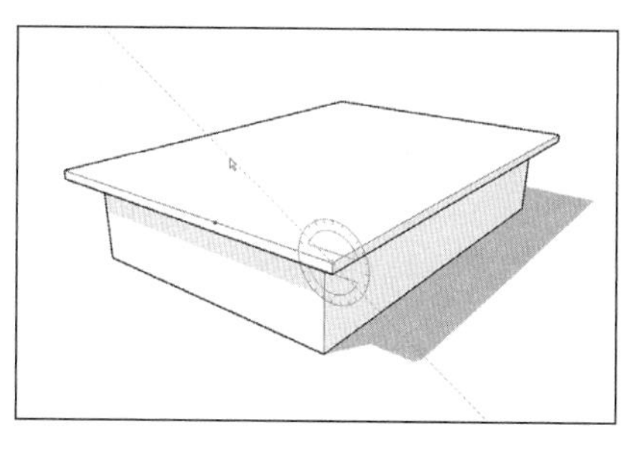

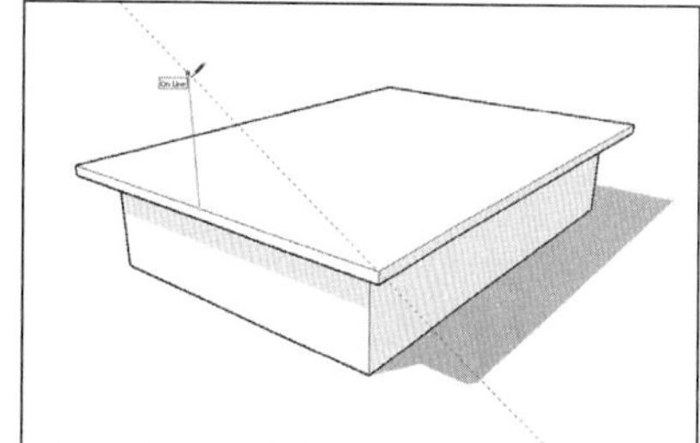

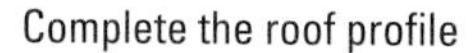

Complete the roof profile

Push/pull it back

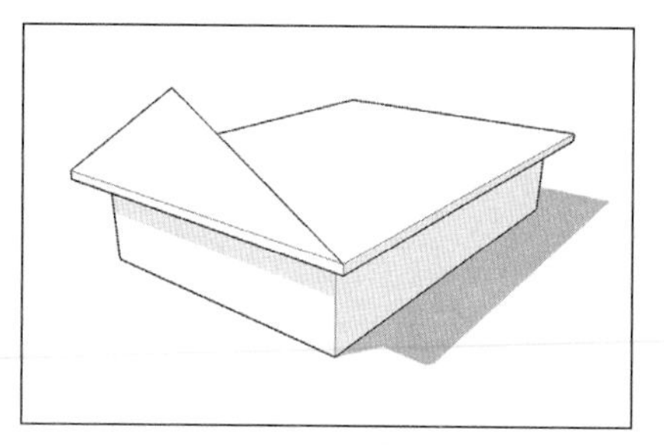

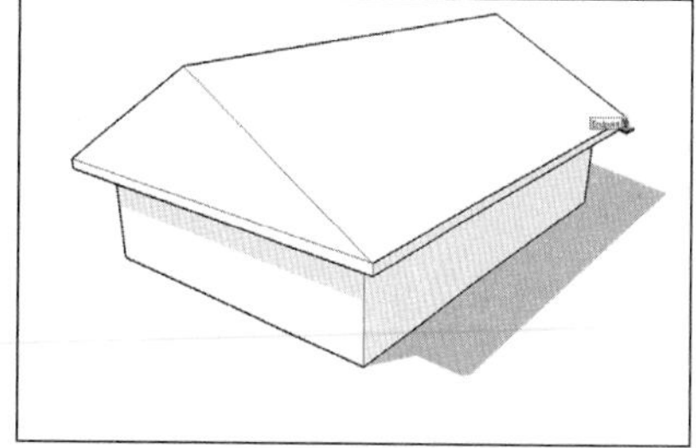

Gabled roofs are relatively easy to make in SketchUp.

1. **Create a roof overhang, following the steps in the previous section.** Most gabled roofs have eaves, so you'll probably need to do this for your building.
2. **Use the Protractor tool to create an angled guide at the corner of your roof.** See the previous section of this chapter for more information about drawing angled guides with the Protractor.

> **FOR EXAMPLE**
>
> RISE OVER RUN RATIOS
>
> Architects and builders often express angles as *rise over run ratios*. For example, a **4:12** (pronounced "four in twelve") roof slope rises 4 feet for every 12 feet it runs. Accordingly, a 1:12 slope is very shallow, while a 12:12 slope is very steep. When using the Protractor tool, SketchUp's Measurements Box understands angles expressed as ratios, as well as those expressed in degrees. Thus, typing **6:12** yields a slope of 6 in 12.

3. **Use the Line tool to draw a vertical edge from the midpoint of your roof to the angled guide you created in step 1.** The point at which your edge and your guide meet is the height of your roof ridge.
4. **Draw two edges from the top of your vertical line to the corners of your roof.** This should cause two triangular faces to be created.
5. **Erase the vertical edge you drew in step 3 and the guide you drew in Step 2.**
6. **Push/Pull your triangular gable back.** If your gabled roof extends all the way to the other end of your building, push/pull it back that far. If your roof runs into another section of roof (as in Figure 4-36), extrude it back until it's completely "buried." The section "Putting Your Roof Together," found later in this chapter, has more information on what to do when you're making a complex roof.
7. **Finish your eaves, fascia, soffit, and rake(s) however you want.** There are many different kinds of gabled roof details, so we can't cover them all here. However, Figure 4-37 shows a few common features of gabled roofs.

Figure 4-36

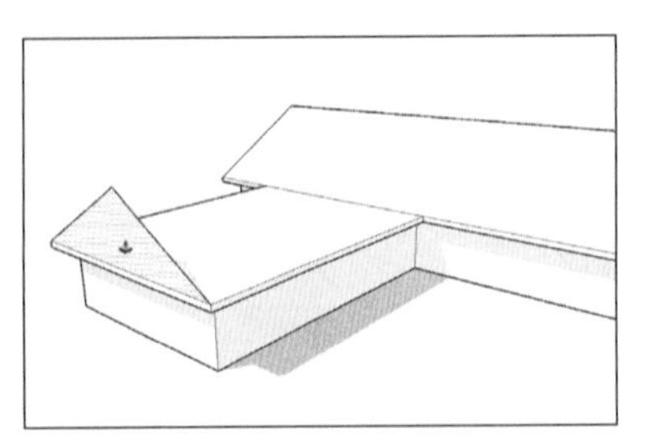

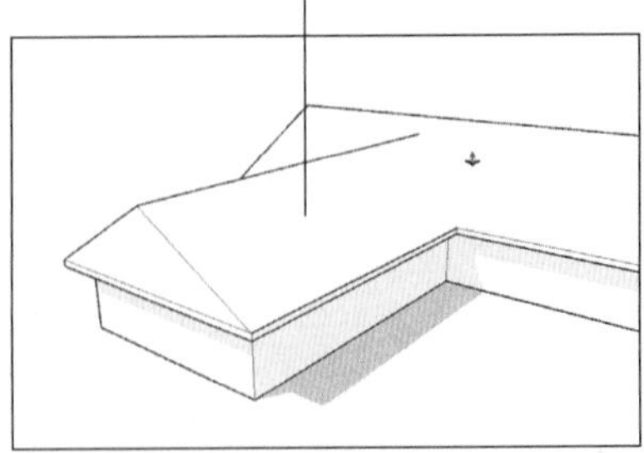

If your gabled roof is part of a larger roof structure, it might run into another roof pitch.

Figure 4-37

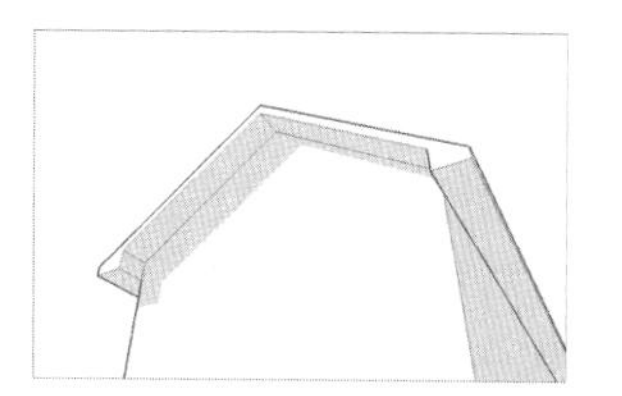
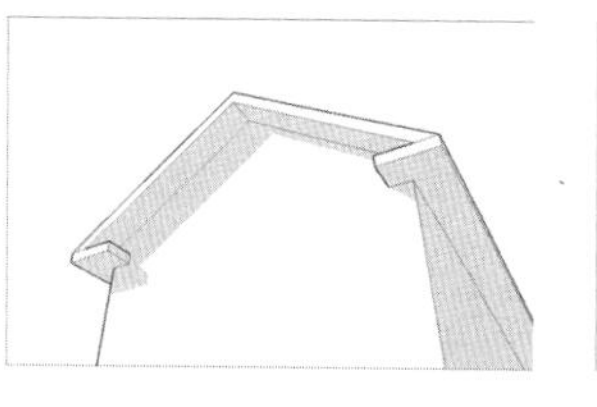
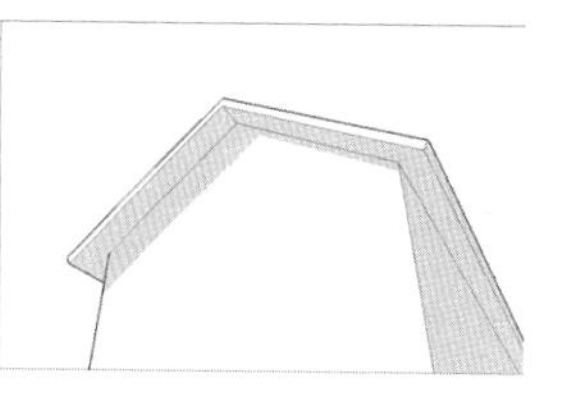

Some common gabled roof details and how to make them.

4.3.4 Building Hip Roofs

Believe it or not, building a hip roof is easier than making a gabled one! Hip roofs don't have rakes, which makes them significantly less complicated to model. Figure 4-38 illustrates the basics of creating a hip roof.

Complex Hip Roofs and the Follow Me Tool

SketchUp also has a tool called Follow Me (described in detail in Chapter 6), which you can use to create complex hip roofs in about one-fifth the time it would normally take to make them. At its core, Follow Me works a bit like

Figure 4-38

Measure half-width of your gable

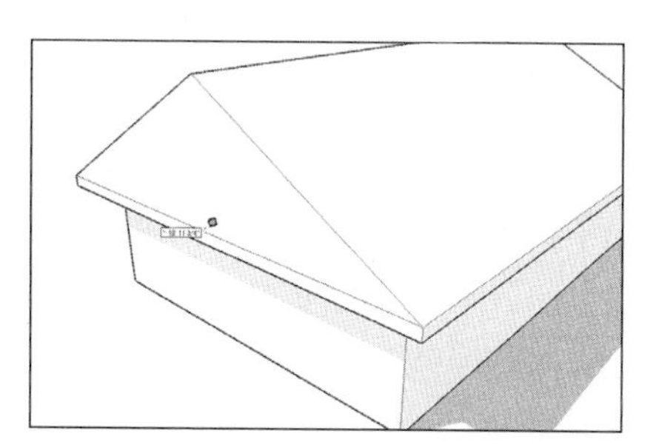

Create a guide that distance from end of gable

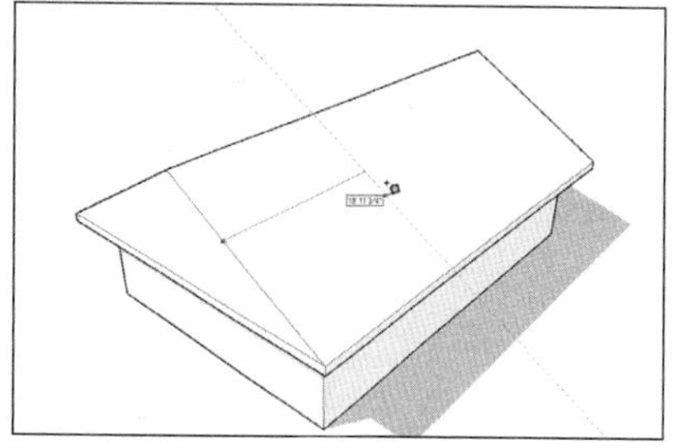

Draw edges connecting ridge and corners and erase three edges that form gable

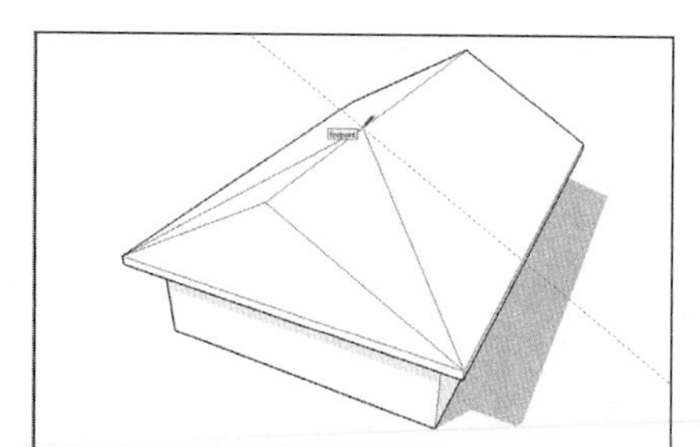

Now you have a hip

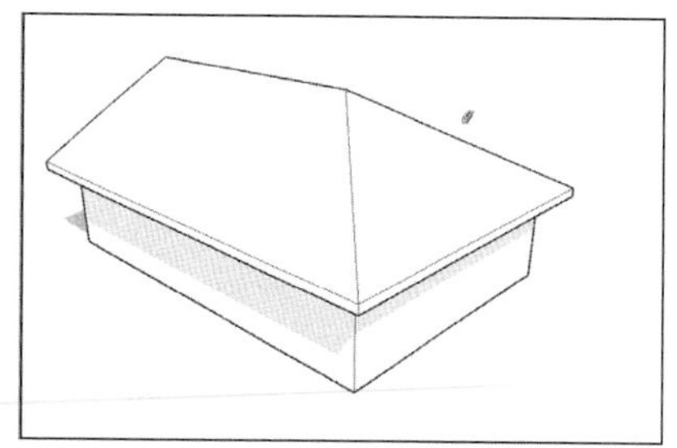

To make a hip roof, start with a gabled one.

Push/Pull, except it lets you extrude faces along predetermined paths. You can use this tool to create very complicated geometry.

When it comes to hip roofs, the Follow Me technique only works if your roof meets the following conditions:

- The pitch needs to be the same on all roof surfaces.
- The roof needs to be "hipped" all the way around.

PATHWAYS TO... BUILDING A HIP ROOF

1. Follow Steps 1 through 5 in the section "Building Gabled Roofs" to begin making a hip roof.
2. **Measure the distance from the midpoint of the gable to the corner of the roof.** Because hip roofs have pitches that are the same on all sides, you can use a simple trick to figure out where to locate the hip in your roof. It's a lot easier than using the Protractor.
3. **With the Tape Measure, create a guide (the distance you just measured) from the end of the gable.**
4. **Draw edges from the point on the ridge you just located to the corners of your roof.** This does two things: It splits the sides of your roof into two faces each and creates a new face (which you can't see yet) under the gabled end of your roof.
5. **Erase the three edges that form the gabled end of your roof, revealing the "hipped" pitch underneath.** Now all three faces of your roof are the same pitch—just the way they should be.
6. If appropriate, repeat the process on the other end of your roof.

Follow these steps to use Follow Me to create a complex hip roof (as shown in the accompanying images) See Figure 4-39:

1. Over the widest part of your building, draw a triangle that represents the slope of your roof. It should only be a half-gable, as shown in the images.
2. Select the top surface of your building.
3. With the Follow Me tool (available on the Tools menu), click the half-roof profile you drew in step 1 once.
4. Select your whole roof by triple-clicking it with the Select tool. (You may want to hide the rest of my building at this point, too.)
5. Right-click anywhere on the roof and choose Intersect Faces ⇨ With Model from the context menu.
6. Use the Eraser to clean up your roof by erasing any geometry that isn't supposed to be part of it (you'll find plenty). You'll probably also have to draw in edges every now and then. If you make a mistake, just use Undo and try again.

Figure 4-39

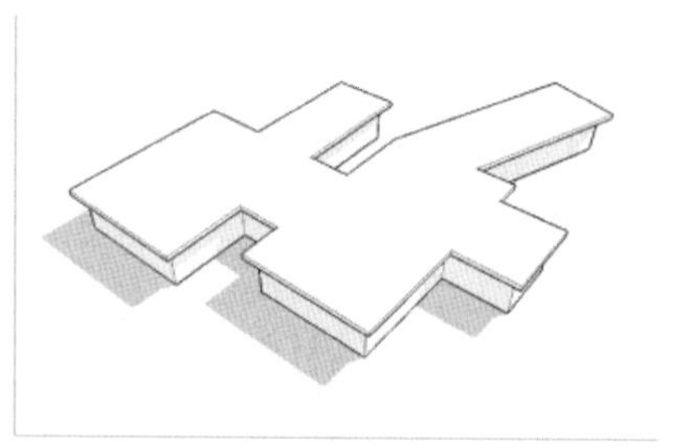

Draw a profile

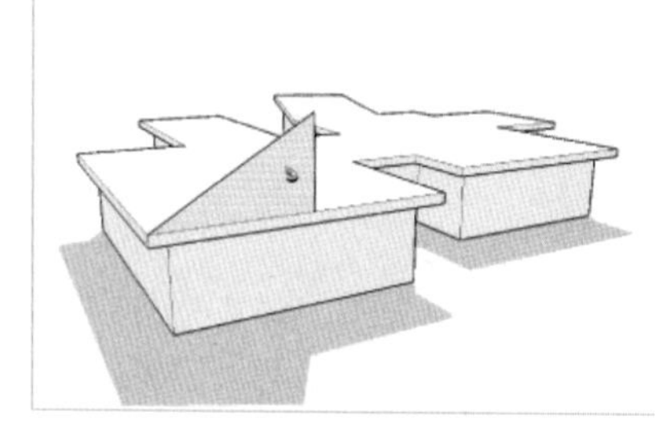

Use Follow Me

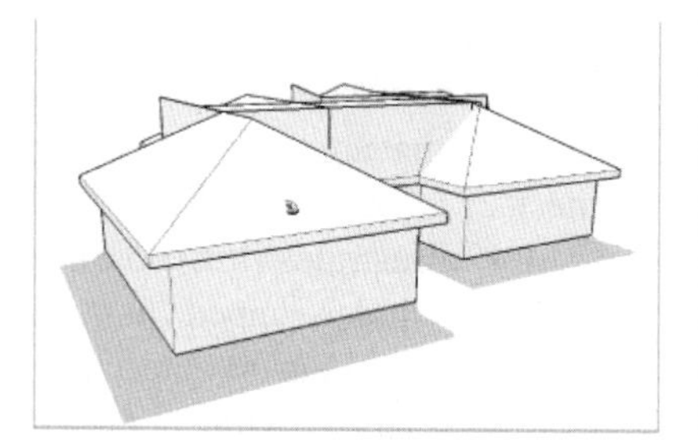

Select the whole roof, right click, and then choose Intersect with Model

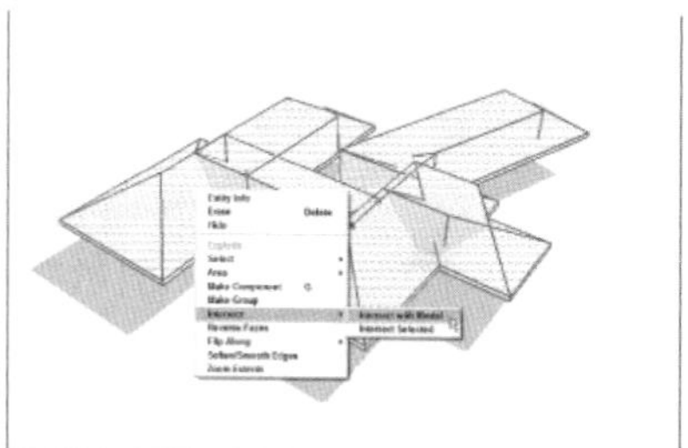

Clean up the mess

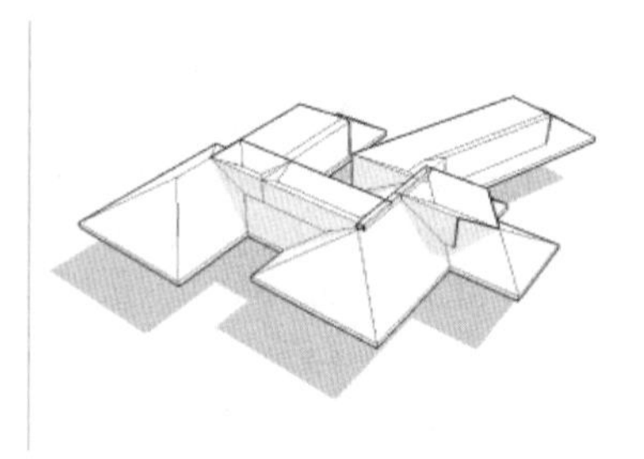

Complex Hip Roof

4.3.5 Putting Your Roof Together

GOOGLE SKETCHUP IN ACTION

Understand and practice putting your roof together.

In general, the newer and more expensive a house is, the more roof slopes it has. This can make modeling difficult! Thankfully, SketchUp has a relatively little-known feature that often helps when it comes to making roofs with numerous pitches: Intersect with Model.

Getting to Know Intersect with Model

Here are the basic things you need to know about the Intersect with Model tool:

- **Intersect Faces with Model makes new geometry from existing geometry.** That's how it works: It takes faces you've selected and creates edges

Figure 4-40

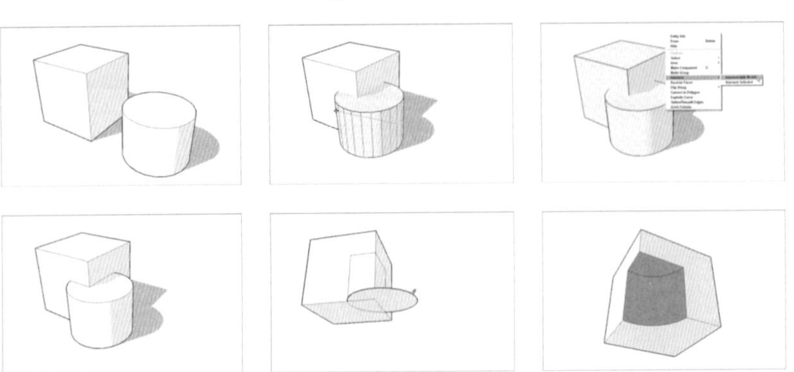

Using Intersect with Model to cut a partial cylinder out of a cube.

wherever they intersect. You use Intersect with Model in cases where you need to create forms that are the *union* (both put together), *difference* (one minus the other), or *intersection* (the part they have in common) of other forms. Figure 4-40 illustrates this. Perhaps you want to make a model that's a cube with a cylinder-shaped chunk taken out of it. To do this, you would model the cube and the cylinder. After positioning them carefully, you could then use Intersect with Model to create edges where the two shapes' faces come together. After that, you can use the Eraser to get rid of the edges you don't want (the rest of the cylinder, in this case).

- **Intersect Faces with Model and the Eraser tool go hand in hand.** Anytime you use Intersect with Model, you need to follow up by spending some time deleting the geometry you don't want. This isn't a bad thing, but it does mean that you need to be good at orbiting, zooming, and panning around your model. It also means that you need to be handy with the Eraser.
- **Most of the time, choose to Intersect with Model.** This tool has three different modes, but most of the time, you'll end up using just the basic one. In any case, here's what all three of the tools do:
- **Intersect with Model:** This creates edges everywhere your selected faces intersect with other faces in your model—whether the other faces are selected or not.
- **Intersect with Selection:** This option only creates edges where *selected* faces intersect with other *selected* faces. This is handy if you're trying to be a bit more precise.
- **Intersect with Context:** This one's a little trickier: Choosing this option creates edges where faces within the same group or component intersect. For this reason, it's only available when you're editing a group or component.
- **Intersect Faces with Model doesn't have a button.** To use it, you have to do either of the following:

  – Right-click and choose it from the context menu.

  – Choose Edit ⇨ Intersect Faces.

Using Intersect Faces with Model to Make Roofs

When it comes to creating roofs, you can use Intersect with Model to combine numerous gables, hips, dormers, sheds, and so on into a single roof. It's not easy, and it requires a fair amount of planning, but it works great when nothing else will.

Figure 4-41 shows a complicated roof with several different elements. Gabled roofs have been pushed/pulled into the main hip roof form at different heights, but edges don't exist where all the different faces meet.

Now, let's use Intersect with Model to create the edges you want, and then the Eraser to clean up the mess:

1. **Select the whole roof.** You can do this using a number of different methods, but one way that often works best is to first hide the group that contains the rest of your building, and then draw a large selection box around the whole roof with the Select tool.
2. **Choose Edit ⇨ Intersect Faces ⇨ Intersect With Selection.** This tells SketchUp to create edges everywhere you have faces that intersect—in other words, everywhere they "pass through" each other without an edge.
3. **Get out your Eraser and *carefully* delete all the extra geometry on the inside of your roof.** This can be a lot of work, but it's much easier than using the Line tool and SketchUp's inference engine to determine where everything should go.

Figure 4-41

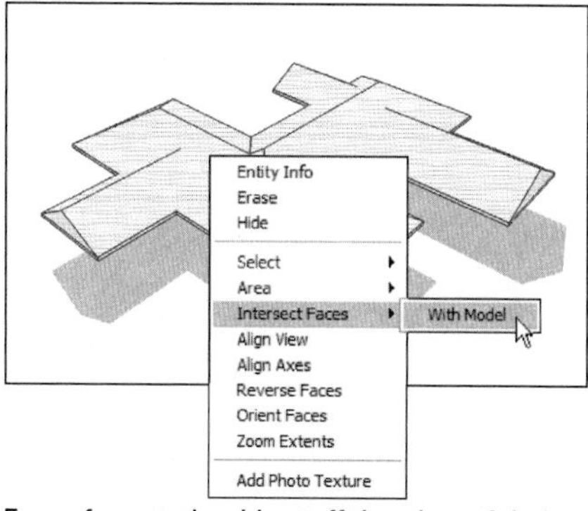

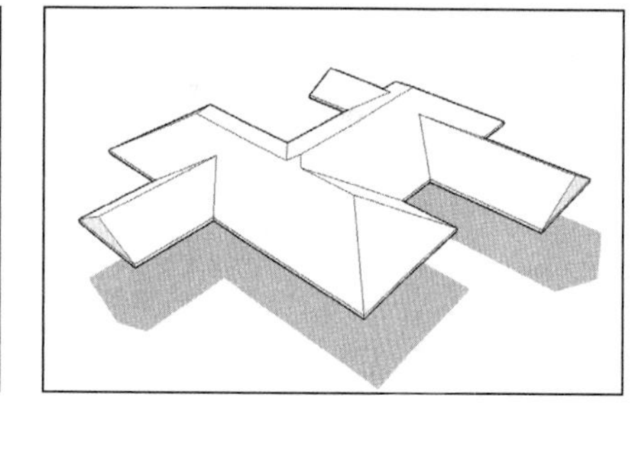

Erase from underside stuff that doesn't belong

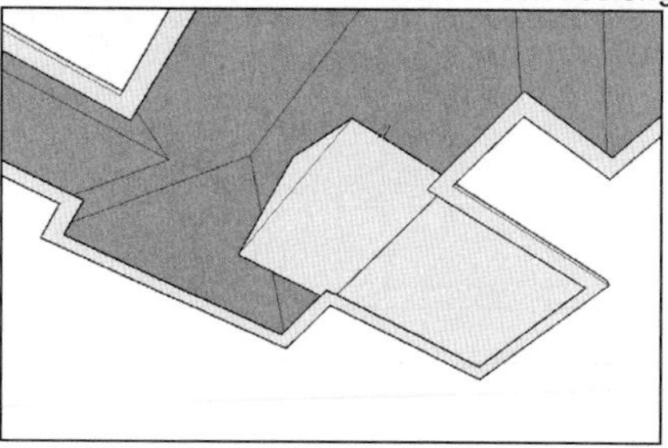

Here's a typically complex roof that could be unified using Intersect with Model.

FOR EXAMPLE

WHEN ALL ELSE FAILS, USE THE LINE TOOL

Fancy tools like Follow Me and Intersect with Model are useful most of the time, but for some roofs, you just have to resort to drawing edges. If that's the case, you'd best become familiar with most of the material at the beginning of Chapter 2, because you're going to be doing a lot of inferencing.

For example, consider the following figure. In it, the Line tool and SketchUp's inference engine are used to add a gabled dormer to a sloped roof surface.

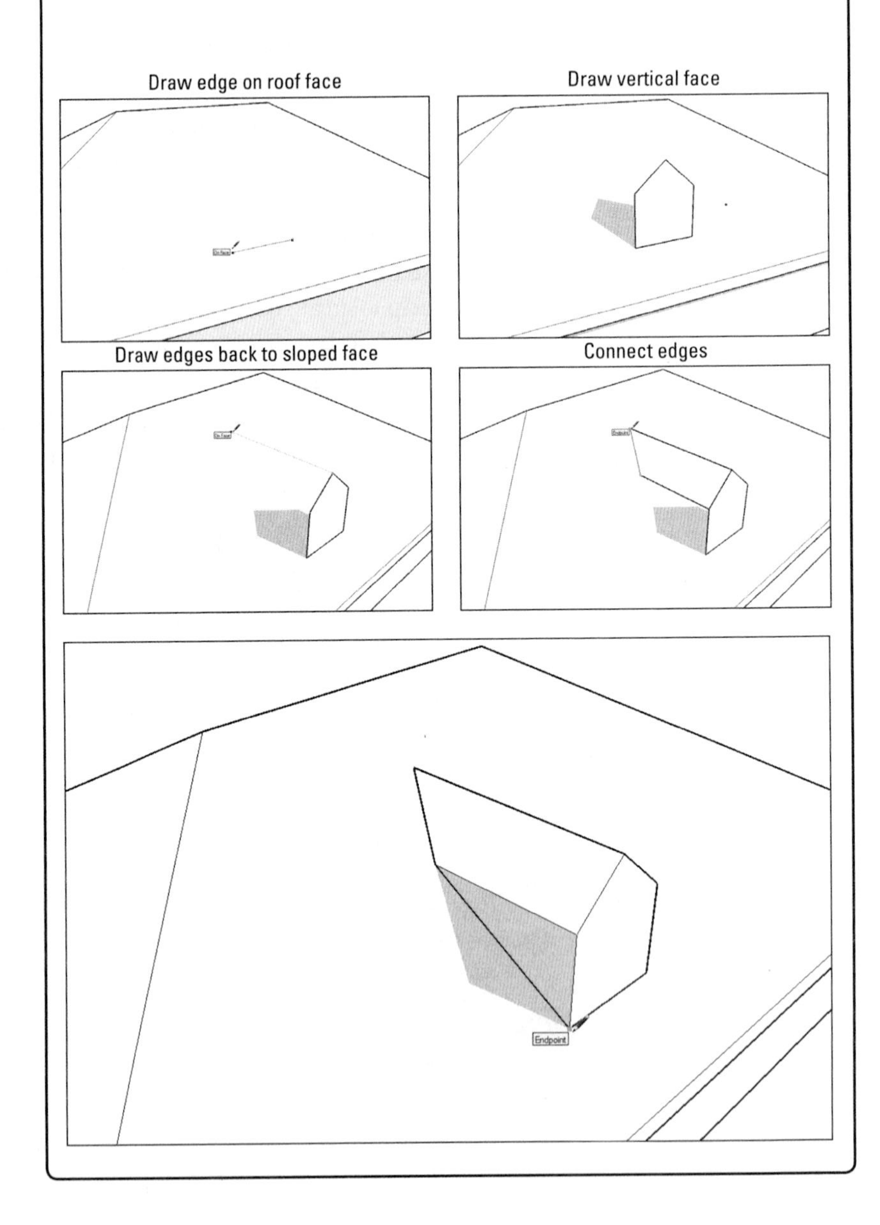

IN THE REAL WORLD

According to Christopher Carley, AIA, LEED AP from Carley Associates, his technique of building staircases depends on the situation. According to him, "my usual method is to draw a vertical line equal to the floor height somewhere outside of the model but aligned with the floors in it. I then use the Divide tool to divide that line into equal segments at an appropriate riser height (<7" in commercial buildings, <7 ¾" in houses). Then, either in our outside of the model, I draw a horizontal line equal to tread depth and connect that to a vertical line equal to riser height using the divided line as a reference. I then copy the riser and tread down the run for as many iterations as there are risers needed. Depending on how precise I need to be, I may adjust the riser to include a nosing before I do this. At this point, I have a jagged line representing the tread and riser profile of the stair, usually in place in the model. Then I integrate the line into a plane and use the push-pull tool to extrude the stair across its width."

Christopher explains that sometimes he would make a 3D extrusion of just one tread and riser and copy that along the run for as far as it needs to go. To make railings, he draws a single line where he needs it and then uses the Follow Me tool or Extrusion tool plugin and then either Follow Me or the Extrusion tool plugin to make the 3D rail from a 2D section. When using the latter method, he makes the stair components into one or more groups to avoid interfering with the rest of the model until he is sure that all is well.

Finally, Christopher explains his technique to raise 3D walls with a thickness: "I generally import a hard-line floor plan as a .dwg file, explode it if it is a group and then lay a rectangle over it. I then clean it up as necessary to get the wall interiors separate from the rest of the rectangle and use the push-pull tool to raise the wall to the height I need. If there is no plan available, I draw the outline of the wall to make a plane of the floor and then use the offset tool to make another line on the interior the same distance from the outside as the wall is thick. I then extrude the wall to the height needed."

SELF-CHECK

1. Name three different types of roofs you can build in SketchUp.
2. What tool would you use to create eaves on a roof line?
3. Explain in a brief sentence how the slope of a roof is described as a rise over run ratio, and the tool we would use in SketchUp to express it.
4. The Intersect with Model tool can be used to combine numerous gables, hips, dormers, and sheds into a single roof. True or false?

Apply Your Knowledge A client comes to you asking you to present three different kinds of roofs for their new home. Mock up three different roof styles. Be sure to use the protractor tool.

SUMMARY

Section 4.1

- An interior model of a building is significantly more complicated than an exterior-only one; dealing with interior wall thicknesses, floor heights, ceilings, and furnishings involves a lot more modeling prowess.
- Use single faces for exterior models and double faces for interior ones.
- It's easiest to draw in 2D when you're directly above your work, looking down at the ground plane.
- All the tools you need to draft in 2D in SketchUp are on the basic toolbar.
- When you are drafting in 2D, whatever you do, *do not* use the Orbit tool.
- Moving from 2D to 3D is basically a one-step process, leveraging one important tool: Push/Pull.

Section 4.2

- The anatomy of a staircase includes the following:
  - Rise and Run
  - Tread
  - Riser
  - Landing
- The subdivided rectangles method is the method most people use to draw their first set of stairs.
- Like the subdivided method, the copied profile method for modeling a staircase relies on Push/Pull to create a three-dimensional form from a 2D face, but it's a lot more elegant.

Section 4.3

- Using a combination of the Offset tool and Push/Pull, allows for creating a parapet in under a minute.
- Modeling pitched roofs can be a complicated and often frustrating process. Consequently, before building such roofs, it is helpful to keep the following tips in mind:
  - Start by making the rest of your building a group.
  - Draw a top view of your roof on paper first.
  - Learn to use the Protractor tool.
  - To make a hip roof, start with a gabled one.

ASSESS YOUR UNDERSTANDING

SUMMARY QUESTIONS

1. It is easier to build two separate models, one showing the interior and one showing the exterior, than to create one model that shows both the interior and the exterior of a building at once. True or false?
2. When creating three-dimensional models, at what point should you add windows and doors?
   - **a.** After drawing an interior outline.
   - **b.** After offsetting exterior walls.
   - **c.** After putting in interior walls.
   - **d.** After extruding the floor plan into a three-dimensional figure.
3. To make an interior model, it is best to measure the exterior of the building and work from the outside in. True or false?
4. Which tool should you never use when drafting in 2D?
   - **a.** Line tool
   - **b.** Orbit tool
   - **c.** Offset tool
   - **d.** Arc tool
5. It is helpful to use guides when putting in interior walls. True or false?
6. A _____ is an individual step.
   - **a.** Tread
   - **b.** Rise
   - **c.** Run
   - **d.** Landing
7. It can be helpful to think of a landing as a really big:
   - **a.** Riser
   - **b.** Step
   - **c.** Tread
   - **d.** Edge
8. The copied profile method for modeling staircases, combined with the Follow Me tool, can produce some elaborate geometry. True or false?
9. In the copied profile method, what would we have to type if we want 12 steps?

10. A ______ is any kind of roof that is not flat.
    a. Pitched roof
    b. Shed roof
    c. Gabled roof
    d. Hip roof
11. Explain when would be a good idea to group your whole building when planning to complete your roof.
12. The process for building hip roofs starts out the same as that for building gabled ones. True or false?

APPLY: WHAT WOULD YOU DO?

1. What tools would you use for the following tasks?
    a. Drawing edges.
    b. Erasing edges.
    c. Drawing circles.
    d. Drawing arcs.
    e. Drawing edges that are a constant distance apart from edges that already exist.
    f. Measuring distances.
2. Why can it be difficult to measure an existing building?
3. What are components and in what circumstances would you use them?
4. List three circumstances in which you would want to use the Push/Pull tool.
5. List and describe four different types of roofs.
6. How can you add doors and windows to a model?

BE AN ARCHITECT

First Steps

You want to create a model of your favorite museum. What are the first three steps you take in creating this model and why?

Creating Staircases

You are creating a set of staircases for a model of your house. Which method (subdivided rectangles method, copied profile method, "treads are components" method) do you use and why? Write two to three paragraphs explaining the method you chose and why.

Creating Roofs

You are creating a house with a roof with four slopes. What would you use to help you accomplish this and why? Open SketchUp and create this roof.

KEY TERMS

As-built
Dormer
Eave
Exterior model
Fascia
Flat roof
Gable
Gabled roof
Hip roof
Interior model
Landing
Nosing
Parapet
Pitch
Pitched roofRak
Rise
Riser
Run
Shed roof
Soffit
Stringer
Tread
Valley

CHAPTER

5 KEEPING YOUR MODEL'S APPEARANCE

Working with Components

Do You Already Know?

- How to group things together?
- Work with Components?
- Create your own Components?
- Use Dynamic Components?

For the answers to these questions, go to **www.wiley.com/go/chopra/googlesketchup2e**

What You Will Find Out	What You Will Be Able To Do
5.1 Why grouping is important.	Understand that making individual items and grouping them together is easier.
5.2 How to work with Components.	Apply various SketchUp tools when working with Components.
5.3 How to work with Dynamic Components.	Access online resources to retrieve Dynamic Components.
5.4 How to take advantage of Components to build better models.	Learn to model symmetrically. Understand how to model with repeated elements.

INTRODUCTION

When it comes to using SketchUp there is nothing more important than using Components. Making a component, or group, is like gluing together geometry in your model. Edges and faces that are grouped together serve as mini-models inside your main model; you use components and groups to more easily select, move, hide, and otherwise work with parts of your model that need to be kept separate. Getting accustomed to using groups and components is the single most important thing you can do to get better at SketchUp. This chapter is about creating and using SketchUp components to make your life a whole lot simpler. The chapter starts by addressing groups which are a lot like small components. After that, component features like finding them, managing them, making your own, and finally, Dynamic Components are discussed.

The final section of this chapter addresses a few modeling techniques that take advantage of component behavior. These techniques are guaranteed to save you time and effort in your 3D modeling activities.

5.1 GROUPING THINGS TOGETHER

Anyone who's worked with SketchUp for even a short time has probably noticed something: SketchUp geometry (the edges and faces that make up a model) is sticky. That is to say, stuff in a model wants to stick to other stuff. The people who invented SketchUp built it this way on purpose; however, the reasons why they did so take a while to explain. In any case, making and using **groups** is the key to keeping elements in your model from sticking together.

Group
Collection of edges and faces that are grouped together and act like a mini-model within the main model.

There are many reasons why you may need to make groups when using SketchUp; the following are a few of those reasons:

GOOGLE SKETCHUP IN ACTION

Understand that making individual items and grouping them together is easier.

- **Grouped geometry doesn't stick to anything.** Perhaps you've modeled a building, and you want to add a roof. You want to be able to remove the roof by moving it out of the way with the Move tool, but every time you try to do this, you end up pulling the whole top part of the house along with the roof (like the middle image in Figure 5-1). Making the roof a separate group allows you to let it sit on top of your house without sticking there, making it easier to deal with, as shown in the right image in Figure 5-1.
- **Using groups makes it easier to work with your model.** For example, you can select all the geometry in a group by clicking it once with the Select tool. You can move groups around and make copies with the Move tool.
- **You can name groups.** If you turn a selection of geometry in your model into a group, you can give it a name. In the Outliner (which is described in Chapter 7) you can see a list of the groups (and **components**) in your model, and if you've given them names, you can see what you have.

Components
Edges and faces grouped together into objects that display certain useful properties.

Figure 5-1

The house is being stretched

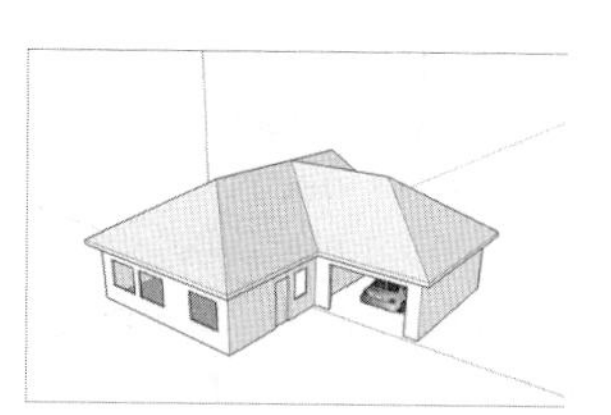

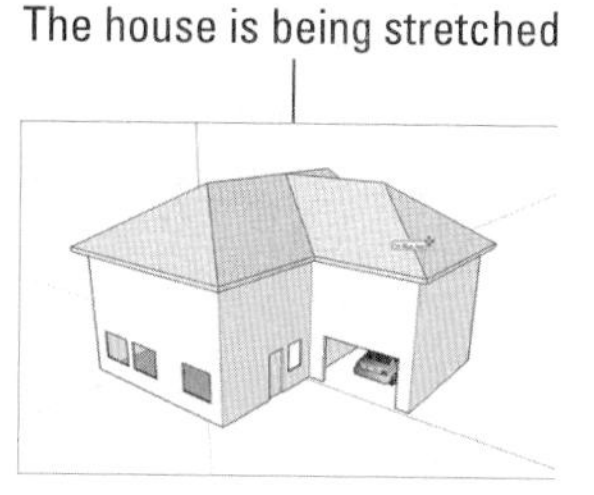

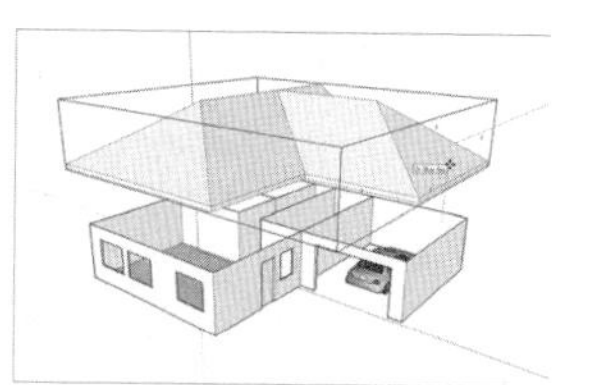

Making the roof into a group means that it won't stick to the rest of your building.

- **Groups can be solids**. Essentially, a *solid* is any group (or component) whose geometry can be thought of as watertight—continuous, with no holes. Solids are important for a couple of reasons:
- If an object is a solid, SketchUp can calculate its volume. You can see any solid's volume by looking in the Entity Info dialog box.
- The Solid Tools let you perform modeling tricks using two or more solids. More details about Solid Tools is covered in Chapter 6.

PATHWAYS TO... CREATING A GROUP

1. **Select the geometry (edges and faces) you'd like to turn into a group.** The simplest way to select multiple entities (edges and faces) is to click them one at a time with the Select tool while holding down Shift. You can also use the Select tool to drag a box around the entities you want to select, but this can be tough, depending on where the entities are. Reference Chapter 2 for more information on making selections.
2. **Choose Edit ⇨ Make Group.** You can also right-click and choose Make Group from the context menu that pops up.

If you want to "ungroup" the geometry in a group, you need to explode it. To do this, right-click the group and choose Explode from the context menu. The edges and faces that were once grouped together won't be grouped anymore.

To edit the geometry inside a group, double-click it with the Select tool. You know you are in edit mode when the rest of your model appears to fade back, leaving only your grouped geometry clearly visible. To stop editing a group, click outside it, somewhere else in your modeling window.

IN THE REAL WORLD

Ronald Schouwink, founder of Ronald Schouwink Design in the Netherlands, started using Google SketchUp four versions ago. What drew him to this software was both the short learning curve and the quick results he could get without having to invest heavily on training. After just a few hours of working with it, he already felt he could use it for his design projects with ease. In his own words "even the price is very friendly."

He designs and fabricates custom-made cabinets and likes to craft his presentations in SketchUp in 3D because he can choose different scenes at different angles and in different viewing styles, ranging from a mere black wireframe perspective or in full color. He finds it rewarding to work directly in three dimensions in SketchUp, and likes to use several layers, scenes, and components.

He could render his SketchUp elevations in other plugins to polish up his screen captures, to create better shop drawings, but his clients are always satisfied with his presentations, as he exports his work into Google layout to create his show drawings.

SELF-CHECK

1. Explain two benefits of grouping our geometry together.
2. What is the simplest way to select edges and faces in SketchUp, before grouping them?

Apply Your Knowledge Select an object and practice grouping and ungrouping it. Then, select a second object and practice grouping it inside the first group. Then, reverse the steps by ungrouping them.

5.2 WORKING WITH COMPONENTS

Even though components are incredibly important, there's nothing too magical about them—they're just groupings of geometry that make working in SketchUp faster, easier, and more fun. In a lot of ways, components are really just fancy groups—they do a lot of the same things. In the following sections, you'll learn about what makes components special and see some examples of what you can do with them. Next, you'll have a quick tour of the Components dialog box, in which you'll discover where components live and how you can organize them. The last part of this section is devoted to making your own components. It's not hard, and once you're able to make components, you're well on your way to SketchUp success. Reference Chapter 11 for more on information on components.

Figure 5-2

These windows are instances of the same component

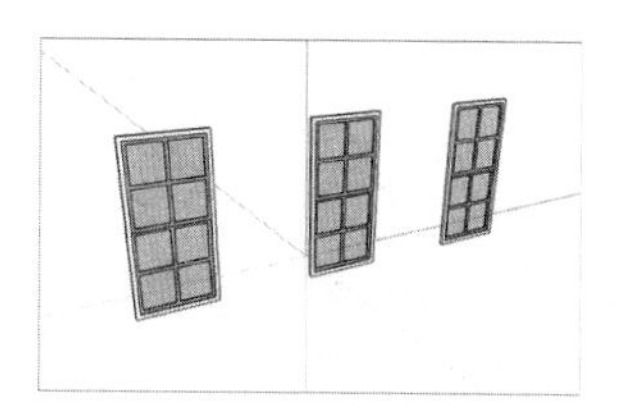

Changing one instance of a component changes all the other instances, too.

GOOGLE SKETCHUP IN ACTION

Apply various SketchUp tools when working with Components.

The following are a few of the reasons why components are important:

- **Everything that's true about groups is true about components.** That's right: Components are just like groups, only better (in some ways, at least). Components don't stick to the rest of your model, you can give them meaningful names, and you can select them, move them, copy them, and edit them easily—just like you can with groups.
- **Components update automatically.** When you use multiple copies (these are called *instances*) of the same component in your model, they're all linked. Changing one makes all of them change, which saves loads of time. Consider a window component that you created and made two copies of, as shown in Figure 5-2. When you add something (in this case, some shutters) to one instance of that component, *all* the instances are updated. Now you have three windows, and they all have shutters.
- **Using components can help you keep track of quantities.** You can use the Components dialog box to count, select, substitute, and otherwise manage all the component instances in your model. Figure 5-3 shows a big (and ugly) building designed to go with the window component from the previous figure. Because the windows are component instances, you have a lot more control over them here than you would if they weren't components.
- **You can make a component cut an opening automatically.** Perhaps you've made a window, and you'd like that window to poke a hole through whatever surface you stick it to. SketchUp components can be set up to cut their own openings in faces. These openings are components that are set up to automatically cut openings can only do so through a single face. As a result, if your wall is two faces thick, your components will cut through only one of the faces.
- **You can use your components in other models.** It's a simple operation to make any component you build available for use whenever you're working in SketchUp, no matter what model you're working on. If you have a group of parts or other things you always use, making your own

Figure 5-3

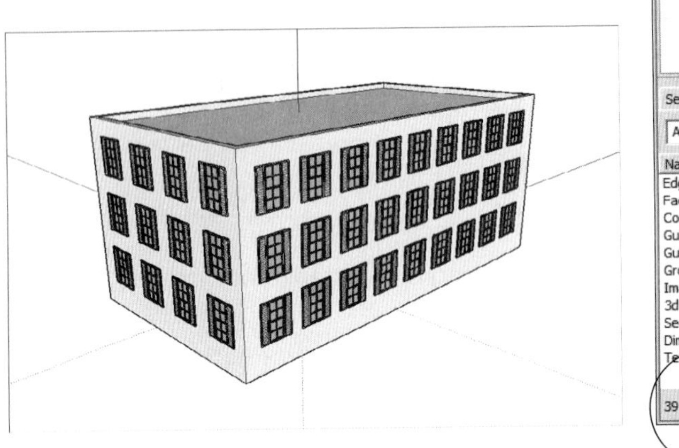

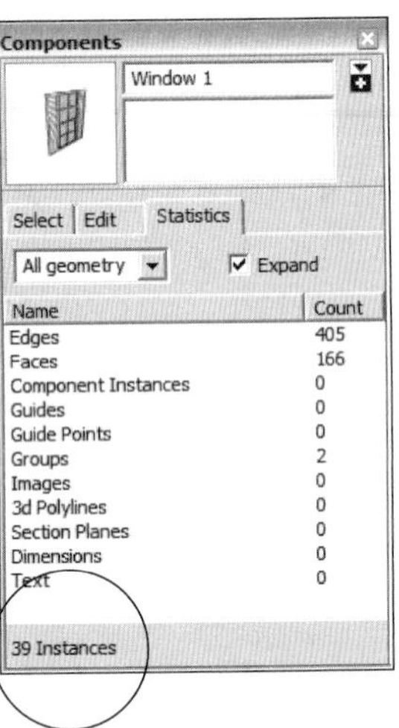

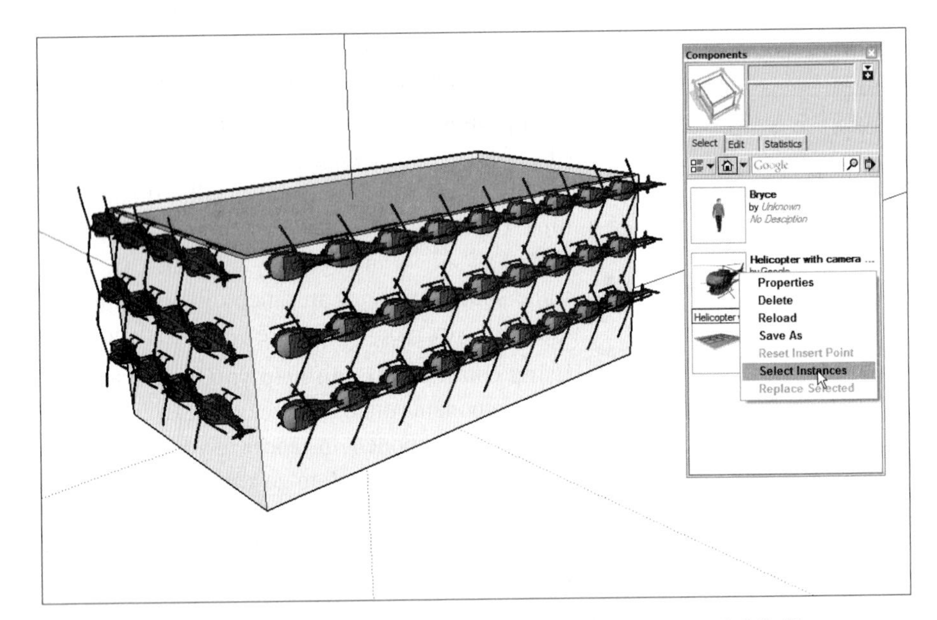

Quickly count all the Window 1 instances in your model (left), or even swap them out for another component.

component library can save you a lot of time and effort. There's more information about creating your own component libraries later in this section.

- **Components are great for making symmetrical models.** Because you can flip a component instance and keep working on it, and because component instances automatically update when you change one of them, using components is a great way to model anything that's symmetrical. And if you look around, you'll notice that most of the things we use are symmetrical.

Figure 5-4

What do all these things have in common? They're symmetrical!

The end of this chapter dives headlong into modeling symmetrical things like couches and hatchbacks; Figure 5-4 shows some examples of symmetrical objects from SketchUp's default component library.

5.2.1 Exploring the Components Dialog Box

It's all fine and well that SketchUp lets you turn bits of your models into components, but wouldn't it be nice if you had someplace to *keep* them? And wouldn't it be great if you could use components made by other people to spiff up your model, instead of having to build everything yourself? As you've probably already guessed, both of these things are possible in SketchUp, and both involve the Components dialog box, which you can find on the Window menu.

You can bring any SketchUp model on your computer into your current file as a component. That's because components are really just SketchUp files embedded in other SketchUp files. When you create a component in your model, you're effectively creating a new, nested SketchUp file.

The Components dialog box is made up of four major areas, which are described in the following sections.

Information and Buttons

This part of the dialog box doesn't have an "official" name, so let's simply refer to it as the Information and Buttons area. Figure 5-5 points out the elements in this area.

Here's what everything in the Information and Buttons area does:

- **Name:** This is where the name of the component you select appears. If it's a component in your model, it's editable. If it's in one of the default collections,

Figure 5-5

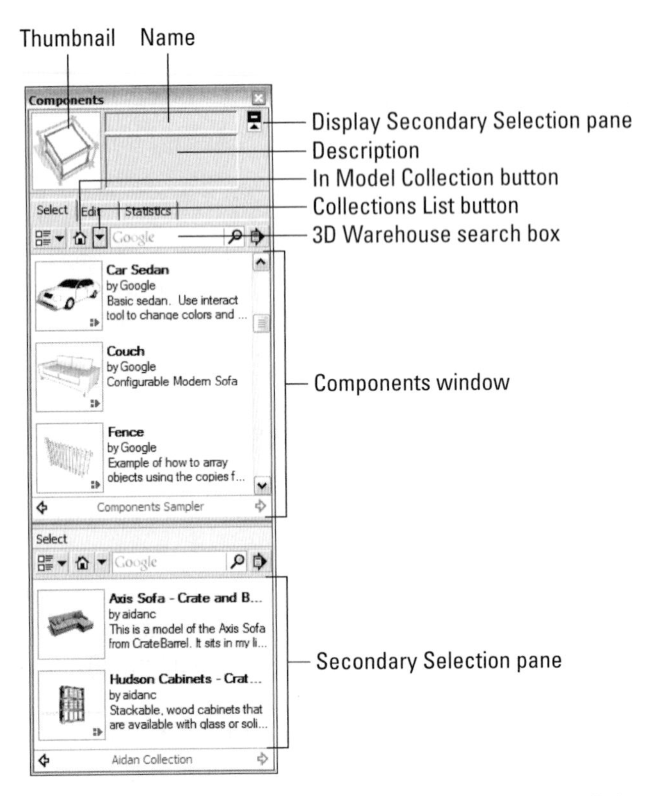

The Information and Buttons area of the Components dialog box.

it's not. A component is considered to be in your model if it appears in your In Model collection, which you can read about later in this chapter.

- **Description:** Some, but not all, components have descriptions associated with them. You can write one when you are creating a new component, or you can add one to an existing component in your model. As with component names, you can only edit descriptions for models if the models are in your In Model library. Sometimes these descriptions include as well tips on how to modify them (for example, we might get a coffee table with many table top options and the description might tell us to find them in the other layers associated with the model).
- **Display Secondary Selection Pane button:** Clicking this button opens a second view of your libraries at the bottom of the Components dialog box. You can use this to manage the components on your computer system.

The Select Tab

This is where your components "live" (if they can be said to live anywhere). You use the Select tab to view, organize, and choose components. Figure 5-6 shows the Select tab in all its glory.

Figure 5-6

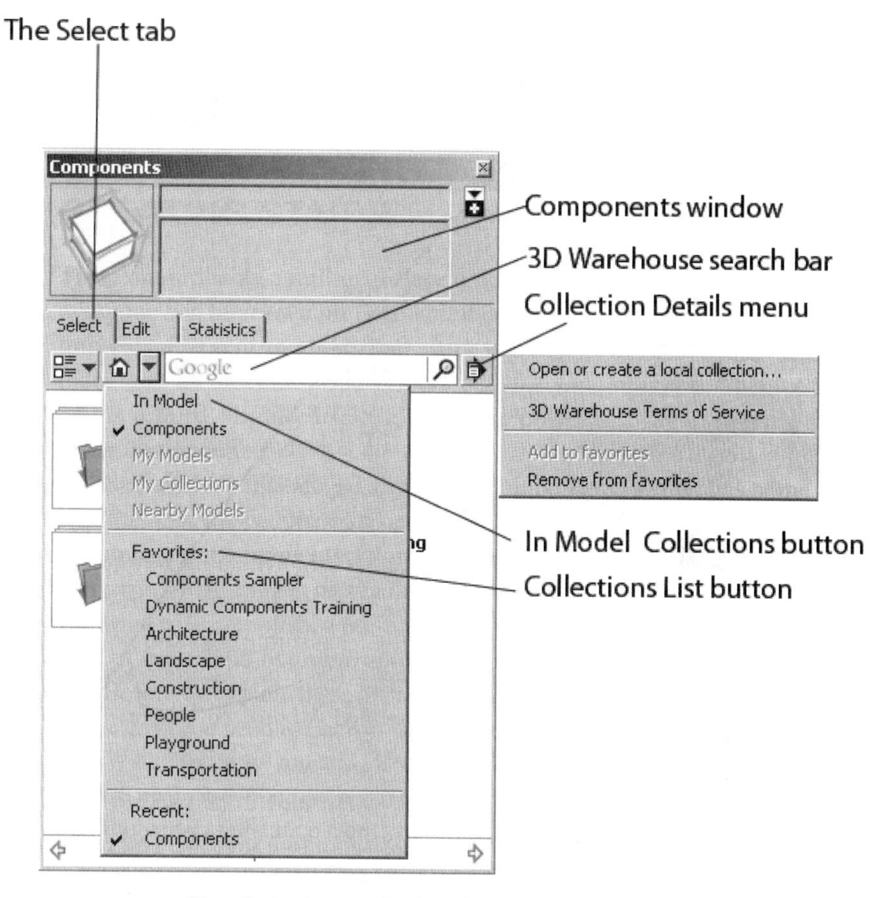

The Select pane in the Components dialog box

The functions of the various parts of the Select pane are as follows:

- **In Model Collections button:** SketchUp automatically keeps track of the components you've used in your model and puts a copy of each of them in your In Model collection. Each SketchUp file you create has its own In Model library, which contains the components that exist in that model. Clicking the In Model Collection button displays the components in your In Model collection, if you have any.
- **3D Warehouse search bar:** It works just like regular Google search; Type what you are looking for and press Enter. Models in the Google 3D Warehouse that match your search terms appear in the Components window below. Naturally, you need to be online for this to work.
- **Components window:** This window displays the components in the currently selected component library or the results of a 3D search you have just performed. Click a component to use it in your model.

FOR EXAMPLE

THE GOOGLE 3D WAREHOUSE

Imagine a place online where everyone in the world can share SketchUp models for free. That's the 3D Warehouse in a nutshell. It's hosted by Google, it's available in more than 40 languages, and it's searchable — just like you'd expect from the world's most popular search engine. You can get to the 3D Warehouse in a couple ways:

- **Through SketchUp:** The Components dialog box is hooked up directly to the 3D Warehouse, as long as you're online. You can also open the 3D Warehouse in a separate window by choosing File ⇨ 3Dwarehouse ⇨ Get Models.

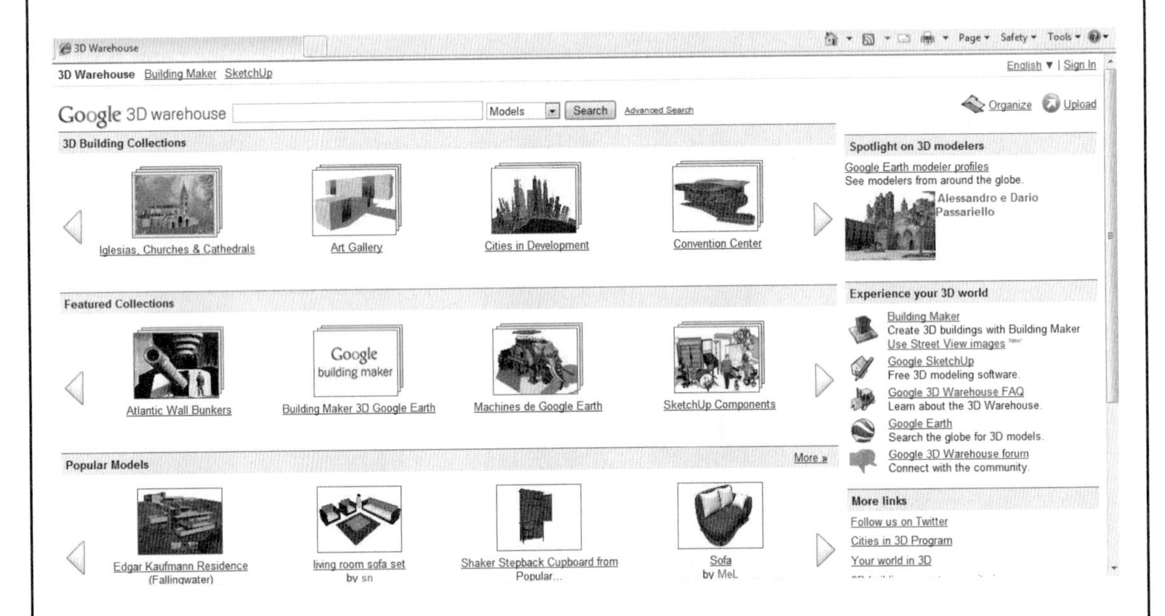

Anything you find on the 3D Warehouse, you can download and use in your own models. You can also upload anything you make so that other people can use it. Find out more about sharing your work on the 3D Warehouse in Chapter 11.

- **On the Web:** Just type http://sketchup.google.com/3dwarehouse into your Web browser. While it does not matter which of the two options you use, it is easier to access the 3D warehouse through the SketchUp, as it is a built-in feature that allows us to work more efficiently, as we can just select the item that we want and drop it in our current model.

- **View Options button:** Simple. This is where you decide how to view the components (or subcollections) in the Components window.
- **Collections List button**: The components listed under the Favorites heading are a mix of two collection types:
  - Local collections are folders of components that live on your hard drive. You can access them anytime because they refer to files on your computer.
  - Online collections are groupings of components that live in the Google 3D Warehouse (which you can read lots more about in this book). Unlike local collections, you can only access online collections when you are online.

    Unfortunately, there's no way to tell just by looking at them in the list which collections are local and which are online. If you click the name of a collection and see a Searching Google 3D Warehouse progress bar before you see any models, that collection is online.

- **Collection Details menu:** Here is where you manage your component collections; it can be found on the small icon to the right of the 3D Warehouse search box (it is the icon with the document with the black arrow behind it). Many options exist and here is an explanation of them:
- **Open or Create a Local Collection**: Allows you to choose a folder on your computer system to use as a component collection. Any SketchUp models in that folder appear in the Components window, ready to be used as components in your model.
- **Create a New Collection**: Allows you to create a folder somewhere on your computer system you can use as a component collection. A collection is handy if you have a number of components that you use all the time, putting them all in one place makes them easier to find.
- **Save as a Local Collection**: When you choose this option, SketchUp lets you save the components that currently appear in your Components window as a brand-new local collection. If the components you are viewing are online, copies of them are downloaded to your computer. If you are vising your In Model collection, the contents are copied and included in a new folder. If you are already viewing a local collection, this option isn't available.
- **View in Google 3D Warehouse**: If you are viewing an online collection, this option opens that collection in a separate window that displays the 3D Warehouse in much more detail.
- **3D Warehouse Terms of Service**: This option is only useful to display information on 3D Warehouse ownership.
- **Add to Favorites**: Choosing this option adds whatever you are viewing in the Components window to the Favorites section of the Collections list. This applies for local collections (folders on your computer); online collections (from the 3D Warehouse); and 3D Warehouse searches; you can save a search as a favorite collection. The models in a Favorite Search collection are always different, depending on what is in the 3D Warehouse.
- **Remove from Favorites**: In reverse, choosing this option will allow you to eliminate whatever you have chosen with a checkmark in the Components window from the Favorites section of the Collections list.

The next two options appear only when you are viewing your In Model collection.

- **Expand:** Because components can be made up of other nested components, a component you use in your model might really consist of *lots* of components. Choosing Expand displays all the components in your model, whether or not they're nested inside other components. Most of the time, you'll probably want to leave Expand deselected.
- **Purge Unused:** Choose this to get rid of any components in your In Model library that aren't in your model anymore. Be sure to use this

before you send your SketchUp file to someone else; it'll seriously reduce your file size and make things a whole lot neater.

On top of all the buttons, menus, and windows you can immediately see in the Select pane of the Components dialog box, some hidden options exist that most people don't find until they go looking for them; they're on the context menu that pops up when you right-click a component in your In Model library, and they are as follows:

- **Properties:** Will show again the Edit tab under Components. It does the same thing as accessing the Edit tab right next to the Select tab.
- **Delete:** Will delete all the instances of that component in your model.
- **Reload:** Replaces the component and all of its instances with another file, or reloads the component if it has been changed since it was used in the 3D model. The difference between Reload and Replace selected, is that Reload will replace ALL of the instances of that component, while Replace Selected will only replace the single instance you have selected.
- **Save As:** It saves the component you selected into a specified folder in your computer.
- **Reset Insert Point:** Allows you to reset any changes done to the insertion point in a component.
- **Select Instances:** Perhaps you have 15 instances (copies) of the same component in your model, and you want to select them all. Just make sure that you're viewing your In Model library, and then right-click the component (in the Components dialog box) whose instances you want to select all of. Choose Select Instances, and your work's done. This can save you tons of time, particularly if you have component instances all over the place.
- **Replace Selected:** Say you want to swap in a different component for one that's currently in your model. Simply select the component instances (in your modeling window) that you want to replace, and then right-click the component (in the Components dialog box) that you want to use instead. Choose Replace Selected from the context menu to perform the swap. Ready for an even better tip? Use Select Instances and Replace Selected together to help you work more efficiently. For example, instead of placing 20 big, heavy tree components in your model (which can seriously slow things down), use a smaller, simpler component instead (like a stick). When you're finished modeling, use Select Instances to select all the stand-in components at once, and then use Replace Selected to swap in the real component. Figure 5-3 earlier in this chapter, shows the mechanics of this operation using windows and helicopters.

The Edit Tab

Because the options in this part of the Components dialog box are similar to the ones you get when you make a new component, you should check out

Figure 5-7

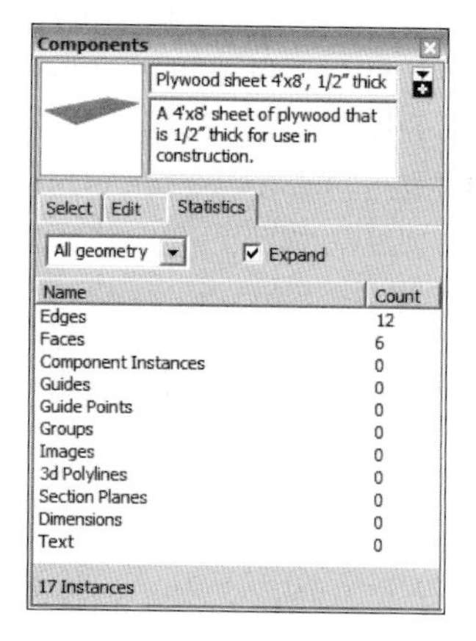

The Statistics tab of the Components dialog box provides a variety of statistics.

the section "Creating Your Own Components," later in this chapter, for the whole scoop. You can only use the options in the Edit tab on components in your In Model collection—everything will be grayed out for components that "live" any other place.

The Statistics Tab

The Statistics tab is a useful place to spend some time. You use it to keep track of all the details related to whatever component you have selected in the Components dialog box. See Figure 5-7 for an example.

This tab is especially useful for doing the following things:

- **Checking the size of your components:** The information in the Edges and Faces areas of this tab lets you know how much geometry is in a component. If you're worried about file size or your computer's performance, try to use small components—ones with low numbers of faces and edges.
- **Seeing what components are inside your components:** The Component Instances line lists how many component instances are in your selected component. If you switch from All Geometry to Components in the drop-down list at the top of the tab, you can see a list of all the constituent components: subcomponents within your main component.

The Statistics tab *doesn't* show details for components you have selected in your actual model; it only shows information about the component that's selected in the Select tab of the Components dialog box. To see information about whatever component (or other kind of object) you have selected in your modeling window, use the Entity Info dialog box (located in the Window menu).

5.2.2 Creating Your Own Components

Now you've learned about the benefits of using components in your models, you're probably ready to start making your own. Thankfully, creating and using components is probably the single best SketchUp habit you can develop. Here's why:

- **Components keep file sizes down.** When you use several instances of a single component, SketchUp only has to remember the information for one of them. This means that your files are smaller, which in turn means that you'll have an easier time emailing, uploading, and opening them on your computer.
- **Components show up in the Outliner.** If you're a person who's at all interested in not wasting time hunting for things you've misplaced, you should create lots of components. Doing so means that you'll be able to see, hide, unhide, and rearrange them in the Outliner, which is described in Chapter 7.
- **Components can save your sanity.** Hooray! You've finished a model of the new airport—and it only took three weeks! Too bad the planning

commission wants you to add a sunshade detail to every one of the 1,300 windows in the project. If you made that window a component, you're golden. If, on the other hand, that window *isn't* a component, you're going to be spending a very long night with your computer.

- **Components can be dynamic.** Dynamic Components are components with special abilities. They can be set up with multiple configurations, taught to scale intelligently, programmed to perform simple animations, and more. Anyone can use existing DCs, but only people with SketchUp Pro can create new ones. Dynamic Components are covered in more detail later in this chapter.

Making a New Component

Creating simple components is a relatively easy process, but making more complicated ones—components that automatically cut openings, stick to surfaces, and always face the viewer—can be a little trickier. Follow these steps, regardless of what kind of component you're trying to make:

1. **Select the edges and faces (at least two) you'd like to turn into a component.** For more information on making selections, see Chapter 2.
2. **Choose Edit ⇨ Make Component.** The Make Component dialog box opens (see Figure 5-8).
3. **Give your new component a name and description.** Of these two, the name is by far more important. Make sure to choose one that's descriptive enough that you'll understand it when you open your model a year from now.
4. **Set the alignment options for your new component.** Wondering what all this stuff means? For a quick introduction to each option and tips for using it, see Table 5-1.

Figure 5-8

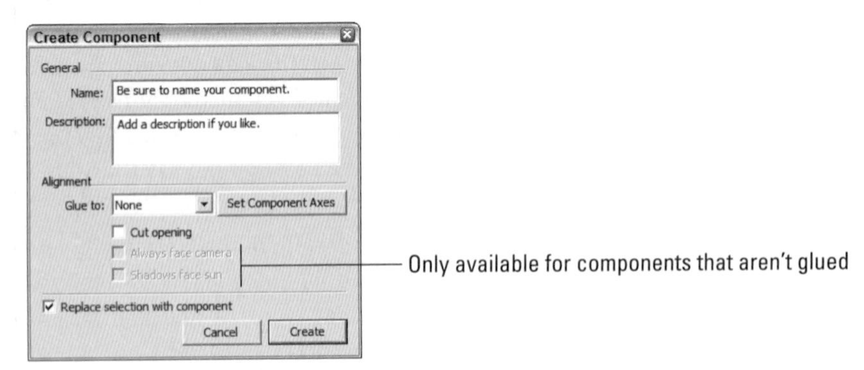

The Make Component dialog box.

Reproduced with the permission of Jorge Paricio, PhD.

Table 5-1 Component Alignment Options

Option	*What It Does*	*Tips and Tricks*
Glue To	This option makes a component automatically stick to a specific plane. For example, a chair will almost always be sitting on a floor. It will almost *never* be stuck to a wall, turned sideways. When a component is glued to a surface, using the Move tool only moves it around on that surface—never perpendicular to it (up and down, if the surface is a floor).	Use this feature for objects that you want to remain on the surface you put them on, especially objects you'll want to rearrange. Furniture, windows, and doors are prime examples. If you want to "unstick" a glued component from a particular surface, right-click it and choose Unglue from the context menu.
Set Component Axes	Sets a component's *axis*, *origin,* and *orientation*. This is important primarily if you have SketchUp Pro and plan to make this into a Dynamic Component. If you aren't, you can safely leave this alone.	Click the Set Component Axes Button to choose where you want your component's axis origin to be (where the red, green, and blue axes meet). Click once to center axes, again to establish the red direction, and again to establish the green and directions. If you are creating a Dynamic Component, this is something you absolutely must know how to do.
Cut Opening	For components "on" a surface, select this check box to automatically cut an opening in surfaces you stick the component to.	As with pre-made components, this opening is temporary. If you delete the component instance, the opening will disappear. If you move the component instance, the opening will move, too.
Always Face Camera	This option makes a component *always* face you, no matter how you orbit around. To make your 2D Face-Me components (as they're called) work, rotate your component to-be so that it's perpendicular to your model's green axis before you choose Make Component.	Using flat, "lightweight" components instead of 3D "heavy" ones is a great way to have lots of people and trees in your model without bogging down your computer.
Shadows Face Sun	This option is only available when the Always Face Camera check box is selected, and it is selected by default.	You should leave this check box selected unless your Face-Me component meets the ground in two or more separate places, as shown in Figure 5-9.

(Continued)

Table 5-1 (*Continued*)

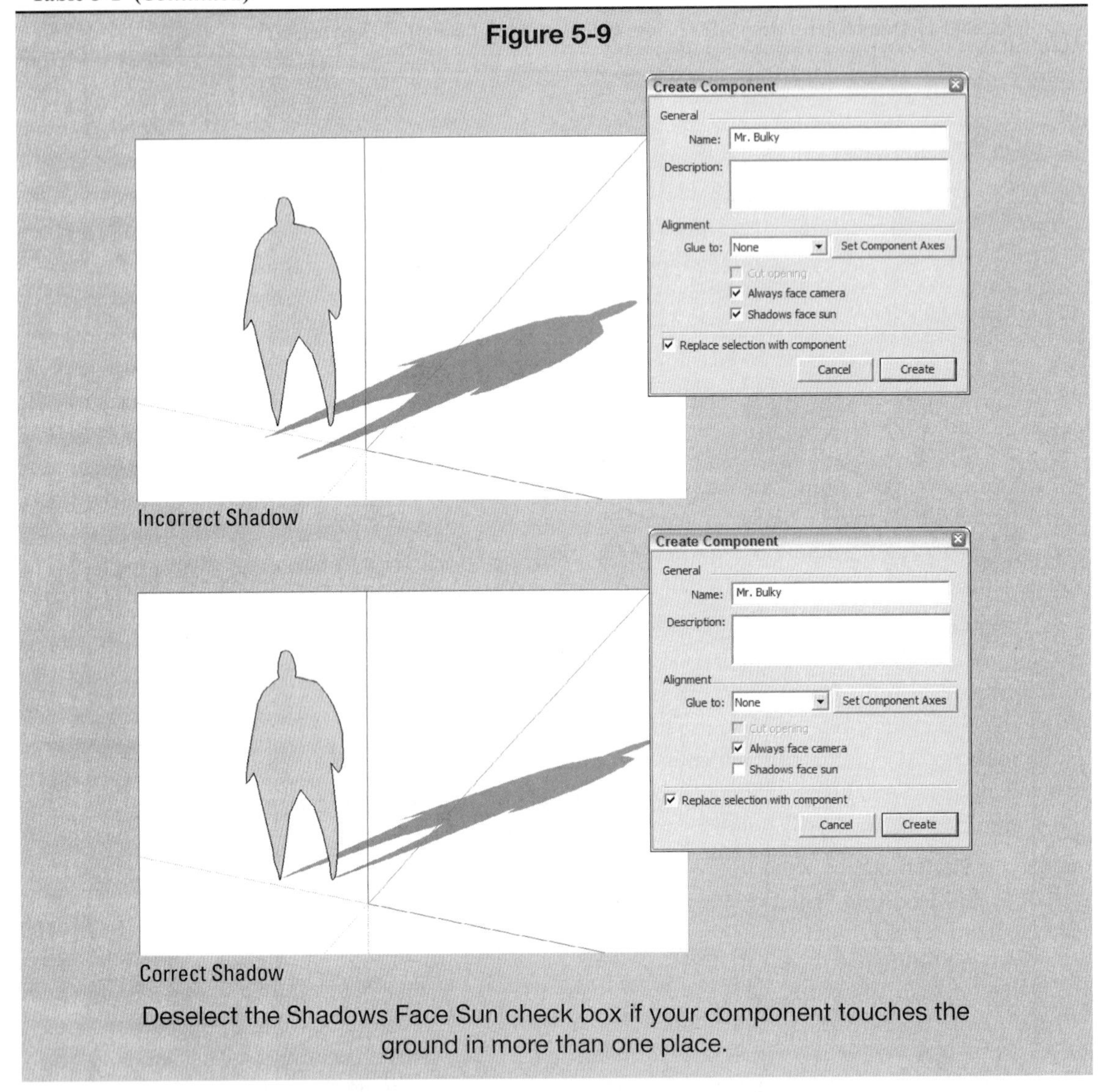

5. **Select the Replace Selection with Component check box, if it isn't already selected.** This drops your new component into your model right where your selected geometry was, saving you from having to insert it yourself from the Components dialog box.
6. Click the Create button to create your new component.

Remember, components can only cut through one face at a time. If your model's walls are two faces thick, you'll have to cut your window and door openings manually (a full reference to this property can be found in the text following Figure 5-5).

5.2.3 Editing, Exploding, and Locking Component Instances

Right-clicking a component instance in your modeling window opens a context menu that offers lots of useful choices. There are 19 different useful choices when we right-click on a component. Here's what some of those choices let you do:

- **Entity Info:** Brings up the information that the component is carrying. It also can be found under WindowàEntity Info, after it has been selected.
- **Erase:** Erases the component from the model.
- **Hide:** Provides a quick way to hide the selected component from view.
- **Lock:** Locking a group or a component instance means that nobody—including you—can mess with it until it's unlocked. You should use this on parts of your model you don't want to change accidentally. To unlock something right-click on it and choose Unlock.
- **Edit Component:** To edit all instances of a component at once, right-click any instance and choose Edit Component from the context menu. The rest of your model will fade back, and you'll see a dashed bounding box around your component. When you're done, click somewhere outside the bounding box to finish editing; your changes have been made in every instance of that component in your model.
- **Make Unique:** Sometimes you want to make changes to only one or a few of the instances of a component in your model. In this case, select the instance(s) you want to edit, right-click one of them, and choose Make Unique from the context menu. This turns the instances you selected into a separate component. Now edit any of them; only those instances you "made unique" will reflect your changes.
- **Explode:** When you explode a component instance, you're effectively turning it back into regular geometry. Explode is a lot like Ungroup in other software programs (in SketchUp, you use Explode to "disassemble" both components and groups).
- **Unglue**: Sometimes components appear to be glued or attached to faces or other objects, and this will allow you to detach it or "unglue" it quickly.
- **Reload**: Let's say that you worked on a table design and made it a component. Then, you placed it into a larger model you are building, let's say, a house and then worked on other areas. Later on, if you his option will allow you to reload it into
- **Save As**: Allows you to save your component as a different name.
- **Share Component**: If you are happy with your work and you want to share it with the rest of the world, you can choose this option to upload it to 3D Warehouse. In order to do that, you would have to create a Google account first, though.

- **Change Axes**: This option will let us modify the axes that are used as a reference for your component. This method would be used if the axis are not aligned with the edges of your object.
- **Reset Scale:** This option will allow you revert your component to the original scale if you modified it at some point.
- **Reset Skew:** Similar to the previous option, this will allow you revert to the original component if you skewed it along the way.
- **Scale Definition:** If you scaled your current component and hit Scale Definition, that component will be duplicated with the current scale, and you will be able to find it in the In Model button.
- **Intersect Faces:** Sometimes you will have to work with two components that are intersecting, and we will have to let SketchUp know that there should be a line that separates the two. This option will let you create a new face where the two intersect, but we will have two more options, With Model or With Selection. The former will intersect all selected that are intersection with the currently selected component, and the latter will allow us to intersect only the components that are selected.
- **Flip Along:** This is a hidden but versatile option if we want to flip horizontally or vertically (or laterally) our components, along each of the three axis, red, blue, and green.

IN THE REAL WORLD

Michael Kanoza, a Graphics Consultant at Steelcase Inc. explains how he uses SketchUp: "It is helping me leverage more than 40,000 AutoCAD 3D drawings from Steelcase's digital assets; in a typical design project we start on AutoCAD, using Steelcase's free application called the Steelcase Furniture Symbol Library (which is a plugin for AutoCAD, http://www.steelcase.com/en/resources/design/fsl/pages/main.aspx) to design typical layouts and then we import the whole thing into SketchUp."

Michael explains that each unique drawing of a piece of furniture (also known as a symbol) that is repeated as Block in AutoCAD, gets imported into Sketchup as a Component. These typical office furniture layouts are then hosted in a flash application on Steelcase's website called "Planning Ideas" at http://www.steelcase.com/en/resources/design/planning-ideas/pages/main.aspx. These office furniture layouts are also shown to students, architect, and designers.

About working with Dynamic Components, Michael thinks that they work very well with repetitive geometric shapes that change size or add features. While they take some time and thought to build, they can save a lot of time in the long run.

- **Zoom Extents:** This is a very simple operation, as it will zoom the selected component so that we can see it magnified to fit our screen.
- **Dynamic Components:** This is a complex button that will be explained in detail later in this chapter—but it basically will allow you to configure and create dynamic components. It has some options available in the free version of SketchUp, which are Component Options and Swap Components, while the Pro version has a few more—Component Attributes, Generate Report, and Redraw.

SELF-CHECK

1. Using components has little effect on file size. True or false?
2. What would be equivalent of Explode command in in other software programs?
3. Name three of the many options that we can access when we right-click on a Component instance.
4. After we are done using components in a SketchUp, what tool could we use to help reduce the size of the file and to get rid of any components in your In Model library that aren't in your model anymore?
5. Where can we see, hide, unhide and rearrange all of the Components that we have in our model?

Apply Your Knowledge Imagine that you work at a furniture gallery and that you need to organize your components' library of the furniture pieces you have for sale. Go to 3D Warehouse and choose 10 different tables and 10 different chairs, and create a new Library Collection of your selected components. Then, practice saving them in your computer.

5.3 DISCOVERING DYNAMIC COMPONENTS

Once upon a time, the smartest thing a component could do was cut its own hole in a surface. "Wow!" all SketchUp aficionados thought, "Components are geniuses!" And so they were—until SketchUp 7 came along. With that release, the folks at Google introduced an entirely new dimension to modeling with SketchUp: Dynamic Components are components with special powers. Until version 7, SketchUp components were basically dumb. If you wanted to make a staircase longer, you had to make copies of the steps and place them in the right spot. If you needed to change the color of a car, you had to dig out the Paint Bucket and dive in to the geometry. The problem was that components didn't know what they were supposed to represent; they were just groupings of faces and edges in the shape of an object.

FOR EXAMPLE

MAKING YOUR OWN DOORS AND WINDOWS

If you really enjoy modeling, nothing beats making your own window and door components. Here's what you need to know (check out the illustration for visual instructions):

1. Start by drawing a rectangle on a vertical surface, like a wall.
2. Erase the face you just created to make a hole in your vertical surface.
3. Select all four edges of the hole you just created. Then right-click one of the edges and choose Make Component from the context menu.
4. Make sure that the Glue option is selected, on the four different options of Glue to Any, Glue to Horizontal, Glue to Vertical, and Glue to Slope. Also, be sure to select Replace Selection with Component, then click the Create button to create your new component.
5. With the Select tool, double-click your new component (in the modeling window) to edit it; the rest of your model will appear to fade back a bit.
6. Use the modeling tools just like you always do to keep building your door or window any way you want.
7. When you're done, click outside your component to stop editing it.

If the opening you create ever closes up, one of two things probably happened:

- **A new surface was created.** Try deleting the offending surface to see whether that fixes things; it usually does.
- **The cutting boundary was messed up.** The cutting boundary consists of the edges that define the hole your component is cutting. If you take away those edges, SketchUp doesn't know where to cut the hole anymore. Drawing them back in usually sets things straight.

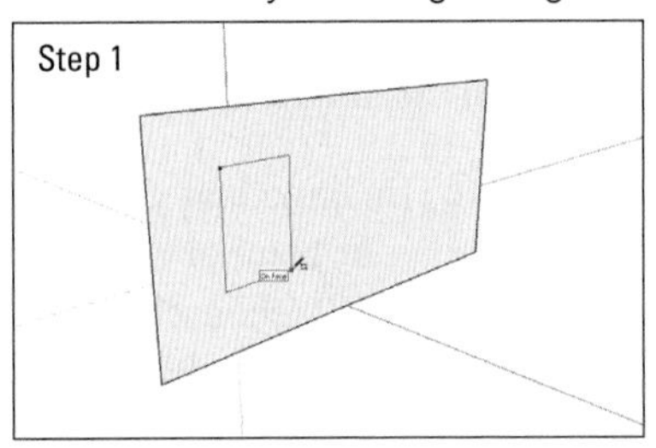

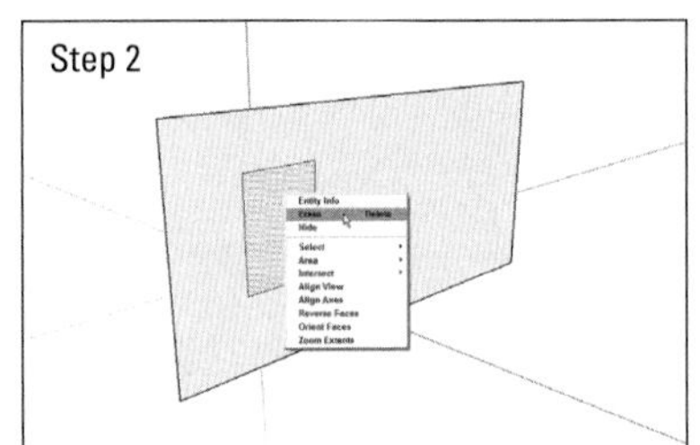

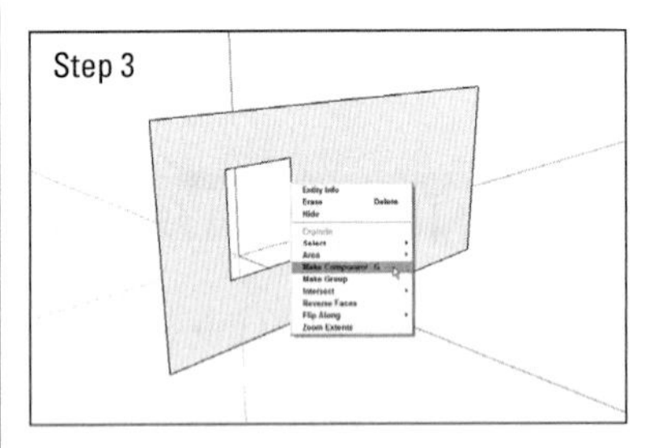

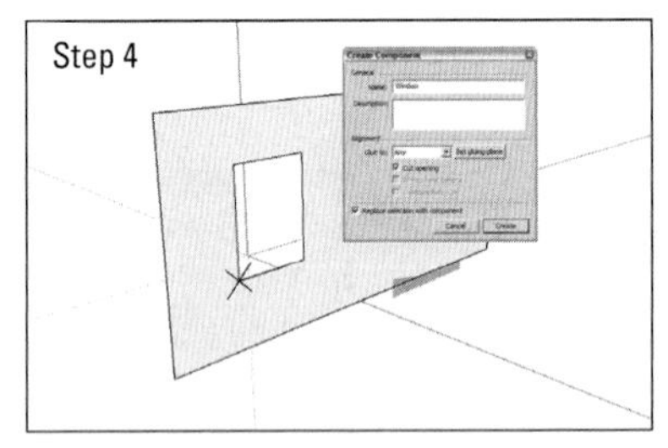

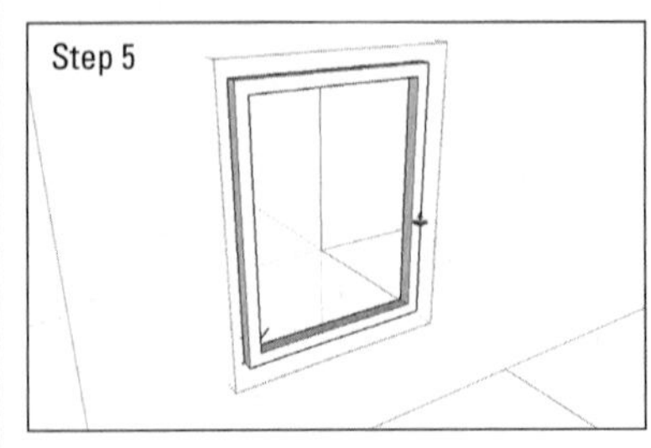

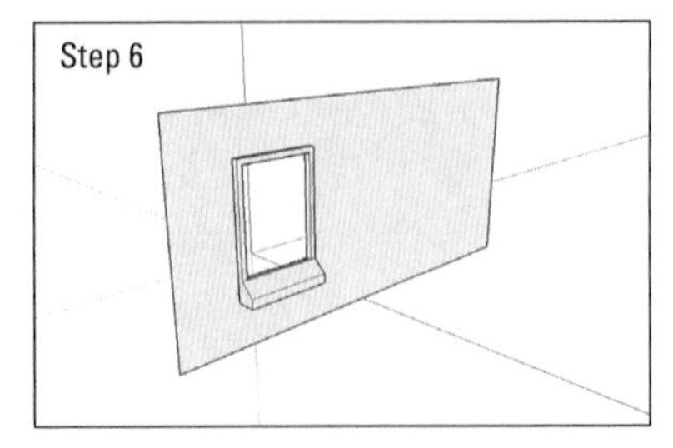

5.3.1 Working with Layout to Manage your Components

When time comes to create a very large model in SketchUp with many dozens or hundreds of Components, SketchUp's Outliner dialog box will provide you with the necessary tools to manage your components as needed. The trick here is to name them (and group them too) in such a way that you will be able to find a particular component without much trouble. For example, if you are trying to pick up a side table component from a perspective view and you happen to be using a ghosted view or a wireframe view, you might end up picking the sofa right next to it, or any object behind it. If you want to be certain that you picked the side table and not anything else, you can select it by name in the Outliner and it will be highlighted for you in the Modeling window.

5.3.2 Getting Acquainted with Dynamic Components

Dynamic Components (DCs) are models that have an idea of what they are; they know what to do when you interact with them. This section outlines what DCs represent for SketchUp modelers and how to use them.

Here's what you need to know about Dynamic Components (DC):

GOOGLE SKETCHUP IN ACTION

Access online resources to retrieve Dynamic Components.

- **DCs are just like regular components, but with extra information added.** That extra information makes them easier to deal with than other components because they know how they're supposed to behave when you need to use them.
- **They can do all sorts of things.** Describing what DCs do is tricky because they're all different. The simple (but totally unsatisfying) answer is that they do what they've been programmed to do. I think some examples are in order (see Figure 5-10):
  - A dynamic door component may be set up to swing open when you click it with the Interact tool.
  - The same dynamic door may also be configured into different sizes, styles, and finishes by using simple drop-down lists in the Component Options dialog box.
  - A dynamic chair may be scaled into a sofa, but without stretching the arms—it would also add cushions as you make it longer.
  - A dynamic stair component may automatically add or remove steps as you use the Scale tool to make it taller or shorter.
  - Susan (the little person who appears by default when you start a new SketchUp file) is also dynamic: Click her shirt with the Interact tool to cycle through various colors. You can replace Susan with another character, too, and his or her shirt also changes color.

Figure 5-10

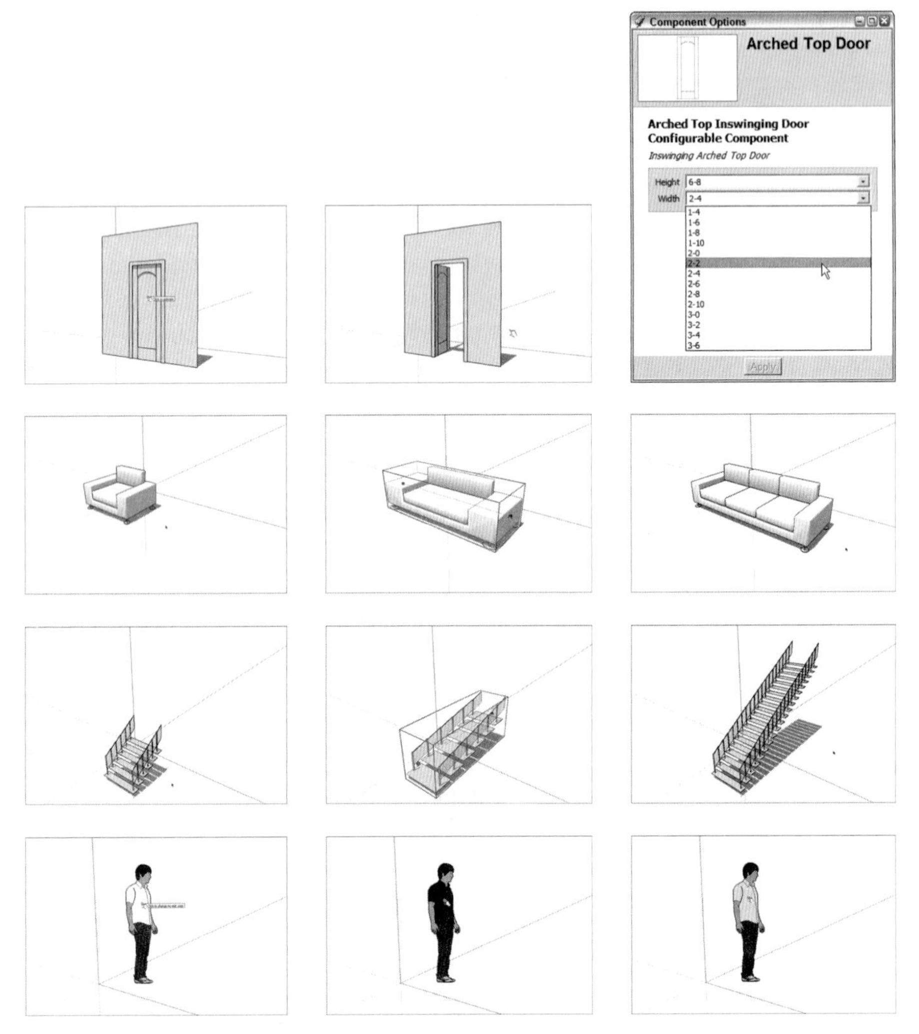

Dynamic Components can do all sorts of things.

- **Anyone can use DCs.** Both the free and Pro versions of SketchUp can read and use Dynamic Components. The SketchUp team invented them (at least partially) to make SketchUp easier for new modelers to pick up.
- **You need Pro to make your own DCs.** If you need to build your own Dynamic Components (or modify ones that other folks have made), you need a copy of SketchUp Pro. Download a free trial of Pro from the SketchUp Web site: http://sketchup.google.com.
- **DCs are free.** People are adding new DCs to the Google 3D Warehouse every day. As you can imagine, companies that make things like furniture and building products (windows, kitchen cabinets, and flooring) are really

excited about the possibilities that DCs offer. Many of them are in the process of producing DCs of everything in their catalogues and posting them to the 3D Warehouse. That's good news for you; soon you can download and use a configurable model of almost anything you need.

- **They're in the 3D Warehouse.** When you download SketchUp, you find a few sample DCs in the Components dialog box. They are the ones with the little, green dynamic icon next to them (that looks kind of like an arrow). The best way to get more is to visit the 3D Warehouse and do a special search:
  1. Choose File ⇨ 3D Warehouse ⇨ 3D Get Models to open a window into the 3D Warehouse from inside SketchUp.
  2. Add the following advanced search operator to any search for models you do: is:dynamic. For example, if you were looking for a dynamic door, you'd search for door is:dynamic.

5.3.3 Using Dynamic Components

In SketchUp, you can interact with Dynamic Components in three basic ways. Depending on what a particular DC has been set up to do, it may respond to one, two, or all three of the following interactions.

Smart Scaling

DCs designed to react intelligently to the Scale tool are the closest things to true magic that SketchUp offers. Instead of stretching and getting all distorted when you scale them, the parts that are supposed to change dimensions, do; the other parts don't.

Take a look at Figure 5-11. The first image shows what happens when scaling a non-dynamic window component to make it wider. See how the frame stretches? Not so nice. The image on the bottom shows the dynamic version of the same window. It gets wider when scaled, but the frame stays the same thickness. It's smart enough to know that only some parts of it should get wider when it's scaled.

Figure 5-11

Original

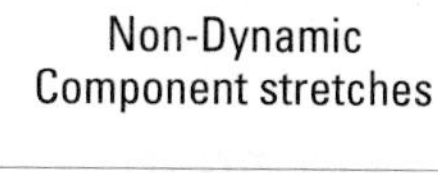

Dynamic Component resizes correctly

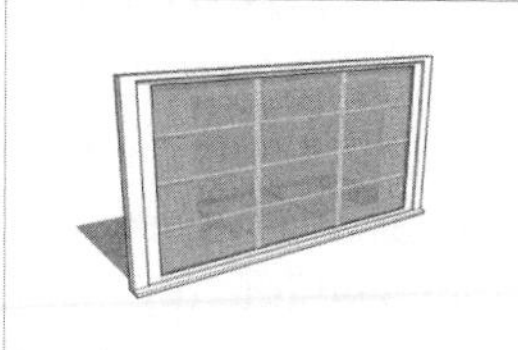

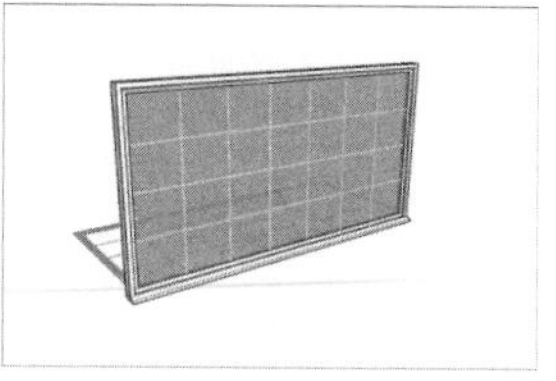

Scaling a non-dynamic window (middle) stretches the entire thing and the DC version scales properly.

Figure 5-12

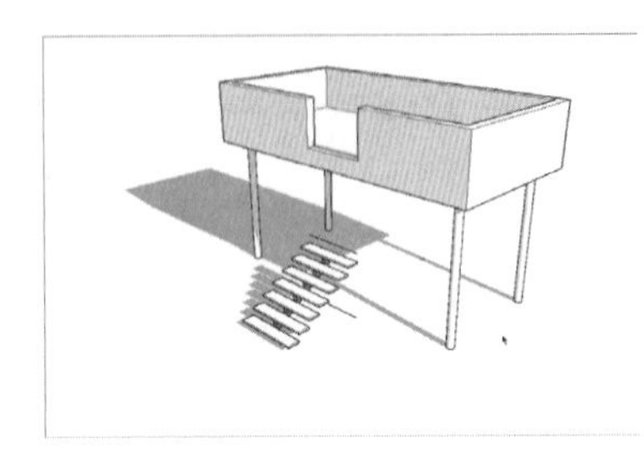

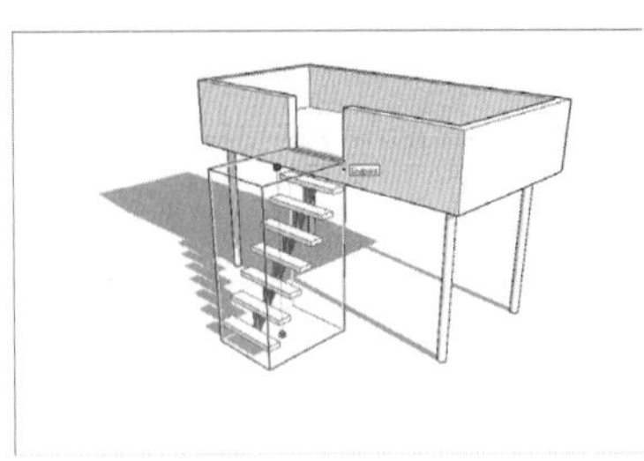

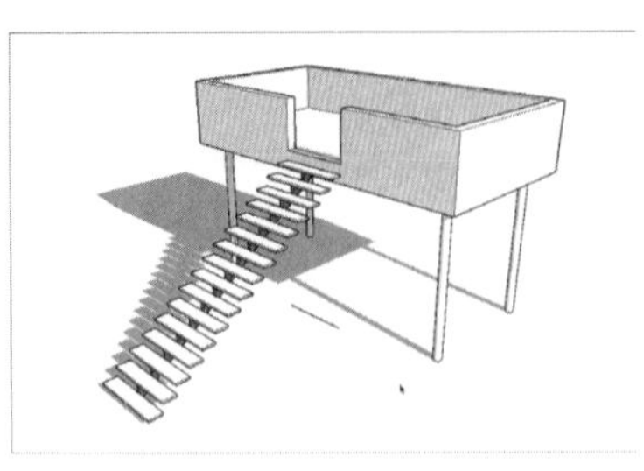

When you make the staircase taller, this dynamic staircase adds steps instead of stretching.

There's another way that DCs can scale smartly: by adding or subtracting pieces as they get bigger or smaller. Dynamic stairs are a perfect example of this, as shown in Figure 5-12. When I use the Scale tool to make the staircase taller, the staircase adds steps instead of stretching.

You can turn on the Dynamic Component toolbar, which is a quicker way to work with DCs than constantly using the menu bar. Just choose View ⇨ Toolbars ⇨ Dynamic Components, and you're all set.

Component Options

SketchUp 7 added the Component Options dialog box, which is on the Window menu. You can configure DCs that have been hooked up to this dialog box by choosing options from drop-down lists, typing dimensions, and performing other simple tasks. When you change a setting in Component Options, the DC you've selected updates to reflect the change, kind of like modeling by remote control.

The Component Options dialog box looks different for every DC. The first image in Figure 5-13 shows the Component Options dialog box for a simple, straight staircase. It is set up so you can choose a riser height and a tread depth from preprogrammed lists. The dialog box also displays the total height (rise), total length (run), and number of steps in the staircase as it currently appears.

The second image in Figure 5-13 shows the Component Options for a circular-stair DC. It is displayed to provide a lot of configuration options, so it looks a lot different. The dialog box lets you enter a size, structure type, and other information and then redraws the staircase based on your specifications.

The Interact Tool

Activate the Interact tool by choosing it from the Tools menu. Using this tool couldn't be simpler: When a DC is set up to react to the Interact tool, it does stuff when you click it. Its actions depend on what you've programmed it to do.

Examine the truck in Figure 5-14. It has been designed to react to the Interact tool in a few ways:

Figure 5-13

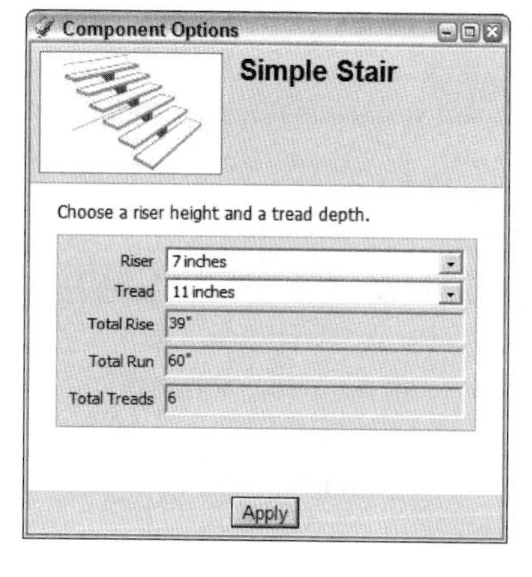

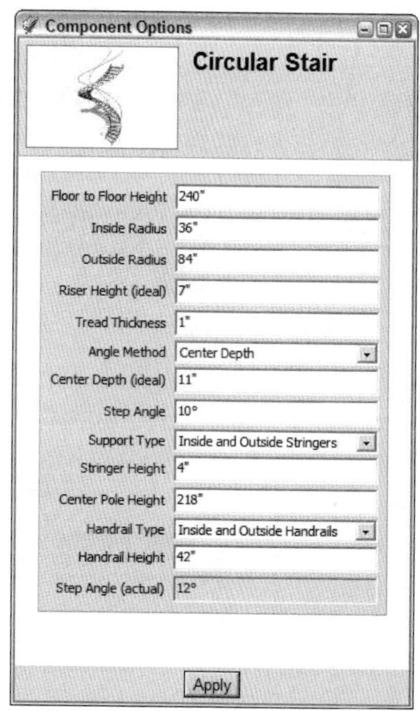

The Component Options dialog box looks different for every Dynamic Component.

- Clicking the back of the truck cycles through the following options: box, flatbed, or flatbed with rails.
- Clicking the front wheels turns them from side to side.
- Clicking the doors makes them open and close.

When you are hovering over a DC after the Interact tool has been turned on, your Select cursor will turn into a small hand with a starburst at the tip of the index finger, signaling that your model is ready for some interaction.

Figure 5-14

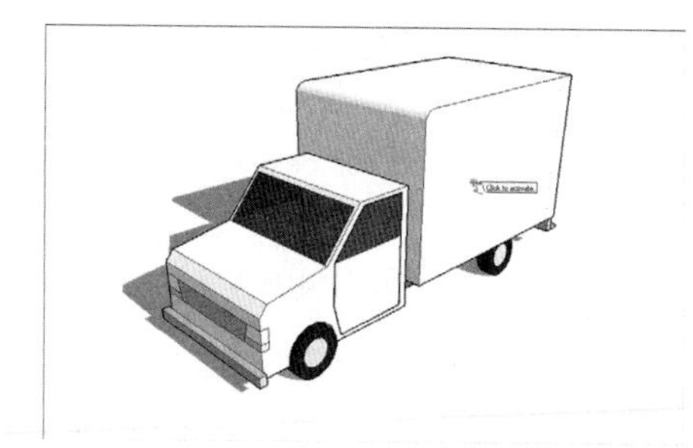

Clicking with the Interact tool makes things happen.

Experimenting with the Interact Tool

You can't know which interactions you can use with any particular DC just by looking at it. If you know you're dealing with a DC, the best way to figure out what it does is to experiment:

PATHWAYS TO... WORKING WITH DYNAMIC COMPONENTS

1. Select it and open Component Options to see whether anything's there.
2. Hover over it with the Interact tool to see whether a glow appears at the end of your cursor.
3. Click it with the Scale tool (it looks like a small gray box inside a larger white box, with a red arrow pointing toward the opposite corner). Then, activate its scale grips (little green boxes located on some corners and center points). Depending on how the Dynamic Component has been created, different scale grips will show up, limiting how it can be scaled. Grab one of the green boxes to see what happens and practice scaling it. If no green box shows up, your DC can't be scaled with the Scale tool.

Groups can be dynamic, too. Deep down in the dark recesses of SketchUp's programming, groups and components are pretty much the same thing—groups are just components that behave differently. This means that a group can be assigned dynamic abilities. What does this matter to you? Not much, but it's worth mentioning. It's good to know, especially if you plan to build your own DCs with SketchUp Pro. You can find more details about creating DCs on this book's companion Web site. DC-creation is tricky, and the extra room that cyberspace affords means more images, color images, and direct links to working examples in the 3D Warehouse.

FOR EXAMPLE

WORKING WITH DYNAMIC COMPONENTS

Working with Dynamic Components can be rewarding if we dedicate enough time to craft and test them, If you are more of a visual person that likes immediate results, you might find the process of DCs time-consuming and tedious, but the rewards of your labor can be huge later on, especially if you will be a recurring user of a particular object.

For instance, we can create fully adjustable doors, in height, width, wall thickness, handle style, or finish. Also, we can make the interact tool cycle through three positions, closed, half-open, and full-open, if we so desire. This can be also applied to a window; we can change the basic dimensions or how it is shown in a model, open or closed. What other models could benefit from this dynamic option?

SELF-CHECK

1. Describe two examples of what two Dynamic Components might be able to do.
2. The free and Pro versions of SketchUp can read and use DCs but only the Pro version will let you make your own DCs. True or false?
3. Explain the advantage of using Smart Scaling if we want to make a staircase taller.
4. What visual clue do we get when we hover over a DC that has been connected to the Interact tool?

Apply Your Knowledge You are managing your company's component library and want to add some Dynamic Components. Visit Google's 3D Warehouse and download some DCs. Then practice with them in your model.

5.4 TAKING ADVANTAGE OF COMPONENTS TO BUILD BETTER MODELS

The fact is a huge amount of the galaxy is made of some kind of repeated element. In the case of bilaterally symmetrical objects (like most furniture), that element is a mirrored half; for things like staircases, it's a step or tread. The whole is composed of two or more instances of a single part. This makes modeling a heck of a lot easier because you don't often have to model things in their entirety—especially if you use components. The following is a list of reasons why you need to work with components whenever you build an object that's made up of repeated elements:

- **It's faster.** This one's obvious. Not having to model the same things twice provides you with more time to play golf or answer e-mail, depending on what you prefer to do.
- **It's smarter.** Everybody knows that things change, and when they do, it's nice not to have to make the same changes more than once. Using component instances means only doing things once.
- **It's appealing.** Modeling something and then watching it repeat in a bunch of other places are fun to do, and the overall effect impresses the heck out of a crowd. Somehow, people will think you're smarter if they see things appearing "out of nowhere."

In this section, two methods are described for modeling with components. The first involves symmetrical objects, and it covers about 50 percent of the things you might ever want to model. The second technique applies to things like stairs and fences, which are both perfect examples of why components were invented in the first place.

5.4.1 Modeling Symmetrically

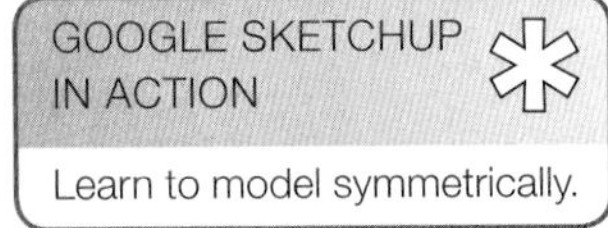

First off, take a hard look at the shape of the things you may want to model. Think about all the objects in the universe. Everything in the world (as I'm sure you realized) can be categorized as either of the following formal types:

- **Symmetrical:** Objects that exhibit bilateral symmetry are made of mirrored halves. You are (more or less) bilaterally symmetrical, and so is your car. Another kind of symmetry is radial symmetry; starfish are good examples of this, as are umbrellas and apple pies. If you were going to build a model of something that exhibits some form of symmetry, building one part and making copies would be a smarter way to do it.
- **Asymmetrical:** Some things—puddles, oak trees, and many houses—aren't symmetrical. There's no real trick to making these things; you just have to make them. You can take advantage of both bilateral and radial symmetry with SketchUp components. To do so, assemble those components as follows, depending on what type of symmetry your object has (reference Figure 5-15):
  - **Bilateral symmetry:** To make a model of something that's bilaterally symmetrical, build half, make it into a component, and flip over a copy.
  - **Radial symmetry:** Radially symmetrical objects can be (conceptually, anyway) cut into identical wedges that all radiate from a central axis. You can use components to model things like car wheels and turrets by building a single wedge and rotating a bunch of copies around a central point.

Figure 5-15

Axis of symmetry

Multiple axes of symmetry

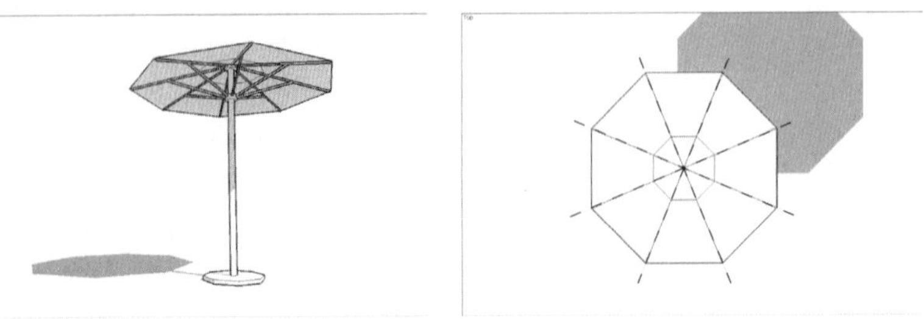

Bilateral symmetry (top) and radial symmetry (bottom) make working in SketchUp easier.

Working Smarter by Only Building Half

Bilaterally symmetrical forms are everywhere. Most animals you can name, the majority of the furniture in your house, and your personal helicopter— they can all be modeled by building half, creating a component, and flipping over a copy.

Follow these steps to get the general idea of how to build a bilaterally symmetrical model in SketchUp (see Figure 5-16):

PATHWAYS TO... BUILDING A BILATERALLY SYMMETRICAL MODEL

1. **Make a simple box.** The easiest way is to draw a rectangle and push/pull it into 3D.
2. **Draw a diagonal edge on the corner of your box.** The point of this step is to mark one side of your box so that when it is flipped over, you don't get confused about which side is which.
3. **Turn your box into a component.** See "Creating Your Own Components," earlier in this chapter, if you wonder how to do this.
4. **Make a copy of your new component instance.** The last part of Chapter 2 has information about moving and copying objects in SketchUp, but here's a simple version:
   a. Choose the Move tool and then press the Ctrl key (Option on a Mac) to toggle from Move to Copy mode. *A little plus sign* (+) *appears next to your cursor.*
   b. Click your component instance, move your copy beside the original, and click again to drop it. Make sure that you move in either the red or the green direction; it makes things easier in the next step.
5. **Flip over the copy.** To do this, right-click the copy and choose Flip Along from the context menu. If you moved your copy in the red direction in the previous step, choose Flip Along ⇨ Component's Red. Choose Component's Green if you moved in the green direction.
6. **Stick the two halves back together.** Using the Move tool (this time without Copy toggled on), pick up your copy from the corner and move it over, dropping it on the corresponding corner of the original. Take a look at the last image in Figure 5-15 to see this illustrated. Doing this precisely is important, if you want your model to look right.

Now you're set up to start building symmetrically. If you want, you can do a test to make sure things went smoothly (see Figure 5-17). Follow these steps:

1. **With the Select tool, double-click one of the halves of your model to edit it.**

Figure 5-16

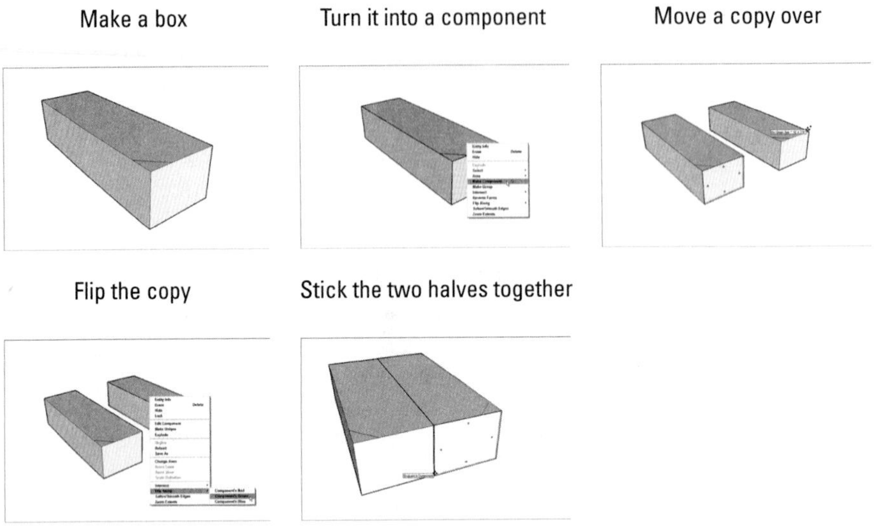

Make a box
Turn it into a component
Move a copy over
Flip the copy
Stick the two halves together

Getting set up to build a bilaterally symmetrical model.

Figure 5-17

Whatever you do on this side should happen on the other side, too

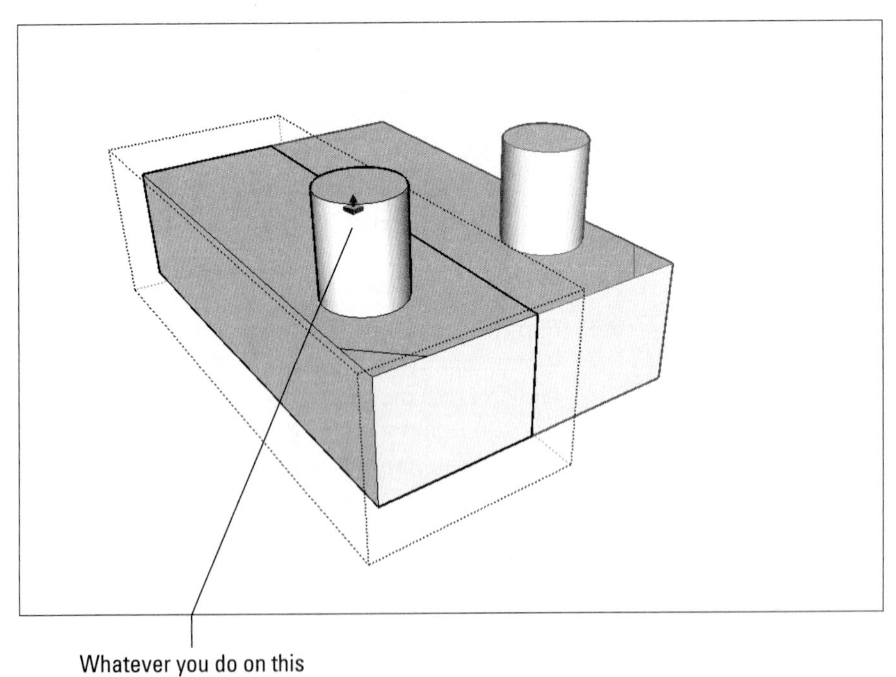

Test your setup to ensure that everything works.

2. **Draw a circle on the top surface and push/pull it into a cylinder.** If the same thing happens on the other side, you're good to go. If the same thing doesn't happen on the other side, it's possible that:
   - **You're not really editing one of your component instances.** If you aren't, you're drawing on top of your component instead of in it. You know you're in edit mode if the rest of your model looks grayed out.
   - **You never made a component in the first place.** If your halves don't have blue boxes around them when you select them, they're not component instances. Start a new file and try again, paying particular attention to Step 3 in the previous steps.

Radial Symmetry

You can model objects that exhibit radial symmetry just as easily as those with bilateral symmetry; you just start slightly differently. The only thing you have to decide before you start is how many wedges—how many identical parts you need to make the whole object.

To model something with radial symmetry, start with one wedge, make it into a component, and then rotate copies around the center. Follow these steps to get the hang of it:

1. **Draw a polygon with as many sides as the number of segments you need for the object you're modeling**. Here's the easiest way to draw a polygon in SketchUp, as shown in Figure 5-18:
   a. Choose Draw ⇨ Polygon to select the Polygon tool.
   b. Click once to establish the center (I like to do this on the axis origin), move your cursor and then type the number of sides you want your polygon to have and press Enter. For example, if we want to have a five-sided polygon, we would need to type 5s and press Enter.
   c. Before you do anything else, move your cursor again and type the radius you want and press Enter.

Figure 5-18

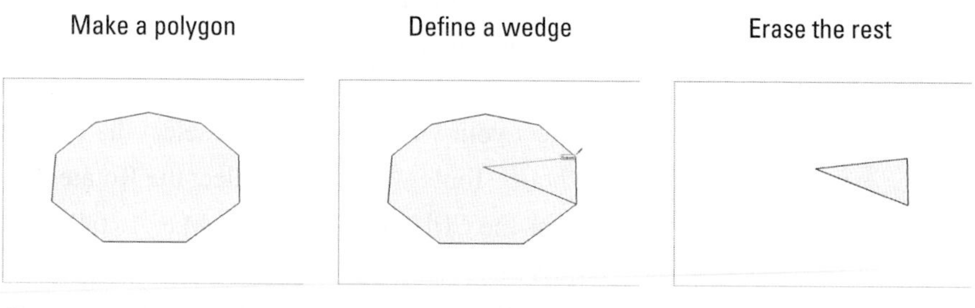

Make a polygon

Define a wedge

Erase the rest

Draw a polygon to start, draw two edges to create a wedge, and erase the rest of your polygon.

Figure 5-19

Click to define center of rotation

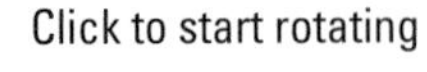

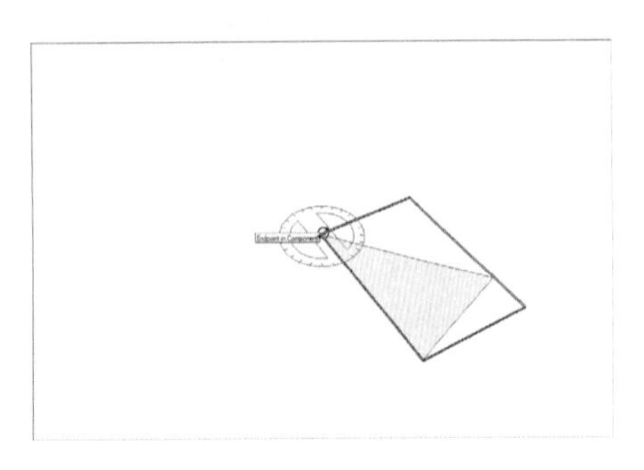

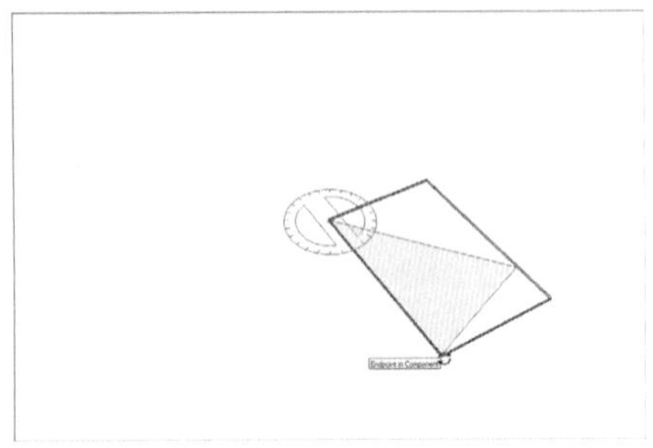

Press Ctrl (Option on Mac) to rotate copy

Make more copies

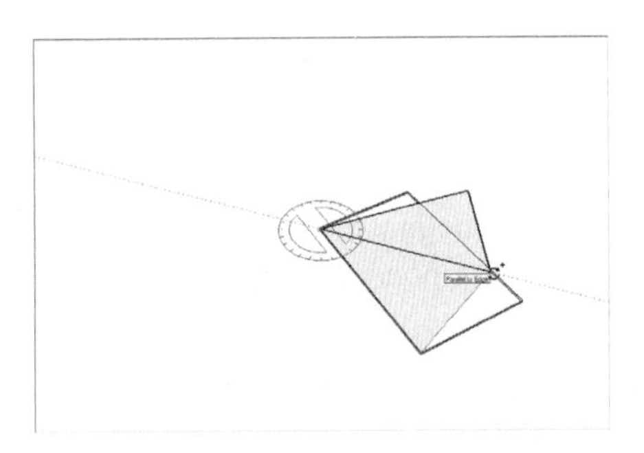

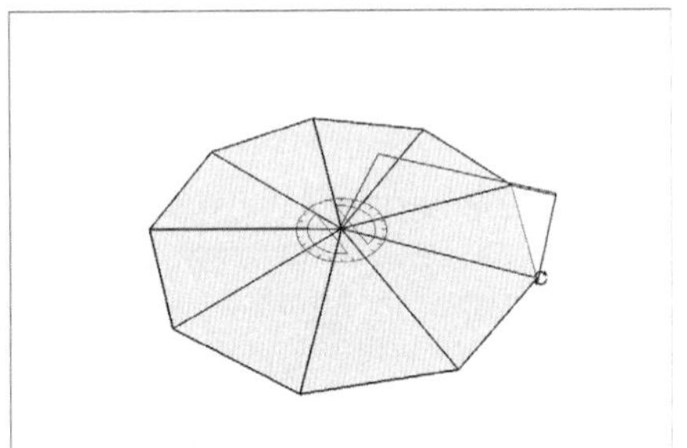

Use the Rotate Tool to make copies of your wedge component instance.

2. **Draw edges from the center of your polygon to two adjacent vertices (endpoints) on the perimeter, creating a wedge.** To find the center of a polygon (or a circle), hover your cursor over the outline for a couple seconds and move the cursor toward the middle; a center inference point appears.
3. **Erase the rest of your polygon, leaving only the wedge.**
4. **Turn your wedge into a component.** Check out "Creating your own components," earlier in this chapter, if you're unsure of how to do this.
5. **Make copies of your wedge component instance with the Rotate tool (see Figure 5-19).**

   Just like with the Move tool, you can use the Rotate tool to make copies. You can even make an array (more than one copy at a time). Here's how:

   a. Select your wedge's edges and select the face, too.
   b. Choose Tools ⇨ Rotate to select the Rotate tool.
   c. Press the Ctrl key (Option on a Mac) to tell SketchUp you want to make a copy. *A + appears next to your cursor.*
   d. Click the pointy end of your wedge to set your center of rotation.
   e. Click one of the opposite corners of your wedge to set your rotation start point.

f. Click the other corner to make a rotated copy of your wedge.

g. Type the number of additional wedges you want, followed by the letter x, and then press Enter.

6. **(Optional) Test your setup.** Follow the steps associated with Figure 5-17 to test whether updates to a single component in your new object updates all instances of the component.

Hiding the edges in your component instances makes your finished model look a whole lot better. Take a look at the sidebar "Making two halves look like one whole," earlier in this chapter, to read how.

5.4.2 Modeling with Repeated Elements

A staircase is a perfect example of an object that is composed of several identical elements. If, when you hear the phrase "several identical elements," a big, flashing neon sign that screams "COMPONENTS!" doesn't appear in your head, you are not using SketchUp enough.

In the following example, is demonstrated how you might use components to model more efficiently and to show readers of Chapter 4 the smartest way to build a set of stairs.

The Treads Are Components method involves making each tread (step) in your staircase into an instance of the same component. Basically, you build one simple tread that is the right depth, make it into a component, and copy a bunch of instances into a full flight of stairs. Because every step is linked, anything you do to one automatically happens to them all.

Go through these steps to build a staircase using the Treads Are Components *method:*

1. **Model a single step, including the tread and the riser.** You can make this very simple at this stage, if you want to; all that matters is that the tread depth and the riser height are correct (the standard for a tread depth is 10" and the standard for a riser height is 7 5/8").

   - Nosing: The overhanging lip of the tread. Standard nosing is 1 1/4 inches.
   - Tread depth: The cut dimension of the stringer; depth of the horizontal walking surface less the nosing dimension. Standard tread depth is 10 inches.
   - Riser height: The vertical dimension between two treads; this dimension must be equal throughout the total flight. Standard rise height is 7 5/8 inches and should not exceed 7 3/4 inches.

   You can try everything else later. Figure 5-20 shows a simple example of this.

2. **Make a component out of the step you just built.** Take a look at "Creating Your Own Components," earlier in this chapter, if you need help.

3. **Move a copy of your step into position, above the first one (see Figure 5-21).**

Figure 5-20

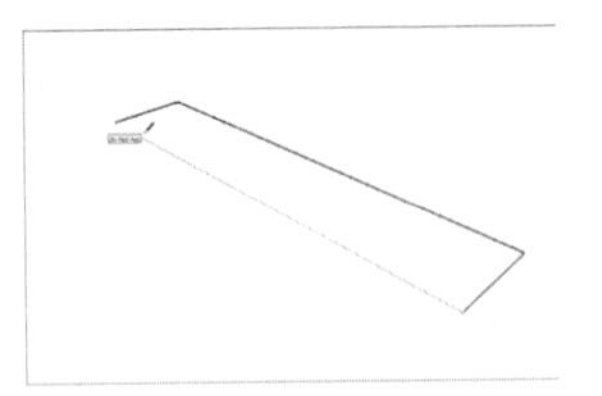

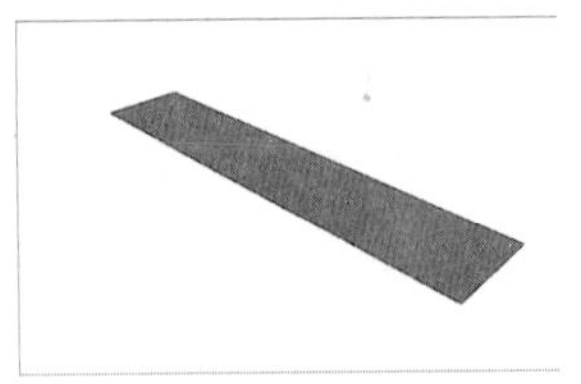

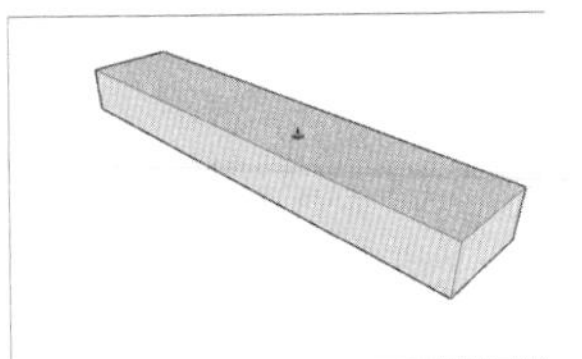

Model a single step, making sure that the depth and height are accurate.

Figure 5-21

1. Create a component

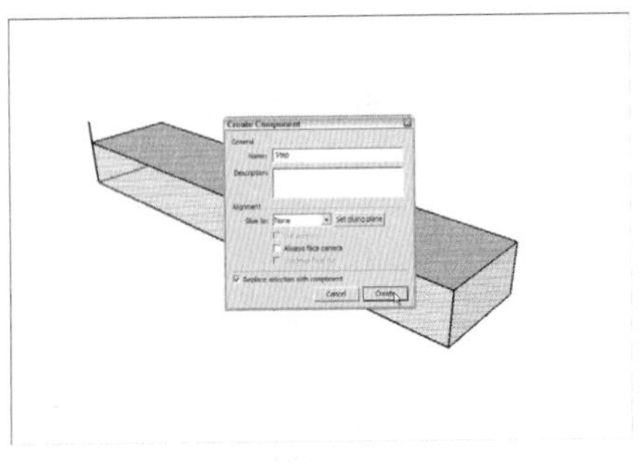

2. Move up a copy

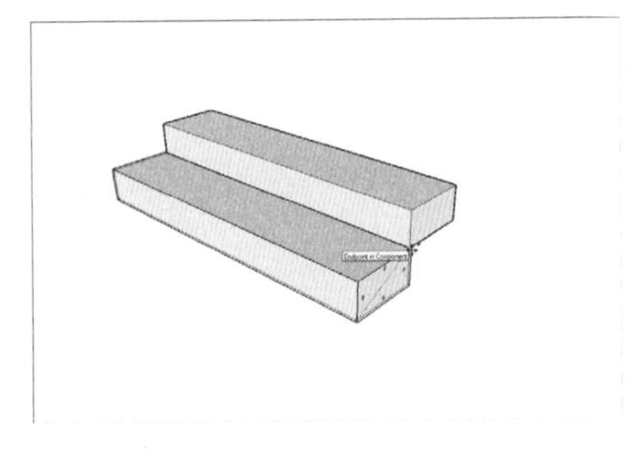

3. Type the number you want, then **x**, and press Enter

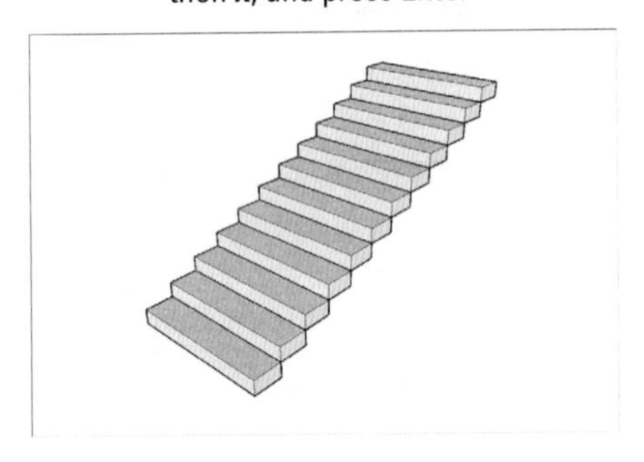

4. Edit one instance

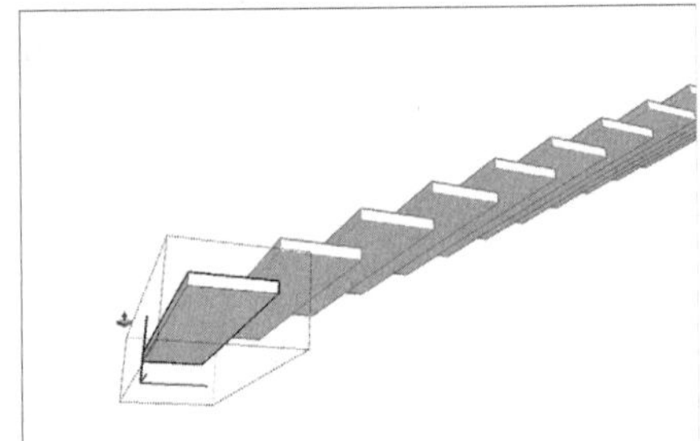

5. All components change

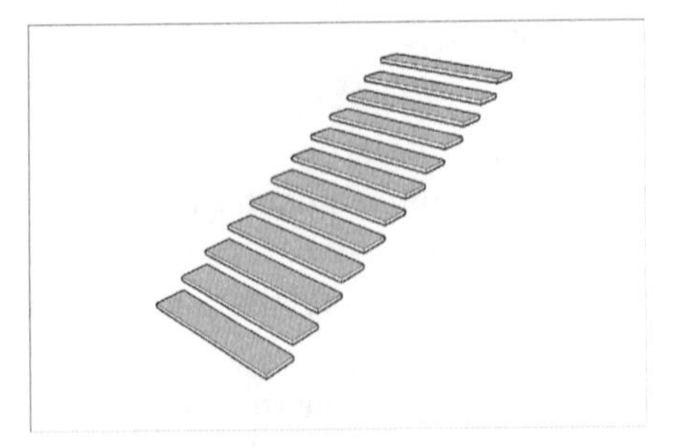

Make you step into a component instance, move a copy into position above the original, and then create an array.

GOOGLE SKETCHUP IN ACTION

Understand how to model with repeated elements.

4. **Type the total number of steps you want, type an x, and then press Enter.** You're creating a linear array, meaning that you're making several copies at regular intervals, in the same direction you moved the first one. Typing **12x** generates 12 steps the same distance apart as the first step and its copy. The last image on the right in Figure 5-21 illustrates this.
5. **With the Select tool, double-click any one of your steps to edit all instances of your component**. Everything besides the component instance you're editing fades out a little.
6. **Have fun.** Having your staircase made up of multiple component instances means that you have all the flexibility to make drastic changes to the whole thing without ever having to repeat yourself. Add a nosing (a bump at the leading edge of each tread), a stringer (a diagonal

TIPS FROM THE PROFESSIONALS

Making two halves look like one whole

Looking carefully at the little boat in the figure that follows, notice how the edges in the middle clearly show that it's made out of two halves? If I were to erase those edges, my whole model would disappear, because those edges are defining faces, and without edges, faces can't exist. Instead of erasing those unwanted edges, I can hide them by using the Eraser while pressing the Shift key. See the second and third images of the boat? When I hold down Shift as I drag over the edges I want to hide with the Eraser, they disappear. Two things are important to know about hidden edges:

- **Hidden edges aren't gone forever.** Actually, this applies to any hidden geometry in your model. To see what's hidden, choose View ⇨ Hidden Geometry. To hide it again, just choose the same thing.
- **To edit hidden edges, you have to make them visible.** If you need to make changes to your model that involve edges you've already hidden, you can either view your hidden geometry (see the previous point) or unhide them altogether. Just show your hidden geometry, select the edges you want to unhide, and choose Edit ⇨ Unhide ⇨ Selected.

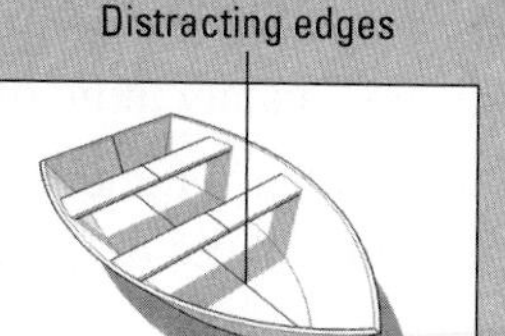

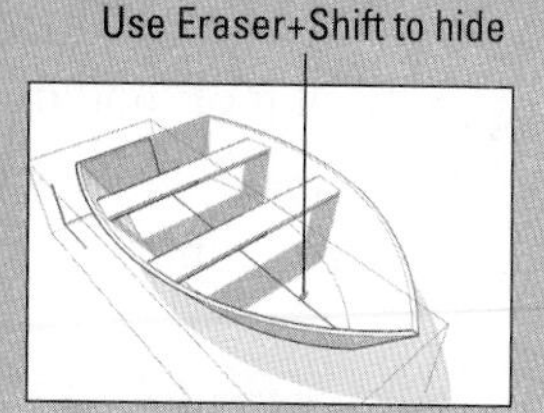

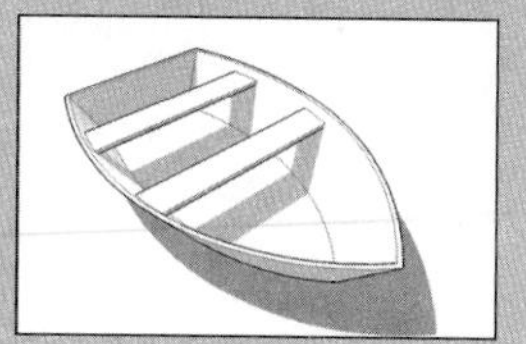

Figure 5-22

Series of component instances

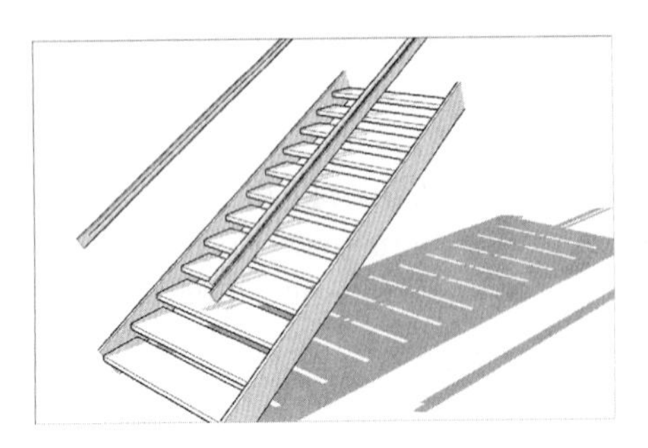

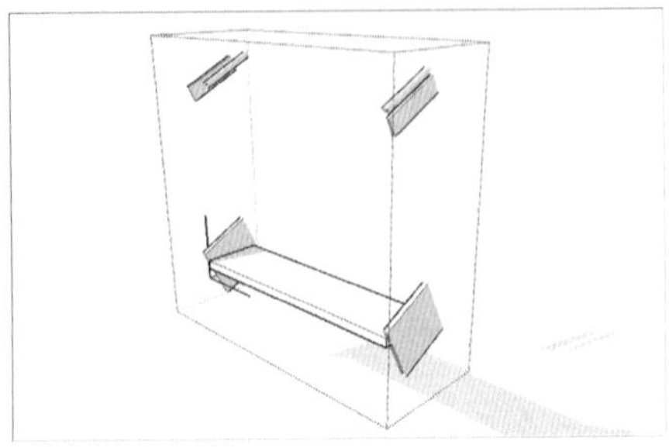

A flight of stairs with side stringers and a handrail. On the right, a single component instance.

piece of structure that supports all your steps), or even a handrail by getting creative with how you modify a single component instance. Figure 5-22 shows some of what you can do. The color insert in this book shows the Treads Are Components method applied to building a spiral stair.

> **FOR EXAMPLE**
>
> GOOGLE SKETCHUP FACT
>
> When you check components done by someone else, it is always a good idea that you open them independently before you insert them in your file. There is no way to verify how each one is built until you open them and check how they are built or how many groups they have, if there are too many GROUPS nested inside other groups or if they come with loose lines that do not connect to anything else.

SELF-CHECK

1. To quickly make two halves to look like one, we have to:
   a. Edit each half and hold the Shift key and hit erase on the edges that we want to hide.
   b. Edit each half and right-click and select erase.
   c. Edit each half and hit erase on the edges that we want to hide.
2. When we are about to make several identical elements in our model, what tool should we immediately think of?
3. To make an object with radial symmetry we would have to build half, make it into a component and flip over a copy. True or false?

Apply Your Knowledge Working efficiently with Components can cut down your modeling time significantly, plus it will allow you to move faster, seeing results even before you have locked the design of a shape. Quickly build a simple four-walled house with a pitched roof that overhangs in the front. Now, imagine that you are including eight columns in the front. Practice by placing a simple cylinder in lieu of a more complex, fully designed column, and make it a Component. Then, copy it seven more times and space out each column in the front. Once you are done right-click and edit one of them and modify its design, adding a capital and a base. Whatever design you made on that edited column will be carried through the other seven columns and you would have completed your design in a fraction of the time.

SUMMARY

Section 5.1

- Using groups makes it easier to work with your model.
- SketchUp geometry (the edges and faces that make up a model) is sticky.

Section 5.2

- Components work much like groups.
- When you use multiple copies (these are called *instances*) of the same component in your model, they're all linked. Changing one makes all of them change, which saves modeling time.
- The 3D Warehouse hosted by Google, is available in more than 40 languages, is searchable and you can get it from the Web or through SketchUp.

Section 5.3

- Dynamic Components are just like regular components, but with extra information added. That extra information makes them easier to deal with than other components because they know how they're supposed to behave when you need to use them.
- Dynamic Components are free to use. However, in order to create new Dynamic Components, you need to use the Pro version of SketchUp.

Section 5.4

- Forms come in the following ways and can be worked in SketchUp as:
  - Symmetrical
  - Asymmetrical
  - Bilateral symmetry
  - Radial symmetry
- The Treads Are Components method involves making each tread into an instance of the same component. Basically, you build one simple tread that is the right depth, make it into a component, and copy a bunch of instances.

ASSESS YOUR UNDERSTANDING

SUMMARY QUESTIONS

1. Components are used when a model includes several copies of the same thing. True or false?
2. Which of the following is a characteristic of groups?
   a. Grouped geometry sticks to everything.
   b. Groups have no names.
   c. Groups cannot be moved.
   d. Ungrouping geometry requires that it be exploded.
3. Explain what happens when you use multiple instances (copies) of the same component, and you create an update. What happens to all the other instances in your model?
4. You can place components inside other components. True or false?
5. Libraries are folders on your computer that contain SketchUp files. True or false?
6. The Statistics pane shows details for components you have selected in your actual model. True or false?
7. Which of the following is true with regard to components?
   a. They keep file sizes down.
   b. They don't appear in the Outliner.
   c. Component instances must be updated manually.
   d. They cannot cut openings.
8. One benefit of using layers is that they can be used to organize large groups of similar items. True or false?
9. Using components can help you keep track of what?
10. Which of the following options in the Outliner displays a list of all the groups and components in your model?
    a. Display
    b. List
    c. Expand
    d. Library list
11. Which option should you use to make changes only to a few of the instances of a component in your model?
    a. Alter selectively
    b. Make unique
    c. Edit component
    d. Select one

12. What can we do to the Outliner so that you only see top-level groups and components?
13. The option Sort by Name lists the groups and components in your model alphabetically. True or false?
14. Explain how you can check the size of your components.

APPLY: WHAT WOULD YOU DO?

1. Name and describe three reasons why you would need to make groups.
2. List and describe the four major areas of the Components dialog box.
3. How do you move entities to a different layer?
4. Name two reasons you should use components and why you would want to create your own components.
5. What are libraries, and why would you use them?
6. Explain what is Layer0 and when would you use it.

BE A DESIGNER

Doors and Windows

Within the model of the house that you've been working on, make your own doors and windows. What happens if the cutting boundary is messed up?

Organize Your Model

Create a model of a simple doghouse surrounded by trees, with one dog outside the doghouse. Using groups, components, the Outliner, and layers, organize the model so that you can easily show or hide certain subsets of objects.

KEY TERMS

Components

Group

TOOL KIT

Save time—Go shopping

Why spend hours modeling an oak tree when you can buy a fantastic one for a reasonable price? If you can't find what you need among the millions of components that you can download for free from the 3D Warehouse, here are a couple great, paid options online:

- Form Fonts: Form Fonts (www.formfonts.com) is a Web site that sells components "all-you-can-eat, buffet style." You pay a (surprisingly low) monthly fee, and you have access to thousands of high-quality models of just about anything. Form Fonts' international team of modelers even takes requests—if you need something that they don't have, they can probably make it. Even if you're not interested in signing up, it's worth checking out the Web site just to see the beautiful models Form Fonts makes.

CHAPTER

CREATING EVERYDAY OBJECTS

Tools, Techniques, and Tips

Do You Already Know?

- How to extrude shapes?
- How to make lathed forms?
- How to work with different kinds of terrain?
- How to create and work with solids?

For the answers to these questions, go to **www.wiley.com/go/chopra/googlesketchup2e**

What You Will Find Out	What You Will Be Able To Do
6.1 What tools and techniques to use for extrusion.	Understand how to work the Follow Me tool. Create lathed forms.
6.2 How to scale objects.	Learn how to operate the Scale tool.
6.3 Learn tools and techniques to create and work with different terrains.	Create new and edit existing terrains.
6.4 About how solids affect your model.	Understand and work with solids.

INTRODUCTION

Here's something you already know: There's more to life than modeling buildings. Even though SketchUp is really good at letting you make models of built structures, you can use it to build just about anything you can think of—all it takes is time, ingenuity, and the ability to take a step back and break down things into their basic parts. SketchUp provides fantastic tools for creating forms that aren't the least bit boxy, but they're not as obvious as Push/Pull and Rectangle, so most people never find them. This chapter is devoted to helping you discover SketchUp's "rounder" side. In this chapter, I present tools, techniques, and other tips for creating forms that are distinctly unbuilding-like—my hope is that you'll use them to push the limits of what you think SketchUp can do.

6.1 EXTRUDING SHAPES WITH FOLLOW ME

Follow Me is an excellent example of a powerful SketchUp tool with an underwhelming name. The problem that faced the software designers when they were trying to determine what to call this feature was this: it does what other 3D modeling programs dedicate two or three other tools to doing. Thus, the designers chose an unconventional name because Follow Me is a wholly unconventional tool.

In the following sections, you'll learn how to use Follow Me to create a number of different types of shapes. Examples of these shapes are shown in Figure 6-1 and are as follows:

Lathed form
3D form created by spinning a 2D shape around a central axis.

GOOGLE SKETCHUP IN ACTION

Understand how to work the Follow Me tool.

- **Bottles, spindles, and spheres:** These are all examples of **lathed forms**, which are 3D models created by spinning and extruding a 2D profile (shape) around a central axis.
- **Pipes, gutters, and moldings:** If you look closely, all three of these things are basically created by extruding a 2D face along a 3D path; the result is a complex 3D form.
- **Chamfers, fillets, and dados:** Without explaining what all these things are know this: You can use Follow Me to cut away profiles, too.

Figure 6-1

Follow Me lets you create all kinds of different shapes.

IN THE REAL WORLD

Industrial Designer Dennis Bostwick designs wine cellars for a small company in Colorado. He started using SketchUp in school and has been using it one way or another of the past four years. He likes it because it is easy software to learn but most importantly, the company he works for can afford to buy a copy of the software. AutoCAD was far too expensive. He works combining images from SketchUp and Illustrator to create his presentations, and interestingly enough, he quotes "the thing I get most comments about are the people I put in my models for scale. Clients find them either cute or creepy."

His workflow usually follows the following pattern: first, he draws a basic sketch of the wine cellar and creates a floor plan of the room he is designing. Once that is complete, he finds out the size of the cabinets he is planning to use in the room and places them. Then, he completes the scene by building in 3D any custom cabinets that he needs and finally adds components like trim, doors, and assignments materials. Finally, he exports his views as JPEG files into Illustrator and composes his board for his presentations.

About the use of Follow Me, Dennis mentions that "there is no other tool I can use to give me the desired results I'm looking for when designing wine cellars. Often enough, the room calls for a complex crown molding and this tool is the only way I can complete my design efficiently. The molding is an intricate design that must follow a combination of straight and curved elements throughout the spaces and it allows me to do a high quality design in very little time."

6.1.1 Using Follow Me

At its core, Follow Me lets you create forms that are extrusions. It's a bit like Push/Pull, except that it doesn't just work in one direction. You tell Follow Me to follow a path, and it extrudes a face all along that path. This means that you need three things to use Follow Me:

- **A path:** In SketchUp, you can use any edge, or series of edges, as a path. All you have to do is make sure that they're drawn before you use Follow Me.
- **A face:** Just like with Push/Pull, Follow Me needs a face to extrude. You can use any face in your model, but it needs to be created before you start using Follow Me.
- **Undo:** Imagining what a 2D face will look like as a 3D shape isn't easy—it usually takes several tries to get a Follow Me operation right. That's what Undo is for, after all.

Follow these steps to use Follow Me, if you want to extrude a shape along an axis; refer to Figure 6-2 to see how the steps work:

Figure 6-2

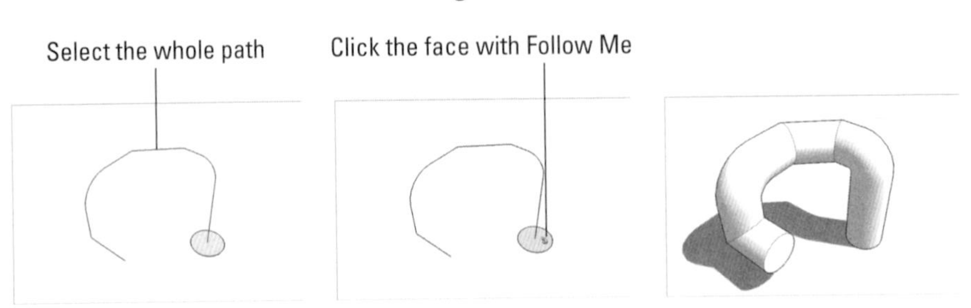

Using Follow Me to create a simple extruded shape.

PATHWAYS TO... USING FOLLOW ME TO CREATE A SIMPLE EXTRUDED SHAPE

1. **Draw a face to use as an extrusion profile.** In this example, we are creating a pipe, so our extrusion profile is a circular face.
2. **Draw an edge (or edges) to use as an extrusion path.** Although the edge (or edges) is touching the face in this case, it doesn't have to for Follow Me to work.
3. **Select the complete extrusion path you want to use.** Refer to the section on making selections in Chapter 2 for pointers on using the Select tool effectively.
4. **Activate the Follow Me tool.** To do this, choose Tools ⇨ Follow Me.
5. **Click the face you want to extrude once.** Magic! Your face (extrusion profile) is extruded along the path you chose in step 3, creating a 3D form (in this case, a section of pipe).

If you want to use Follow Me all the way around the perimeter of a face, you don't need to spend time selecting all the individual edges. Just select the face and then use Follow Me; the tool automatically runs all the way around any face you have selected. The key to success in using this tool is to have all the edges or shapes that conform your path, touching. If we have a gap in between two segments, the Follow Me tool will stop right at the first gap.

You can use Follow Me another way, too: Instead of preselecting a path (as in Step 3 of the preceding list), you can click any face with Follow Me and attempt to drag it along edges in your model. While this works on simple things, preselecting a path works a lot better—it's really the only option for using Follow Me in a predictable way.

Figure 6-3

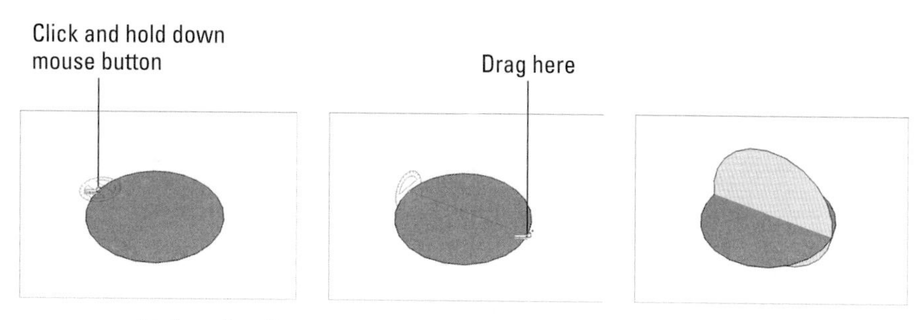

Using the Rotate tool to make a rotated copy of a circle.

6.1.2 Making Lathed Forms Like Spheres and Bottles

A surprising number of things can be modeled by using Follow Me to perform a *lathe* operation. A lathe is a tool that carpenters and machinists use to spin a block of raw material while they carve into it—this is how baseball bats are made, for instance.

A simple example of a lathed object is a sphere. Here's how you might make one with Follow Me:

1. **Draw a circle on the ground.**
2. **Rotate a copy of your circle up by 90 degrees, as shown in Figure 6-3.** If you are wondering how to do this, follow these steps:
   - Select the face of your circle with the Select tool.
   - Choose Tools ⇨ Rotate to activate the Rotate tool.
   - Press Ctrl (Option on a Mac) to tell SketchUp you want to make a copy.
   - Click a green endpoint inference along the edge of your circle *and hold down* your mouse button to drag. Don't let go just yet.
   - Still dragging, move your cursor over to the endpoint on the *exact opposite* side of your circle; then release your mouse button. Now your *axis of rotation* is a line right through the center of your circle.
   - Click anywhere on the edge of your circle, and then move your mouse over a little bit.
   - Type in **90** and press Enter.
3. **Make sure that one of your circles is selected.**
4. **With the Follow Me tool, click the circle that's not selected once (see Figure 6-4).** Now you have a sphere. The Follow Me tool "lathed" your circular face around the path you selected—the other circle.

Note, however, that if you really need a sphere, the easiest way to get one is in the Components dialog box. The Shapes library that comes installed with SketchUp has a selection of spheres (and cones and other things) you can choose from.

Figure 6-4

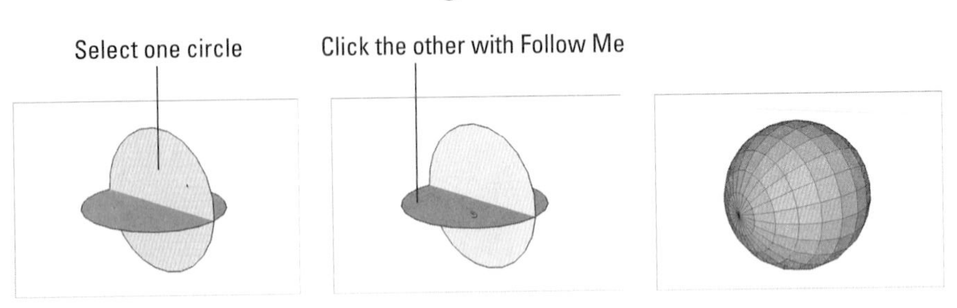

Clicking one circle with Follow Me while the other one is selected produces a sphere.

Under normal circumstances, you only have to model half a profile to use Follow Me to make it three-dimensional. Let's say that you are building a bottle of wine; you can draw half of the profile and then you can rotate it 360 degress along its axis to obtain the full shape. Figure 6-5 shows a few different examples of 3D objects.

6.1.3 Creating Extruded Shapes Like Gutters and Handrails

Much of the time, you will want to use Follow Me to create geometry (edges and faces) that's attached to another part of your model. An example of this might be modeling a gutter that runs all the way around the roof of a house. In this case, you're likely to already have the path along which you want to extrude a profile (the edge of the roof).

When you're using Follow Me to extrude a face along a path that consists of edges that already exist as part of your model, there are two things you should *always* do:

- **Before using Follow Me, make the rest of your model a separate group.** Follow Me can sometimes mess things up, so you want to be able to keep the geometry that it creates separate, just in case.

Figure 6-5

A few examples of lathed objects created with Follow Me.

- **Make a copy of your extrusion path outside your group.** There's a consequence to working with Follow Me on top of a group: The edge (or edges) you want to use as an extrusion path will no longer be available, because you can't use Follow Me with a path that's in a separate group or component. What to do? You need to make a copy of the path *outside* the group, and then use the *copy* to do the Follow Me operation. Here's the best way to make a copy of the path:
  - With the Select tool, double-click your group to edit it.
  - Select the path you want to use for Follow Me.
  - Choose Edit ⇨ Copy.

FOR EXAMPLE

WHY YOUR COMPUTER IS SO SLOW

When you use Follow Me with an extrusion profile that's a circle or an arc, you're creating a piece of 3D geometry that's very big. In other words, this geometry has lots of faces, and faces are what make your computer slow down. Without going into detail about how SketchUp works, keep this in mind: The more faces you have in your model, the worse your computer's performance will be. At a certain point, you'll stop being able to orbit, your scenes (which are explained in Chapter 10) will stutter, and you'll be frustrated.

The first pipe in the figure that follows has been extruded using Follow Me. It was made with a 24-sided circle as an extrusion profile, and it has 338 faces. Hidden Geometry is turned on (in the View menu) so that you can see how many faces the pipe has.

The second pipe uses a 10-sided circle as an extrusion profile. As a result, it only has 116 faces. What an improvement!

The third pipe also uses a 10-sided circle as an extrusion profile, but the arc in its extrusion path is made up of only 4 segments, instead of the usual 12. It has a total of 52 faces. Even better!

The second image in the figure shows all three pipes with Hidden Geometry turned off. Is the difference in detail worth the exponential increase in the number of faces? Most of the time, the answer is no.

To change the number of sides in a circle or an arc, just before or just after you create it, follow these steps:

1. Type in the number of sides you'd like to have.
2. Type an *s* to tell SketchUp that you mean "sides."
3. Press Enter.

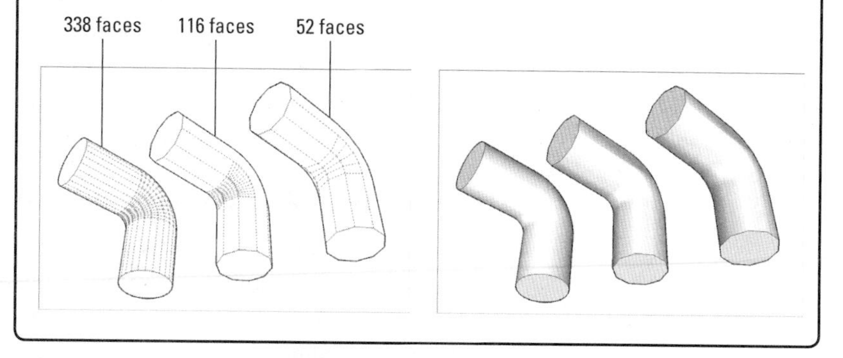

- Exit (stop editing) your group by clicking somewhere else in your modeling window.
- Choose Edit ⇨ Paste in Place. Now you have a copy of the path you want to use, and it's outside your group.

When you are using an existing edge (or series of edges) as an extrusion path, the hard part is getting your profile in the right place. You have a choice of two ways to proceed; which one you use depends on what you need to model:

- **Draw the profile in place.** Do this only if the extrusion path is parallel to one of the colored drawing axes.
- **Draw the profile on the ground and then move it into position.** If your extrusion path doesn't start out parallel to a colored drawing axis, you should probably draw your profile somewhere else and move it into place later.

Drawing Your Profile in Place

Consider that you have a model of a house. You want to use Follow Me to add a gutter that goes all the way around the perimeter of the roof. You decide to draw the profile in place (right on the roof itself), because the edges of the roof are drawn parallel to the colored drawing axes. This means that you'll have an easier time using the Line tool to draw "in midair."

The trick to drawing an extrusion profile that isn't on the ground is to start by drawing a rectangular face. You then draw the profile on the face and erase the rest of the rectangle. Figure 6-6 shows how you would draw the profile of a gutter directly on the corner of a roof; the steps that follow explain the same things in words:

1. **Zoom in on what you're doing.** Get close, and fill your modeling window with the subject at hand.
2. **Using the Line tool, draw a rectangle whose face is perpendicular to the edge you want to use for Follow Me.** This involves paying careful attention to SketchUp's inference engine; watch the colors to make sure that you're drawing in the right direction.
3. **Use the Line tool (and SketchUp's other drawing tools) to draw your profile directly on the rectangle you just created.** The important thing here is to make sure that your extrusion profile is a single face; if it's not, Follow Me won't work the way you want it to.
4. **Erase the rest of your rectangle, leaving only the profile.**

Drawing Your Profile Somewhere Else

The awful thing about handrails is that they're almost always at funny angles, and thus not parallel to a colored axis. In such cases, it's not convenient to draw your extrusion profile in place. Here, it's best to draw the profile on the ground and move it into position afterward.

Figure 6-6

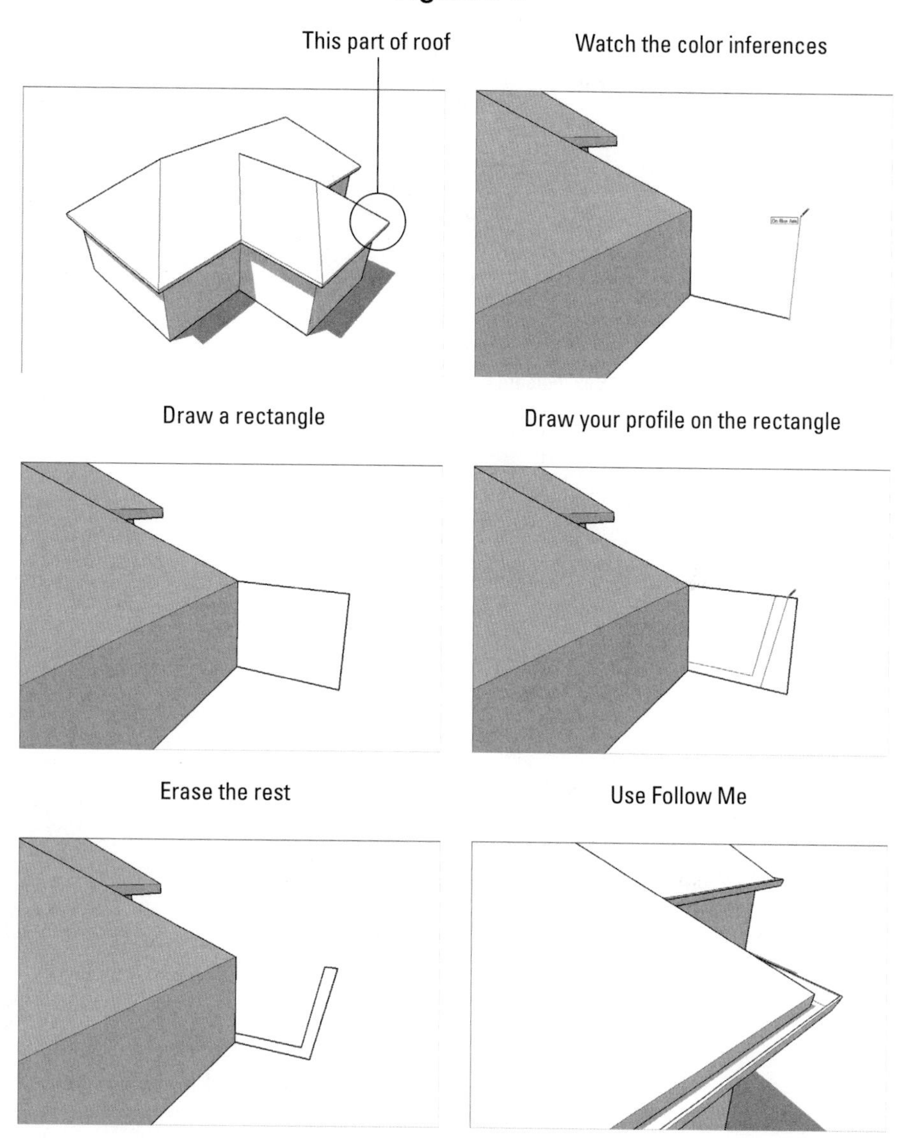

Drawing an extrusion profile in place by starting with a rectangle.

Here's the trick: Draw a "tail"—a short edge—perpendicular to your extrusion profile. You can use this tail to help you line up your profile with the edge you want to use as an extrusion path for Follow Me. The following steps, and Figure 6-7, describe how you would draw and position a profile for a handrail:

1. **Draw your extrusion profile flat on the ground.**
2. **Draw a short edge perpendicular to the face you just drew.** This "tail" should come from the point where you want your profile to attach to the extrusion path.

Figure 6-7

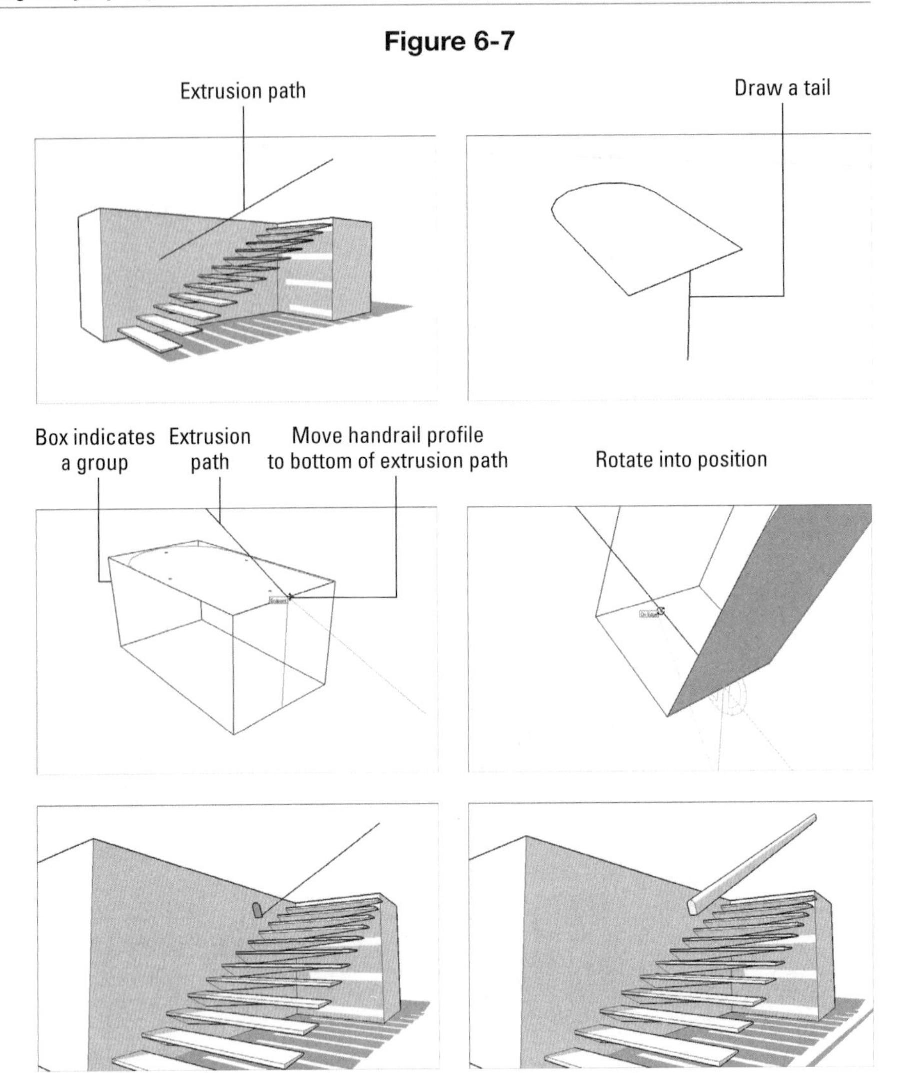

Draw a short tail on your extrusion profile to help you position it with the Move and Rotate tools.

3. **Make your profile and its tail into a group.** This makes it easier to move and rotate around all at once. See Chapter 5 for information on creating and using groups, if you need it.
4. **Using the Move tool, place your profile at the end of the extrusion path.** To make sure that you position it accurately, pick it up by clicking the point where the tail meets the face, and drop it by clicking the end of the extrusion path.
5. **With the Rotate tool, rotate your profile into position.** Here's where you need to use a bit of skill. (See Chapter 2 for guidance.) The Rotate tool is easy to use—after you get the hang of it.

6. **Right-click the group you made in Step 3 and choose Explode; delete your tail.**

6.1.4 Subtracting From a Model with Follow Me

What if you want to model a bar of soap? Or a sofa cushion? Or any object that doesn't have a sharp edge? The best way to round off edges in SketchUp is to use Follow Me. In addition to using this tool to *add* to your model, you can also use it to *subtract* from your model.

Here's how it works: If you draw an extrusion profile on the end face of a longish form, you can use Follow Me to "remove" a strip of material along whatever path you specify. Figure 6-8 demonstrates this concept on the top of a box.

If the extrusion path you want to use for a Follow Me operation consists of the entire perimeter of a face (as is the case in Figure 6-8), you can save time by just selecting the face instead of all the edges that define it.

But what if you want to create a corner that's rounded in *both* directions, as so many corners are? That one's a little trickier to do in SketchUp, but because it's such a common problem, let's take a closer look at how to do it. The basic technique involves using Follow Me on a corner you've already

Figure 6-8

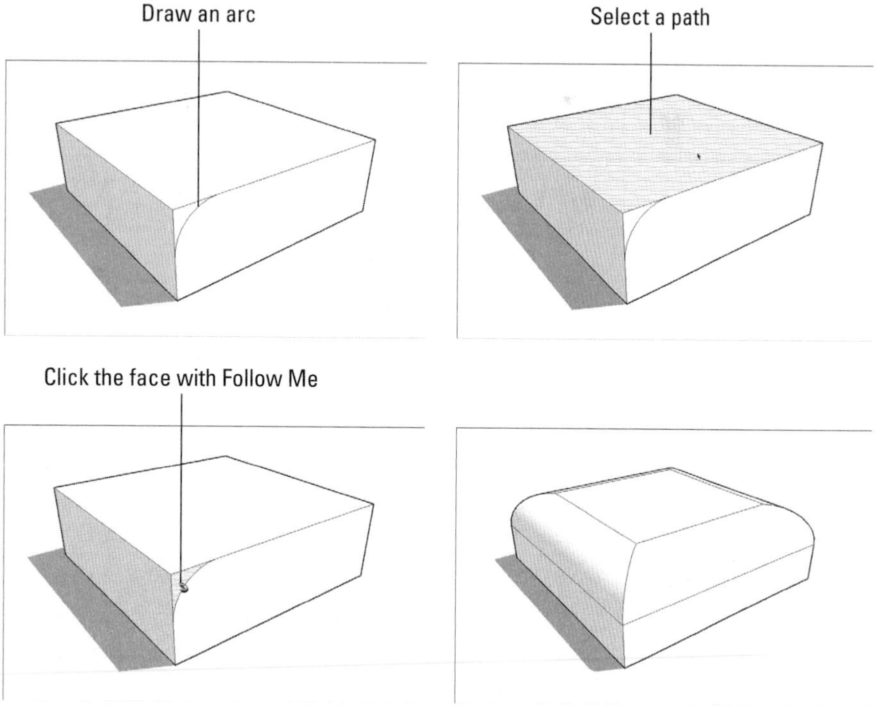

Creating a filleted edge with Follow Me.

Figure 6-9

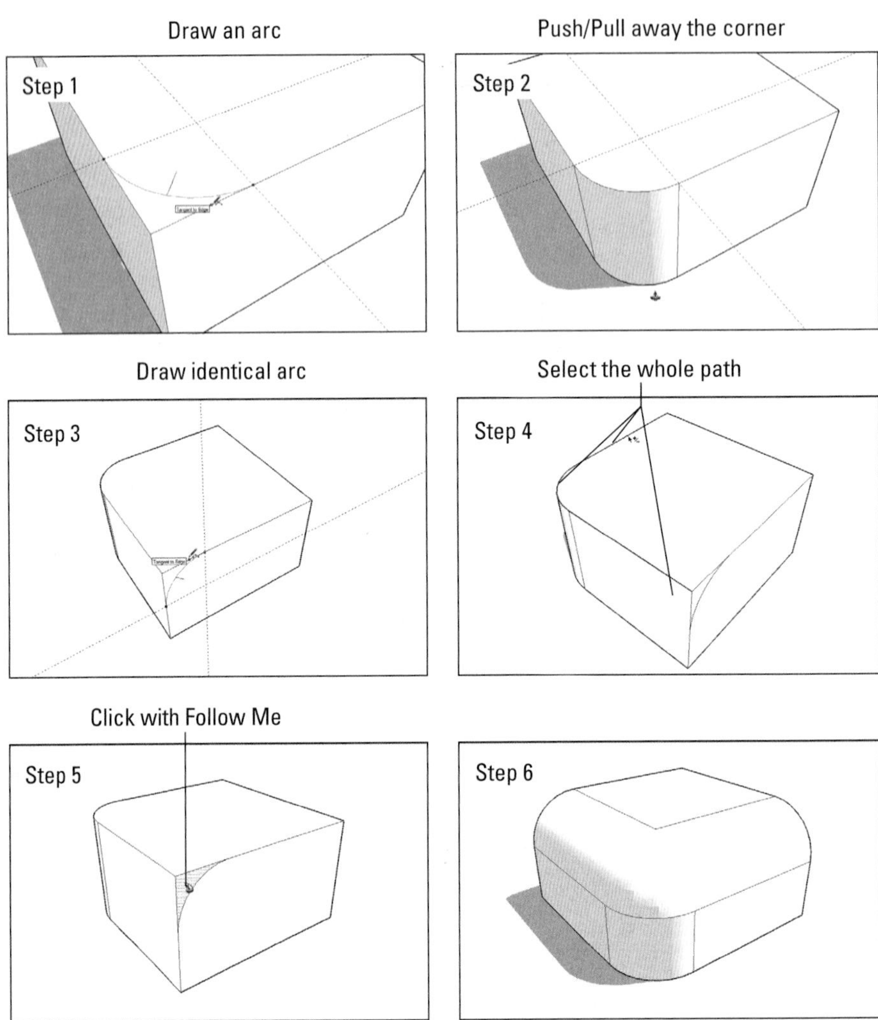

Making a corner that's filleted in both directions.

rounded with the Push/Pull tool. After you have a corner that's filleted with an arc of the correct radius, you can use copies (or component instances, if you're clever) of that corner several times, wherever you need them. It's not an elegant solution, but it's possible, and it works when you need it to.

Figure 6-9 gives a step-by-step, visual account of this process. First draw a box. It doesn't really matter how big the box is, as long as it's big enough for the fillet you want to apply.

1. **With the Arc tool, draw an arc on the corner of the box.** When you're drawing an arc on a corner, keep an eye out for the inferences that help you draw properly:
   - After clicking to place one endpoint of your arc, as you cut across the corner, the point at which your line turns magenta is where your

TIPS FROM THE PROFESSIONALS

Smoothing Out Those Unsightly Edges

If you're wondering how to get rid of all the ugly lines you end up with when you use Follow Me, the answer is simple: You can *smooth* edges, just like you can hide them. The difference between hiding and smoothing is illustrated by the images of the cylinders in the figure that follows:

- When you *hide* an edge between two faces, SketchUp still treats those faces as though your edge is still there—it just doesn't show the edge. Materials you've applied to each face stay separate, and each face is "lit" separately by SketchUp's sun. The latter fact is the reason why simply hiding the edges between faces that are supposed to represent a smooth curve doesn't make things look smooth; you still end up with a faceted look.
- When you *smooth* an edge between two faces, you're telling SketchUp to treat them as a single face, with a single material and smooth-looking shading. The difference is significant, as you can see in the second cylinder.

You can smooth edges in two different ways:

- **Use the Eraser.** To smooth edges with the Eraser tool, hold down Ctrl (Option on the Mac) while you click or drag over the edges you want to smooth.
- **Use the Soften Edges dialog box.** Located on the Window menu, this dialog box lets you smooth multiple selected edges all at once, according to what angle their adjacent faces are at. It's a little complicated at first, but here's what you need to know to get started: Select the edges you want to smooth, and then move the slider to the right until things look the way you want them to.

To unsmooth edges, you need to make them visible first; turn on Hidden Geometry to do just that. Then, do the following:

1. Turn on hidden geometry to make edges visible.
2. Select the edges you want to unsmooth.
3. In the Soften Edges dialog box, move the slider all the way to the left.

Visible edges Hidden edges Smoothed edges

endpoints are *equidistant* (the same distance) from the corner across which you're cutting.

- After clicking to place your second endpoint, you will see a point at which the arc you're drawing turns magenta—this means your arc is *tangent to* (continuous with) both edges it's connected to. You want this to be the case, so you should click when you see magenta.

2. **Push/Pull the new face down to round off the corner.**
3. **Draw another *identical* arc on one of the corners directly adjacent to the corner you just rounded.** This is where you'll have to refer to Figure 6-9. Pictures are better than words when it comes to explaining things such as which corners are adjacent to which.
4. **Select the edges shown in Figure 6-9.**
5. **Activate the Follow Me tool. Click the arc corner face to extrude it along the path you selected in Step 4.**
6. **Hide or smooth out any edges that need hiding or smoothing.** For a review of hiding edges, see Chapter 5. Also, review the previous Tips from the Professionals box on "Smoothing Out Those Unsightly Edges."

After you have a fully rounded corner, you can use a number of them to make anything you want; it just takes a little planning. Figure 6-10 shows a simple bar of soap that was created out of eight rounded corners that were copied and flipped accordingly. The text (in case you're wondering) was created with SketchUp's 3D Text tool, which you can find on the Tools menu.

Figure 6-10

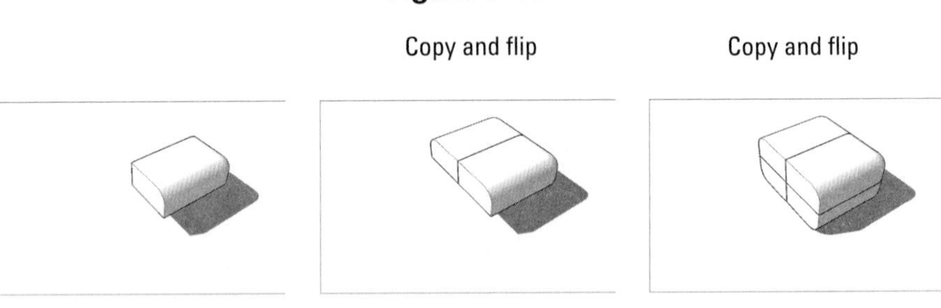

Copy and flip   Hide and smooth edges; then add text

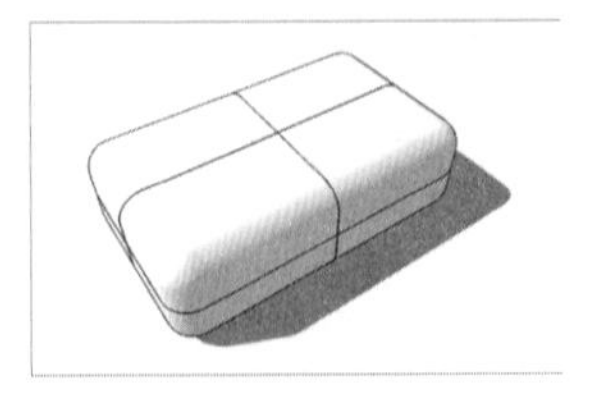

Assembling multiple rounded corners to make objects is relatively easy.

TIPS FROM THE PROFESSIONALS

One of the most frustrating moments using the Follow Me tool is when making a revolving object the profile does not want to revolve around a circle. Despite multiple efforts, we keep on getting error messages. The most probable cause of such notices would be that the profile that we are trying to revolve is not coplanar. What does it mean to be coplanar? The term refers to a shape that belongs to a single plane; to make an equivalent comparison, would be drawing a profile on a flat piece of paper. As long as the profile that we have belongs to that virtual single sheet of paper, the shape will revolve without trouble. Be sure to use the Orbit tool to make sure that your profile is in fact, flat.

SELF-CHECK

1. Name three shapes that we can obtain by spinning a 2D shape around central axis.
2. If you just used Follow Me and the computer seems to be sluggish, what would be the most probable cause?
3. What tool and what key stroke would we have to use to smooth out the edges of an object?

Apply Your Knowledge You have been contracted to design a wine cellar in the basement of a home. Create the cellar, build a floor and use the Follow Me tool.

6.2 MODELING WITH THE SCALE TOOL

> GOOGLE SKETCHUP IN ACTION
>
> Learn how to operate the Scale tool.

As SketchUps hero, the Scale tool is frequently misunderstood. New modelers assume that Scale is for resizing things in your model. That's technically true, but most only use it to resize whole objects. The real power of Scale happens when you use it on parts of objects to change their shape (see Figure 6-11.)

Understanding Scale

The basic principle of this technique is pretty simple: You select the geometry (edges and faces) in your model that you want to resize, activate the Scale tool, and go to town.

Figure 6-11

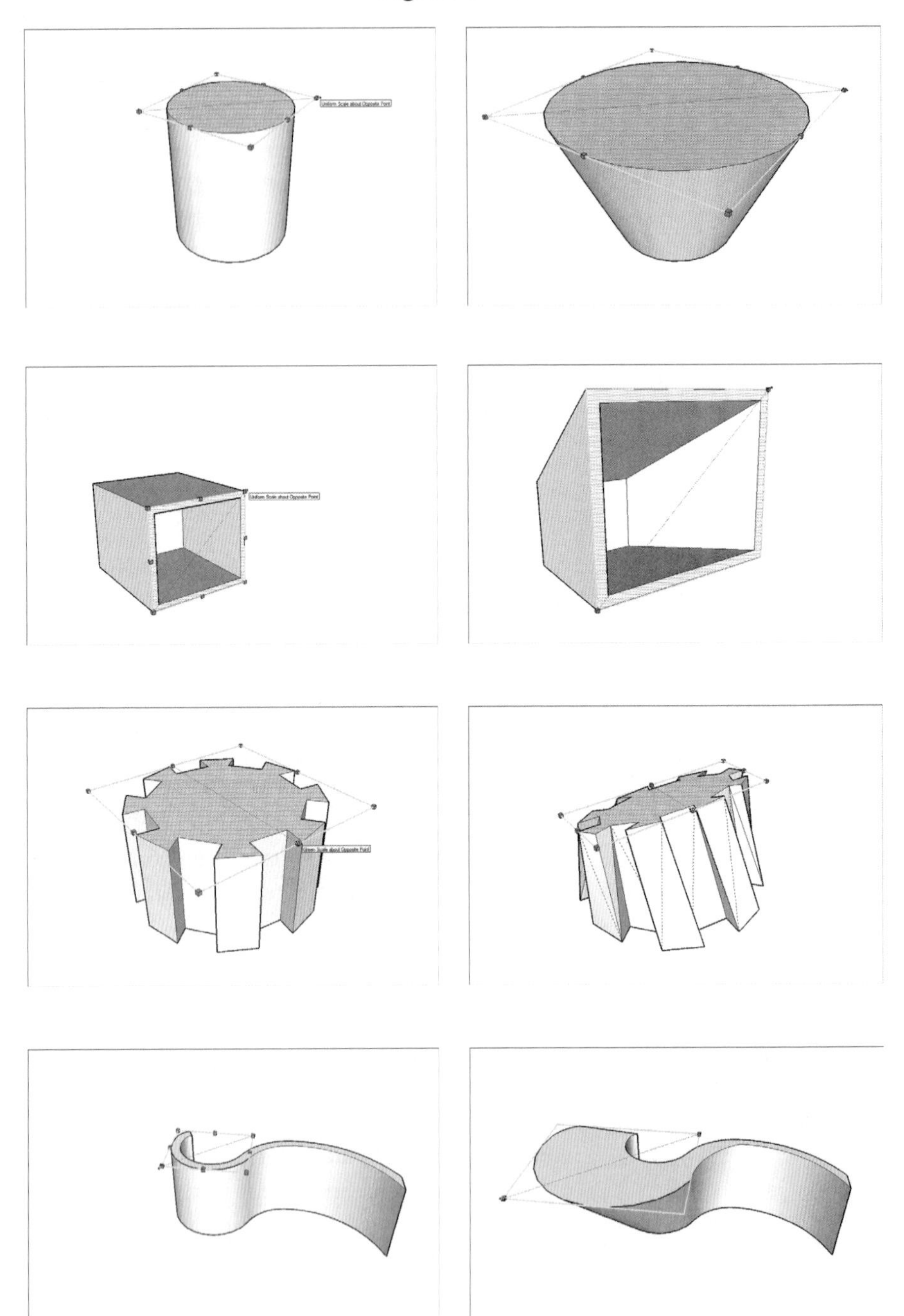

Using the Scale tool on parts of objects changes their shape.

Figure 6-12

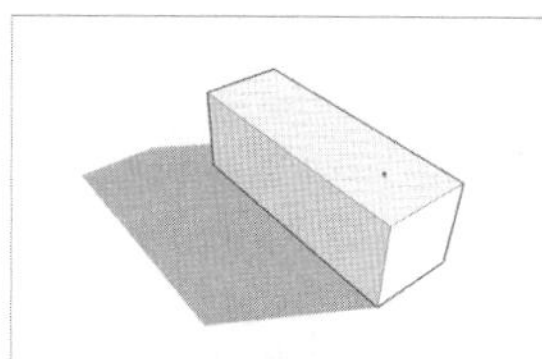
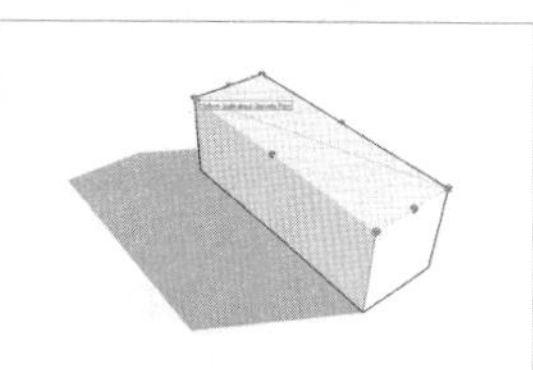
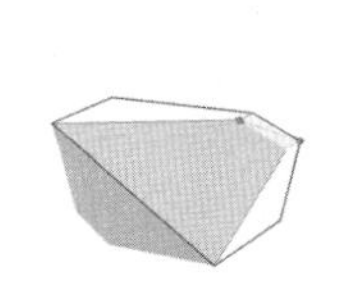

The Scale tool is easy to use.

Here is a list of steps: (Figure 6-12 tells the story in pictures):

PATHWAYS TO... USING THE SCALE TOOL

1. **Select the part of your model that you want to scale.** Use the Select tool to do this; check out the latter part of Chapter 2 if you need a refresher.
2. **Activate the Scale tool by choosing Tools ⇨ Scale from the menu bar.** You can also make Scale active by clicking its button on the toolbar or by pressing the S key on your keyboard. After you activate Scale, the geometry you selected in Step 1 should be enclosed in a box of little green cubes, or grips.
3. **Click a grip and then move your mouse to start scaling your selected geometry.** Take a look at the next part of this section for details on all the different grips.
4. **When you are finished scaling, click again to stop.** Here are a few more things you should know about scaling:
   - **Use different grips to scale different ways.** Which grip (the little green boxes that appear when you activate the Scale tool) you use determines how your geometry scales:
     - Corner grips scale proportionally—nothing gets distorted when you use them.
     - Edge and side grips distort your geometry as you scale—use them to squeeze what you are scaling.
5. **Hold down the Shift key to scale proportionally.** This happens automatically if you are using one of the corner grips, but not if you are using any others. If you don't want what you are scaling to be distorted, hold down Shift.
6. **Hold down the Ctrl key (Option on a Mac) to scale about the center of your selection.** You might find yourself using this more often than not.

(Continued)

PATHWAYS TO... USING THE SCALE TOOL *(continued)*

7. **Type a scaling factor to scale accurately.** To scale by 50 percent type 0.5. Typing 3.57 scales your geometry by 357 percent, and typing 1.0 doesn't scale it at all. See Chapter 2 to read more about using numbers while you work.
8. **Which grips appear depend on what you're scaling.** Reference Figure 6-13.
   - Most of the time, you see a scaling box enclosed by 26 green grips.
   - If you are scaling flat, coplanar geometry (faces and edges that all lie on the same plane) and that plane is perfectly aligned with one of the major planes in your model, you get a rectangle consisting of 8 grips instead of a box composed of 26.
   - If you are scaling a Dynamic Component, you may see anywhere from 0 to all 26 grips; it depends on how the builder set up the component. Take a look at Chapter 5 for more information about Dynamic Components.
9. **You can't make a copy while you scale.** Both the Move and Rotate tools let you make copies by holding down a button on your keyboard while you are using them, but Scale doesn't work this way, unfortunately. If you need to make a scaled copy, try this instead:
   a. Select the geometry that you want to scale and copy, and then make it into a group. See Chapter 5 for more information on making groups.
   b. Choose Edit ⇨ Copy from the menu bar and then choose Edit ⇨ Paste in Place from the menu bar.
   c. Scale the copied group as you would anything else.

Figure 6-13

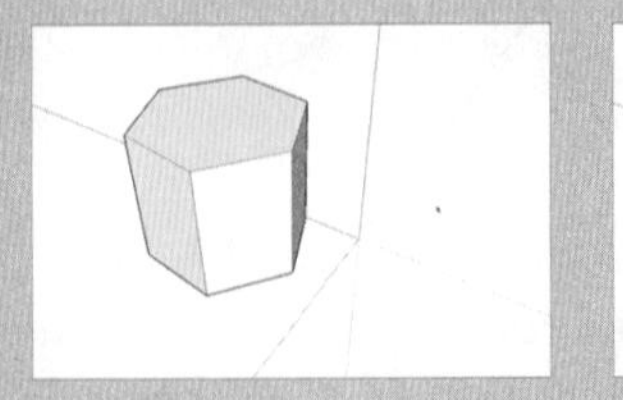

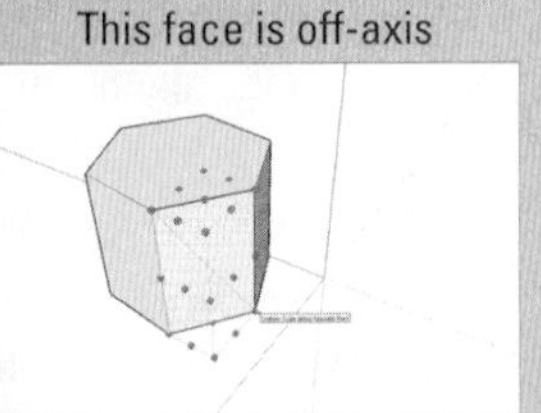

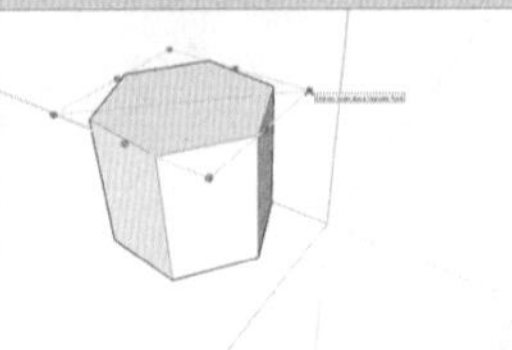

Grips depend on what you are trying to scale.

6.2.1 Scaling Profiles to Make Organic Forms

Now it gets really interesting. Thanks go to SketchUp super-user, Justin Chin, for demonstrating the power of scaling profiles to make organic forms; it's great because it's easy to understand and powerful enough to be applied in many areas.

So what is this method? It involves using the Scale tool in combination with a series of 2D profiles to create curvy, un-boxy 3D shapes. An awful lot of the stuff in the universe fits squarely in this category: humans, animals, bananas, just about everything that wasn't made by a machine, can be modeled using the scaled profiles method of 3D modeling.

Combining Scale and Push/Pull

The simplest way to use this method is in association with Push/Pull. Below is a very simple example of how it works (reference the first column of Figure 6-14 for an illustrated view of the steps below):

1. **Create a 2D shape.** This shape may be something simple (such as a circle) or something more complex; it all depends on what you are trying to model. The shape may also be a half-shape if what you are trying to make exhibits bilateral symmetry. Take a look at the last section in Chapter 5 for more information on using components to build symmetrical models.
2. **Push/pull your 2D shape into a 3D form.**
3. **Scale the new face you created so it's slightly bigger (or slightly smaller) than the original 2D shape from Step 1.** See the previous section in this chapter for more specifics about using the Scale tool. Pay special attention to the points about using modifier keys (keyboard buttons) to scale proportionally or about the center of what you're working on.
4. **Push/pull the face you scaled in the preceding step.** Try to make this extrusion about the same as the one you made in Step 2. You can usually double-click a face with the Push/Pull tool to repeat the last Push/Pull operation you did.
5. **Repeat Steps 3 and 4 until you're done.** You can add skillful use of the Rotate tool into the mix if you like; doing so allows you to curve and bend your form as you shape it.

Keep the following tidbits in mind as you explore this technique:

- **Watch your polygon count.** Polygons are faces, basically—the more you have, the "heavier" your model becomes, and the worse it performs on your computer. Try to minimize the number of faces you're working with by reducing the number of edges in your original 2D shape. Look at the sidebar "Why Your Computer is so Slow," earlier in this chapter, for the whole scoop.

Figure 6-14

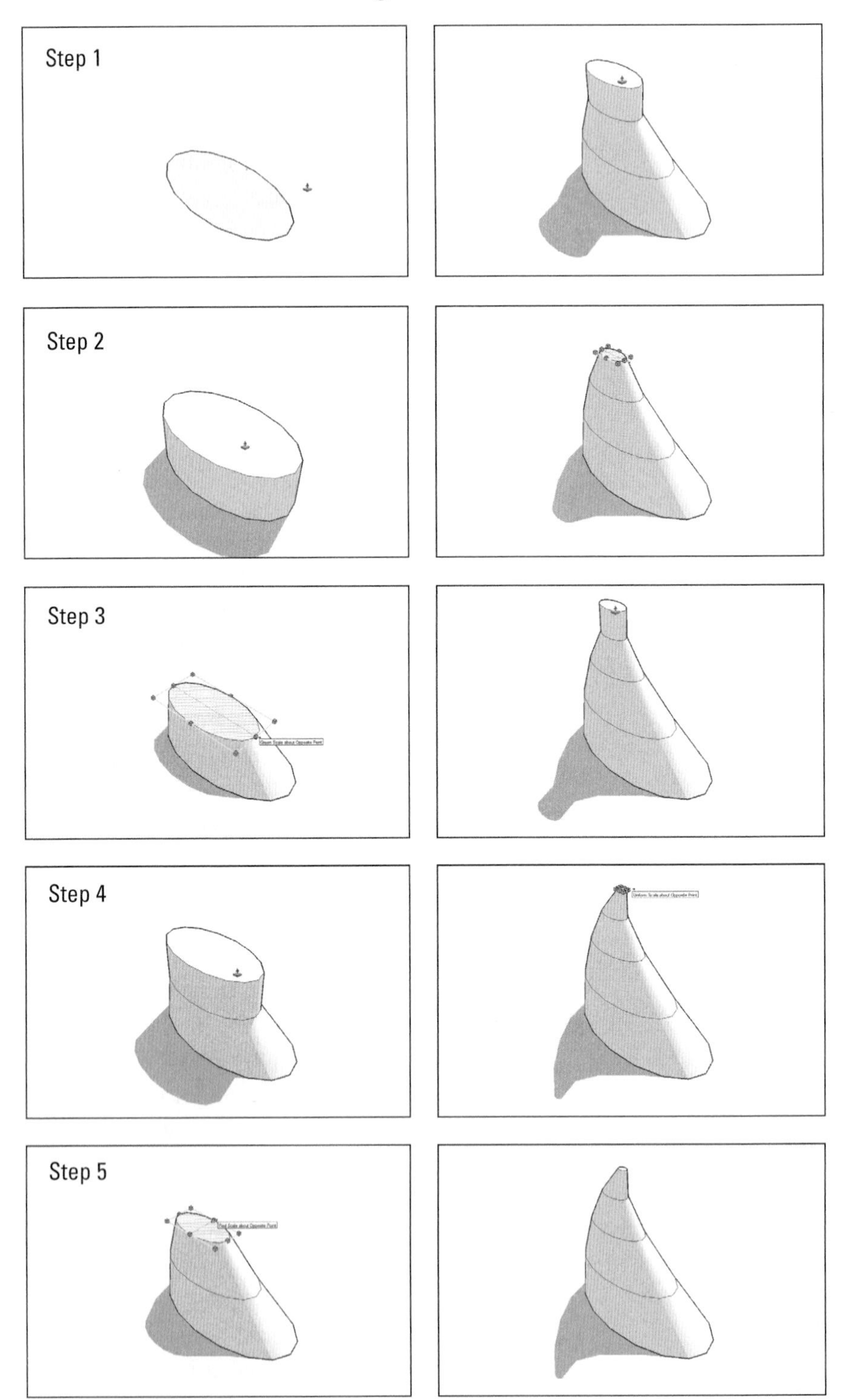

Using Scale and Push/Pull together is a simple way to make organic forms.

Figure 6-15

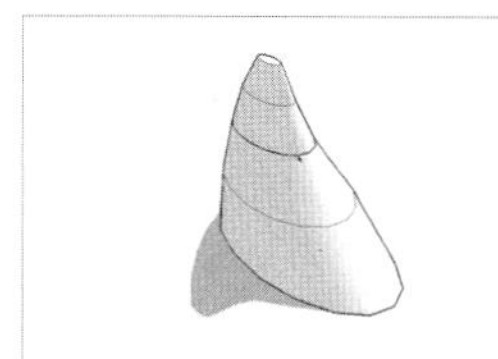
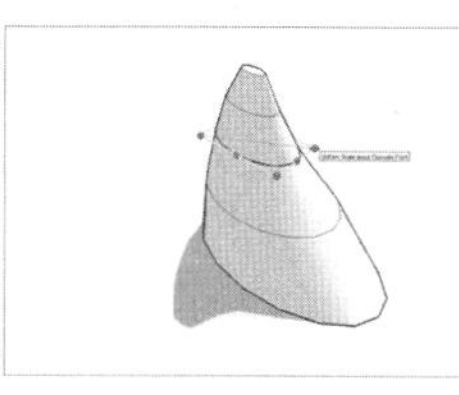
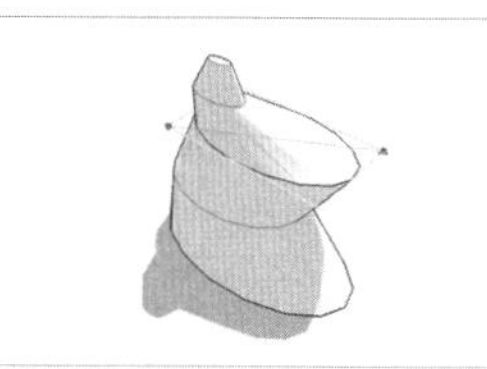

You can go back and scale any profile at any time.

- **Don't be afraid to go back and tweak.** The beauty of this method is its flexibility. While you are working, you can select any of the 2D profiles (shapes) in your model and use the Scale tool to refine them. Just select the loop of edges along the perimeter of the profile you want to scale and take it from there. Check out Figure 6-15 to see an illustration of this.

PATHWAYS TO... CREATING EXTRUDED FORMS USING FOLLOW ME

1. **Draw a circle.** This is the extrusion profile for Follow Me. Strongly consider reducing the number of sides in your circle from the standard 24 to something more like 10 or 12. See the sidebar "Why Your Computer is so Slow" (earlier in this chapter) to find out how and why you should do this.
2. **Draw a 10-sided arc that starts perpendicular to the center of the circle you drew in Step 1.** Type **10s** and press Enter right after you click to finish drawing your arc. This tells SketchUp to make sure your arc has 10 sides (instead of the default 12). Why 10 sides? It makes the math easier a few steps from now. The easiest way to create a halfway-accurate arc in 3D space is to start by drawing a rectangle. When you are sure this rectangle is properly situated, use the Arc tool to draw on top of it and then delete everything but the arc.
3. **Select the arc you just drew.** This is the extrusion path for Follow Me.
4. **Activate the Follow Me tool by choosing Tools ⇨ Follow Me from the menu bar.**
5. **Click the circle you drew in Step 1 to extrude it along the path you drew in Step 2.**
6. **Choose View ⇨ Hidden Geometry from the menu bar.** Showing the hidden geometry in your model lets you select the edges that were automatically smoothed, made hidden, when you used Follow Me in Step 4.

(continued)

PATHWAYS TO...
CREATING EXTRUDED FORMS *(continued)*

7. **Scale the face at the end of your new extrusion by a factor of 0.1.** See "Understanding Scale," earlier in this chapter, for instructions on how to do this. Use any of the four corner grips on the scaling box, and don't forget to hold down the Ctrl key (Option on a Mac) while you're scaling—this forces SketchUp to scale about the center of the face you're resizing.
8. **Select the edges that define the next-to-last profile in your extruded form.** Depending on the angle of your arc, making this selection can get tricky. Here are some considerations that may help:
   - See Chapter 2 for tips on making selections.
   - Choose View ⇨ Face Style ⇨ X-Ray from the menu bar to make it easier to see what you've selected.
   - Hold down the Ctrl key (Option on a Mac) while you orbit to turn off SketchUp's "blue is up/down gravity bias." While orbiting this way, try drawing lots of tight, little circles with your mouse to get your view to tilt in the direction your want. This is by no means easy, but getting the hang of temporarily disabling the Orbit tool's tendency to keep the blue axis straight up and down is a very nice way to work. Doing so makes it infinitely easier to get just the right angle for making a window selection. This in turn makes selecting the edges that define profiles a whole lot easier, and that's what becoming master of the Orbit tool is all about.
9. Scale the edges you selected in the preceding step by a factor of 0.2.
10. **Repeat Steps 8 and 9 for each of the remaining profiles in your form, increasing the scaling factor by 0.1 each time.** Of course, you can absolutely choose to sculpt your form however you like, but this method (counting up by tenths) yields a smooth taper.

Combining Scale and Follow Me

Another way to create extruded forms is to use Follow Me. (See the first part of this chapter if you need a refresher.) This technique is ideally suited to taking long, curvy, tapered things like tentacles and antlers; it's a little time consuming but works like a charm.

Modeling a simplified bull's horn is a good, straightforward illustration of how the Follow Me variation of this method works. Here's how to go about it; reference Figure 6-16, starting with the left column for an illustration:

Figure 6-16

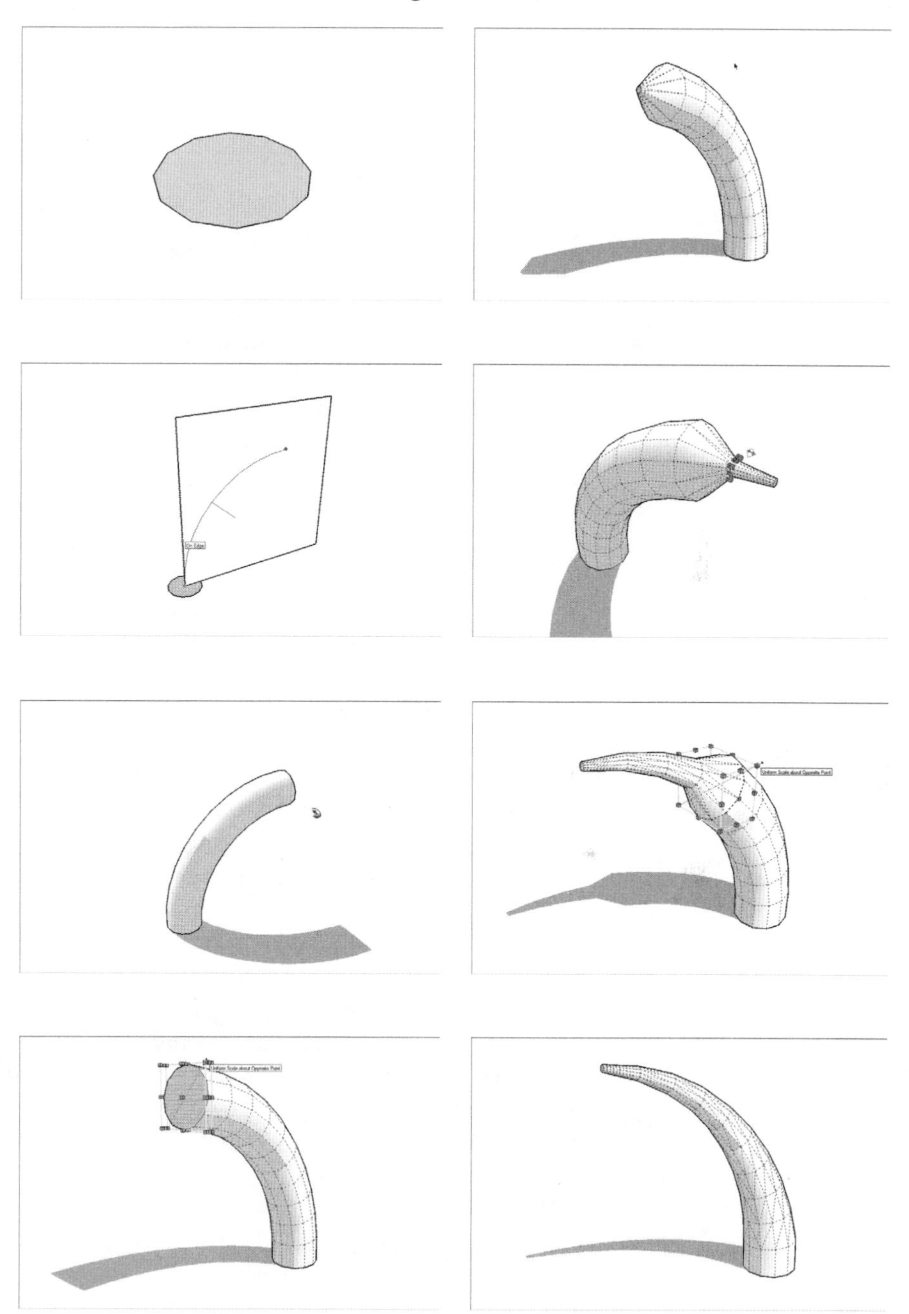

Use Scale with Follow Me to create long, tapered forms like this bull's horn.

Look at the Santa and reindeer project in the color insert (in the center of this book) to get an idea of the kind of fancy, not-a-box models you can build after you master the Scale tool. It's not beginner-level material, but it's worth the time when you're ready for it.

IN THE REAL WORLD

Justin Pohl, a student in the Architecture program at Kansas State University, likes using SketchUp because "it allows me to get concepts of my designs out faster. Believe it or not, I am also able to draft out plans quicker in SketchUp than in AutoCAD. It has an easy interface and is very useful in crafting most forms—especially if one understands how the program operates."

He likes to do most of his design work in SketchUp and to produce presentable plans and sections, he exports 2D .dwg files to alter them further in AutoCAD. Sometimes he rebuilds the model entirely in Revit and exports it to 3Dstudio Max if he wants to obtain photorealistic materials in his renderings. For his last project, he used V-Ray for SketchUp to produce the renderings that included exterior views, interior views, and section perspectives.

He mentions that there are few SketchUp tools that he doesn't use, but on the other hand he has downloaded some plugins for renderings such as Vray, Podium, and Kerkythea (if rendering out of SketchUp). He does not use Google Layout because he feels it is not as customizable as he would like it to be. Consequently, he composes his boards in Adobe InDesign. The response of his professors tends to be positive, "my professors' critiques tend to be that of praise, but they usually comment more on design rather than the use of specific software."

Justin uses Follow Me extensively in many projects. In one specific instance, he used it to produce a curved (complex) fabric roof and he likes the fact that the tool is really limitless in what you can produce. He is quick to admit that, "Follow Me allows a way to create complex shapes easily, such as any sort of domed shape or any volumetric shape with curves."

He always uses the terrain tools, such as Contour when generating topography. He usually takes CAD plans of the site and orients the contours in SketchUp (if there are no Z depths in CAD) and the uses the Sandbox tool to smooth them out. In his own words, "most recently it was used effectively in a winery project last semester." He also likes to work with the Intersect, Union, and Subtract tools similarly to the command Intersect with Model; Justin says, "I can intersect two components or groups and either subtract the overlap, join the two objects together, or cut away areas that I don't need."

FOR EXAMPLE

When creating extruded or revolving surfaces using Follow Me sometimes there are unpredictable results and it is important to practice a few times until we get it right. At this point you have two possibilities: you either make multiple copies of your path and profile and try to create your shape in a variety of ways, or you practice hitting Undo several times if it did not turn out right the first time (To Undo, hold down the Ctrl key + Z or Option+Z on a Mac). Also, this tool is particularly useful when we are creating tubular designs.

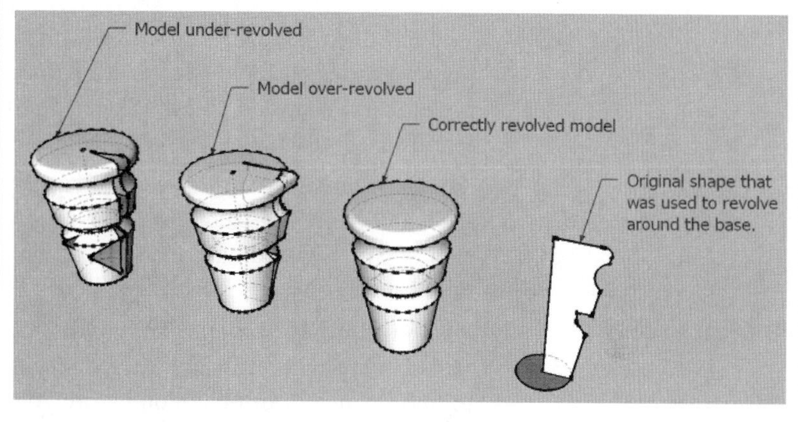

Reproduced with the permission of Jorge Paricio, PhD.

SELF-CHECK

1. What two tools do we use to create an organic shape?
2. How do we control the scale factor on a model?
3. Choose the description that is correct about long, curvy, tapered objects:
   a. Those shapes cannot be made in SU.
   b. The process is laborious but it works like a charm.
   c. It is recommended that we start with a curved profile that has a standard number of 24 sides.

Apply Your Knowledge You are designing a wooden side table with soft edges and revolving legs. Combine the use of Push/Pull with Scale to make the table top and then practice extruding the form of the legs using Follow Me. If you are having trouble understanding what a revolving leg is or how it is made industrially, it is a leg that is made in a machine called a lathe. A long piece of wood is secured by the two ends and made spin rapidly. Then, a sharp chisel is applied to the wood block and the shape is modeled. When you are deciding what shape to give to your revolving leg, think of a very long, inverted truncated cone that is split in half, similar to the profile that we can see in the figure above.

6.3 MAKING AND MODIFYING TERRAIN

SketchUp introduced the Sandbox in version 5. It is being introduced here because it helps people to model terrain—the land your buildings sit on (or in, if what you are making is underground).

The Sandbox isn't new, but because its location isn't evident, most SketchUp users have never used it. Here are the facts:

- **The Sandbox is a collection of tools.** Each tool serves a fairly specific purpose and is meant to be used at a particular stage of the terrain modeling process. That said, like all SketchUp's tools, they are incredibly flexible. You can use them to model anything you want.
- **The Sandbox is in both free and Pro:** Despite what many people think, the Sandbox tools are not just for Pro users; people who use the free version of SketchUp can use them, too. They are just hidden, which brings up the next point.
- **The Sandbox is hidden.** The reasons for this are complicated, but the tools in the Sandbox are a little bit special; they're extensions—you have to find them and turn them on before you can use them. If you're using SketchUp Pro, you can skip the first two steps in the following list—they're already turned on.

Follow these steps to switch on the Sandbox tools:

1. Choose Window ⇨ Preferences from the menu bar to open the Preferences dialog box. Choose SketchUp ⇨ Preferences if you're on a Mac.
2. In the Extensions panel, make sure the Sandbox Tools check box is selected and then close the Preferences dialog box.
3. Choose View ⇨ Toolbars Sandbox from the menu bar to show the Sandbox toolbar.

GOOGLE SKETCHUP IN ACTION

Create new and edit existing terrains.

Rather than continuing to describe what each element of the Sandbox does, tools that are necessary are covered. Don't worry about figuring out everything all at once—pick up new tools as you need them.

6.3.1 Creating a New Terrain Model

Whether you are modeling a patch of ground for a building or redesigning Central Park, you need one of two terrain-modeling methods:

- **Starting from existing data:** This existing data usually arrives in the form of contour or topo lines; see the next section to read more about them.
- **Starting from scratch;** If you don't have any data to start or if you're beginning with a perfectly flat site, you can use SketchUp's From Scratch tool to draw a grid that's easy to form into rolling hills, berms, and valleys. Skip ahead to the next section, "Modeling Terrain from Scratch," for more information.

There's a neat trick for modeling small (yard-sized) amounts of terrain—the piece of land immediately surrounding a building, for example. You could use the From Scratch tool to start with a flat site, but there's a better way: See "Roughing out a Site" a little later in this chapter for more information.

Modeling Terrain from Contour Lines

You know the curvy lines on topographical maps that show you where the hills and valleys are? They are contour lines (or contours) because they represent the contours of the terrain; every point on a single line is the same height above sea level as every other point on that line. Where the lines are close together, the ground between them is steep. Where the lines are far apart, the slope is less steep. Cartographers, surveyors, engineers, and architects use contour lines to represent 3D terrain in flat formats like maps and site drawings.

Sometimes, you have contour lines for a building site that you want to model in 3D. You can use the From Contours tool in the Sandbox to automatically generate a three-dimensional surface from a set of contour lines. See Figure 6-17 for an illustration.

Figure 6-17

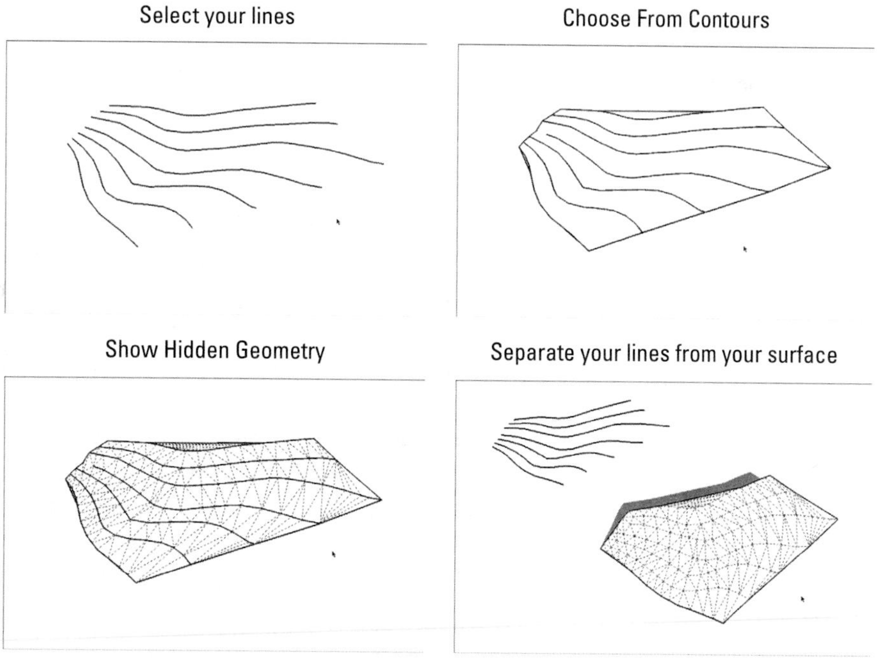

Select your lines

Choose From Contours

Show Hidden Geometry

Separate your lines from your surface

Use the From Contours tool to turn a set of contour lines into a 3D surface.

Here are some things to keep in mind about the From Contours tool:

- **It's a two-step tool.** Using From Contours is simple after you get used to it:
  - **a.** Select all the Contour lines you want to use to create a surface.
  - **b.** Choose Draw ⇨ Sandbox ⇨ From Contours from the menu bar (or click the From Contours tool button, if the Sandbox toolbar is visible).

  If you can't see the Sandbox tools in your menus, you haven't turned them on yet. See the beginning of this section, "Making and Modifying Terrain," to rectify the situation.
- **Your contour lines need to be lifted up.** The From Contours tool creates a surface from contour lines that are already positioned at their proper heights in 3D space. Most of the time you work with contours that were part of a 2D drawing, and that means you probably have to lift them up yourself using the Move tool—one at a time. It's tedious but necessary. For a refresher on selecting things, take a look at the last part of Chapter 2.
- **Download and install Weld.** The Weld Ruby script (which you can read about in Chapter 16) turns selections of individual line segments into polylines—this makes them much, much easier to work with. If you work with contour lines imported from a computer-aided drawing (CAD) file, using Weld makes your life a little easier.
- **You end up with a group.** When you use From Contours, SketchUp automatically makes your new surface (the one you generated from your contour lines) into a group. It leaves the original lines themselves completely alone; you can move them away, hide them, or delete them if you want. I recommend making another group out of them, putting them on a separate layer (see Chapter 7 for more on this), and hiding that layer until you need it again.

To edit the faces and edges inside a group, double-click it with the Select tool. Chapter 5 has all the details on groups and components.

- **To edit your new surface, turn on Hidden Geometry.** The flowing, organic surface you just created is actually just a bunch of little triangles. The From Contours tool smoothes the edges that define them, but they are there. To see them, choose View ⇨ Hidden Geometry from the menu bar.
- **Try to keep your geometry reasonable.** The From Contours tool is super useful, but it has its limits. The trouble is that it's too easy to use it to create enormous amounts of geometry (faces and edges) that can really bog down your system. If it takes forever for your contours to turn into a surface, or if that surface is so big that your computer can't seem to handle it, you need to go back a few steps and do one (or perhaps all) of the following:
  - **Work on a smaller area.** As nice as it'd be to have the whole neighborhood in your SketchUp model, you may have to narrow your scope. Creating only what you need is good modeling policy.

Figure 6-18

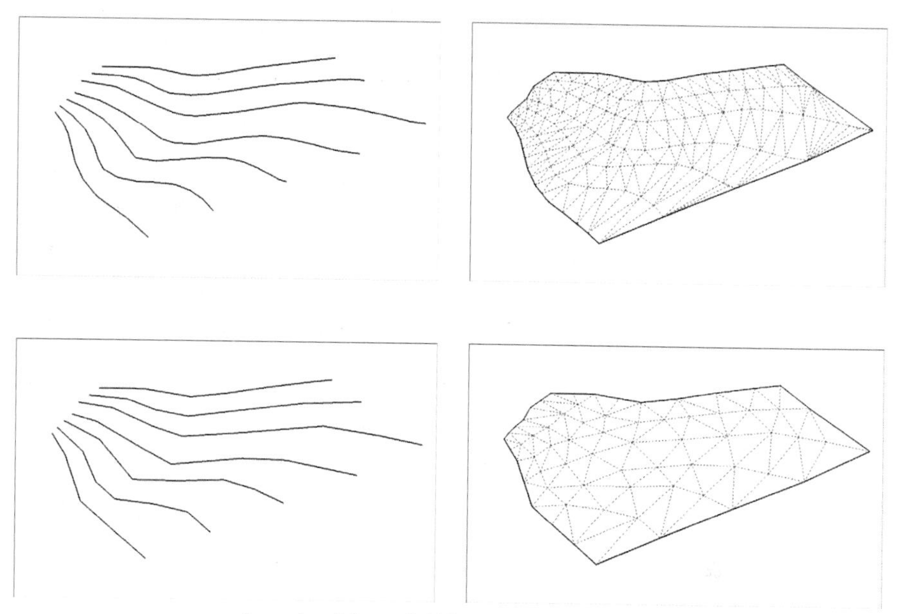

Low-detail lines yield fewer triangular faces

How many triangles are created depends on the number of edge segments in the contour lines you start with.

- **Use only every other contour line**. Doing this effectively halves the amount of geometry in your resulting surface.
- **Simplify the contour lines**. This is a little bit hard to explain, but here goes: The From Contours tool works by connecting adjacent contour lines together with edges that form triangles. How many triangles it creates depends on how many individual edge segments are in each contour line; Figure 6-18 provides an illustration. Unless you created the contour lines to begin with—there's a good chance you imported them as part of a CAD—you have no control over how detailed they are. Redrawing each contour line is not ideal, but luckily, you can download a great Simplify Contours Ruby script that makes the process much simpler. Reference Chapter 16 for more information on using Ruby scripts.

- **You don't have to start with existing contour lines.** In fact, drawing your own edges and using From Contours to generate a surface from them is one of the most powerful ways to create organic, non-boxy forms in SketchUp. Take a look at the next part of this chapter for more information.
- **Get ready to do some cleanup.** The surfaces that From Contours creates usually need to be cleaned up to some extent. Use the Eraser to delete extra geometry (you'll find lots along the top and bottom edges of your surface). Use the Flip Edge tool to correct the orientation of your triangular faces. See the nearby sidebar "Sandbox's Flip Edge Tool" for the lowdown.

FOR EXAMPLE

SANDBOX'S FLIP EDGE TOOL

The sandbox's flip edge tool is simple, but it's indispensable if you're working with the From Contours tool. Basically, you use flip edge to clean up the surfaces that From Contours creates. When you turn contour lines into a surface, lots and lots of triangular faces appear. Sometimes, the From Contours tool decides to draw an edge between the wrong two line segments, creating two triangular faces that form a "flat spot" in your surface. The following image illustrates this.

- You get rid of these flat spots manually by flipping the edges that create them. Doing so changes the resulting triangular faces, usually making them end up side-by-side (instead of one-above-the-other).
- To use the flip edge tool (choose Tools ⇨ Sandbox ⇨ Flip edge), just click the edge you want to flip. If you're not sure about an edge, go ahead and flip it; then see if things look better. If they don't, you can always undo or flip it back.

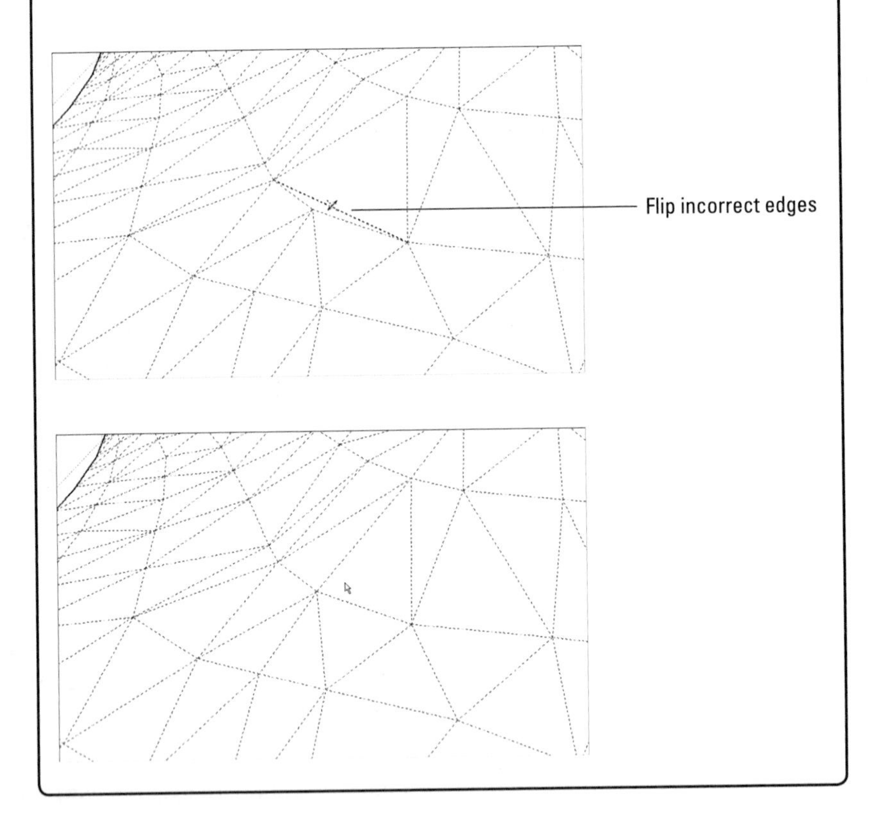

Modeling Terrain from Scratch

Without contour lines that define the shape of the terrain you want to model, you have to start with a level surface. Use the From Scratch tool to create a big, flat rectangle that represents a chunk of ground. Because the rectangle is already divided into triangular faces, it's easy to use the Smoove tool, addressed next in this chapter, to shape the rectangle into hills, valleys, sand traps, and whatever else you have in mind.

Here's the thing, though: It's a very rare occasion that you have carte blanche with a piece of land. Unless you design something like a golf course in the middle of a dry lake bed or terra form a new planet for colonization, you probably have pre-existing terrain conditions to contend with. And if that's the case, you're probably better starting off with a set of contour lines that describe those conditions that were addressed in the From Contours tool earlier in this chapter.

So although the From Scratch Tool works great, it's doubtful you'll need to use it much. All the same, here's how to do so, just in case. Follow these steps to create a new terrain surface with the From Scratch tool and take a look at Figure 6-19 while you're at it:

PATHWAYS TO... CREATING A NEW TERRAIN SURFACE WITH THE FROM SCRATCH TOOL

1. Choose Draw ⇨ Sandbox ⇨ From Scratch from the menu bar to activate the From Scratch tool.
2. **Type a grid spacing amount and press Enter.** The default grid spacing amount is 10 feet, which means the tool draws a rectangle made up of squares that are 10 feet across. The grid spacing you choose depends on how big an area you're planning to model and how detailed you plan to make the terrain for that model. If modeling a single-family house on a reasonably sized lot, use a grid spacing of 2 feet, which provides enough detail for elements like walkways and small beams without creating too much geometry for the computer to handle. If laying out an 18-hole golf course, on the other hand, choose grid spacing closer to 50 feet and then add detail to certain areas later.
3. Click to position one corner of your new terrain surface where you want it.
4. Click to determine the width of the surface you're drawing.
5. Click to establish the length of your new terrain surface. When you are done, the great big rectangle you have created will automatically be a group. Double-click with the Select tool to edit it and good luck. You will probably decide to use the Smoove tool next; move ahead to "Making Freeform Hills and Valleys with Smoove" (later in this chapter) to find out how.

Figure 6-19

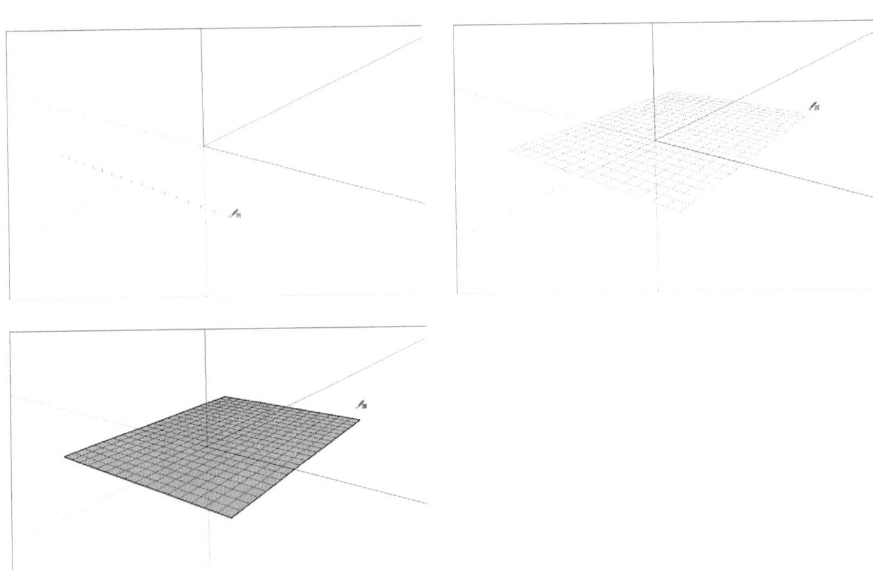

Use the From Scratch tool to create big swatches of flat terrain.

Roughing Out a Site

Perhaps you want to model a small piece of non-flat terrain that surrounds a building. Maybe you are trying to reproduce existing site conditions, or maybe you are in the process of designing the landscape for a project. There is a great technique for cases like this one: You can use From Contours to quickly generate a surface from just a few simple outlines. Follow these steps and Figure 6-20 starting on your left to model a simple terrain surface with the From Contours too:

1. **Extend the bottom of your building down so the exterior walls drop below ground level.**
2. **Make your building into a group.** See Chapter 5 if you need help.
3. **Use the Tape Measure and Line tools to draw the outline of the chunk of terrain you want to model around the building.** Keep in mind that the resulting horizontal face is flat; just imagine you are drawing in 2D space. It doesn't matter if the outline you draw is below, above, or in line with the building, as you see in the next step.
4. **Use the Push/Pull tool to extrude the face you drew in Step 3 into a box that extends above and below your building, and then delete the top and bottom faces of the box you just drew.**
5. **Paint the walls of your box with a translucent material.** You can find some in the Translucent library, in the Materials dialog box.
6. **Draw edges on the sides of the box that represent where the ground should intersect them.**

Figure 6-20

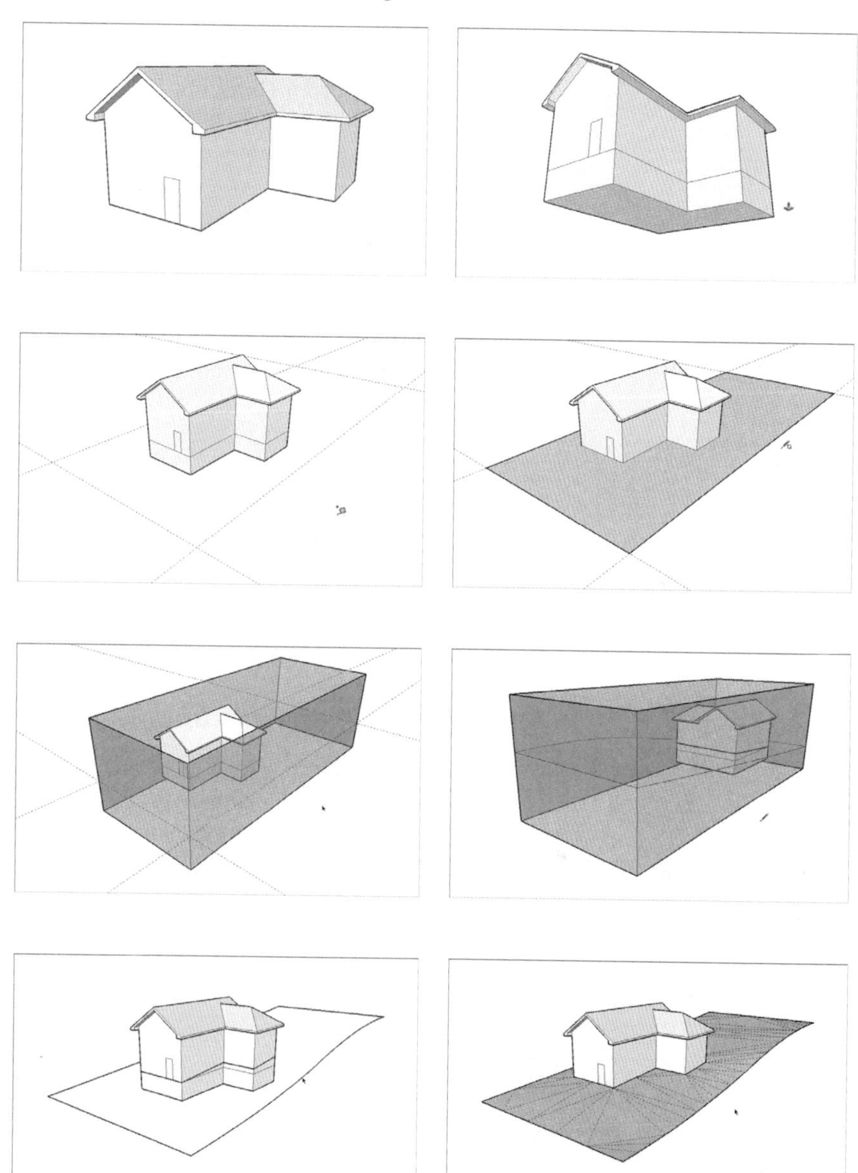

You can create irregular terrain surfaces very quickly with the From Contours tool.

7. **Draw edges on the sides of the building that represent where the ground meets the building.**
8. **Delete the box you created in Step 4, leaving only the edges you drew in Step 6.**
9. **Select all the edges you drew in Steps 6 and 7.**

10. **Choose Draw ⇨ Sandbox ⇨ From Contours from the menu bar to generate a surface based on the edges you selected in the previous step.** Take a look at the section "Modeling Terrain from Scratch" for tips on using From Contours; at this point, you need to use the Flip Edge tool and the Eraser to clean up your terrain model—particularly where your building is supposed to go.

6.3.2 Editing an Existing Terrain Model

No matter how you make a terrain model, there's a 99-percent chance that it consists of lots and lots of triangles. Switch on Hidden Geometry (choose View ⇨ Hidden Geometry) to see them. As long as you have triangles, you can use the Sandbox's terrain editing tools. This section shows you how to do the following:

- Shape (or re-shape) your terrain with the Smoove tool.
- Create a flat spot for a building with the Stamp tool.
- Draw paths and roads with the Drape tool.

Keep in mind that both From Contours and From Scratch create terrain objects that are groups. To edit a group, double-click it with the Select tool. When you are done, click somewhere else in your modeling window.

Making Freeform Hills and Valleys with Smoove

Smoove is a tool for moving smoothly. Smooth + Move = Smoove. Smoove is actually one of the coolest tools in SketchUp; it lets you shape terrain (or any horizontal surface that is made up of smaller, triangular faces) by pushing and pulling bumps and depressions of any size. Smoove is fun to use and yields results that would be difficult to produce with any other tool in SketchUp. Figure 6-21 shows what Smoove can do.

PATHWAYS TO...
SHAPE A SURFACE WITH SMOOVE

1. Double-click the group containing your terrain to edit it. If your terrain is not part of a group, then ignore this.
2. Choose Tools ⇨ Sandbox ⇨ Smoove from the menu bar to activate the Smoove tool.
3. Type a radius and press Enter. Smoove creates lumps, bumps, and dimples that are circular. The radius you enter here determines how big those lumps, bumps, and dimples should be.
4. Click somewhere on your terrain surface to start smooving.
5. Move your mouse up or down (to create a bump or a depression, respectively), and then click again to stop smooving.

Figure 6-21

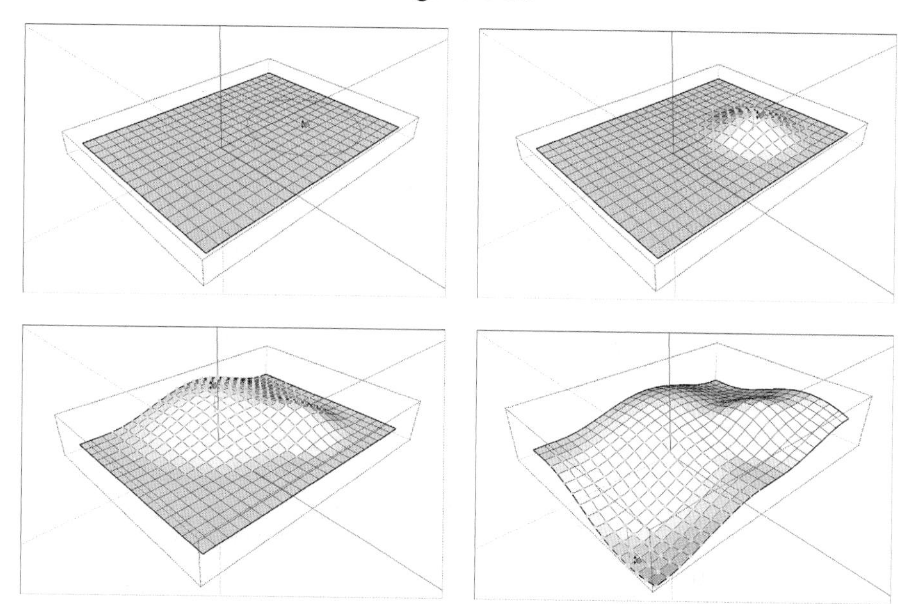

Smoove creates shapes that are unlike anything else can create with SketchUp.

Here are some more things to keep in mind when you use Smoove:

- **Use the From Scratch tool beforehand.** You don't have to, but creating a surface with the From Scratch tool (which is described earlier in this chapter) is by far the easiest way to end up with terrain that you can smoove easily.
- **Try smooving to edit other terrain surfaces.** You can also use Smoove after you create a terrain surface with the From Contours tool.
- **Double-click to repeat your previous Smoove.** As with Push/Pull, double-clicking tells SketchUp to carry out the same operation as you did the last time you used the tool.
- **Preselect to smoove shapes other than circles.** Any faces and edges you select before you use the Smoove tool will move up (or down) by a constant amount. This means you can use Smoove to create things like ridges and ditches by selecting the right geometry beforehand. Figure 6-22 provides a picture of this.

Placing a Building on Your Terrain with Stamp

Eventually, you may need to insert a building (or some other structure) on the terrain you have crafted. The Stamp tool provides an easy way to stamp a building footprint into a terrain surface, creating a flat "pad" for something to sit on. This tool also provides a way to create a gently sloping offset around the perimeter of your stamped form. This creates a smoother transition between the new, flat pad and the existing terrain.

Figure 6-22

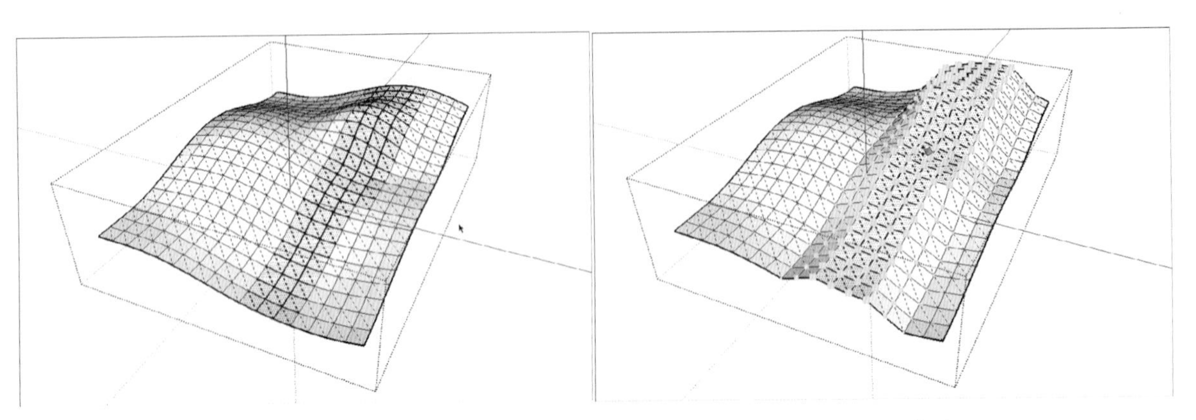

Preselect faces and edges to smoove shapes other than circles.

FOR EXAMPLE

NEED MORE TRIANGLES? ADD DETAIL

Like the Flip Edge tool, Add Detail is one of a kind. Use it to add triangles to areas of your terrain surface that need more detail. That way, you can save geometry (and file size, and waiting) by having lots of faces only in the areas of your terrain that require it. If a golf course, use very big triangles for the vast majority of it. Use the Add Detail tool to add triangles to areas where you planned to have smallish things like sand traps.

You can use the Add Detail tool in two ways:

- Add detail to faces one at a time. You can activate the tool (see the next bullet) without having any geometry selected. Then click faces or edges on your terrain to divide them into more faces. This comes in handy when you model something very precisely.
- Add detail to an area all at once. It's quick and easy to understand. Simply select the faces on your terrain you want to subdivide and choose Tools ⇨ Sandbox ⇨ Add Detail from the menu bar. Take a look at the figure to see what happens when you do.

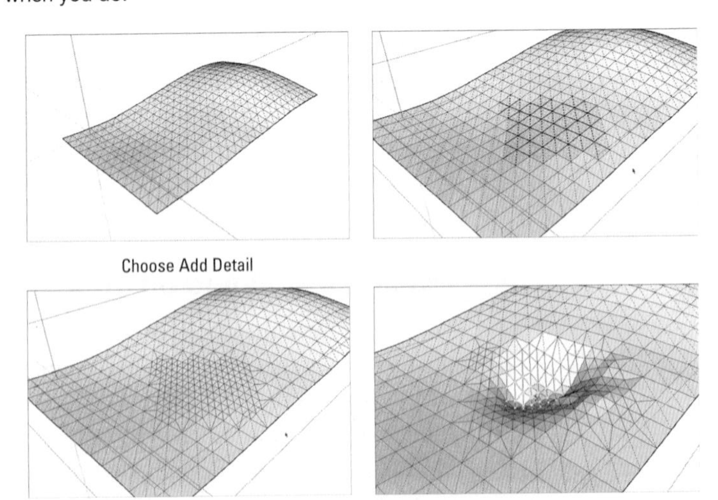

Follow these steps to use the Stamp tool; reference Figure 6-23 to see the corresponding pictures:

Figure 6-23

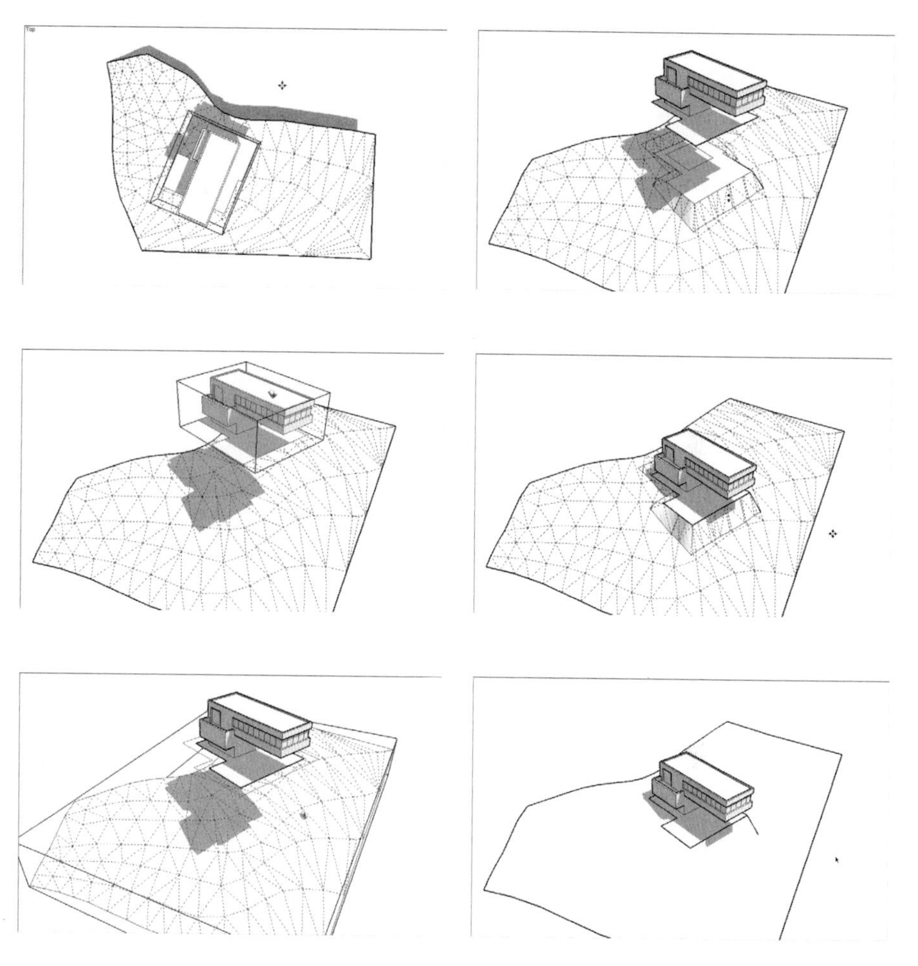

Use the Stamp tool to create a nice, flat spot for your building.

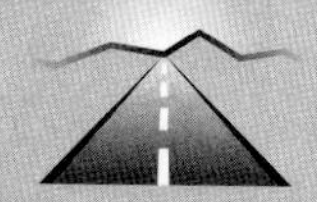

PATHWAYS TO... USING THE STAMP TOOL

1. **Move the building you want to stamp into position above your terrain surface.** The building should not touch the terrain but float in space directly above it. Also, turn the building into a group before you start moving anything; look at Chapter 5 to find out all

(continued)

PATHWAYS TO... USING THE STAMP TOOL *(continued)*

about groups and components. If you are having trouble moving your building into position accurately, move it to the correct height first and then switching to a top, no-perspective view to finish the job. Look in the Camera menu for both these commands.

2. **Choose Tools ⇨ Sandbox ⇨ Stamp from the menu bar to activate the Stamp tool.**
3. **Click the floating object to tell SketchUp what you want to use as the stamp.**
4. **Type an offset distance and press Enter.** The offset distance is the amount of space around the perimeter of whatever you are stamping that SketchUp uses to smooth the transition between the flat spot it's creating and the existing terrain. The offset amount you choose depends entirely on what you are stamping. Play with this and practice using Undo.
5. **Move your cursor over your terrain surface and click again.**
6. **Move (but don't drag) your mouse up and down to position the flat pad in space.** Click again to finish the operation.

Here are a couple things that are useful to know when you use Stamp:

- SketchUp uses the bottommost face in your stamp object as the template for the flat pad it creates in your terrain.
- Read the "Sandbox's Flip Edge Tool" sidebar, earlier in this chapter; Stamp creates triangular faces that sometimes need cleaning.

Creating Paths and Roads with Drape

The Drape tool works a little like a cookie cutter; use it to transfer edges from an object down onto another surface, which is directly beneath it.

Perhaps you have a gently-sloping terrain and you want to draw a meandering path on it. The path has to follow the contours of the terrain, but because you want to paint it with a different material, it needs to be a separate face. In this case, you'd draw the path on a separate face and use the Drape tool to transfer it to your terrain surface.

Taking the preceding example, follow these steps to use the Drape tool to draw a path on a non-flat terrain surface (Figure 6-24 illustrates the steps, starting with the left column.):

Figure 6-24

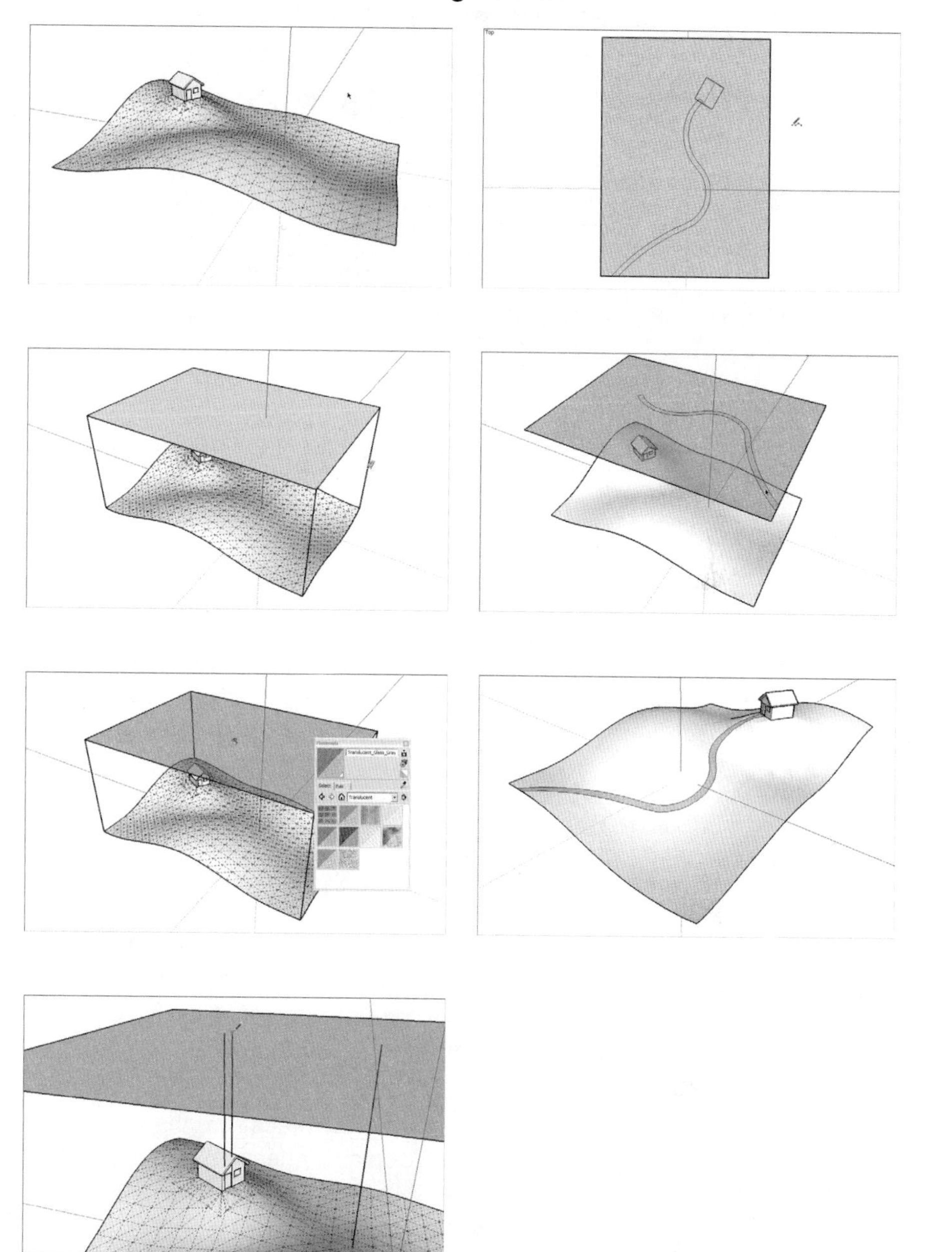

Use Drape to transfer edges onto your terrain surface.

PATHWAYS TO... USING THE DRAPE TOOL TO DRAW A PATH ON A NON-FLAT TERRAIN SURFACE

1. **Use the Line tool (see Chapter 2) to draw a flat face somewhere directly above your terrain surface.** If you can, make your flat face exactly the same size as your terrain. Just make sure it's big enough for whatever you plan to draw next (in this example, a path).
2. **Paint the face you just created with a translucent material.** A light gray works well; there is a good one in the Translucent library, inside the Materials dialog box.
3. **Use the Line tool to carry up any important points on your terrain surface.** In this case, make sure the path begins precisely at the door of the building, so draw vertical lines from the sides of the door to the flat face directly above. That way, you have something to inference to in Step 6.
4. **Choose Camera ⇨ Standard Views ⇨ Top from the menu bar to switch to a top view.**
5. **Choose Camera ⇨ Parallel Projection from the menu bar to turn off perspective.**
6. **On the upper face, draw the edges you want to drape.** Make sure that your edges form closed loops to create faces. If they don't, you'll have a difficult time trying to paint the path (in this case) after it's draped onto your terrain surface.
7. **Orbit your model so you can see both the upper and lower surfaces.**
8. **Soften/smooth the edges of the triangles in your terrain surface (if they aren't already).** To do this, follow these steps:
   a. Select all the edges and faces in your terrain, and then choose Window ⇨ Soften Edges from the menu bar.
   b. In the Soften/Smooth Edges dialog box, move the slider to the far right and make sure that both the Smooth Normals and Soften Coplanar check boxes are selected.
9. **Select the edges you want to drape.** If your edges define closed faces, you can select those faces instead; sometimes that's easier than selecting a bunch of individual edges. Reference Chapter 2 for tips on selecting things.
10. **Choose Draw ⇨ Sandbox ⇨ Drape from the menu bar to activate the Drape tool.**
11. **Click once on your terrain surface to drape the edges you selected in Step 9.** It doesn't matter if your terrain is inside a group—the Drape tool works anyway.

IN THE REAL WORLD

Csaba Pozsárkó is an archaeologist, 3D modeller, and SketchUp trainer and started using Google SketchUp when the free V.5 came out, and has become a Pro user since. He mentions, "I am not a designer but an archaeologist and use SketchUp to reconstruct archaeological finds as accurately as possible. I need to 'exclude' my design 'instincts' as much as possible, too. So when there is the field with the walls for instance, we take an accurate survey and then try to find out how that building may have looked like. Finally, we render the result with all bells and whistles so that the general audience can imagine what it was like."

Since most of Csaba's models "do not exist" any more, he cannot export them into Google Earth as a historic 3D Layer. However, he regularly uses the terrain tools in his workflow. He usually starts by reconstructing the contemporary terrain and modifies it as needed using SketchUp tools and many other plugins that are available. Then, he creates scenes and layers to jump from one view to the next and to gain access to different visible options. Next he creates animations and 360-degree panoramic shots that can be exported into "game like" engines or into 3ds max for further processing or rendering (in a team work) etc. Generally these projects are completed in at least two months, which allows for continuous feedback from different sources.

CAREER CONNECTION

Interior Designer Nancy Lin started using SketchUp with version 6. She states "not only was it free and easy to use, but I loved how we could apply the materials and textures to make the drawing look realistic. For people who don't have great hand-drafting skills, Google SketchUp is a wonderful tool in order to show quick sketches in different styles and views."

Her process always begins with a drawing in AutoCAD and then it gets imported into SketchUp to build the 3D model. After she builds the shell, she starts plugging in all the components, then she applies the finishes and finally renders the views with SUPodium. When the 3D symbols are not available, she uses the already-made components in the 3D warehouse that we can modify to meet our requirement.

For a presentation, she normally renders different views of the space layout or the custom furniture and quite often an animation, or a walk through, tool are included in the presentation as well. Often clients have difficulty describing what they have in mind, but with SketchUp the client can see close to what the end result will look like, thus eliminating their fears. Having these renderings, or "walk-throughs," within the space can help the client not only visualize the space but also catch anything the client would like to change before the project is completed. According to Nancy, "by incorporating SketchUp with a presentation, if the client dislikes anything or wants to change something we can easily change the finishes or stretch/shrink a component with just a 'click.' For most of our custom or AI projects, clients love this tool because it can avoid a lot of 'surprises' afterwards."

SELF-CHECK

1. What do we call the curvy lines on topographical maps?
   a. Topography outlines
   b. Contour lines
   c. Terrain lines
2. What can the Weld Ruby script do the individual line segments that we need to use to build our terrain?
3. What tool do we use to create large swatch of flat terrain and what do we do to edit it?
4. Pick the correct answer about the Smoove tool:
   a. It allows you to shape terrain by pushing and pulling the triangular faces.
   b. It sharpens a smooth terrain by reducing the bumps and depressions.
   c. It allows you to view an animation of the terrain in slow motion.
5. What tool would allow us to create a flat surface on a terrain, where we can situate a building?

Apply Your Knowledge Imagine that you are an architect that needs to build a 15' × 15' structure on an uneven terrain. Create a 100' ×100' terrain with the Smoove tool and be sure to make it slightly curved. Practice with Stamp to create a flat pad where you would be building the structure, and then create a path leading to one side of that pad, using Drape.

6.4 BUILDING A SOLID TOOLS FOUNDATION

Brand new for SketchUp 8, the Solid Tools provide a completely new way for SketchUp modelers to work. Solid modeling operations (otherwise known as Boolean operations) give you the ability to create the shapes you need by adding or subtracting other shapes to or from each other. This type of modeling is actually pretty common in other 3D apps; now SketchUp can do it too. In the next few pages, you are shown how to use all six of SketchUp 8's new solid Tools, giving detailed examples for the three that are the most useful.

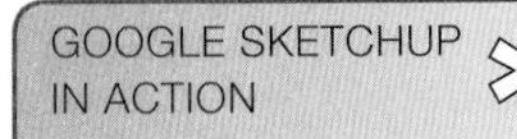

Understand and work with solids.

Important to mention right away is that five of the six Solid Tools are only in the Pro version of SketchUp 8. Read Table 6-1 (later in this chapter) to see what is available.

6.4.1 Understanding Solids

Before you can use the new Solid Tools, you need solids. Here are six things you need to know about solids; you can think of them as the Solid Rules:

- **A solid is nothing more than an object that's completely enclosed.** It has no holes or other gaps; if you filled it with water, none would leak out. For this reason, solids are sometimes referred to as being watertight. Here is another way to think about it: Every edge in a solid must be bordered by two faces.
- **No extra edges or faces allowed.** You wouldn't think that one or two edges or faces would make much of a difference, but it does—solids can't contain any extra geometry. Figure 6-25 shows some examples of

Figure 6-25

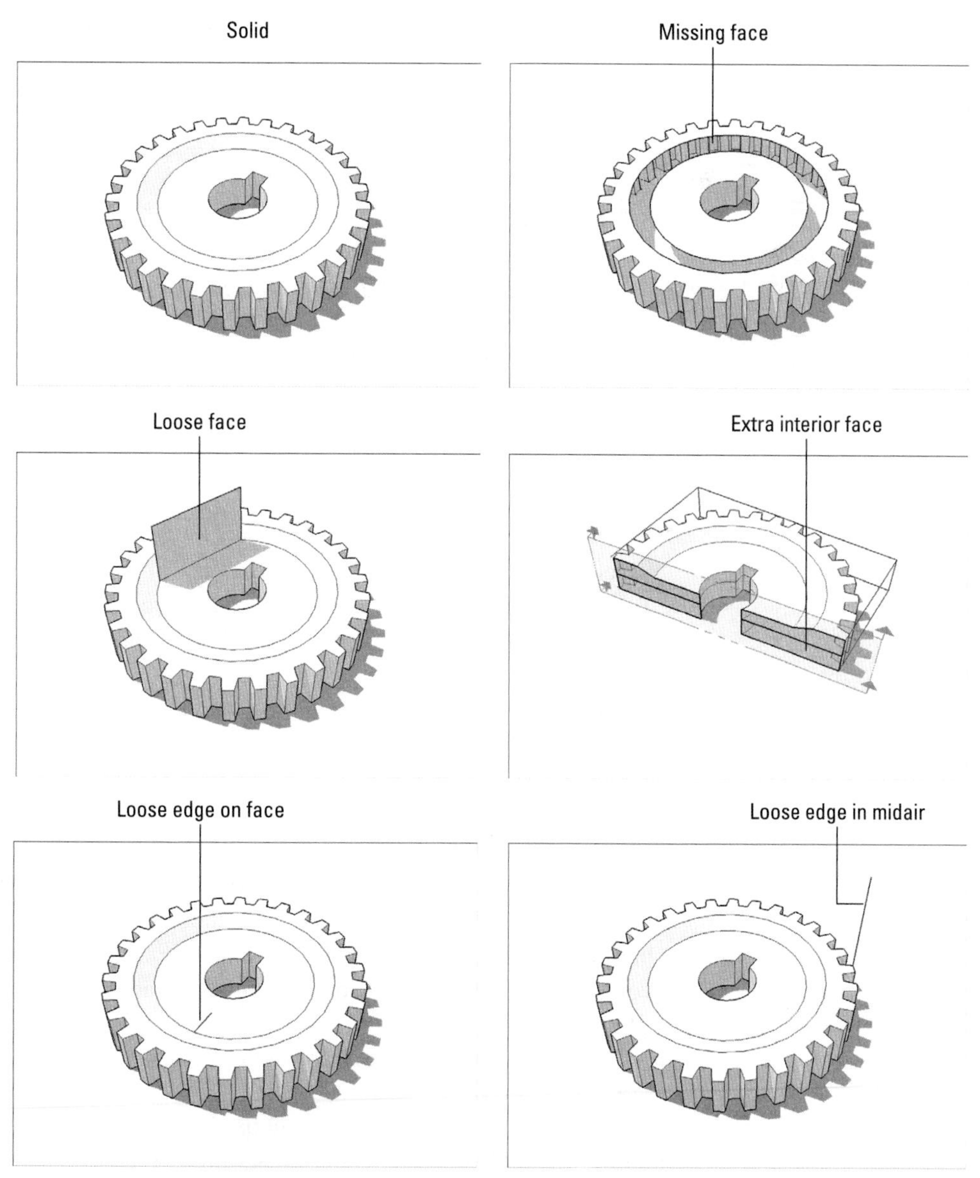

Solids cannot contain any extra edges or faces.

things that can disqualify otherwise completely enclosed shapes from being solids.

- **Only groups and components can be solids.** This one is very important. For SketchUp to realize an object is a solid, you have to make it into either a group or a component first. Another thing: Solid groups and components can't have other groups and/or components nested inside them.
- **Making a solid does not require any special tools.** You do not have to pick from a special list of objects to create solids; you make them with the same SketchUp tools you use all the time. Case in point: Every time you have push/pulled a rectangle into a box, you have created a solid.
  - **Solids have volumes.** The easiest way to tell whether a group or component is a solid is to select it and choose Window ⇨ Entity Info. If the dialog box includes a value for Volume, you have a solid on your hands. Reference Figure 6-26 for an illustration.
  - **Solids can be multiple objects.** This one is a bit confusing. As long as each individual cluster of geometry within a group or component is completely enclosed and is selected without loose lines or faces that are missing, then SketchUp considers it to be a solid. It doesn't matter that they are not connected or touching in any way; what is important is that an area of space is fully surrounded by faces. Only then, the Volume reference will appear.

Figure 6-26

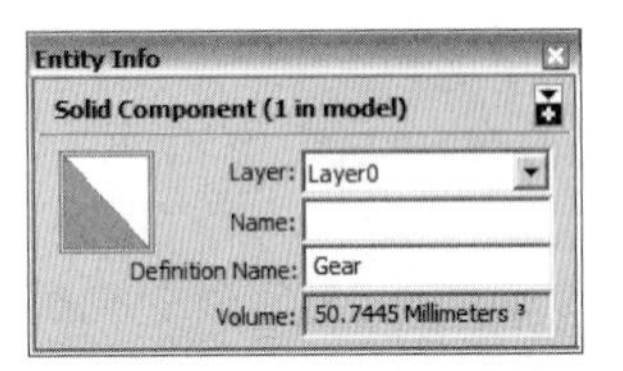

Check the Entity Info dialog box to see whether your selection is a solid.

6.4.2 Checking Out the Solid Tools

When you have a solid object or objects, you can use SketchUp 8's Solid Tools in powerful ways to create shapes that would otherwise be very complicated and time-consuming to make. For example:

- Add two solids together to create a new one.
- Use one solid to cut away part of another one.

With the SketchUp Intersect Faces tool (formerly Intersect with Model), you can achieve many of the same things that the Solid Tools do. Intersect Faces takes longer because it requires a lot of cleanup; however, it's still useful for two very important reasons: It's available in both the free and Pro versions of SketchUp, and it works on any face in your model—not just on solids. You can read about Intersect Faces in Chapter 4.

Two things you need to know before you start using the Solid Tools:

- **Open the dedicated toolbar.** Choose View ⇨ Toolbars ⇨ Solid Tools to open the toolbar that contains all six tools. You can also find them on the

Tools menu. Keep in mind that five of them—all but the Outer Shell tool—are available only if you have SketchUp Pro 8.

- **To use the Solid Tools, preselect—or don't.** Pick the tool you want to use either before or after you have told SketchUp which solid objects you want to affect. Like most "order of operations" issues, like the Follow Me tool, this can be confusing for some. It's easiest to use the Solid Tools by preselecting the solids you want to use and then choose the tool to carry out the operation. The glaring exceptions to this rule are the Subtract and Trim tools; both of these depend heavily on the order in which you pick your solids. Read Table 6-1 for more specifics and reference Figure 6-27 for an illustration.

Figure 6-27

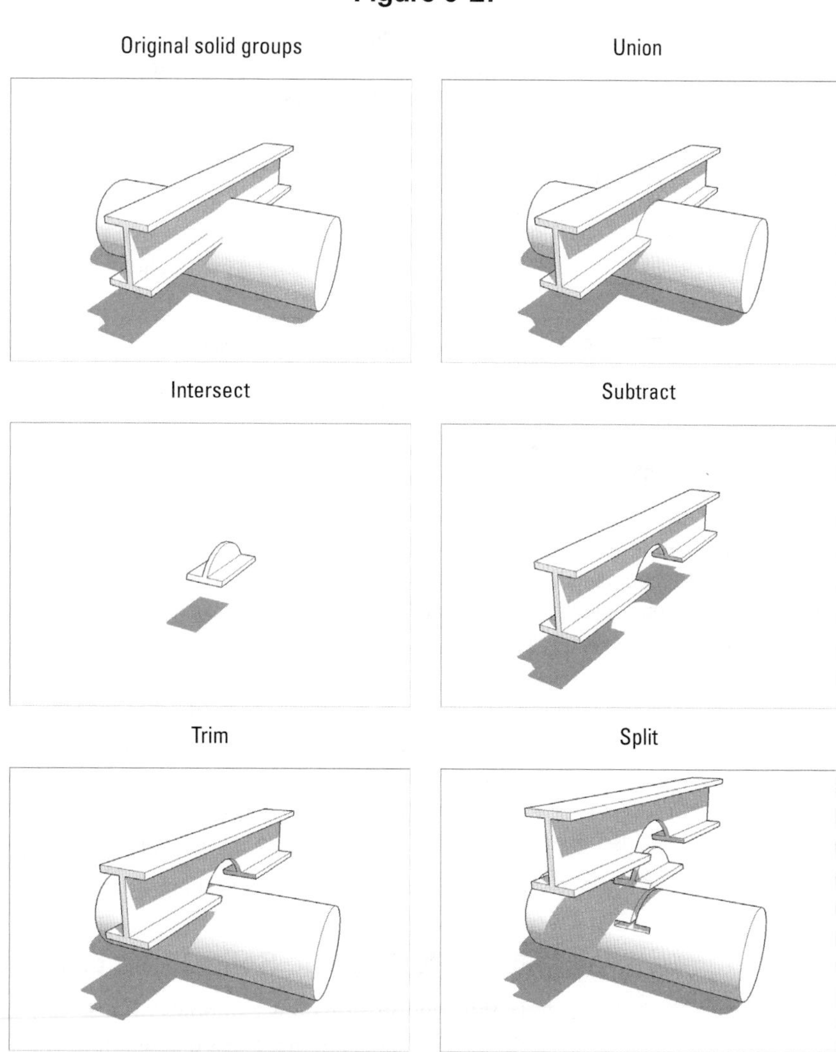

The Solid Tools lets you do additive and subtractive modeling operations.

Table 6-1: The Solid Tools

Tool	*Free or Pro?*	*What It Does*	*How to Use It*	*Start With*	*End With*
Union	Pro only	Combines two or more solids into a single solid. Deletes overlapping geometry. Preserves internal pockets.*	Select the solids you want to use and then activate the tool.	Two + solids	One solid
Outer Shell	Free and Pro	Combines two or more solids into a single solid. Deletes overlapping geometry, including internal pockets.*	Same as Union tool	Two + solids	One solid
Intersect	Pro only	Makes a single solid where two or more solids overlap. Deletes everything else.	Same as Union tool	Two + solids	One solid
Subtract	Pro only	Uses one solid to cut away part of another solid. Deletes the first solid when it's done.	Activate the tool, click "cutting" solid, and then click solid to be cut.	Two solids	One solid
Trim	Pro only	Cuts two solids where they overlap and creates a new solid from the overlap. Doesn't delete anything.	Same as Union tool.	Two solids	Three Solids

*An internal pocket is like a solid within a solid—it's a completely enclosed volume that happens to be located inside the main volume of a solid. Picture a SketchUp model of a tennis ball. Because tennis balls have a thickness, you'd need two surfaces to model one: one for the inside, and one for the outside. If you selected both and made a group, you'd have a solid with an internal pocket inside.

Note that the Split tool actually does three operations every time you use it: It yields two subtractions and an intersection. That is to say, using Split is like using both Subtract and Intersect on your solids. For this reason, replace both of these tools with Split full-time. It's easier to keep track of what's going to happen, and the only downside is that you have to delete a couple extra objects when you are done.

6.4.3 Putting the Solid Tools to work

In this section, are a few examples of everyday modeling challenges that the Solid Tools can help make less challenging. You are almost certain to encounter these tricky situations while you continue your exploration of SketchUp.

Figure 6-28

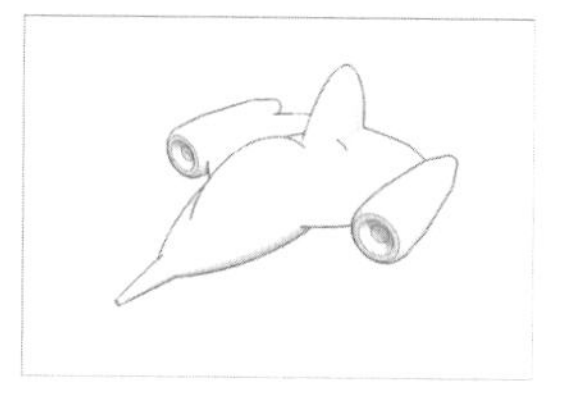

Using Union or Outer Shell to combine several solids into one produces edges where their faces intersect and gets rid of excess geometry.

Assembling Complex Objects with Union or Outer Shell

Chapter 4 has a section about using the Intersect Faces tool to combine multiple roof pitches into a single, solitary roof. If all those gables, hips, dormers, and other roof elements are solids, you can absolutely use SketchUp's Union or Outer Shell tools to make quick work of the problem.

The same goes for anything that is composed of several disparate elements that you have assembled by moving them together until they overlap. In the spacecraft in Figure 6-28, the hull (or body) of the craft is a combination of different pieces modeled separately. Notice the lack of edges where the components intersect? Edges add detail and definition, especially when the model is displayed using a lines-only style (as it is here). There is also the issue of all the geometry hidden inside the hull. Combining everything together into a single solid helps it shed weight and look better, all at the same time.

Using Intersect in Combination with Front, Top, and Side Views

Anyone who has ever tried to model a car with SketchUp knows it's a tricky undertaking. The problem is that cars (and most other vehicles) are kind of curvy; worse yet, they are curvy in several directions.

One trick lots of modelers use to block out a basic shape for things like cars is to start with orthographic (straight-on top, front, and side) views of the thing they are trying to model. Here's how the method works:

1. **Position each 2D view where it belongs in 3D space.**
2. **Push/pull them all so their extrusions overlap.**
3. **Use the Intersect tool (Tools ⇨ Solid Tools ⇨ Intersect) to find the object the extrusions all have in common.** This method doesn't always produce perfect results, but it's a lot better than guessing. Figure 6-29 shows the technique in action.

Figure 6-29

1. Create solids from front, top, and side views

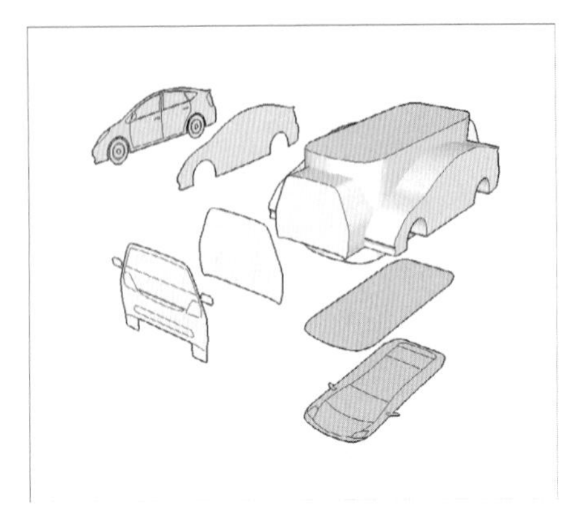

2. Position them precisely

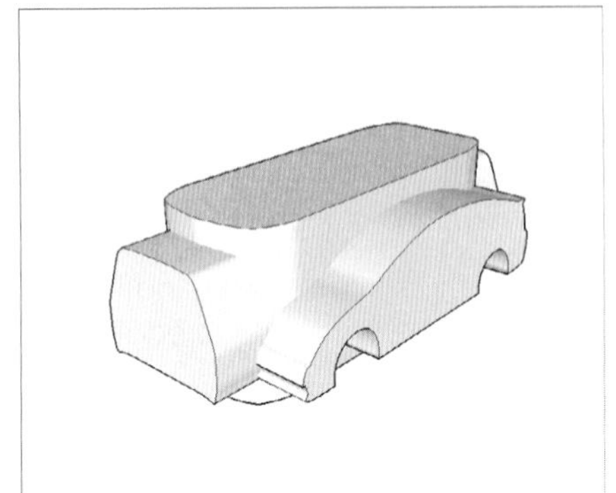

3. Use Intersect to find shape they have in common

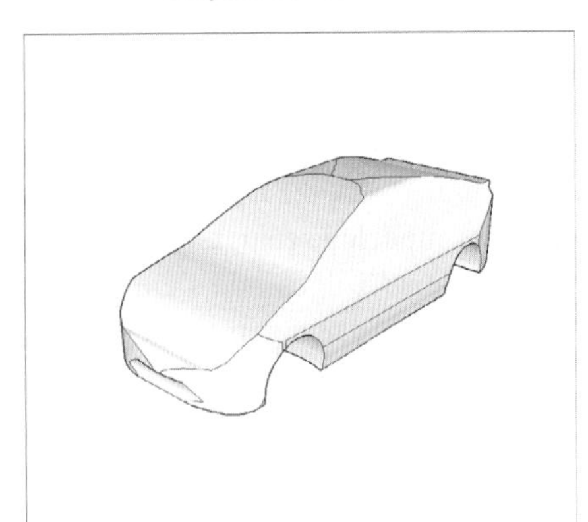

4. Smooth edges and use Intersect Faces with Model to transfer details

If you have orthographic views of the thing you are trying to model, you can use Intersect to give yourself a head start.

Modeling Close-Fitting Parts with Trim

Woodworkers and industrial designers, take heed: SketchUp Pro 8's Trim tool saves you literally hours of work. Any time you need to build a model with parts that interlock or otherwise fit together closely, Trim is where you should look first.

Trim basically tells one part to "take a bite" out of another, which is perfect for joinery (dovetails, finger joints, dadoes, and so on), machine parts, ball-and-socket joints, and any other positive/negative conditions where two parts meet. In Figure 6-30, there is illustrated a small wooden box with dovetailed sides and a dadoed bottom.

Figure 6-30

1. Start with sides as solid components

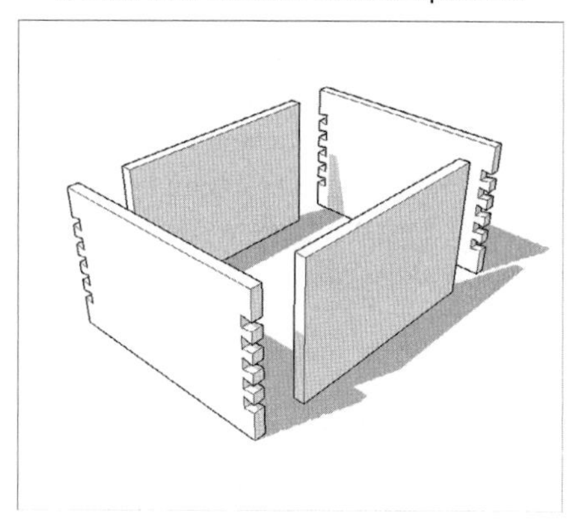

2. Fit everything together

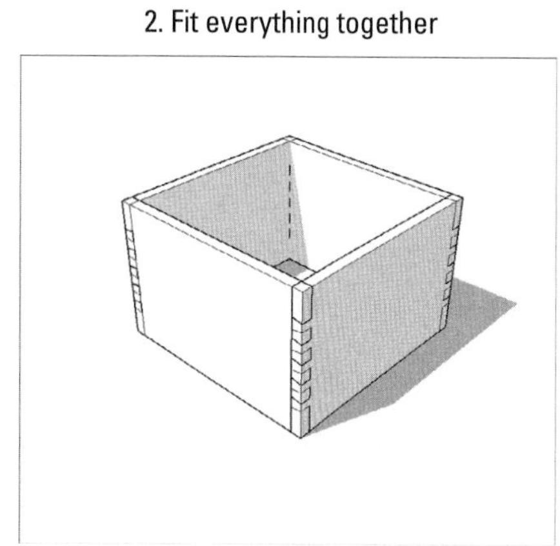

3. Use Trim to cut dovetails into other two sides

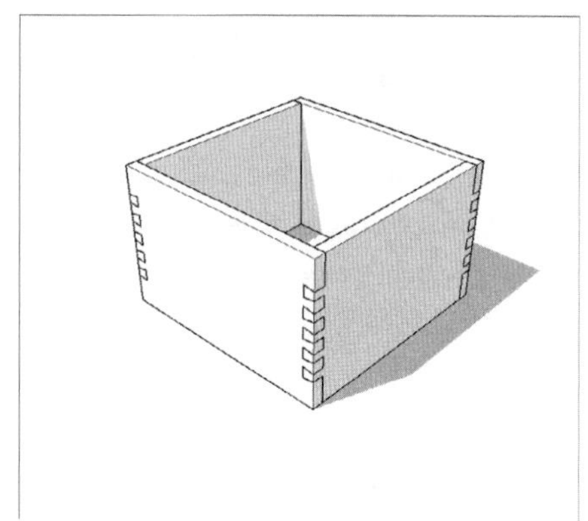

4. Create bottom to fit

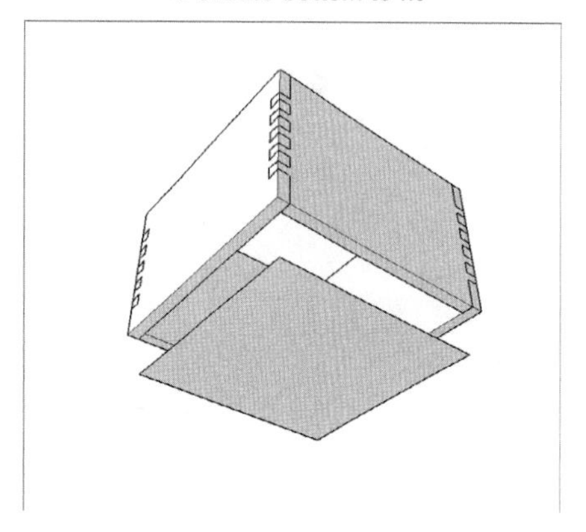

5. Position bottom and use Trim to cut dadoes in sides

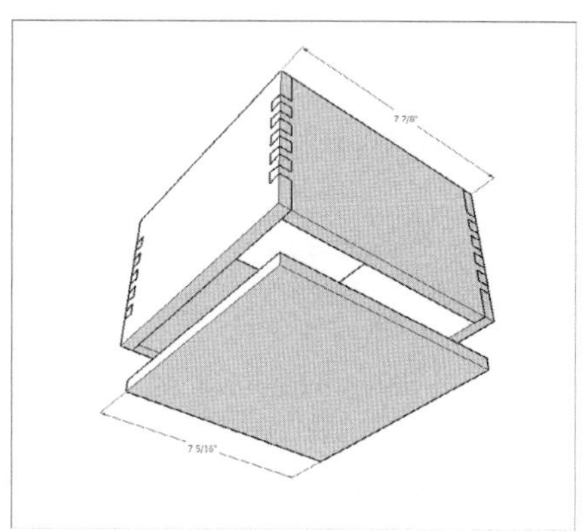

6. Take everything apart to make sure it looks right

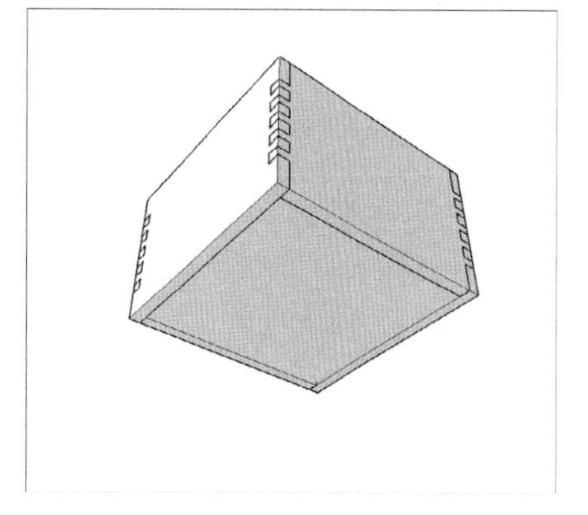

The Trim tool is perfect for modeling joinery and other close-fitting parts.

The only tricky thing about using the Trim tool is remembering which solid to pick first. Remember that the first thing you pick (or click) is the one you want to use to cut with. In the case of the box in Figure 6-30, that would be side with the dovetails. When you select the dovetails and then select the blank side, the Trim tool cuts the dovetails into the second piece. You get the hang of it after a few tries.

The Trim tool is nice to use: You can keep using your cutting solid on multiple other solids. To cut the dado (or groove) into the sides of the box in Figure 6-30, follow these steps:

PATHWAYS TO... USING THE TRIM TOOL

1. **Choose Tools ⇨ Solid Tools ⇨ Trim to activate the Trim tool.** Your cursor has the number 1 on it.
2. **Select the box bottom.** Your cursor changes to show the number 2.
3. **Select one side on the box.** You just cut a dado using the box bottom you picked in Step 2. Your cursor still says 2.
4. **Select another of the box's sides to create another dado.**
5. **Select the remaining two sides to cut dadoes in them, too.**

A question that comes up pretty frequently concerns what happens when you use one of the Solid Tools on a component instance. Why doesn't the effect of what you just did affect all the other instances of that component? It should, shouldn't it? Chapter 5 of this book speaks to this.

As soon as you use a Solid Tool on a component instance, SketchUp makes that instance unique; it's still a component, it just isn't connected to the other instances anymore.

SELF-CHECK

1. Three dimensional objects, regardless if they are grouped or not, can be solids. True or false?
2. What is the only operation that can be performed under the Solid Tools, in the free version of Google SketchUp?
3. Imagine that we are building a car in three dimensions from three extruded profiles—front, top, and side. What operation would be used under the Solid Tools, to obtain an approximation of the shape?

4. Choose the correct answer about the Trim tool:
   a. The last thing you pick or click is the one you want to cut with.
   b. The first thing you pick or click is the one you want to cut with.
   c. It does not matter which part we select first, to create our cuts.

Apply Your Knowledge Practice using the solid tools an object such as an oscillating electric heater or room dehumidifier. If you analyze the shape carefully, you will discover the following:

- If your object does not have straight lines and it has soft edges, you might want to consider extruding the front view and then the side view of the main body. After that you can use Trim to discern the overlapping shape.
- If your object has a clearly defined base or a front panel that would easier to build alone from the main body, you can build those parts separately and then join them later with the Union, Outer Shell, or Intersect tools, depending on your interest in preserving internal pockets or not.
- If you have a grill where the air comes in or out, you can build the shape of one of the grill lines and Subtract it from the body of the object easily. You would have to repeat the same operation as many times as grill lines you have.

SUMMARY

Section 6.1

- Follow Me lets you create all kinds of different shapes.
- Bottles, spindles, and spheres are all examples of **lathed forms,** or 3D forms created by spinning a 2D shape around a central axis.
- You need three things to use Follow Me:
  - A path
  - A face
  - Undo
- When you're using Follow Me to extrude a face along a path that consists of edges that already exist as part of your model, there are two things you should *always* do:
  - Make the rest of your model a separate group.
  - Make a copy of your extrusion path outside your group.

Section 6.2

- To resize objects with the scale tool, you select the geometry (edges and faces) in your model that you want to resize, then activate the Scale tool.
- You can't make a copy while you scale.

Section 6.3

- The Sandbox helps to model terrain.
  - The Sandbox is a collection of tools.
  - The Sandbox is in both free and Pro.
  - The Sandbox is hidden.
- Both From Contours and From Scratch create terrain objects that are groups.
- The Drape tool works a little like a cookie cutter; use it to transfer edges from an object down onto another surface, which is directly beneath it.

Section 6.4

- Brand new for SketchUp 8, the Solid Tools provide a completely new way for SketchUp modelers to work. Solid modeling operations (otherwise known as Boolean operations) give you the ability to create the shapes you need by adding or subtracting other shapes to or from each other.
- Six things you need to know about solids; you can think of them as the Solid Rules:
  - A solid is nothing more than an object that's completely enclosed.
  - No extra edges or faces allowed.
  - Only groups and components can be solids.
  - Making a solid does not require any special tools.
  - Solids have volumes.
  - Solids can be multiple objects.

ASSESS YOUR UNDERSTANDING

SUMMARY QUESTIONS

1. At its core, Follow Me lets you create forms that are:
   a. Lathes
   b. Extrusions
   c. Paths
   d. Faces
2. When using Follow Me along a path that consists of edges that are already part of your model, you should always:
   a. Make sure the rest of your model is not a separate group.
   b. Delete these edges first.
   c. Rotate the edges around a central axis.
   d. Make a copy of your extrusion path outside your group.
3. Imagine that we want to build an armchair with the following description: four revolving legs, softened edges on the upholstered seat and back, and a bent metal tubing armrest. What single tool would we use to make those shapes, after using the Push/Pull and the Scale tools?
4. What two keys would we have to hold down to either scale proportionally or scale about the center of our selection?
5. What combination of tools can we use to make organic forms, such as a banana or an animal?
6. The Sandbox Tool can only be found and used in the Pro version. True or false?
7. After we activate the Smoove tool, what controls the size of the lumps, bumps, and dimples of our terrain?
8. What tool would be use to create paths and roads on a terrain, just like a cookie cutter would do?
9. The tool that we would use to "take a bite" out of another is:
   a. Subtract
   b. Trim
   c. Intersect
   d. Outer shell
10. What does the Stamp tool do in an undulating terrain?

APPLY: WHAT WOULD YOU DO?

1. Name and describe the three things that you need to use Follow Me.
2. You've been asked to create a model of a wine bottle. Describe the steps that would take you to do it, naming the tools that you would use on each phase
3. You've made a model of a vase, but you have many visible edges in your model. How would you smooth these edges out to make them invisible?
4. What is the trick to draw an extrusion profile that is not on the ground?
5. Where can we find the Solid Tools?
6. Explain the meaning of the offset distance of the Stamp tool.
7. What tool would we use to add detail to a terrain surface, only on the areas where we really more precision?

BE A DESIGNER

Balcony Railing

Take your model of a house and create a balcony that consists of a flat platform and a handrail. Write down the list of steps you take to accomplish this as you are completing this task.

Add to Your House

Within the model of your house, add four wine bottles, three trash baskets, and one bar of soap. Write down the steps you take to create each one of these objects.

KEY TERMS

Lathed form

CHAPTER

7 KEEPING YOUR MODEL ORGANIZED

Using the Outliner and Layers

Do You Already Know?

- How to work with the Outliner and Layers?
- That you could use the Outliner to control visibility?
- That you could move one layer to another through the Entity Info Dialog box?

For the answers to these questions, go to **www.wiley.com/go/chopra/googlesketchup2e**

What You Will Find Out	What You Will Be Able To Do
7.1 How to understand your organizing options.	Differentiate between Outliner and Layers.
7.2 How to work with Outliner and Layers.	Apply Outliner and Layers to your model.
7.3 How to put it all together.	Apply Outliner and Layers on roofs and walls.

INTRODUCTION

As everybody knows, living life can be a messy ordeal, and modeling in SketchUp is no exception. As you crank away at whatever it is you're building, you will reach a time when you stop, orbit around, and wonder how your model got to be such a mess. It's inevitable.

Luckily, SketchUp includes a number of different ways to keep your geometry (edges and faces) from getting out of control. Because big, unwieldy, disorganized models are a pain to work with—they can slow your computer, or even cause SketchUp to crash—so you should definitely get in the habit of "working clean."

In this chapter, you will compare and contrast two main tools that SketchUp provides for organizing your model. In the first section, both tools are outlined. Later, the details of each one are presented, describing how to use them and how *not* to use them. This chapter ends with a detailed example of how you can use both tools together for a better SketchUp work experience through the modeling of a house.

7.1 TAKING STOCK OF YOUR ORGANIZATION OPTIONS

GOOGLE SKETCHUP IN ACTION

Learn the definitions of Outliner and Layers.

When it comes to sorting out the thousands of edges and faces in your model, it's all about lumping things together into useful sets. After you've separated things out, you can name them, hide them, and even lock them so that you (or somebody else) can't mess them up.

If you haven't read about groups and components yet, now would be a good time to take a look at Chapter 5. The content in this chapter is best understood if you have a firm grasp on the information provided in Chapter 5.

Outliner
A dialog box that lists all of the groups and components in a model.

Layers
A collection of geometry (including groups and components) that can be made visible or invisible all at once.

- **Outliner:** The **Outliner** dialog box is basically a fancy list of all the groups and components in your SketchUp model. It shows you which groups and components are nested inside other ones, lets you assign names for them, and gives you an easy way to hide parts of your model that you don't want to see. If you use a lot of components (and you should), the Outliner may well become your new best friend.
- **Layers:** All of this process is done with the idea to have an organized file with components and elements that can be easily turned on and off and that can be easily found. You will also avoid untimely computer crashes right when you are ready to share your model with the client.

A word of caution: In SketchUp, using layers the wrong way can seriously mess up your model. If you plan to use them, be sure to read the section "Discovering the Ins and Outs of Layers," later in this chapter.

TIPS FROM THE PROFESSIONALS

Working Efficiently with the Outliner Dialog Box

Getting to know how to work effectively with the Outliner dialog box can be a deal breaker when you are working with a complicated model. For example, if you are designing an entire interior of a residence and you are placing hundreds of components in it, in the living room, kitchen, bathroom, etc., you have to find a way to organize your Components effectively; the Outliner will give you that flexibility.

As you keep on placing one component after another in your model, your file will progressively slow down your computer, especially if you are constantly orbiting and zooming in and out in full color and with the texture mapping visible. So in order to streamline your workflow, you should do the following:

1. Open the Outliner and right-click and rename each component so that each one has a name that you can identify clearly.
2. Group your components by room—you can right-click on them and rename each one using room names.
3. To turn them on and off, right-click and choose hide or unhide—depending on what room you are working on.

This technique will work if your number of components is not over, let's say, thirty elements. For larger numbers you will have to start considering moving all of the furniture pieces into a different layer (that gets explained later in this chapter).

All this process is done with the idea to have an organized file with components and elements that can be easily turned on and off and that can be easily found, and also you will avoid untimely computer crashes right when you were to share your model with the client.

SELF-CHECK

1. What dialog box do we use to control groups and components?
2. Layers can be turned on and off whenever we want. True or false?

Apply Your Knowledge Taking into consideration the information provided in the "Tips from the Professionals" above practice placing furniture in a room and naming each piece using the Outliner dialog box.

7.2 USING THE OUTLINER

Most halfway-complicated SketchUp models consist of dozens, if not hundreds, of groups and components. These groups and components are nested inside each other and a lot of them are heavy, computer-killing behemoths like 3D trees and shrubs.

Apply Outliner and Layers to your model.

Without a list, how are you going to manage all your groups and components? How are you going to keep track of what you have, hide what you don't want to see, and (more importantly) *unhide* what you *do* want to see? The answer to these questions lies in the Outliner dialog box.

7.2.1 Taking a Good Look at the Outliner

You can open the Outliner dialog box by choosing Window ⇨ Outliner. Figure 7-1 shows what this box looks like when a model consists of a simple room with some furniture in it. Each piece of furniture is a separate component that was downloaded from the Google 3D Warehouse. Read Chapter 5 for more information on the Google Warehouse.

Figure 7-1

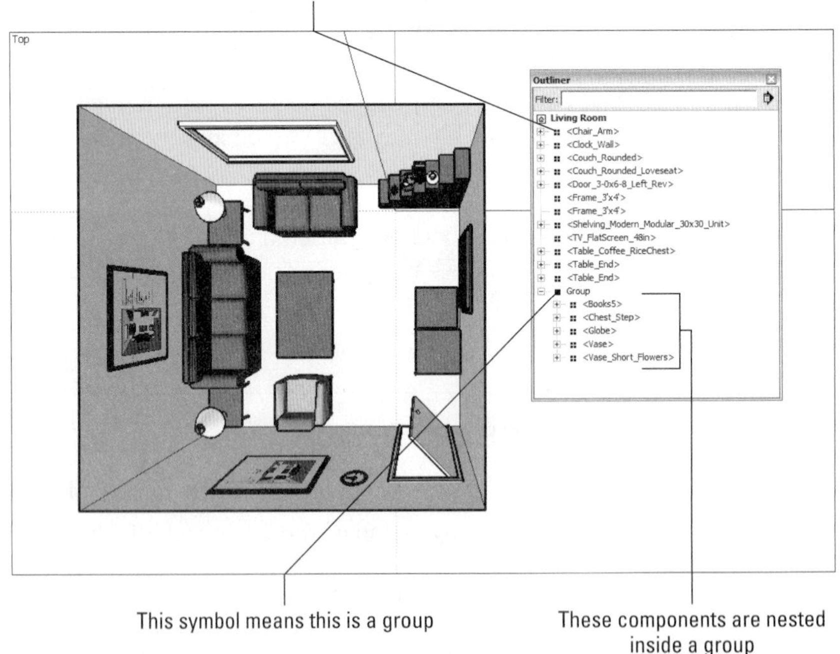

The Outliner, when there are a few components within a model.

The Outliner dialog box has the following features:

- **Search filter box:** If you type a word or phrase into this box, the Outliner will only show the items in your model that include that word or phrase in their name. For example, if you were to type in *coffee*, only the coffee table component would be visible.
- **Outliner Options flyout menu:** This handy little menu contains three options:
  - **Expand All:** Choose this option to have the Outliner show *all* the nested groups and components in your model (provided they're on visible layers).

    The Outliner shows only groups and components that exist on layers that are visible in your model. In other words, anything on a hidden layer does not appear in the Outliner, so be extra careful if you are using both the Outliner and layers to organize your model. You can read more about layers in the "Discovering the Ins and Outs of Layers" section later in this chapter.
  - **Collapse All:** This option collapses your Outliner view so that you only see *top-level* groups and components—ones that aren't nested inside other groups and components.
  - **Sort by Name:** Select this option to make the Outliner list the groups and components in your model alphabetically.
- **Outliner List window:** This is where all the groups and components in your model are listed. Groups and components that have nested groups and components inside them have an Expand/Collapse toggle arrow next to their names. When they're expanded, their constituent groups and components appear as an indented list below them.

CAREER CONNECTION

Thom Freeman, AIA, LEED, likes to use Google SketchUp at BSA+A Architecture/Interior Design because he can quickly create and present designs to clients. He says, "It has become my sketch book." He likes to capture still images, but sometimes he ventures into animations. In his own words, "Clients love it, as it helps them visualize; which is a skill most people do not have."

The process of organizing the material for his presentation often varies from project to project. Sometimes he imports CAD files into Google SketchUp and into Artlantis and/or Adobe Photoshop, but it depends on the specific needs and the desired result in each individual project. Other times he exports SketchUp files into CAD for further refinement. Recently, he collaborated in a project that was completely designed in SketchUp. His design included large polycarbonate panels that were cut from a CAD file that was originally a SketchUp file.

7.2.2 Making Good Use of the Outliner

If you're going to use lots of groups and components (and you should), having the Outliner open on your screen is one of the best things you can do to model efficiently. Here's why:

- **Use the Outliner to control visibility.** Instead of right-clicking groups and components in your model to hide them, use the Outliner instead. Just right-click the name of any element in the Outliner and choose Hide. When you do, the element is hidden in your modeling window, and its name is grayed out and italicized in the Outliner. To unhide it, just right-click its name in the Outliner and choose Unhide.
- **Drag and drop elements in the Outliner to change their nesting order.** Don't like having the component you just created nested inside another component? Simply drag its name in the Outliner to the top of the list. This moves it to the top level, meaning that it's not embedded in anything. You can also use the Outliner to drag groups and components into or out of other ones, too.
- **Find and select things using the Outliner.** When you select something in the Outliner, its name gets highlighted, and it gets selected in your modeling window. This is a much easier way to select nested groups and components, especially if you're working with a complex model.

SELF CHECK

1. How do we control the nesting order of our components in the Outliner?
2. When we highlight a component in the Outliner box it gets deselected in the modeling window. True or false?
3. Outliner shows groups and components that exist on layers that are either visible or invisible in your model. True or false?

Apply Your Knowledge How would you control visibility of the components in one of your design?

7.2.3 Discovering the Ins and Outs of Layers

As was mentioned in the introduction, layers are a very useful part of SketchUp, and they can make your life a lot easier. Layers can also be a major source of heartache because they can really mess up your model of you're not careful. This section helps you avoid problems when using layers.

In a 2D program like Photoshop or Illustrator, the concept of layers makes a lot of sense: You can have content on any number of layers, sort of like a stack of transparencies. You find a distinct order to your layers, so anything that's on the top layer is visually "in front of" everything on all the other layers. Figure 7-2 shows what this means.

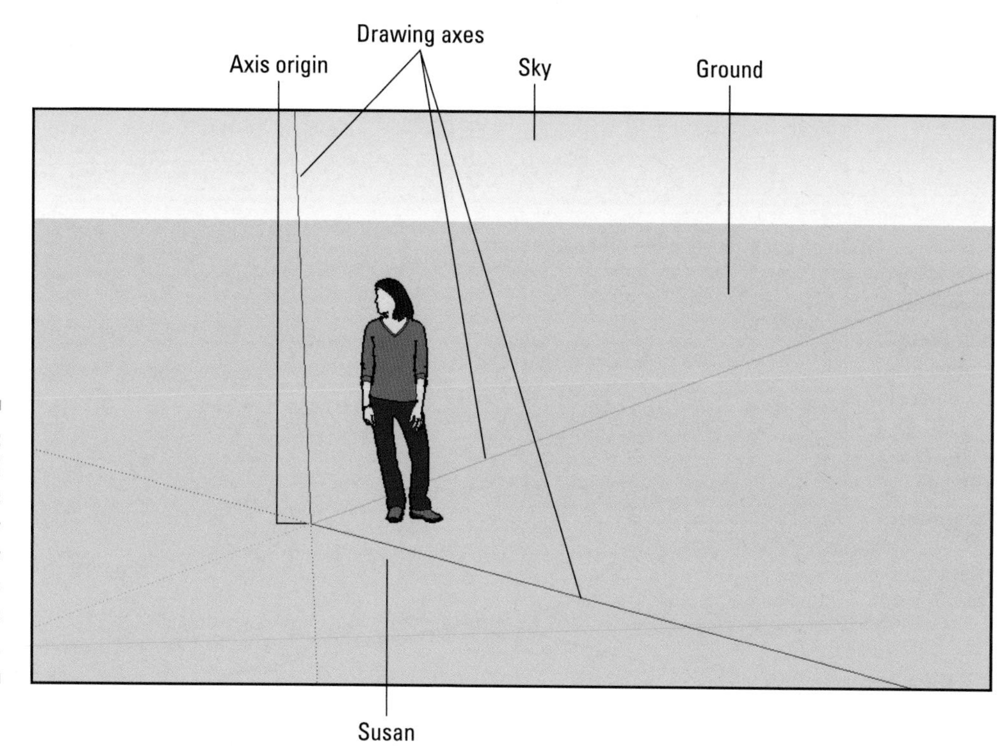

Color Plate 1: The colored drawing axes are the key to how SketchUp works (Chapter 2).

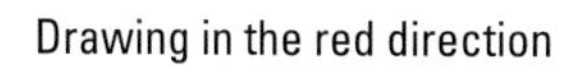

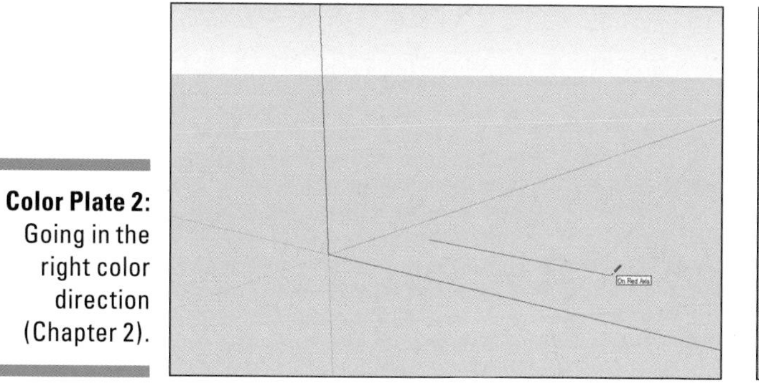

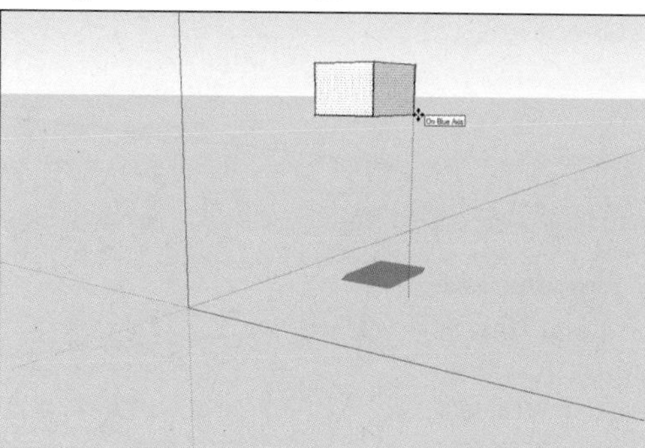

Color Plate 2: Going in the right color direction (Chapter 2).

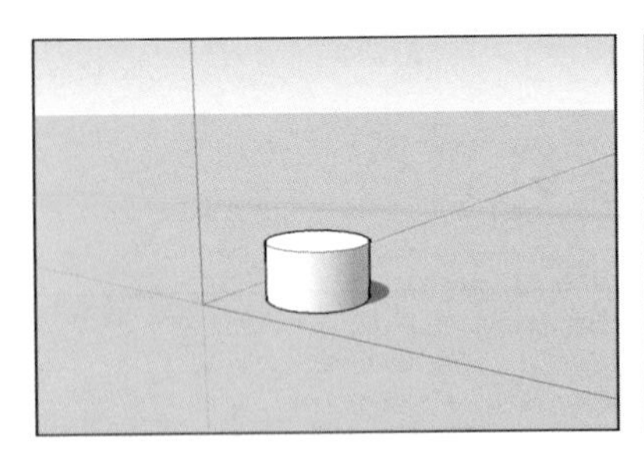

To go up, move in the blue direction

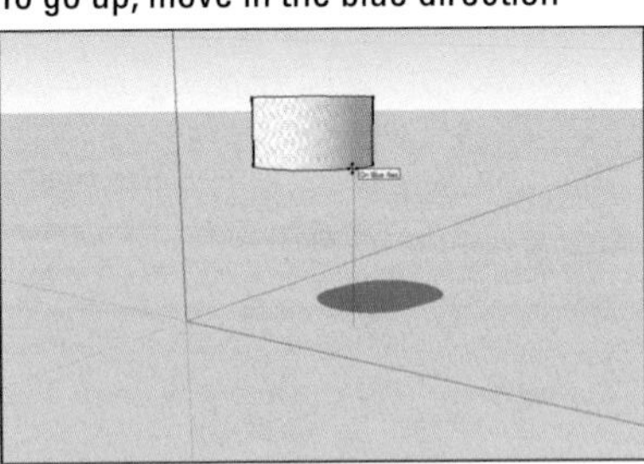

To go back, move in the green direction

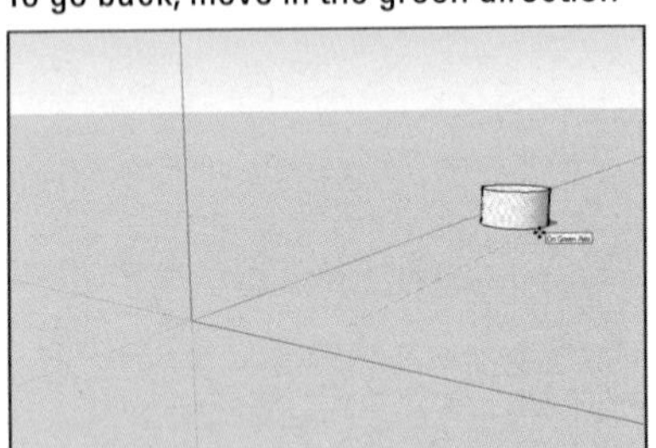

Color Plate 3: Moving an object in 3D on a 2D screen (Chapter 2).

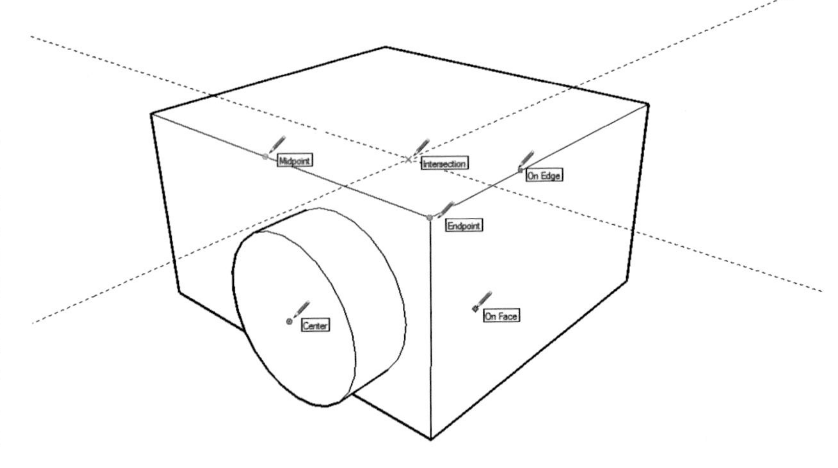

Color Plate 4: Point inferences are colored by type and help you model more precisely (Chapter 2).

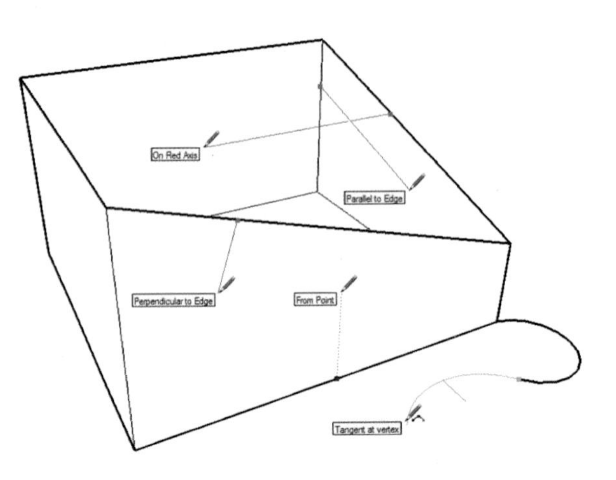

Color Plate 5: Linear inferences are color coded and help you draw in 3D, too (Chapter 2).

Hold down Shift to lock yourself in the blue direction

Hover over the point to which you want to infer; then click to end your edge

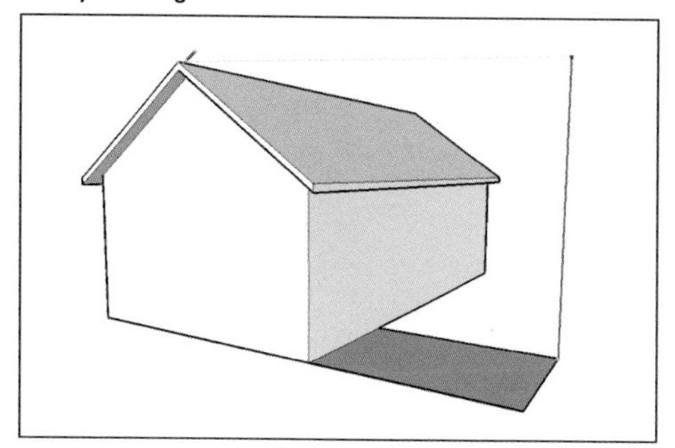

Color Plate 6: Locking an inference helps you draw in the right direction (Chapter 2).

Hover over the center point inference

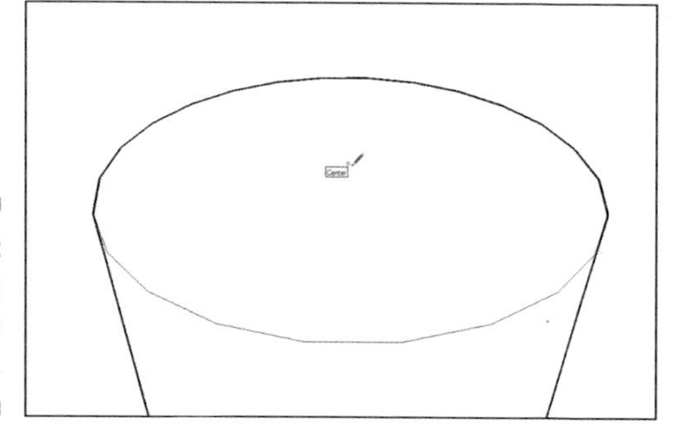

Move slowly away in the red direction

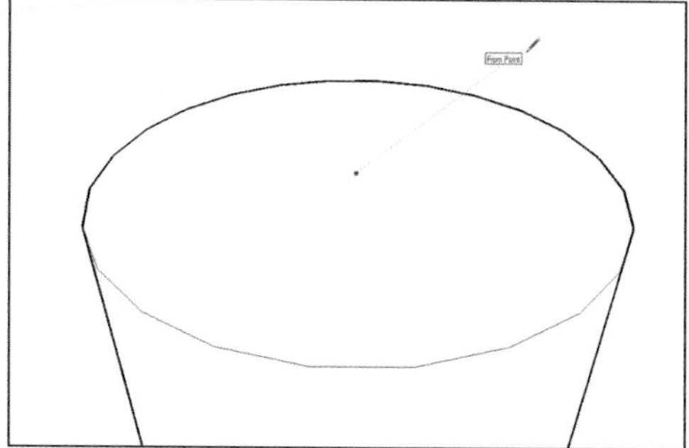

Color Plate 7: Encouraging an inference (Chapter 2).

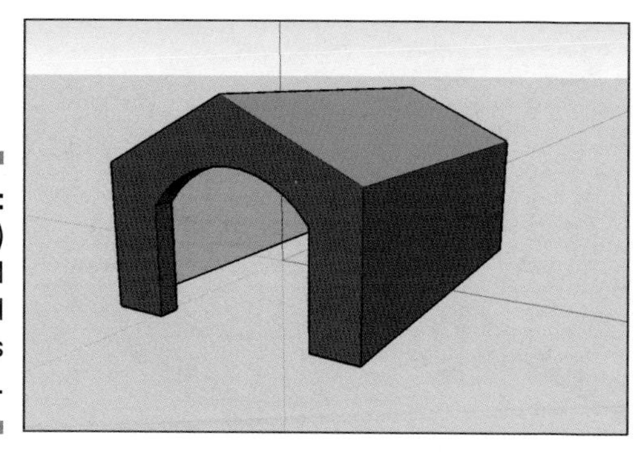

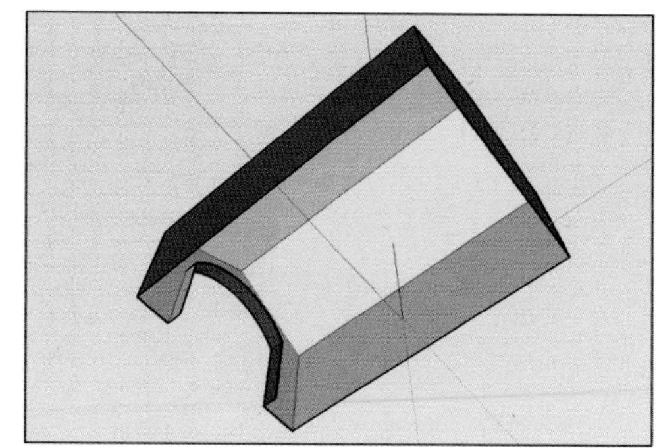

Color Plate 8: Orbit (spin) your model to paint all the faces (Chapter 3).

Anatomy of a Completed House Model

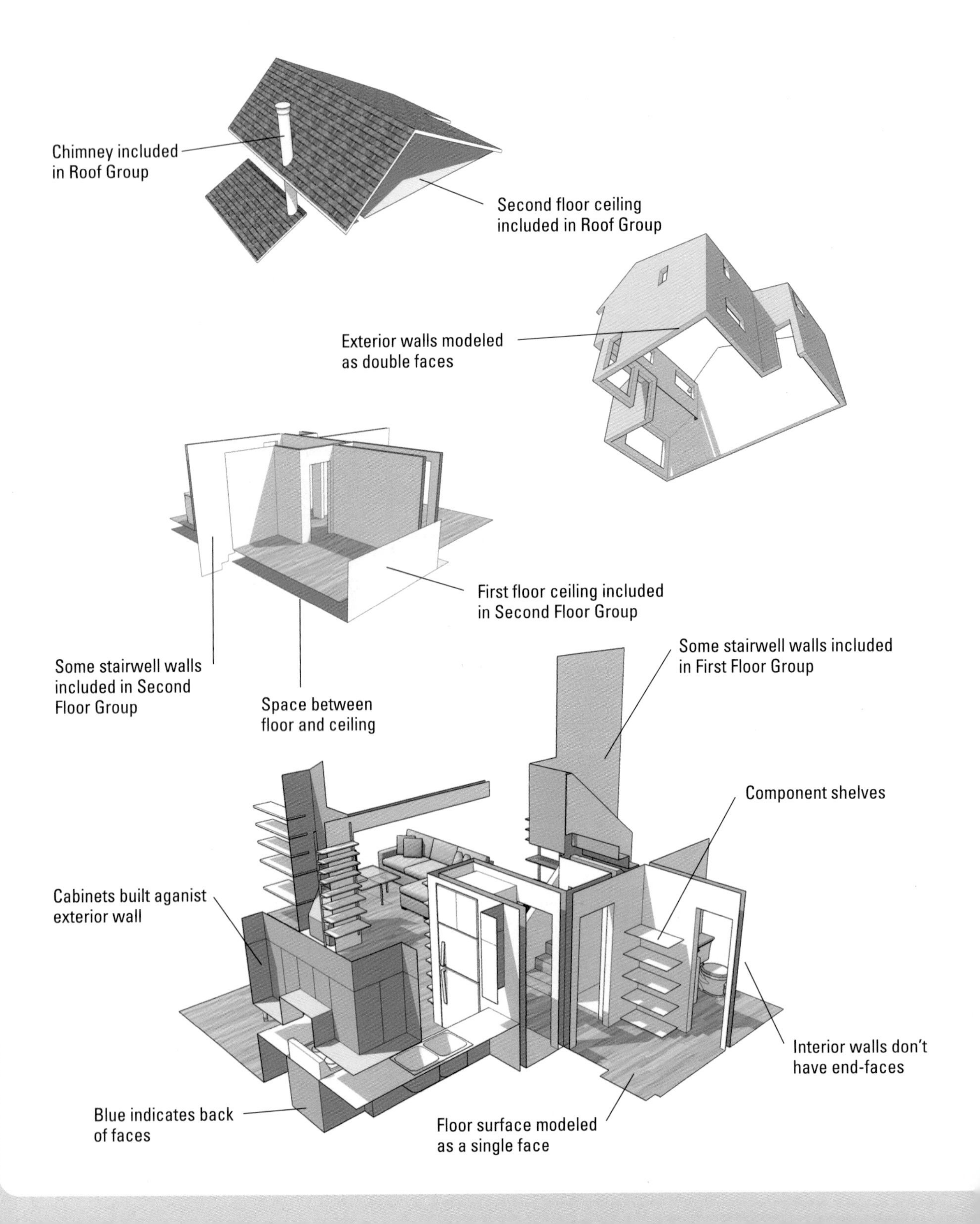

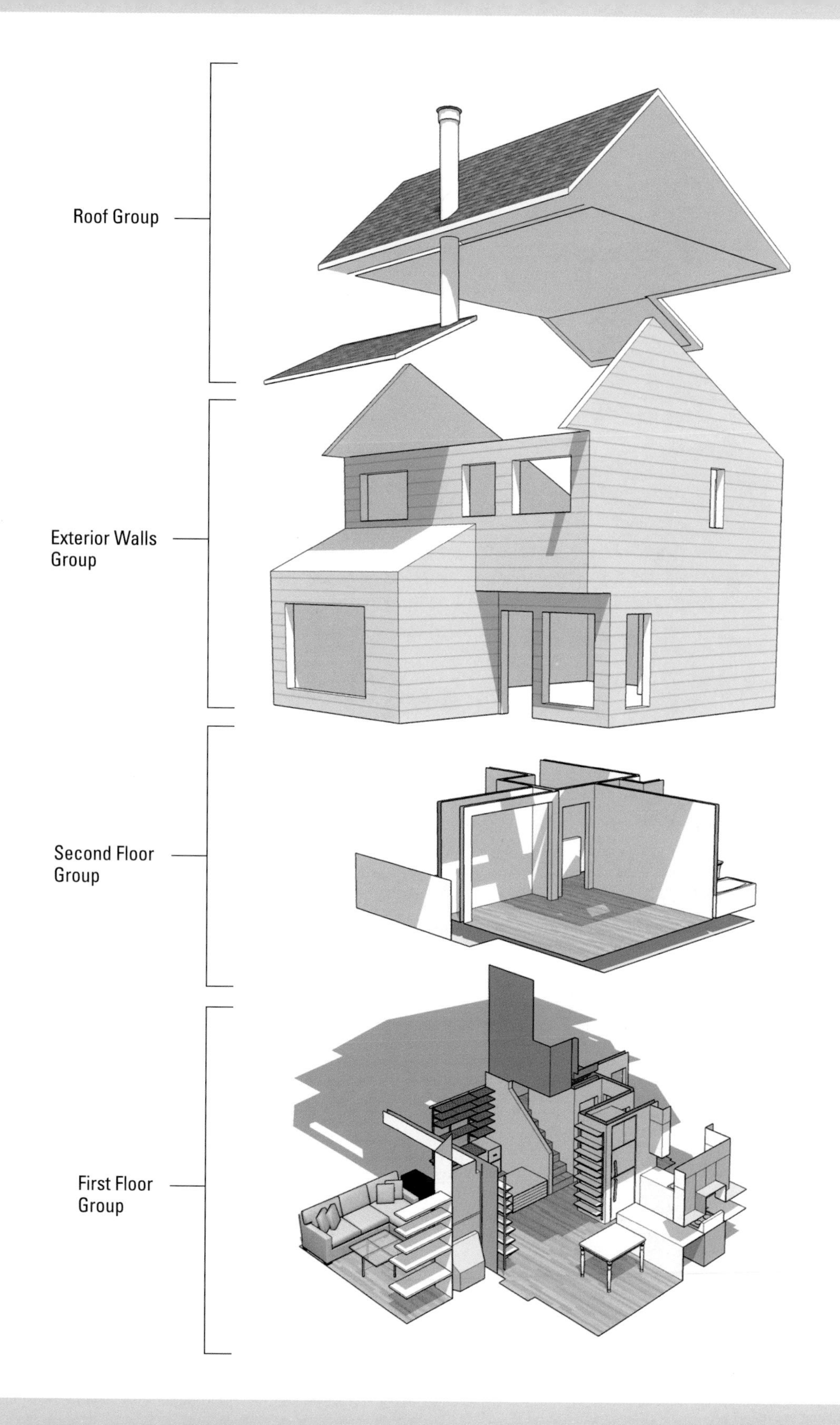
Roof Group
Exterior Walls Group
Second Floor Group
First Floor Group

Building a Circular Stair with Components

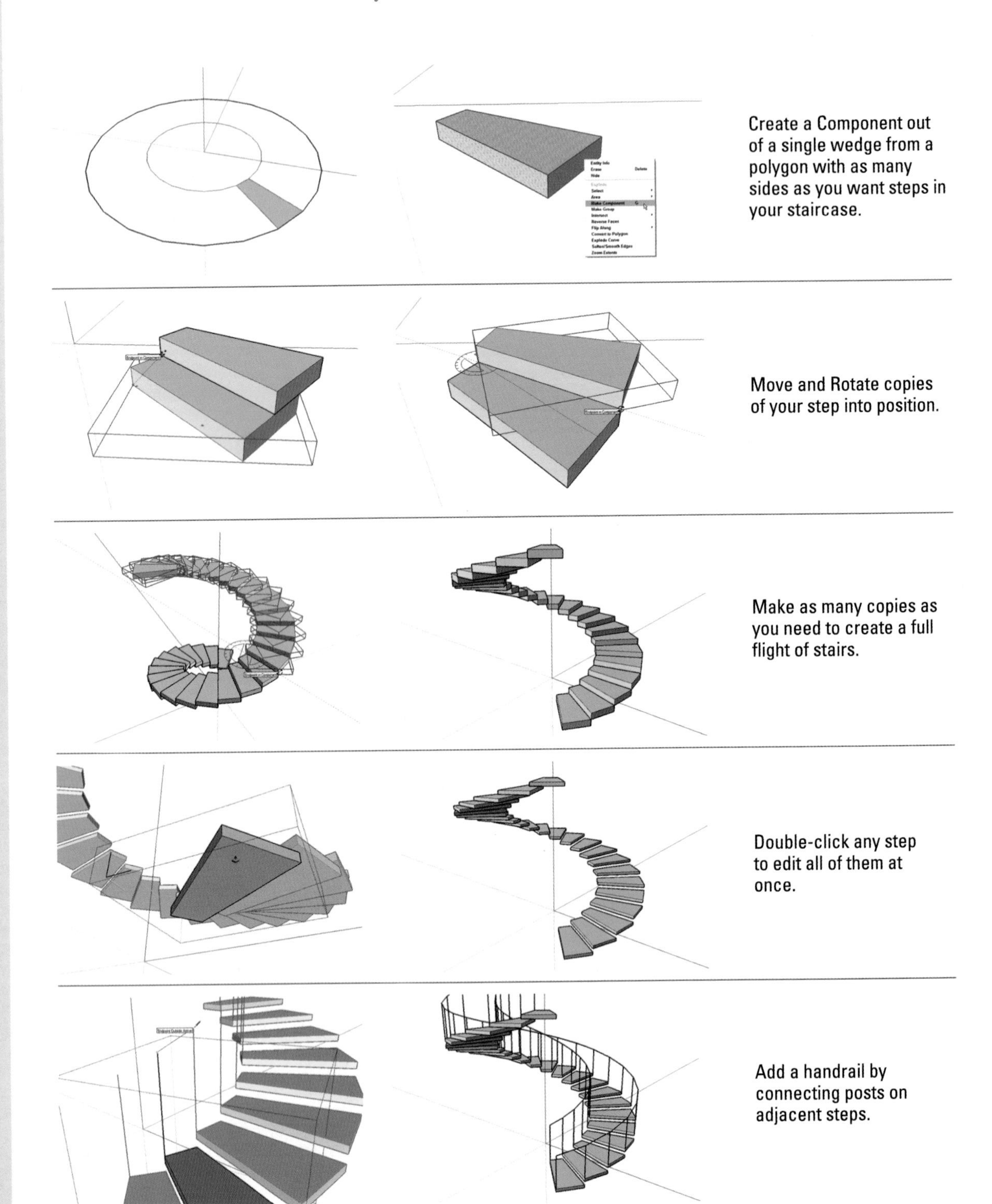

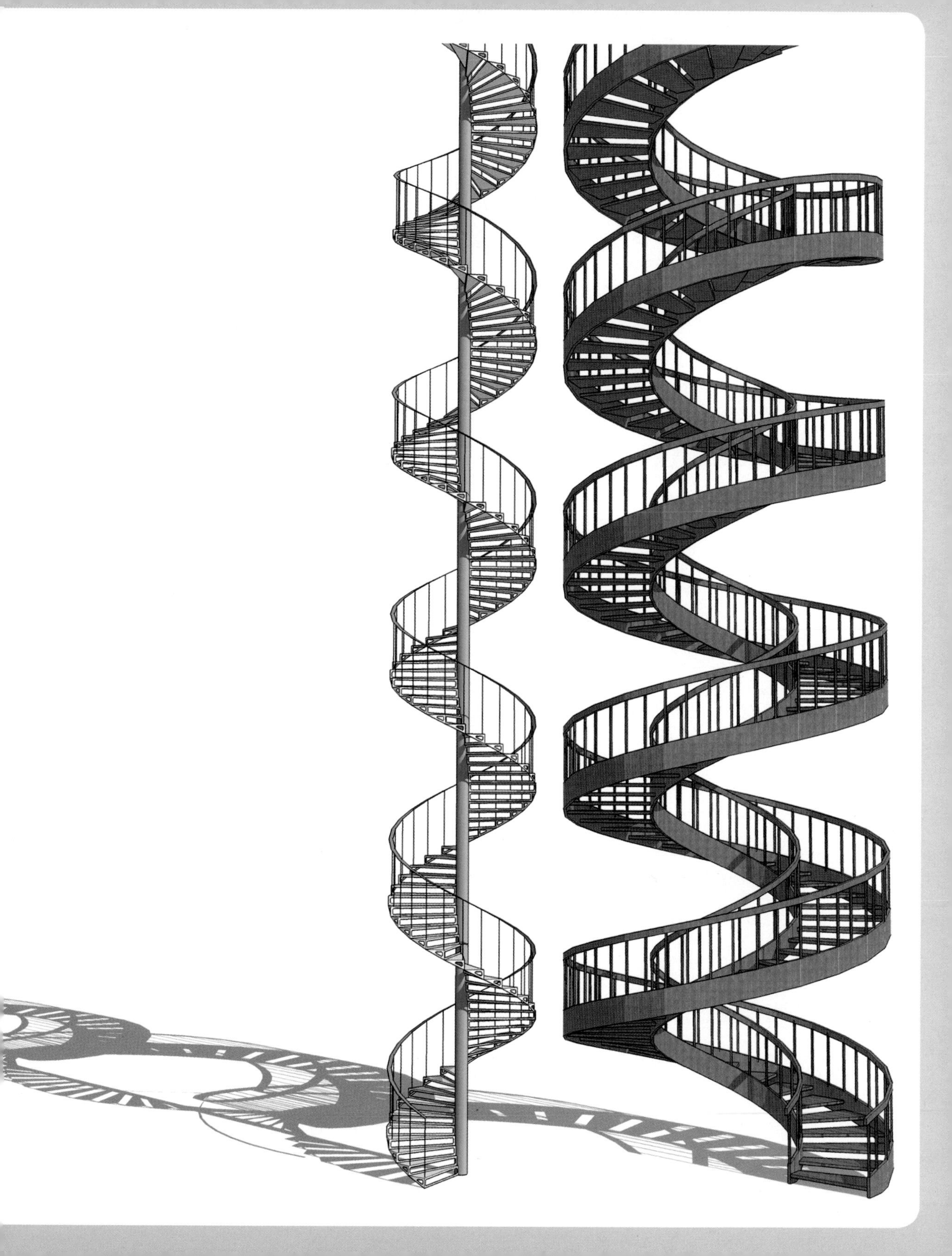

Building a Simple Boat with Mirrored Components and the Scale Tool

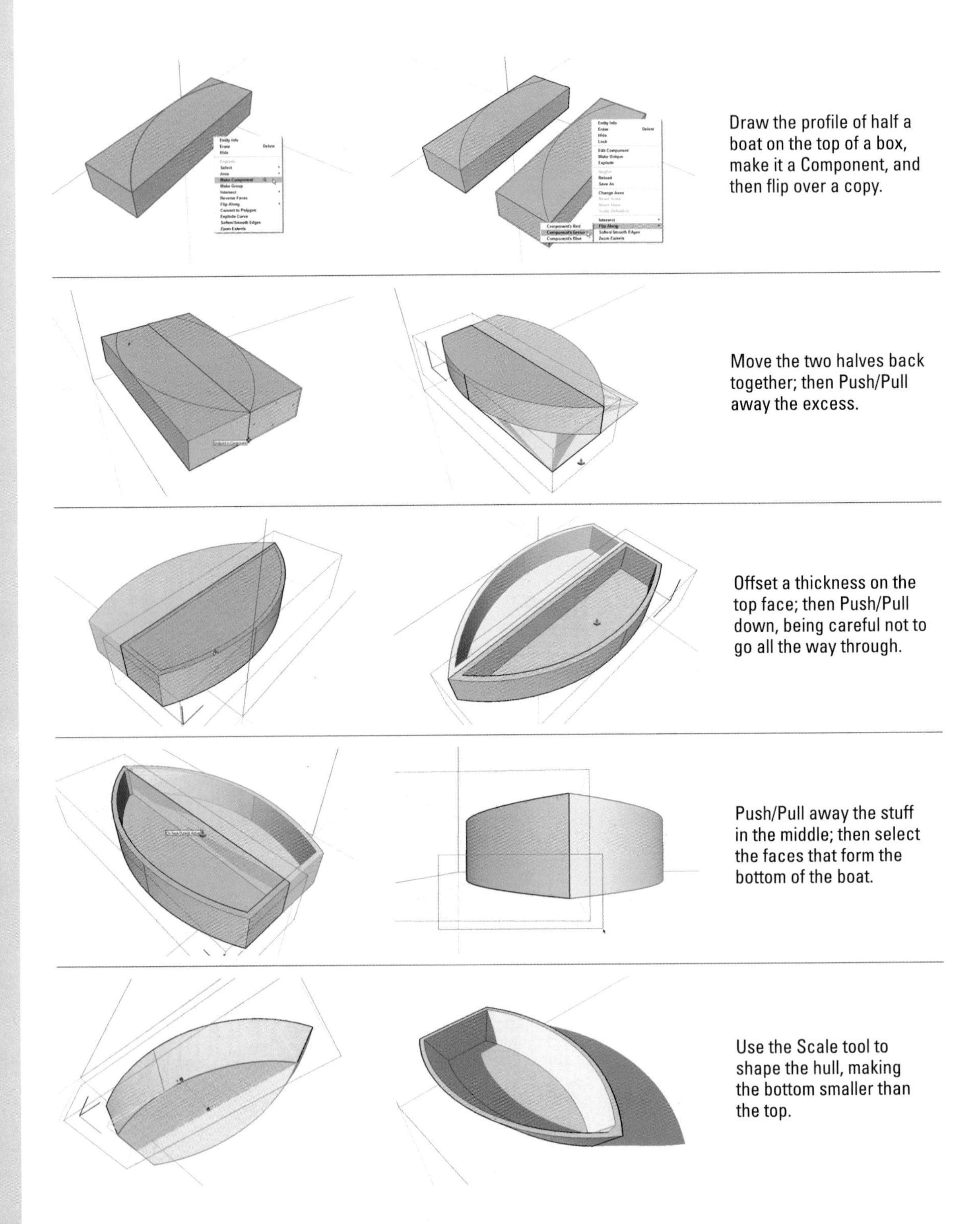

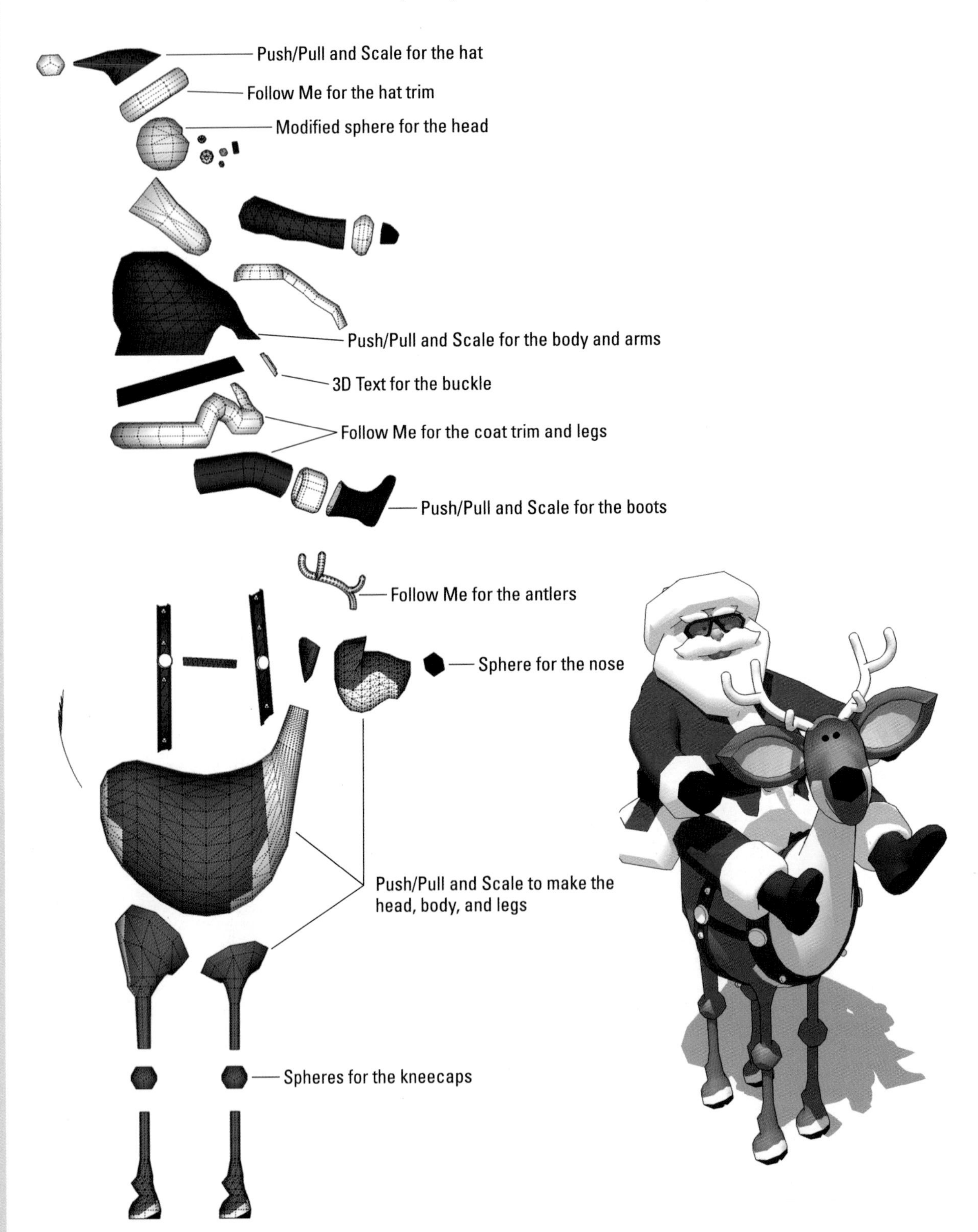

Push/Pull and Scale for the hat
Follow Me for the hat trim
Modified sphere for the head
Push/Pull and Scale for the body and arms
3D Text for the buckle
Follow Me for the coat trim and legs
Push/Pull and Scale for the boots
Follow Me for the antlers
Sphere for the nose
Push/Pull and Scale to make the head, body, and legs
Spheres for the kneecaps

Color Plate 9: A transparent version of your image, along with four colored pins, appear when editing textures (Chapter 8).

Color Plate 10: The Match Photo interface shows your picture, plus tools to create a model from it (Chapter 8).

Color Plate 11: Lining up the Perspective bars (Chapter 8).

Color Plate 12: All four Perspective bars, properly lined up with edges in the picture (Chapter 8).

Color Plate 13: Placing the Axis Origin in a good spot (Chapter 8).

Color Plate 14: Using the grid lines to give your picture an approximate scale (Chapter 8).

Color Plate 15: Tracing an edge in one of the three main directions (Chapter 8).

Color Plate 16: Creating a face to match a surface in the photograph (Chapter 8).

Color Plate 17: Projecting textures from a picture onto a face; then Orbiting around to see the result (Chapter 8).

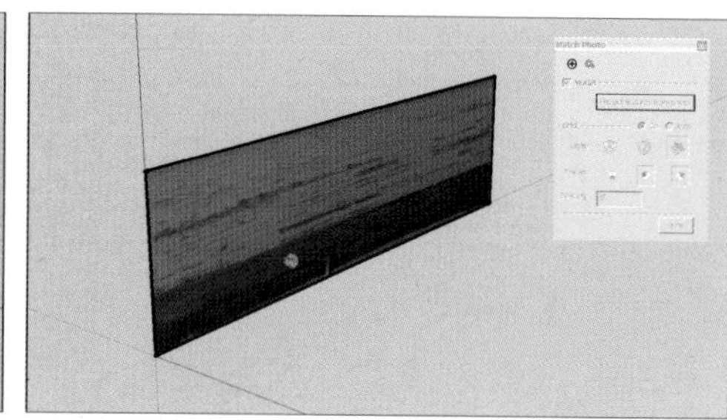

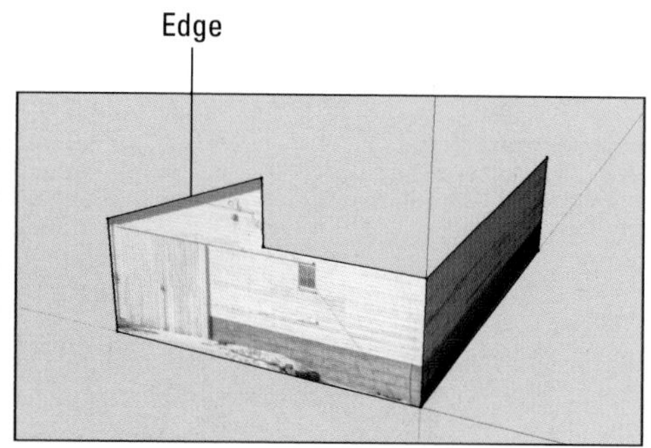

Color Plate 18: Using the endpoints of perpendicular edges to draw a diagonal (Chapter 8).

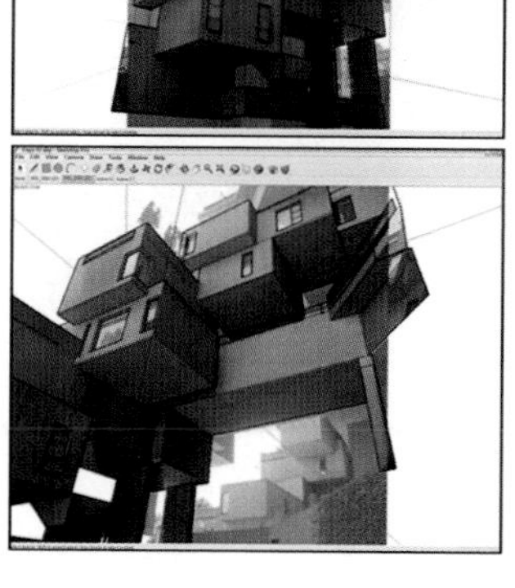

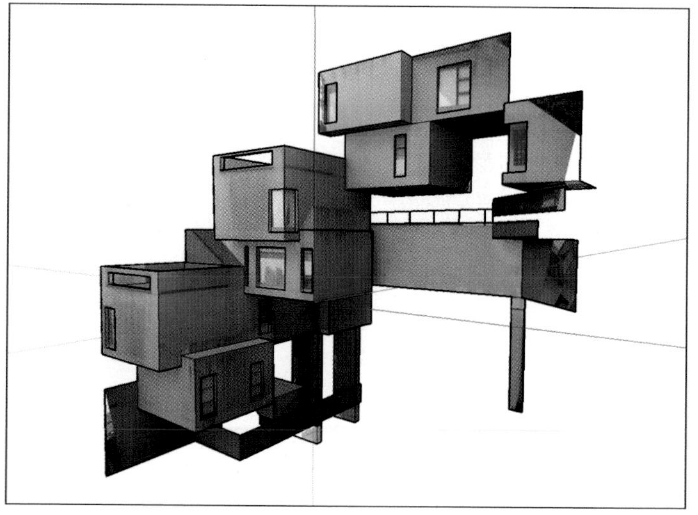

Color Plate 19: A model created from two matched photos (Chapter 8).

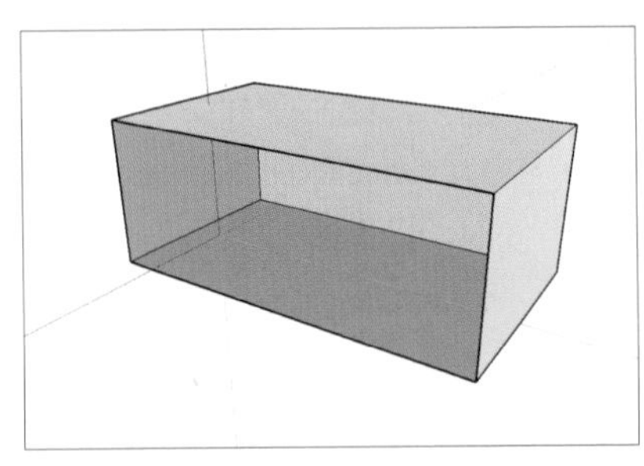

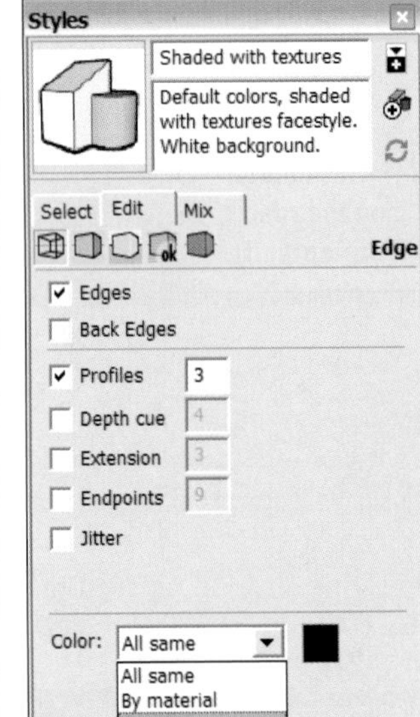

This edge should be blue

Their color tells me these edges aren't right

This edge should be green

Color Plate 20: Turning on Color by Axis helps you track down edges that aren't what they seem (Chapter 16).

Color Plate 21: Use Styles to make your model look any way you want (Chapter 9).

Figure 7-2

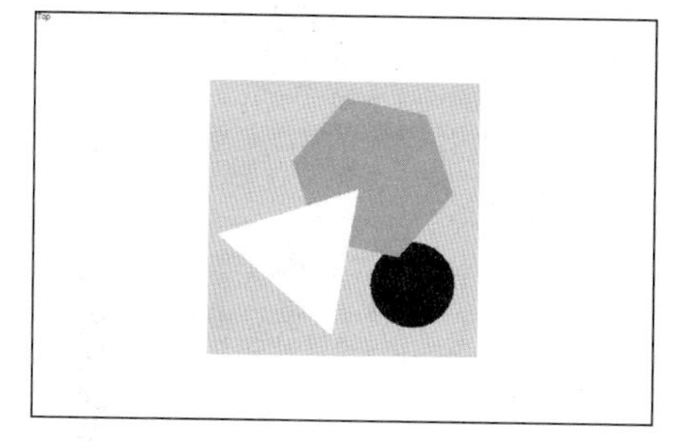

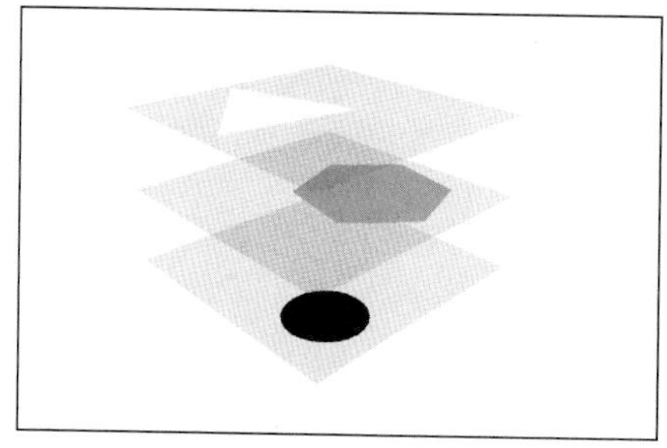

In 2D software, layers are pretty straightforward.

But hold on a second—SketchUp isn't a 2D program; it's a 3D program. So how can it have layers? How can objects in 3D space be "layered" on top of each other so that things on higher layers appear "in front of" things on lower ones? In short, they can't—it's impossible. This means that layers in SketchUp are different from layers in most other graphics programs, and that's confusing to lots of people.

SketchUp has a layers system because some of the very first SketchUp users were architects, and many, many architects and interior designers use drawing software called AutoCAD. Because AutoCAD uses layers extensively, layers were incorporated into SketchUp to maximize compatibility between the two products. When you import a layered AutoCAD file into SketchUp, its layers show up as SketchUp layers, which is rather convenient.

So what are SketchUp layers for? Layers are for controlling visibility. You use them to gather particular kinds of geometry so that you can easily turn that geometry on (make it visible) and turn it off (make it invisible) when you need to. That said, layers *don't* work the same way as groups and components; your edges and faces aren't isolated from other parts of your model, which can cause major confusion if you're not careful.

TIPS FROM THE PROFESSIONALS

Importing for the first time an AutoCAD file into SketchUp can be intimidating, as chances are you might encounter many layers. Fortunately, SketchUp will not merge all the layers automatically and with some patience you will be able to turn on and off just the ones you really need. In a typical workflow you will be using only your AutoCAD files to help you build and raise the walls on the perimeter of your structure and on the inside if you are creating a building in 3D.

7.2.4 Using Layers in SketchUp

You can find the Layers dialog box on the Window menu. It's a pretty simple piece of machinery, as you can see in Figure 7-3.

Figure 7-3

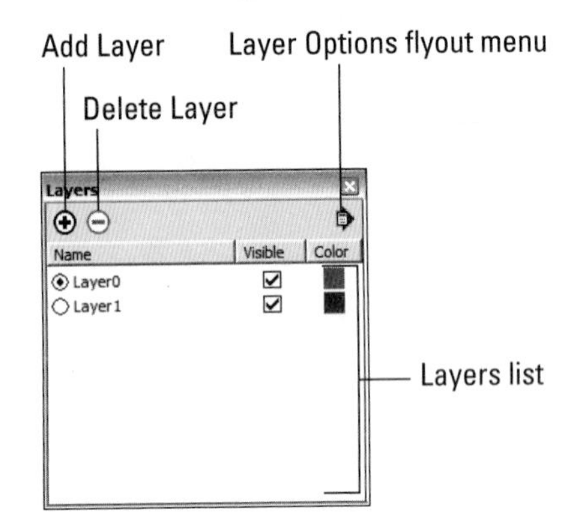

The Layers dialog box.

Here's what everything does:

- **Add Layer:** Click this button to add a new layer to your SketchUp file.
- **Delete Layer:** Click this button to delete the currently selected layer. If anything is on the layer you're trying to delete, SketchUp will ask you what you want to do with it; choose an option and select Delete.
- **Layer Options flyout menu:** This contains the following useful options:
  - **Purge:** When you choose Purge, SketchUp deletes all the layers that don't contain geometry. This is a handy way to keep your file neat and tidy.
  - **Color by Layer:** Notice how each layer in the list has a little color swatch next to it? Choosing Color by Layer temporarily changes all the colors in your SketchUp model to match the colors assigned to each layer. To see what's on each layer, this is the way to go.
- **Layers list:** This is a list of all the layers in your SketchUp file. You need to know the following about the three columns in this list:
  - **Name:** Double-click a layer's name to edit it. Giving your layers meaningful names is a good way to quickly find what you're looking for.
  - **Visible:** This check box is the heart and soul of the Layers dialog box. When it's selected, the geometry on that layer is visible; when it's not, it's not.
  - **Color:** You can choose to view your model using Color by Layer, which is described in the previous list. You can choose which color to assign to each layer by clicking the Color swatch.

Adding a New Layer

Follow these steps to add a layer to your SketchUp file:

1. **Choose Window ⇨ Layers.** This opens the Layers dialog box.
2. **Click the Add Layer button to add a new layer to the Layers list.** If you want, you can double-click your new layer to rename it.

Moving Entities to a Different Layer

Moving things from one layer to another involves using the Entity Info dialog box. Follow these steps to move an entity (an edge, face, group, or component) to a different layer:

1. **Select the entity or entities you want to move to another layer**. Keep in mind that you should only be moving groups and components to other layers; have a look at the next section in this chapter to find out why.
2. **Choose Window ⇨ Entity Info**. This opens the Entity Info dialog box. You can also open it by right-clicking your selected entities and choosing Entity Info from the context menu.
3. **In the Entity Info dialog box, choose a layer from the Layer drop-down list.** Your selected entities are now on the layer you chose from the list.

7.2.5 Staying Out of Trouble

Layers can be really helpful, but you need to know how to use them; if you don't, bad things can happen. Here's some more detail:

- **Do all your modeling on Layer0.** Always make sure that Layer0 is your current layer when you're working. Keeping all your "loose" geometry (that's not part of a group or component) together in one place is the *only* way to make sure that you don't end up with edges and faces all over the place. SketchUp, unfortunately, lets you put geometry on whatever layer you want, which means that you can end up with a face on one layer, and one or more of the edges that define it on another. When that happens, it's next to impossible to work out where everything belongs; you'll spend literally hours trying to straighten things out. This property of SketchUp's layers system is a major stumbling point for new SketchUp users; knowing to keep everything on Layer0 can save you a lot of anguish.
- **Only put groups and components on other layers.** If you're going to use layers, follow this rule: *Never* put anything on a layer other than Layer0 unless it's a group or a component. Doing so ensures that you don't end up with stray edges and faces on separate layers.
- **Use layers to organize big groups of similar things.** More complicated SketchUp models often include trees, furniture, cars, and people. These elements (or objects) are almost always already components, so they're perfect candidates for being kept on separate layers. For instance, you might make a layer called Trees and put all your tree components on it. This makes it easy to hide and show all your trees all at once. This will also speed your workflow by improving your computer's performance (trees are usually big, complicated components with lots of faces).
- **Don't use layers to organize interconnected geometry; use the Outliner instead.** "Interconnected geometry" means things like building floor levels and staircases. These are parts of your model that aren't meant to be physically separate from other parts (like vehicles and people are).

When you put Level 1 on one layer and Level 2 on another, more often than not, you'll get confused about what belongs where: Is the staircase part of Level 1 or Level 2? Instead, make a group for Level 1, a group for Level 2, and a group for the staircase.

Iteration
The process of doing multiple versions of the same thing.

- Feel free to use layers to iterate. **Iteration** is the process of doing multiple versions of the same thing. Lots of designers work this way to figure out problems and present different options to their clients. Using layers is a great way to iterate: You can move each version of the thing you're working on to a different layer, and then turn the layers on and off to show each version in turn. Just remember to follow the rule about only using groups and components on separate layers (mentioned previously), and you'll be fine.

Check out the sidebar "Using Scenes to Control Layers" (later in this chapter) for a nice way to quickly flip through layers that represent design iterations in your model.

IN THE REAL WORLD

Carlo Ferrando, head designer and founder of Mostaza Design in Madrid, Spain, designs stores, restaurants, and duty free shops. He states that he likes using SketchUp, "because it is quick, easy to use and the user interface is great." Quite often his clients ask him specifically to use this program and to avoid using photo-realistic images that you can obtain in other programs. Some other times he shares with his clients renderings done directly from SketchUp, because of budget constraints. In any case, the response from his clients is always positive as they understand his designs created in SketchUp.

Usually, Carlo imports an AutoCAD drawing into SketchUp and then he models it in three dimensions. Sometimes he uses 3ds Max to render high quality images if he has the budget for it.

SELF-CHECK

1. How do we change one or multiple entities from one layer to another?
2. When you import a layered AutoCAD file into SketchUp, its layers show up as Components. True or false?
3. What should we use to organize interconnected geometry, Outliner or Layers?

Apply Your Knowledge What layer should we use to do all of our modeling on?

7.3 PUTTING IT ALL TOGETHER

Apply Outliner and Layers on roofs and walls.

In this chapter (and in Chapter 5), is presented SketchUp's organizational methods in isolation: discussing how they work, why they're special, and when to use them. When you are actually working in SketchUp, you probably use a combination of them all, so an example of all the organization tools in action is especially helpful.

Figure 7-4 shows a model of a small house that we're building in SketchUp. Here, we're using all of SketchUp's organizational tools to help manage our model's complexity while we're working:

- **Each floor level is a group.** By working with each floor level as a separate group, we're able to use the Outliner to hide whichever one we're not working on. This makes it easier to see what we're doing. We're including the house's only staircase in the first floor group, because that turns out to be the easiest thing to do. We've also decided to include the interior walls on each level of the house in that level's group. Because we don't think we'll ever have to hide them, it wasn't worth making them a separate group. For what it's worth, the same thing probably applies to most buildings, unless you plan to study different floor plans with different interior wall arrangements.
- **The roof and exterior walls are groups inside of another group.** We want to be able to "remove" the roof and the exterior walls separately, so

Figure 7-4

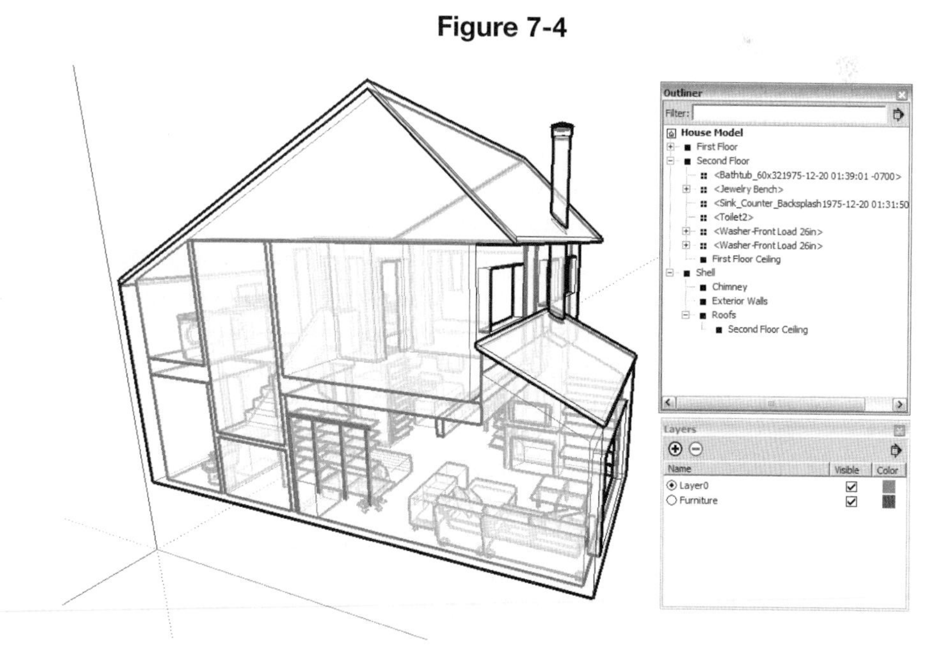

All of SketchUp's organizational tools are used to build this model.

Figure 7-5

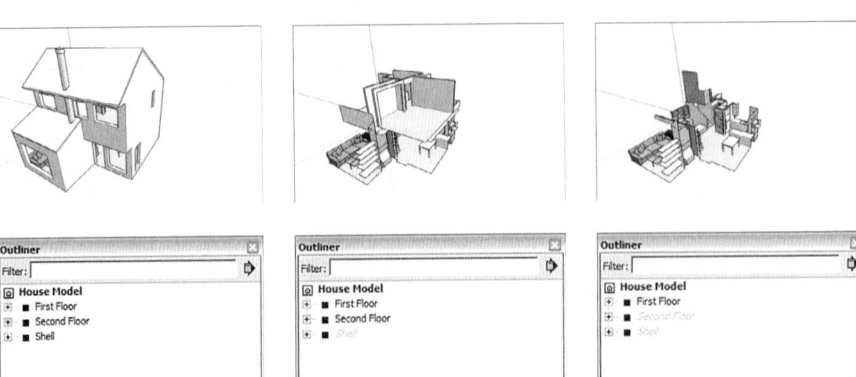

Each floor of the house, as well as the roof and the exterior walls, is a group.

we've made each of them a group. We also want to be able to hide and unhide them both at the same time, so we made a group called "Shell" that includes both of them. Using the Outliner, we can selectively show or hide just the geometry we want (see Figure 7-5). The floor levels, roof, and exterior walls of the house are groups instead of components because they're *unique*—in other words, because we only have one first floor, it doesn't need to be a component.

- **All the furniture and plumbing fixtures are components.** All the components used to furnish the house are ones we built ourselves, took from the Components dialog box, or found in the 3D Warehouse. But we only have one couch: Why make it a component instead of a group? By making every piece of furniture in the model a component, we're able to see a list of our furniture in the In Model collection of the Components dialog box. We can also save that as a separate component collection and use it in future models.
- **All the furniture is on a separate layer.** Because furniture components can be a little heavy (they're taxing on a computer system), and because we want to be able to see the house without furniture, we created a new layer (called "Furniture") and moved all the furniture onto it. Using the Layers dialog box, we can control the visibility of that layer with a single click of the mouse. But why not just create a group from all the furniture components and use the Outliner to hide and unhide them all, instead of bothering with layers? Good question. The answer is because it's easier to change a component's layer than it is to add it to an existing group. To add something to a group, we would need to use the Outliner to drag and drop it in the proper place; with complex models, this can be a hassle. Changing a component's layer is just a matter of using the Entity Info dialog box to choose from a list.

FOR EXAMPLE

USING SCENES TO CONTROL LAYERS

If you're reading this book from front to back, you haven't yet encountered any mention of SketchUp's Scenes feature—Chapter 10 is where you can go to read all about it. Without diving into too much redundant detail, scenes are basically saved views of your model. Instead of fiddling with navigation tools and dialog boxes every time you want to return to an important view, you can click a scene tab.

Scenes are relevant in this chapter because scenes don't just save different camera positions; you can also use them to control layer visibility. Being able to click a scene tab to instantly change which layers are showing is a crazy-powerful way to do iterative design: creating and presenting different options within the same design.

A very simple example: you've modeled a living room and want to try three different furniture configurations:

1. **Make three layers—Option 1, Option 2, and Option 3.**
2. **Do three separate furniture arrangements, one per layer.**

   Of course, this means that you have three copies of each object you move.
3. **Use the Layers dialog box to show Option 1, and hide Option 2 and Option 3.**
4. **Create a new scene and name it Option 1 using the Scenes dialog box.**
5. **Repeat Steps 3 and 4 for the other two configurations.**

Now all you have to do is click a scene tab to switch between the three options; this is much more elegant than having to fiddle with the Layers dialog box during a presentation. See the image below to get an idea of the setup.

The key to making this technique really sing is a working knowledge of how to use the Properties to Save check boxes in the Scenes dialog box. *Be forewarned*: This isn't beginner-level work. Hooking up scenes and layers takes practice, but after you get the hang of it, it's an elegant way to work. Pick your way through Chapter 10, when you're ready.

One more useful tidbit: The Layers Ruby script plugin lets you (among other things) create a new layer that isn't visible in any of the scenes you've made previously. This plugin comes in handy when you need to add a new iteration after you've already made a bunch of scenes; without the plugin, your new layer is visible in every scene, forcing you to manually go through and hide it in each one. Take a look at Chapter 16 for more information about the Layers Ruby script.

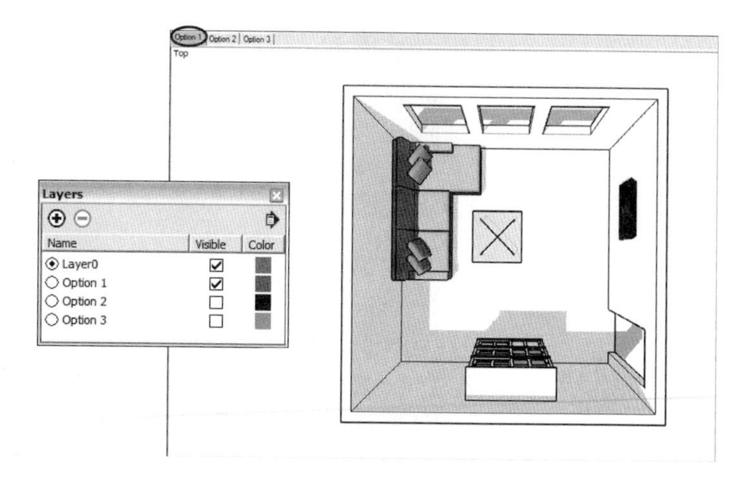

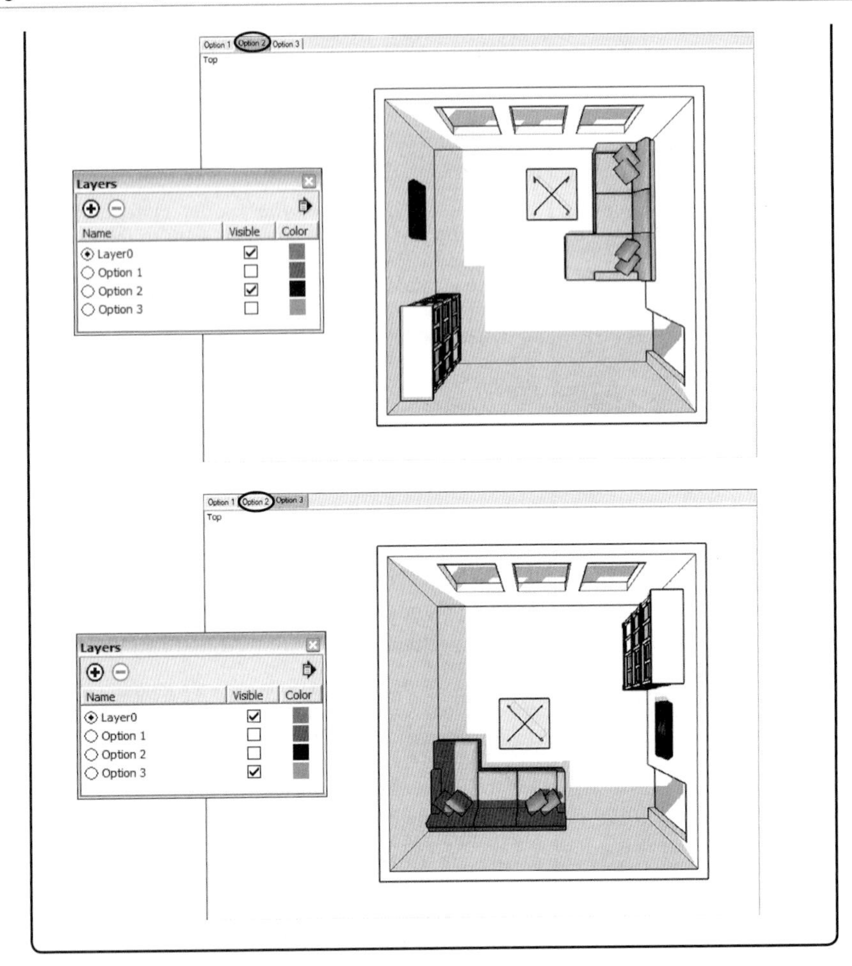

TIPS FROM THE PROFESSIONALS

Whenever you are building a structure with walls, it is a good idea to think for a moment if you want to group all of your walls as a single entity or make each segment of a wall a separate group, and then group those independent segments together into a larger group.

If you are dealing with a structure with small rooms, you might need to "step back through one wall" to gain enough distance to capture a good perspective view. If that is the case:

1. Right-click and hide the wall that you walked through
2. Leave the rest of the walls intact, to provide the framework for your perspective.

IN THE REAL WORLD

Rolf Carlsen works for a playground manufacturer and has always loved SketchUp because it has a low learning curve and it is quite affordable in comparison to other 3D CAD software products. He also likes that it has an active community of users that constantly develop a lot of handy add-ons.

Rolf likes to import AutoCAD files into SketchUp when he is involved in architectural visualizations and he uses many components to get his models ready. He employs the Photomatch option, or measures, as he goes on. After the whole model is complete, he applies textures and renders the final views with Podium with some post processing done in Adobe Photoshop. For presentations he uses Podium, under his "render scene" file, which contains different styles and shadow settings. In his own words, "At meetings I can quickly change the model which is a big time saver, and it helps me avoid any misunderstandings between the client and me." Rolf manages a very large model library with more than 4,000 product pictures, all done in SketchUp.

SELF-CHECK

1. Why is it usually best to gather the floor levels, roof, and exterior walls of a house in groups instead of components?
2. It is easier to change a component's layer than it is to add it to an existing group. True or false?
3. If we are building a house, why would we want to group the floor levels, roof, and exterior walls and not make them components instead?
4. What two tools would we use to for example create three different furniture configurations easily on a floor plan?

Apply Your Knowledge Using the previous exercise with the three rooms and the fifteen components, rebuild the outer walls so that each one is grouped separately. Then, right-click and hide one or two walls to capture a scene taken from farther away.

SUMMARY

Section 7.1

- The Outliner dialog box shows you which groups and components are nested inside other ones, lets you assign names for them, and gives you an easy way to hide parts of your model you don't want to see.
- Put different kinds of elements on different layers, name the layers, and then turn them on and off when you need to.

Section 7.2

- If you are going to use lots of groups and components, having the Outliner open on your screen is one of the best things you can do and here's why:
  - Use the Outliner to control visibility.
  - Drag and drop elements in the Outliner to change their nesting order.
  - Find and select things using the Outliner.
- Layers are for controlling visibility.
- You use layers to gather particular kinds of geometry so that you can easily turn that geometry on (make it visible) and turn it off (make it invisible) when you need to.

Section 7.3

- Scenes are basically saved views of your model. Instead of fiddling with navigation tools and dialog boxes every time you want to return to an important view, you can click a scene tab.

ASSESS YOUR UNDERSTANDING

SUMMARY QUESTIONS

1. Components are used when a model includes several copies of the same thing. True or false?
2. Which of the following is a characteristic of groups?
   a. Grouped geometry sticks to everything.
   b. Groups have no names.
   c. Groups cannot be moved.
   d. Ungrouping geometry requires that it be exploded.
3. When you use multiple instances (copies) of the same component, what happens if we update only one instance?
4. We should only be moving groups and components to other layers. True or false?
5. What tool do we use to search for a specific component from a long list in the Outliner?
6. Which of the following is true with regard to components?
   a. They keep file sizes down.
   b. They don't appear in the Outliner.
   c. Component instances must be updated manually.
   d. They cannot cut openings.
7. What would be the benefit of including all of the furniture elements of an office, in a single, dedicated layer?
8. Using components can help you keep track of _______ of items.
9. Which of the following options in the Outliner displays a list of all the groups and components in your model?
   a. Display
   b. List
   c. Expand all
   d. Library list
10. Which option should you use to make changes only to a few of the instances of a component in your model?
    a. Alter selectively
    b. Make unique
    c. Edit component
    d. Select one

11. How do we make the Outliner list all the components and groups alphabetically?
12. Just like we would do in many other CAD programs, it is recommended that we leave Layer0 intact and that we build in separate layers. True or false?
13. Why would be the benefit of working with Scenes and Layers together?

APPLY: WHAT WOULD YOU DO?

1. How do you move entities to a different layer?
2. Name two reasons you should use components and why you would want to create your own components.
3. What is Layer0 and when would you use it?
4. Build a simple, one-storey house or continue working on the house that you have been working on previous exercises, and design the living room and kitchen with two possible scenarios. Place each scenario in one separate layer and your walls on a separate one. Then, create different scenes to showcase easily both options.

BE A HOME DESIGNER

Doors and Windows

Within the model of the house that you've been working on, make your own doors and windows. What happens if the cutting boundary is messed up?

Organize Your Model

Create a model of a simple doghouse surrounded by trees, with one dog outside the doghouse. Using groups, components, the Outliner, and layers, organize the model so that you can easily show or hide certain subsets of objects.

KEY TERMS

Iteration
Layers
Outliner

CHAPTER

MODELING WITH PHOTOGRAPHS

Using Photo Match to Build Models

Do You Already Know?

- That you can paint faces in your model using photographs?
- About applying photo texturing on curved surfaces?
- How to build a model from scratch with SketchUp's photo-matching tools?
- That you can model on top of photo-textured faces?

For the answers to these questions, go to **www.wiley.com/go/chopra/googlesketchup2e**

What You Will Find Out	What You Will Be Able To Do
8.1 About the many techniques in working with photos.	How to paint faces, add, edit and optimize faces, photos, and textures.
8.2 The usage of photo-matching	Learn how to work with photo-matching.
8.3 What's important in modeling on top of photo textures.	Understand the importance of texture being projected.

INTRODUCTION

It's almost impossible to meet someone who doesn't take pictures. Aside from the millions of digital cameras out there, lots of mobile phones have cameras in them, too. In addition, more and more people are scanning existing photographs and saving them in digital form. In SketchUp, you can use photos from all of these sources in a couple of different ways:

- If you have a model you'd like to "paint" with photographs, you can do that in SketchUp. You can apply photos to faces and then use the information in the pictures to help you model; building windows is a lot easier when they are painted right on the wall.
- If you want to use a photo to help you model something from scratch, you can do that in SketchUp as well. Photo-matching makes it relatively simple to bring in a picture, set things up so that your modeling window view matches the perspective in the photo, and then build what you see by tracking with SketchUp's modeling tools.

One great application for *photo-textured* (painted with photos) models is Google Earth. You can make buildings and see them in Earth—you can even contribute models to Google Earth's default 3D Buildings layer where everyone can see them. This chapter talks about the techniques to use if that's where your models are headed. Read Chapter 11 for details about actually putting your models in Google Earth.

8.1 PAINTING FACES WITH PHOTOS

Mapping
Painting surfaces with pictures using 3D software.

Technically, painting surfaces with pictures using 3D software is called **mapping,** as in "I mapped a photo of your face to the underside of the pile-driver model I'm building." Different software programs have different methods for mapping pictures to faces, and luckily, SketchUp's are pretty straightforward.

This section deals with mapping photos to two different kinds of faces: flat ones and curved ones. The tools are similar, but the methods aren't so both methods are described

SketchUp uses lots of different terms to refer to the tools you can paint faces with; generically, they are all materials. Materials can be colors or textures; textures are image-based, and colors are a single, solid hue. When you import an image to map it to a face, it becomes a texture—just like any of the other textures in your Materials dialog box. Read more about using materials in SketchUp in Chapter 2.

GOOGLE SKETCHUP IN ACTION

How to paint faces, add, edit and optimize faces, photos, and textures.

8.1.1 Adding Photos to Faces

When mapping photos onto flat faces, you can choose the easy way or the hard way. Unfortunately, the hard way is the method you end up using the vast majority of the time, so it describe first. Importing images using the File menu lets you take any image and map it to any flat face in your model.

Figure 8-1

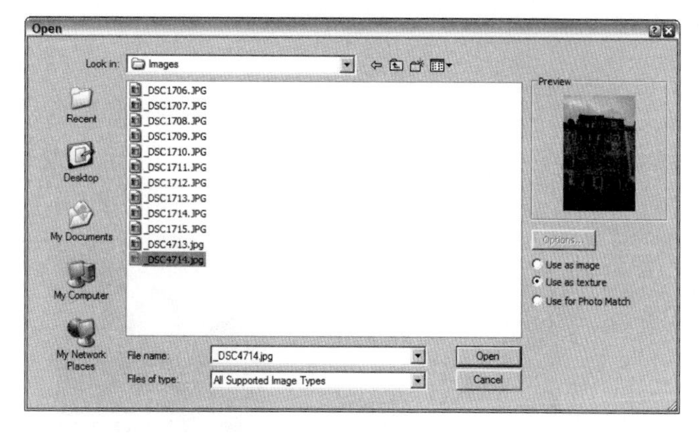

Make sure to pick "Use as texture."

The easy way, which is described later, is designed for one particular case: It gives you access to Google's vast collection of Street View imagery, letting you paint your models with building facades photographed by Google's roving fleet of specialized vehicles. The feature is cool, but also very specific.

Importing Images: Use Your Own Photos

Follow the steps in the Pathways to… section to map an image to a face (and find additional help on this book's companion Web site; see the beginning of this chapter for details):

Before you follow these steps, have at least one face in your model because you map your texture to a face.

Importing Images: Map an Image to a Face

1. **Choose File ⇨ Import.** The Open dialog box opens.
2. **Select the image file you want to use as a texture.** You can use JPEGs, TIFFs, PNGs, and PDFs as textures in SketchUp. All of these are common image-file formats.
3. **Select the "Use as texture" option (see Figure 8-1).**
4. **Click the Open button.** This closes the Open dialog box, switches your active tool to Paint Bucket, and "loads" your cursor with the image you chose to import.
5. **Click once in the lower-left corner of the face you want to "paint" (see Figure 8-2).** Where you click tells SketchUp where to position the lower-left corner of the image you're using as a texture. You can click anywhere on the face you're trying to paint, but using the lower-left corner keeps things simple.
6. **Click somewhere else on the face you're painting to locate the upper-right corner of the image texture (see Figure 8-2).** Image textures in

Figure 8-2

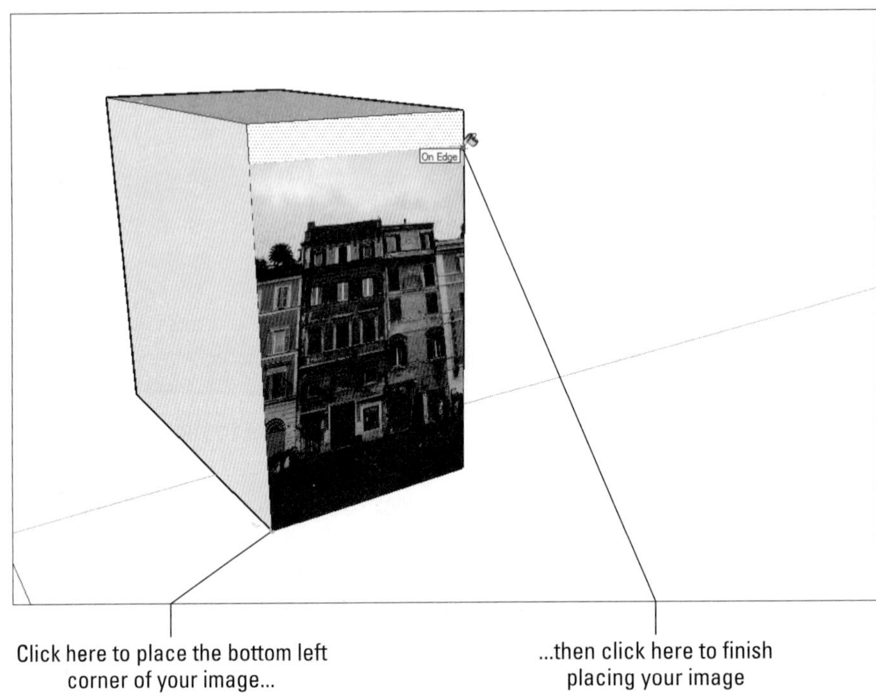

Click once to locate the lower-left corner of the image you're using as a texture, and then again to locate the upper-right corner.

SketchUp are made up of *tiles*. To make a large area of texture (like a brick wall), SketchUp uses multiple tiles right next to each other. In the case of a brick wall, it may look like there are thousands of bricks, but it's really just the same tile of about 50 bricks repeated over and over again. Because SketchUp treats imported image textures just like any other texture, what you're really doing when you click to locate the upper-right corner of your image is this: You're telling SketchUp how big to make the tile for your new photo texture. Don't worry too much about getting it right the first time, though—you can always tweak things later on.

Unless the proportions of your image perfectly match the face onto which it was mapped, you should see your image repeating. Don't worry—that's normal. SketchUp automatically tiles your image to fill the whole face. If you want to edit your new texture so that it doesn't look tiled (and you probably do), keep reading. You can scale, rotate, skew, or even stretch your texture to make it look however you want.

Get Photo Texture: Use Online Imagery

A few years ago, Google undertook a Street View project to enhance Google Maps. The company built special photography units, strapped them onto the roofs of vehicles, and drove them down every public highway, street, and lane it could, taking pictures. The result is an immersive and interesting way to experience the outside world from the comfort of your computer screen.

Google recently wired together Street View and SketchUp, making it possible to grab imagery from the former and use it in the latter. If your goal is to build photo-textured models of real-world buildings, you're in luck.

Building Maker, which Google debuted around the middle of 2009, is an incredibly useful tool for creating photo-textured models based on aerial imagery which it provides. See the sidebar "Introducing Building Maker" to find out more.

To use this feature, you must meet two important prerequisites:

- **Your model must be geo-located.** You have to have already told SketchUp precisely where it is by adding a **geo-location** snapshot to your file. Consult Chapter 11 if you need more clarification on the preceding sentence.
- **Street View data must exist for the thing you are trying to texture.** Google's photographed many places, but it is always possible that wherever you are working isn't one of them.

Geo-location
The act of telling SketchUp the precise coordinates where the model is situated in the world.

Follow the steps to paint a flat face in your model with Google Street View imagery (Figure 8-3):

1. **Select the face you want to paint with Street View imagery.** Selecting a rectangle-shaped face helps. You see why in a couple steps.
2. **Choose Edit ⇨ Face ⇨ Add Photo Texture.** The Photo Textures window pops up. If Street View data is available for the location where you are modeling, you can choose Select Region to mark the portion of the photograph you want and then choose Grab. If otherwise Street View data isn't available for the location where you're modeling, this is when you will find out.
3. **Frame the imagery you want to use in the window:**
   a. Click and drag to swivel the "camera."
   b. Click the arrows superimposed on the photo to move up and down the street.
   c. Zoom in and out using the + and – buttons.
   d. If you need to, try resizing the whole window to get a better view.
4. Click the Select Region button in the upper-right corner of the window. A rectangle with blue pins at the corners appears.
5. Drag the blue pins to define an area to paint on the face you selected in Step 1.
6. Click the Grab button to paint the face you selected in Step 1 with the imagery you defined in Step 5.
7. **Close the Photo Textures window.** The photo textures you apply using Add Photo Texture are just like any other photo textures in your model; you can edit them exactly the same way, as explained in the next section.

Figure 8-3

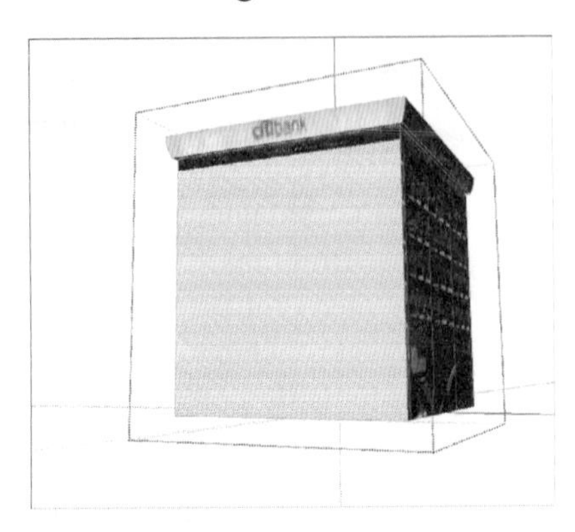

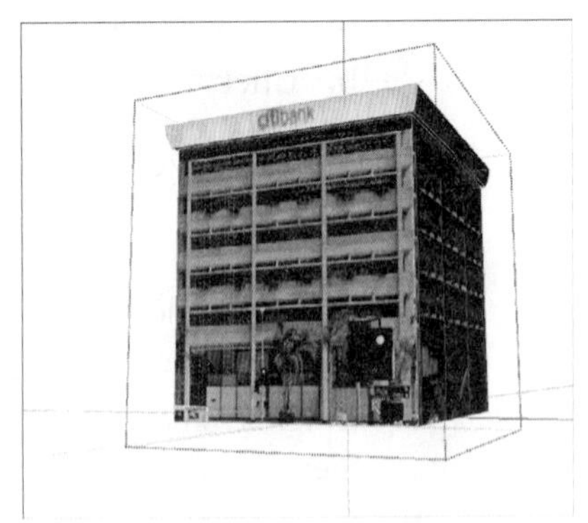

You can use Google's Street View imagery to phototexture your model.

FOR EXAMPLE

INTRODUCING BUILDING MAKER

Sometime between the release of SketchUp 7 and SketchUp 8, Google launched a brand new application—Building Maker; it's free, runs inside a Web browser (such as Firefox, Chrome, Safari, or Internet Explorer), and is hyper-specialized. Building Maker's sole raison d'être is to make it easier for people to create photo-textured building models that will be displayed in Google Earth (and the new Earth view in Google Maps).

In areas where Google's gathered the right kind of aerial imagery, all you have to do is pick a building and start modeling. Just type **Building Maker** into your favorite search engine and see what happens. It is not described how to use it in this book for two main reasons:

It's not SketchUp. Building Maker works absolutely nothing like SketchUp. To build a model, line up primitives (basic shapes like boxes and pitched roofs) with bird's eye images of the building you are working on. Outline the shape, and Building Maker takes care of the rest: It automatically geo-locates, photo-textures, and adds your building to the Google 3D Warehouse.

It's a Web app. And Web apps change weekly. Anything written now will be obsolete by the time you read it. So-called cloud computing is great for users but not conducive for books. SketchUp 8 is tightly integrated with Building Maker. There is even a brand-new Add New Building button on the Getting Started toolbar. Clicking that button opens Building Maker in a dedicated browser window and lets you use it right away. If you have already added a geo-location (see Chapter 11), you can drop anything you make right into your model—perfect for adding context to your design.

If geo-modeling (modeling for Google Earth) is your objective, the ability to start buildings in Building Maker and touch them up in SketchUp is really nice. The first app provides free, easy imagery; the second gives you the freedom to model anything you like.

IN THE REAL WORLD

Mikel from www.sketchucation.com enjoys working with scenes and animations and his clients love his presentations. He started using SketchUp with version 1 because it is "head and shoulders above any other 3D application when it comes to ease of use." He likes to use it from the very early "sketchy" stages and likes to evolve those stages all the way to finished concepts using Layout. He recently used SketchUp in conjunction with Google Earth to complete a golf holiday village project for a client.

8.1.2 Editing Your Textures

After you've successfully mapped an image to a face, you're probably going to want to change it somehow: make it bigger, flip it over, rotate it around, and so on. This is where the Position Texture tool comes in.

The Position Texture tool is actually more of a mode; thus, let's refer to it as Texture Edit mode. Within this mode, you can be in either of two submodes. The names of these submodes are less important than what they do, so here's a description of how they work:

Move/Scale/Rotate/Shear/Distort Texture mode
Mode to use when manipulating textures in SketchUp; also called Fixed Pin Mode.

- **Move/Scale/Rotate/Shear/Distort Texture mode:** As you'd expect, you use this mode to move, scale, rotate, shear, or distort your texture. Its technical name is *Fixed Pin mode*; you'll learn why a little later in the chapter.

Stretch Texture Mode
Mode to use to edit a texture by stretching it to fit the face it is painted on. Also known as Free Pin Mode.

- **Stretch Texture Mode:** Stretch Texture mode lets you edit your texture by stretching it to fit the face it's painted on. If you want to map a photograph of a building façade to your model, this is the mode you want to use. In SketchUp's Help documentation, Stretch Texture mode is called *Free Pin mode*.

Note that you can only edit textures on flat faces; the Position Texture tool doesn't work on curved faces. To find out more about working with textures and curved faces, see the section "Adding Photo Textures to Curved Surfaces," found later in this chapter.

Moving, Scaling, Rotating, Shearing, and Distorting Your Texture

Moving, scaling, rotating, shearing, and distorting your texture involves Texture Edit mode, which is a little bit hidden, unfortunately.

PATHWAYS TO... MOVING, SCALING, ROTATING OR SKEWING YOUR TEXTURE

1. **With the Select tool, click the face with the texture you want to edit.**
2. **Choose Edit ⇨ Face ⇨ Texture ⇨ Position.** This enables Move/Scale/Rotate/Shear/Distort Texture mode. You should be able to see a transparent version of your image, along with four pins, each a different color. Have a look at Color Plate 9 to see what this looks like. Also note that if all your pins are yellow, you're in Stretch Texture mode. Right-click your textured face and make sure there's a check mark next to Fixed Pins to switch into the correct mode. (A quicker way to get to Edit mode is to right-click the textured face and then choose Texture ⇨ Position from the context menu.)
3. **Edit your texture.** At this point, the things you can do to edit your texture are located in two different places. Right-clicking your texture opens a context menu with the following options:
   - **Done:** Tells SketchUp you're finished editing your texture.
   - **Reset:** Undoes all the changes you've made to your texture, and makes things look like they did before you started altering them.
   - **Flip:** Flips your texture left to right or up and down, depending on which suboption you choose.

- **Rotate:** Rotates your texture 90, 180, or 270 degrees, depending on which suboption you choose.
- **Fixed Pins:** When this option is selected, you're in **Move/Scale/Rotate/Shear/Distort Texture mode** (Fixed Pin mode). Deselecting it switches you over to Stretch Texture mode, which is described later in this chapter.
- **Undo/Redo:** Goes back or forward a step in your working process.

Dragging each of the colored pins has a different effect (see Figure 8-4):

Shearing
Action that keeps the top and bottom edges of an image parallel while making the image lean to the left or right.

- **Scale/Shear (Blue) pin:** This pin scales and shears your texture while you drag it. **Shearing** keeps the top and bottom edges parallel while making the image "lean" to the left or right.
- **Distort (Yellow) pin:** This pin distorts your texture while you drag it; in this case, the distortion looks like kind of a perspective effect.
- **Scale/Rotate (Green) pin:** This pin scales and rotates your texture while you drag it.

Figure 8-4

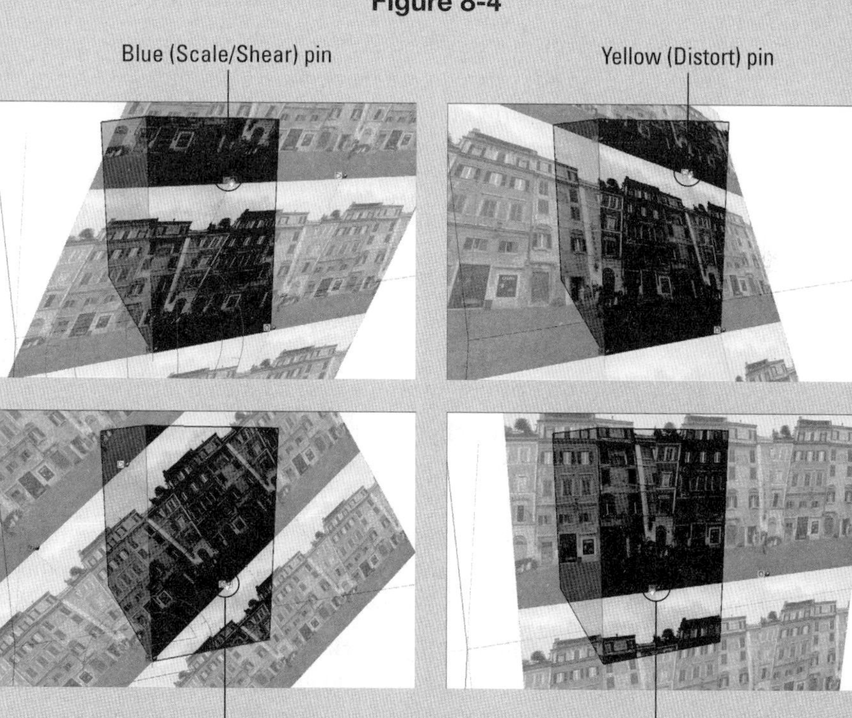

Dragging each of the colored pins does something different.

(continued)

PATHWAYS TO...
MOVING, SCALING, ROTATING
OR SKEWING YOUR TEXTURE *(continued)*

- **Move (Red) pin:** This pin moves your texture around while you drag it. Of all four colored pins, this one may be most useful. You can use it to precisely reposition brick, shingle, and other building material textures in your models. Instead of just dragging around the colored pins, try single-clicking one of them to pick it up; this lets you place it wherever you want (just click again to drop it). This comes in especially handy when you're using the Move and Rotate pins.

4. **Click anywhere outside your texture in your modeling window to exit Edit mode.** You can also right-click and choose Done from the context menu, or press Enter.

Stretching a Photo Over a Face

To better understand how this feature works, think about having a photograph printed on a piece of stretchy fabric. Now, imagine that you stretch the fabric until the photo looks the way you want. Finally, you hold the fabric in place with pins.

In SketchUp, follow these steps to stretch your texture using the Position Texture tool's Stretch Texture mode:

1. **With the Select tool, click the face with the texture you want to edit.**
2. **Choose Edit ⇨ Face ⇨ Texture ⇨ Position.** A quicker way to get to Edit mode is to right-click the textured face and choose Texture ⇨ Position from the context menu.
3. **Right-click your texture and *deselect* the Fixed Pins option (make sure that no check mark is next to it).** Deselecting Fixed Pins switches you to Stretch Texture mode (or Free Pin mode, if you're reading SketchUp's online Help). Instead of four different-colored pins with little symbols next to them, you should see four identical yellow pins. Figure 8-5 shows you what to expect.
4. **Click a pin to pick it up.** Your cursor should clench up into a fist, and the pin should follow it as you move your mouse around. Remember, you can press Esc to drop the pin you're carrying without moving it; pressing Esc cancels any operation in SketchUp.
5. **Place the pin at the corner of the building in your photograph by clicking once.** If the pin you're "carrying" is the upper-left one, drop it on the upper-left corner of the building in your photograph, as shown in Figure 8-6.

Figure 8-5

You know you're in Stretch Texture mode when all the pins are yellow.

Figure 8-6

Place the pin at the corresponding corner (upper-left to upper-left, for instance) of the building in your photo.

Figure 8-7

Drag the pin you just placed to the corresponding corner of the face you're working on.

6. **Click and drag the pin you just moved to the corresponding corner of the face you're working on.** If the pin you just moved is the upper-left one, drag it over to the upper-left corner of the face whose texture you're editing. Check out Figure 8-7 to see this in action.
7. **Repeat steps 4–6 for each of the three remaining pins (see Figure 8-8).** If you need to, feel free to orbit, zoom, and pan around your model to get the best view of what you're doing; just use the scroll wheel on your mouse to navigate without switching tools. A good way to work is to pick up and drop each yellow pin in the general vicinity of the precise spot you want to place it. Then zoom in and use your better point of view to do a more accurate job.
8. **Press Enter to exit Texture Edit mode.**

If you don't like what you see, just go back and edit the texture again; there's no limit to the number of times you can change things around.

Scaling Your Model Until the Photo Looks Right

When you're happy with the way your texture is stretched to fit the face, one of two things will be true:

- **The proportions are correct.** If the proportions are correct, the photo won't look stretched or squashed. This will only be the case if the face

Figure 8-8

Repeat steps 4–6 for each of the other three yellow pins.

to which you applied the photo texture was already exactly the right size.

- **The proportions aren't correct.** If the photo texture you just "tweaked" looks stretched or squashed, the face it's on is the wrong size. No worries—you just need to stretch the whole face until the texture looks right. Better yet, if you know how big the face is *supposed* to be (in real life), you can stretch it until it's correct.

Follow these steps to stretch a face until the texture looks right:

1. **Use the Tape Measure tool to create guides that you can use to accurately stretch your face.** In this case, say you know that the building you're modeling is supposed to be 50 feet wide. (Using the Tape Measure tool and guides is described in Chapter 2, just in case you need a refresher.)
2. **Select the face you want to stretch.** If your model is at a fairly early stage, just select the whole thing. Triple-click the face with the Select tool to select it and everything attached to it. Figure 8-9 shows the whole model selected.
3. **Choose Tools ⇨ Scale to activate the Scale tool.** When the Scale tool is active, everything that's selected in your model should be surrounded by SketchUp's Scale Box. Its 27 little green cubes (they're called grips) and thick, yellow lines are hard to miss.

Figure 8-9

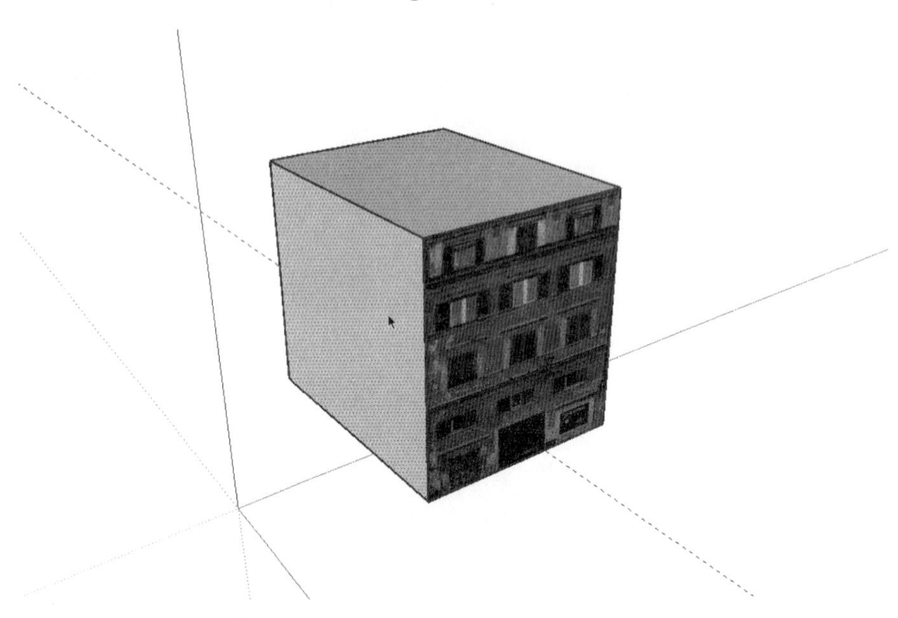

Select everything you want to stretch.

4. **Scale your selection to be the right size (see Figure 8-10).** Use the Scale tool by clicking on the grips and moving your cursor to stretch whatever's selected (including your texture). Click again to stop scaling.

To scale something precisely using a guide, click a scale grip to grab it and then hover over the relevant guide to tell SketchUp that's where you want to scale to. Click again to finish the scale operation.

Figure 8-10

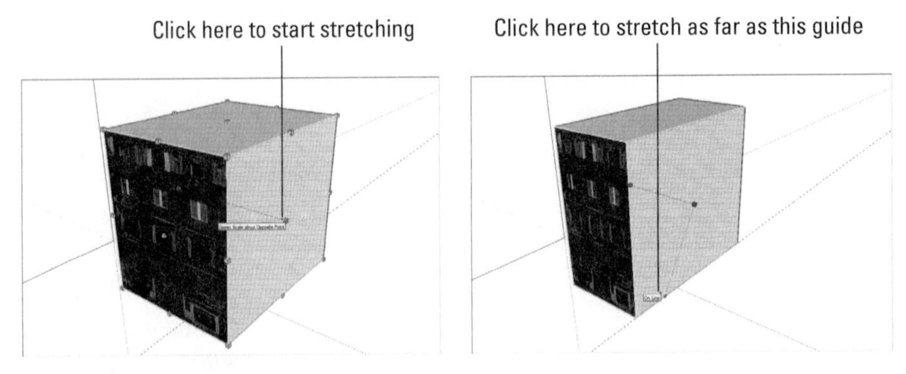

Use the Scale tool's grips to stretch your selection (texture and all).

It is perfectly normal to want to keep modeling with your photo-textured faces; tracing a window and pushing it in a bit with the Push/Pull tool is one of the most satisfying things you can do in SketchUp. Go to the end of this chapter and take a look at "Modeling on Top of Photo Textures" to discover everything you need to know.

8.1.3 Optimizing Your Photo Textures

If your goal is to build models that will eventually show up in Google Earth, one of your primary concerns has to be file size. Big models *are not easily understood by* Google Earth. In fact, anything over 10 megabytes can't even be uploaded to the 3D Warehouse, which is the first step in getting your models accepted into Earth's default 3D Buildings layer.

Optimizing your photo textures goes a long toward reducing your file size. Back in SketchUp 7, the folks at Google added two super-useful new features that make it a whole lot easier to build efficient photo-textured models.

Make Unique Texture

Right-clicking any face in your model and choosing Make Unique Texture do two things: These actions create a copy of the texture you've selected and crop (trim away everything that isn't visible) that copy according to the face it's on. Why is this important? Just because you can't see part of an image doesn't mean it's not there; SketchUp saves the whole photo with the model, even if you use only a little bit of it. In a complex model with dozens of photo textures, all that invisible, extra photo data adds up. Making your textures unique can make your models much, much smaller.

Combine Textures

Take two or more textures in your model and combine them into a single texture. Why? The fewer unique textures in your model, the smaller its file size.

PATHWAYS TO...
COMBINING TEXTURES

1. **Select two or more coplanar (on the same plane) faces with different textures applied to them.** The faces you select must all be adjacent to each other.
2. **Right-click any of the faces you selected in Step 1 and choose Combine Textures.** This creates a new texture in the In Model library of your Materials dialog box. Letting SketchUp delete the interior edges (the ones between the faces whose textures you combined) further reduces your file size because it eliminates faces.

Edit Texture Image

This feature isn't about reducing file size; it's for editing the pixels in your photo textures themselves. Perhaps there's something in a photograph you're using, and you don't want it to be there. You can use Edit Texture Image to open the texture you've selected in an image-editing program, where you can edit the texture directly.

PATHWAYS TO... EDITING A TEXTURE IMAGE

1. **Right-click the texture in your model you want to edit and choose Texture ⇨ Edit Texture Image.**
2. **In the program that opens, make whatever changes you need to make.**
3. **Save (don't Save As and change the filename) the image you're editing and close it if you like.**
4. **Back in SketchUp, check to make sure your edits have been applied.**

Which image-editing program actually opens depends on what you have installed on your computer; you specify which one to use in the Applications panel of the Preferences dialog box. For what it's worth, most designers use Adobe Photoshop, but you can use whatever you have.

Take a look at Figure 8-11. In it, Edit Texture Image *was used* to remove a pesky element from a photograph *used* as a photo texture.

IN THE REAL WORLD

Alex Manuele designs furniture and staircases and likes to use SketchUp from the preliminary idea all the way to the final presentation. He likes this program because it is very fast and in his own words "no other modeler comes close to how fast you can model in SketchUp." He uses Layout for client presentations and their responses are always good; he likes it because it gives his clients a clear understanding of how the project t is evolving and if they raise a concern, he can quickly make changes. He uses photo textures all the time because a lot of his work involves the use of specific wood grain and the texture needs to be shown well.

Figure 8-11

I want to remove this traffic light

Removing the light in Photoshop

Gone!

Use Edit Texture Image to open an image in a photoediting app like Photoshop.

8.1.4 Adding Photo Textures to Curved Surfaces

Notice how the title of this section ends with "surfaces" and not with "faces?" That's because (as you know by now) individual faces in SketchUp are always flat—no exceptions. When you see a non-flat surface, it's actually made up of multiple faces. You can't see the edges between them because they've been smoothed. Choosing View ⇨ Hidden Geometry exposes all curved surfaces for what they really are. Check out Figures 2-3 and 2-5 (in Chapter 2) for a visual reminder.

Figure 8-12

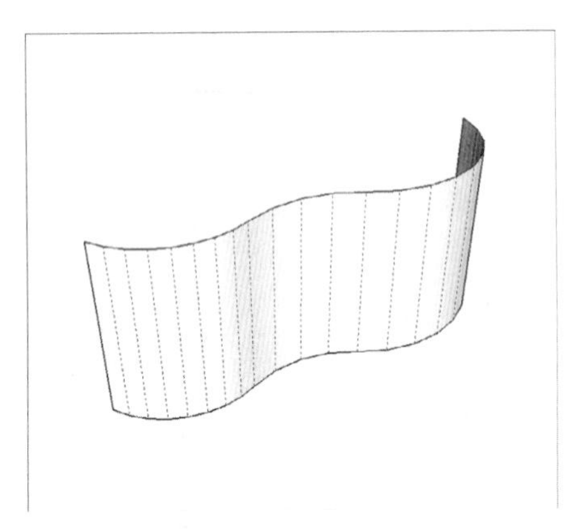

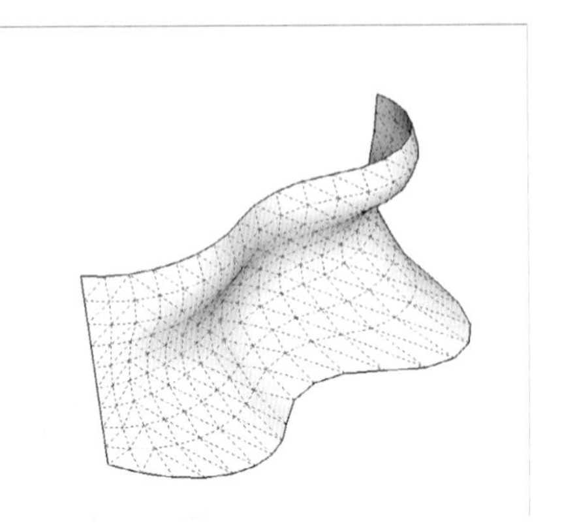

All curved surfaces are either single-direction (left) or multi-direction.

How you go about mapping an image to a curved surface in SketchUp depends on what type you have. With that in mind, curved surfaces fall into two general categories (see Figure 8-12):

- **Single-direction curves:** In this category we can include cylinders or curving outlines that are extruded in one direction only. In SketchUp, a cylinder is understood as a series of small, tall rectangles set side-by-side that make the curve. Most curved walls you see on buildings are the same way, in segments; they don't taper in or out as they rise. Another way to think about single-direction curves is to consider how they might have been made. If the curved surface you're staring at could be the result of a single push/pull operation (such as turning a circle into a cylinder), there's an excellent chance it is single-direction. For mapping an image to a single-curve surface, you can use the Adjacent Faces method; it works well and doesn't stretch your image.
- **Multi-direction curves:** Terrain objects, saddles, and curtains are all prime examples of surfaces that curve in more than one direction at a time. They're always composed of triangles—never basic rectangles. To map an image to this type of curved surface, you must use the Projected Texture method. Skip ahead a couple pages to read all about it. Please keep in mind that *the names are* made up for the Adjacent Faces and Projected Texture methods of mapping images to non-flat surfaces. *They had to be named* something, and these sounded descriptive without seeming too technical.

The Adjacent Faces Method

If you need to paint an image onto a surface that curves only in a single direction (such as a cylinder) you can use this technique. Follow these steps to see how and see Figure 8-13 to see the process in action:

Figure 8-13

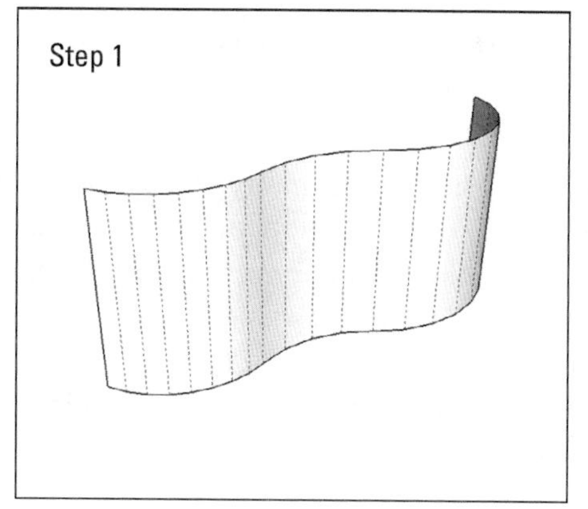

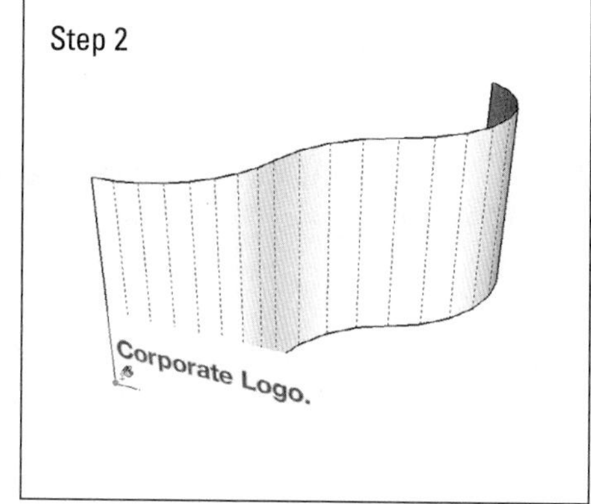

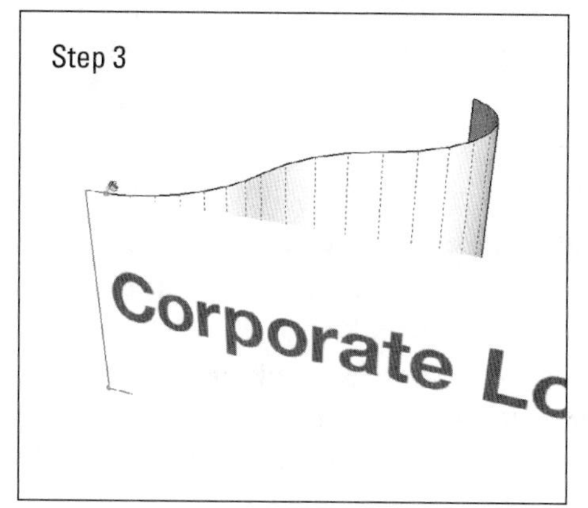

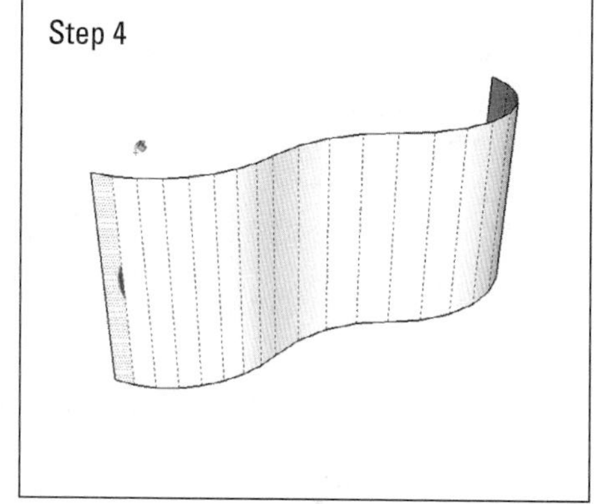

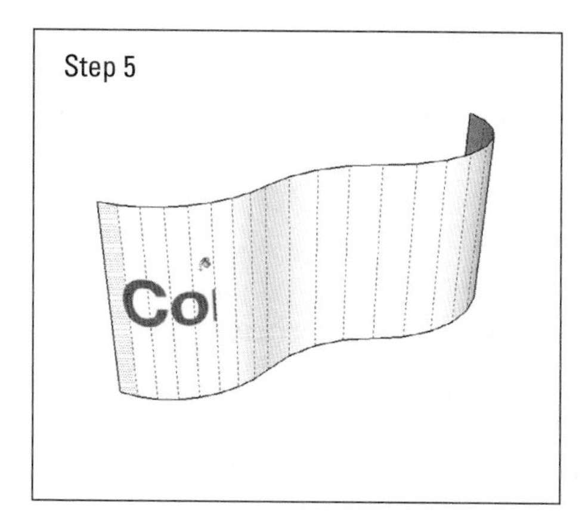

The Adjacent Faces method lets you map images to simple curved surfaces.

PATHWAYS TO... MAPPING IMAGES TO SIMPLE CURVED SURFACES

1. **Choose View ⇨ Hidden Geometry to turn on Hidden Geometry so you can see the individual faces in your model.**
2. **"Load" your cursor with an imported image.** Follow Steps 1–4 in "Importing Images: Use your own photos" (earlier in this chapter) to import an image as a texture.
3. **Paint the leftmost sub-face entirely with the image.** Your curved surface is composed of sub-faces. Here's how to paint the right one:
   a. **Hover your loaded cursor over the lower-left corner of the sub-face farthest to the left. Don't click yet.**
   b. **When the image is oriented in the right direction, click once.**
   c. **Click again on the upper-right corner of the same sub-face. This places the image; it should be cropped on the right.**
4. **Use the Paint Bucket tool with the Alt key (Command on a Mac) held down to sample the texture (image) you just placed.** This "loads" your Paint Bucket tool with the texture.
5. **With the Paint Bucket tool, click once on the face immediately to the right of the face you painted in Step 3.** If everything's working correctly, the image you placed appears on the face you just clicked.
6. **Keep painting sub-faces until you're done.** Remember to work your way from left to right; skipping a sub-face messes up things. To fix a problem, just undo and keep going.

The Projected Texture Method

For painting an image onto a complex curved surface, there's literally no substitute for this method. *Large areas* of terrain are good examples of complex curved surfaces—bumpy, twisted, rippled, and multi-directional. If the curve you are dealing with is more complicated than a simple extrusion, you need to use this image-mapping technique.

The key is to line up a flat surface with the curved surface to which you want to apply the photo texture. You then "paint" the flat surface with the texture, make it projected, sample it, and finally, paint the curved surface with the projected, sampled texture.

Follow these steps to get the basic idea (see Figure 8-14):

PATHWAYS TO... MAPPING PROJECTED TEXTURES TO CURVED SURFACES

1. **Create a flat surface that lines up with your curved surface.** Use the Line tool and SketchUp's inferencing system to draw a flat face that lines up with (and is the same size as) *the* curved surface.
2. **Apply a photo texture to your flat surface and make sure that it's positioned correctly.** You can refer to the earlier parts of this chapter for detailed instructions on how to do this.
3. **Right-click the textured face and choose Texture ⇨ Projected.** This ensures that the texture is projected, which is the key to this whole operation.
4. **Hold down the Alt key (Command on a Mac) while using the Paint Bucket tool to sample the projected texture.** This "loads" your Paint Bucket tool with the projected texture.
5. **Use the Paint Bucket tool without pressing anything on your keyboard to paint the curved surface with the projected texture.** The photo texture is painted on your curved surface, although the pixels in the image look stretched in some places.
6. **Delete the flat surface that you originally mapped the image to; you don't need it anymore.** If you're trying to do this on your own curved surface and things don't seem to be working, your curved surface is probably part of a group or component. Either explode or double-click to edit the group or component before you do Step 5 and see whether that helps.

Figure 8-14

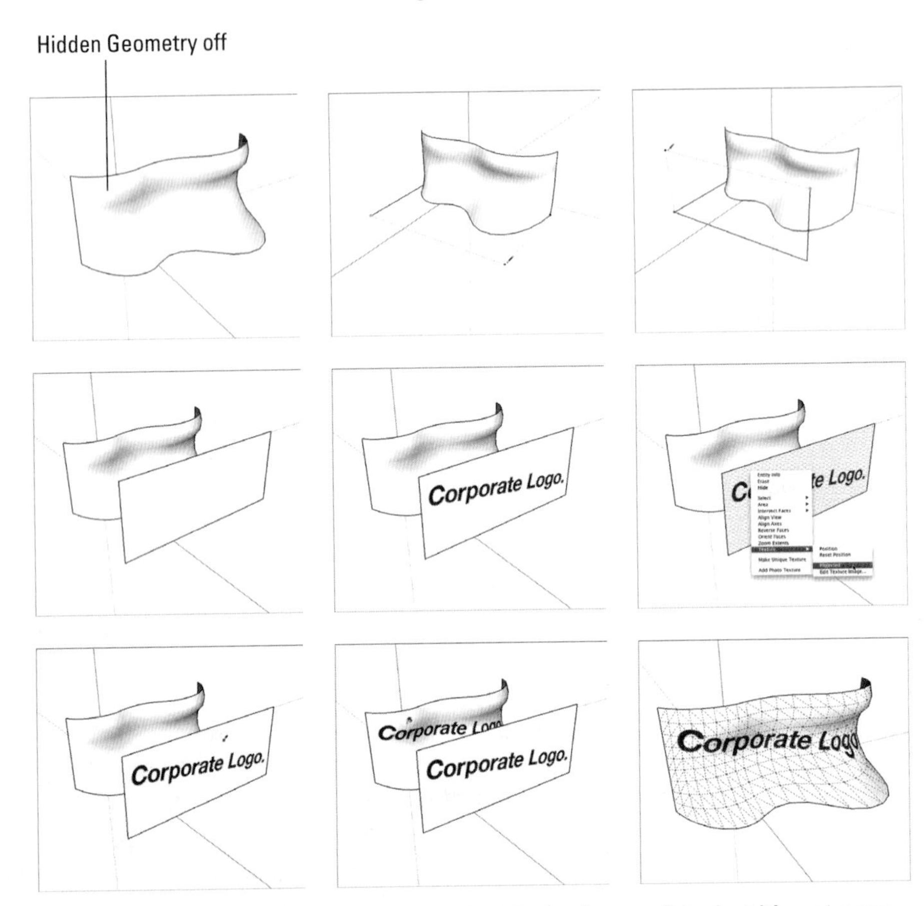

Mapping projected textures to curved surfaces is possible, but it's not easy.

TIPS FROM THE PROFESSIONALS

Photographs

Adding photographs to faces can be rewarding and surely can impress our clients if we are trying to instill a sense of realism to the models we work on. But we need to be careful about the quality of the images we use. Google SketchUp is not a photo editing software and will not retouch your images or make them look better; in other words, if your images are pixilated or do not have enough contrast, they will still look unfinished on the face you are working on. To solve this problem, be sure to have high-quality images available from the very beginning. It would also be a good idea to use a variety of tools in your favorite photo retouching program—Adobe Photoshop or even GIMP would be a good choice here—to enhance them, like cropping unnecessary parts in the scene or applying any of the automatic features available, such as Auto Contrast, Auto Tone, Auto Levels, or Unsharp Mask filter.

SELF-CHECK

1. How can be obtain quickly photographs of specific corners in cities, to be used in Google SketchUp?
2. What can we do if the photo texture that we are using is not proportionately correct?
3. When we right-click on a textured face, what color do we need to have our pins in?
4. SketchUp can project textures only onto flat surfaces; curved surfaces are off limits. True or false?
5. What happens when we hold down the Paint Bucket tool with the Alt key (⌘ Command on a Mac)?

Apply Your Knowledge Take a high-quality picture of two textures that you will find around your neighborhood, such as a particularly interesting brick wall or a highly textured stone wall. Then, retouch the images in a photo-editing program such as Adobe Photoshop, and improve the overall look (think of AutoTone, AutoContrast, or AutoColor settings). Once you are done, build a quick one-story building and apply those two textures on two different sides.

8.2 MODELING DIRECTLY FROM A PHOTO: INTRODUCING PHOTO-MATCHING

The newest version of SketchUp includes a fantastic feature that allows you to model directly from photos. More specifically, you can use this feature, called photo-matching, to do the following:

- **Build a model based on a photograph:** If you have a good photograph (or multiple photographs) of the structure you want to model in SketchUp, photo-matching can help you set things up so that building your model is much easier.
- **Match your model view to a photograph:** Perhaps you have a model of a building and a photograph of the spot where the building will be constructed. You can use photo-matching to position your "camera" in SketchUp to be exactly where the real-life camera was when the photograph was taken. Then, you can create a composite image that shows what your building will look like in context.

GOOGLE SKETCHUP IN ACTION

Learn how to work with photo-matching.

Photo Matching only works on photographs of objects with at least one pair of surfaces that are at right angles to each other. Luckily, this includes millions of things you might want to build. Still, if the thing you want to photo-match is entirely round, or wavy, or even triangular, photo-matching won't work.

8.2.1 Understanding Colors

Like some of SketchUp's other features, photo-matching is more of a *method* than a tool: You use it to set things up, you model a bit you use the Match Photo dialog box a bit, and so on. If you don't know the basics of modeling in SketchUp yet, you won't have any luck with photo-matching—it's really more of an intermediate-level feature, if such a thing exists.

Color Plate 10 shows what your screen might look like when you're in the throes of photo-matching. It's a bit daunting, but after you've used it once or twice, it's not so bad. This image is included in the color section of this book because photo-matching (at least at the beginning of the process) uses color as a critical part of its user interface.

Specifically, the following elements of the photo-matching interface will show up in your modeling window:

- **Photograph:** The photograph you pick to create shows up as a kind of background in your modeling window; it stays there as long as you don't use Orbit to change your view. To bring it back, click the Scene tab (at the top of your modeling window) labeled with the photograph's name.
- **Perspective bars:** These come in two pairs: one green and one red. You use them when you're setting up a new Photo Match by dragging their ends (grips) to line them up with *perpendicular* pairs of *parallel* edges in your photograph. For a clearer explanation of how this works, see the next section in this chapter.
- **Horizon line:** This is a yellow, horizontal bar that, in most cases, you won't have to use. It represents the horizon line in your model view, and as long as you placed the perspective bars correctly, it takes care of itself.
- **Vanishing point grips:** These are found at both ends of the horizon line, and once again, as long as you did a good job of setting up the perspective bars, you shouldn't have to touch them.
- **Axis origin:** This is the spot where the red, green, and blue axes meet. You position it yourself to tell SketchUp where the ground surface is.
- **Scale line/vertical axis:** Clicking and dragging this blue line lets you roughly scale your photograph by using the colored Photo Match grid lines. After you're done, you can always scale your model more accurately using the Tape Measure tool (check out Chapter 2 for more information on how to do this).

You also need to work with a few things that appear outside your modeling window:

- **Matched Photo scene tab:** Every time you create a new photo match, you create a new scene, too (you can read all about scenes in Chapter 10).

Clicking a Match Photo scene tab returns your view to the one you set up when you created (or edited) that matched photo. You need to remember that every time we orbit into another view, the photograph will disappear but if clicking on that scene tab will make the associated photograph reappear.

- **Match Photo dialog box**. This is photo-matching "mission control." It's where you can find almost all the controls you need for creating, editing, and working with your matched photo.
- **Photo visibility settings in the Styles dialog box.** Deep down in the inner-workings of the Styles dialog box, in the Modeling Settings section of the Edit tab, you can control the visibility of your matched photo. See Chapter 9 to find out all about Styles.

8.2.2 Getting Set Up for Photo-Matching

Modeling with SketchUp's photo-matching feature is generally a step-by-step procedure. Whether you are building a new model or lining up an existing model with a photograph, start by getting your modeling window ready. How you do this *depends on which one you are* trying to do:

- **Line up a model you have built already with a photograph:** This case requires you to re-orient your view and then reposition your drawing axes *before* you are ready to begin photo-matching. To do this, follow these steps:

  1. **Orbit around until your model view more or less matches the camera position in your photograph.**
  2. **Choose Tools ⇨ Axes and then click to place your axis origin somewhere on your model.** The axis origin is where your colored axes meet. Try to choose a spot that's also visible in your photograph, if there is one.
  3. **Click somewhere in the lower-left quadrant of your modeling window.** This ensures that the red axis runs from the upper-left to the lower-right corner of your screen.
  4. **Watch your linear inferences to be sure that your repositioned red axis is parallel to some of the edges in your model.** Chapter 2 has more about linear inferences.
  5. **Click somewhere in the upper-right quadrant of your modeling window to make sure that the blue axis is pointing up.**

- **Use a photograph to build a model:** Open a fresh, new SketchUp file, and you are *ready* to go. After your modeling window is set up, follow these steps to create a new matched photo in your SketchUp file:

  1. **Choose Camera ⇨ Match New Photo.** A dialog box opens.
  2. **Select the image on your computer that you want to use and click the Open button.** The dialog box closes, and you see the image you

Figure 8-15

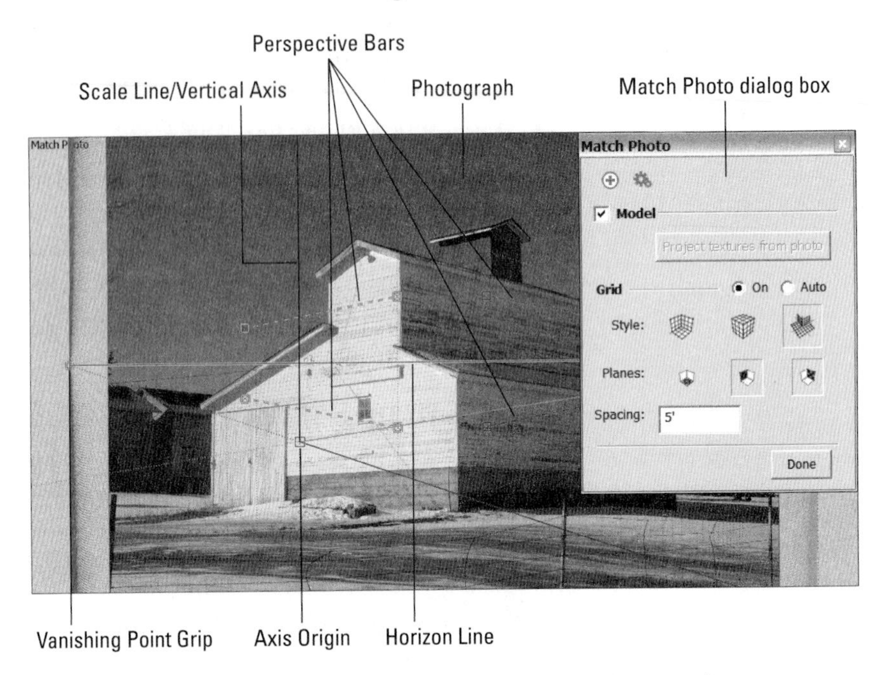

The photo-matching interface includes your picture, plus many other things.

chose in your modeling window. You also see a jumble of colorful *technical things* all over the place. Don't worry—it's all part of the photo-matching interface. Figure 8-15 gives you an idea of this; Color Plate 10 shows the same image in color. SketchUp's photo-matching feature requires that you use certain kinds of photographs for it to work properly. See the sidebar, "Taking the Right Kind of Picture," later in this chapter, for pointers on what kinds of photos you can—and can't—use.

3. **In the Match Photo dialog box (Window ⇨ Match Photo), choose the style that matches your photograph.** The style buttons in the Match Photo dialog box correspond to three types of photographs you may use:

   - Inside if your photo is an interior view.
   - Above if it's an aerial shot.
   - Outside if your photo is an exterior view taken from a human vantage point.

   Figure 8-16 shows examples of each of these scenarios.

4. **Begin positioning the perspective bars, starting with the two green ones, by lining them up with any two parallel edges.** The tops and bottoms of windows are good candidates, as are rooflines,

Figure 8-16

Inside

Above

Outside

Choose the style that best describes your photograph's camera position.

tabletops, and ceiling tiles. This is easier than it looks. Move each perspective bar one at a time, dragging each end into position separately. Color Plate 11 illustrates this in color.

The following tips can help you position the bars correctly:

- Zoom in and out (using the scroll wheel on your mouse) to better view your photograph while you place your perspective bars. The more accurately you place the bars, the better things will turn out.
- Match your perspective bars to nice, long edges in your photograph; you get better results that way.
- If you are working with an existing model, hiding it while you place your perspective bars may help; sometimes a model gets in the way. Just deselect the Model check box in the Match Photo dialog box to temporarily hide it.

5. **Line up the two red perspective bars with a different set of parallel edges—just be sure that these parallel edges are perpendicular (at right angles) to the first pair**. If the parallel edges aren't perpendicular to the first set of edges, photo-matching doesn't work. Color Plate 12 shows what it looks like when all four perspective bars have been positioned properly.
6. **Drag the axis origin (the little square where the axes come together) to a place where your building touches the ground.** This is how you tell SketchUp where the ground plane is. Try to make sure your axis origin is right at the intersection of two perpendicular edges. Color Plate 13 shows what this looks like. If you are photo-matching an existing model, dragging the axis origin moves your model, too. Line up

your model with the photograph so that the spot where you placed the axis origin is right on top of the corresponding spot in your photo. Don't worry about size right now; you deal with that in a moment.

7. **Roughly set the scale of your photograph by clicking and dragging anywhere on the blue scale/vertical axis line to zoom in or out until your photograph looks to be at about the right scale.** Do this by first setting your grid spacing in the Match Photo dialog box and then using the grid lines in your modeling window to "eyeball" the size of your photo until it looks about right. Color Plate 14 shows an example where the grid spacing is set at 5 feet (the default setting). Because we know the barn in the photo is about 20 feet tall, zoom in or out until it's about 4 grid lines high because 4 times 5 feet is 20 feet. If you're trying to match an existing model to your photo, just zoom in or out until your model looks like it's the right size. You don't have to be very exact at this stage of the game. You can always scale your model later by using the Tape Measure tool (Chapter 2 instructs how to do that).
8. **Click the Done button in the Match Photo dialog box.** When you click the Done button, you stop editing your matched photo. All the colorful lines and grips disappear, and you are left with the photo you brought in, your model axes, and your thoughts. It may have seemed like a lot of magic, but what you did was pretty simple: You used photo-matching to create a scene (which is discussed extensively in Chapter 10) with a camera position and lens settings that match the ones used to take the picture that's on your screen. In effect, you are now "standing" exactly where the photographer was standing when the photograph was taken.

8.2.3 Modeling by Photo-Matching

Setting up a new matched photo was just the first step. Now it's time to use SketchUp's modeling tools (with a little help from the Match Photo dialog box) to build a model based on the photograph you "matched." Here are a couple of the basic concepts to keep in mind when doing this:

- **It's not a linear process.** Building a model using a Photo Matched photo entails going between drawing edges, orbiting around, drawing some more edges, going back to your Photo Match scene, and drawing yet more edges. Every photo is different, so the ones you work with will present unique challenges that you will (hopefully) have fun figuring out.
- **Don't forget the photo textures.** By far one of the best features of Photo Match is its ability to automatically photo-texture your model's faces using your photograph as "paint."

FOR EXAMPLE

TAKING THE RIGHT KIND OF PICTURE

Your level of success with photo-matching depends to some extent on the photograph you start out with. Here are some tips for what kind of images are good candidates for this process:

- **Make sure that the edges of two perpendicular surfaces are visible in the shot.** You need to be able to see planes that are at right angles to each other to be able to use photo-matching properly.
- **Shoot at a 45-degree angle if you can.** Because of the way perspective works, you will get a more accurate result if you use a photograph where you can see both perpendicular surfaces clearly. If one of them is sharply distorted, you will have a harder time. The images in the following figure show what this means.
- **Watch out for lens distortion.** When you take a picture with a wide-angle lens, some of the straight lines in the image "bow" a little bit, depending on where they are in the frame. Try to use photos taken with a normal or telephoto lens: 50mm to 100mm is a good bet.

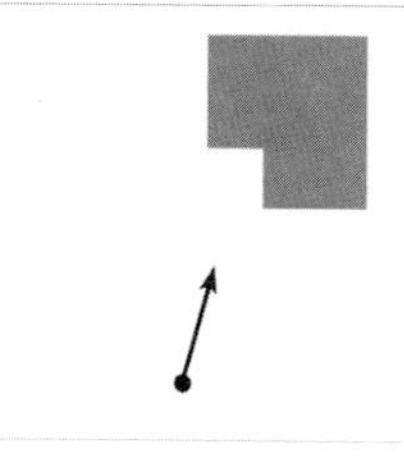

Bad Match Photo candidate: bad angle

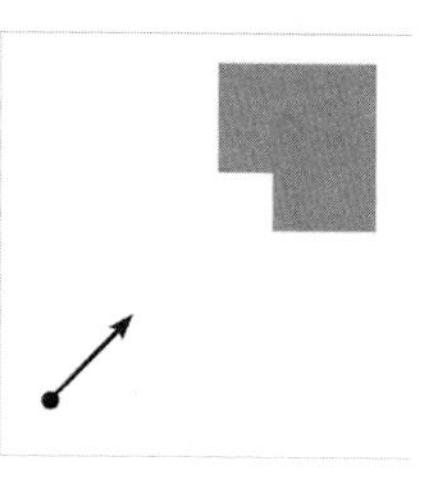

Good Match Photo candidate: you can see both sides clearly

Follow these steps to start building a model with Photo Match:

1. **Click the matched photo scene tab to make sure that you're lined up properly.** If you orbit away from the vantage point you set up with Photo Match, you'll know it; your photograph will disappear. You can easily get back by clicking the scene tab for your Photo Match.

Figure 8-17

Scene tab for this matched photo

Clicking the scene tab for your Photo Match takes you back to that vantage point (and brings back your photograph).

It's labeled with the name of your photo, and it's at the top of your modeling window (see Figure 8-17).

2. **Trace one of the edges in your photograph with the Line tool.** Make sure that you're drawing in one of the three main directions: red, green, or blue. Color Plate 15 shows this in action. It's a good idea to start drawing at the axis origin; it'll help to keep you from getting confused.
3. **Keep tracing with the Line tool until you have a rectangular face.** The key here is to make sure that you keep watching the color of your edges as you draw. You always want to see your lines turn red, green, or blue when you're starting out. Have a look at Color Plate 16 to see what this looks like. Be careful not to orbit while you're drawing—if you do, repeat step 1 and keep going. You can zoom and pan all you want, though.
4. **Use SketchUp's modeling tools to continue to "trace" the photograph in three dimensions.** Here are some pointers for doing this successfully:
   - **Always start awn edge at the end of an edge you've already drawn.** If you don't, your geometry won't make any sense, and you won't end up with what you expect.

- **Never draw an edge in "midair."** This is basically the same as the previous point, but it bears repeating: if you don't draw edges based on other edges, you won't get good results.
- **Orbit frequently to see what's going on.** You'll be surprised what you have sometimes—tracing a 2D image in 3D is tricky business. Get in the habit of orbiting around to check on things and draw certain edges. Click the Photo Match scene tab to return to the proper view (see Color Plate 17).
- **Use other tools (like Push/Pull and Offset) when appropriate.** Nothing prevents you from using the full complement of SketchUp's modeling tools when using Photo Match. However, using only Line and Eraser while drawing the basic skeleton of your model keeps things simple.
- **Pay attention to the colors.** With a photograph as an underlay, it's a little harder to see what you're doing. Watching to make sure that you're drawing the edge you intend to draw is critical.
- **Draw angles by "connecting the dots."** If you need to trace an edge in your photo that doesn't line up with any of the colored axes (an angled roofline, for example), figure out where the endpoints are by drawing perpendicular edges and connecting them with an angled line. Color Plate 18 shows this in detail.
- **Show or hide your photograph.** You can work with the visibility of the picture you are using. Doing so sometimes helps you see what you are working on. You can find the controls in the Modeling Settings section of the Styles dialog box's Edit tab. Reference Chapter 9 for more detail.

If you have more than one photo of your modeling subject, you can have multiple matched photos in the same SketchUp file. Just get as far as you can with the first photo and then start again with next using the geometry you created as an "existing building." See "Getting Set Up for Photo-Matching" earlier in this chapter and follow the steps to line up an existing model with a new photograph.

Color Plate 19 shows the beginnings of a model of Habitat 67 in Montreal. Here, two pictures were used to create two matches in the same SketchUp file, which made it possible to build more of the model than could be seen in a single picture.

You can edit any texture in your model—including ones produced by photomatching—by opening them in image-editing software (such as Photoshop) directly from SketchUp. This is handy for taking out things you might not want in your photos, such as trees, cars, and ex-husbands. Take a look at "Optimizing Your Photo Textures," earlier in this chapter, for more details.

SELF-CHECK

1. Photo Matching works on photographs of objects with surfaces that are at 30-, 45-, and 90-degree angles. True or false?
2. When we use the Orbit tool in Photo Matching, the image that we brought into the scene "revolves" with the view and we never lose sight of it. True of false?
3. Choose the correct answer. What style buttons can we choose from the Match Photo dialog box?
   - **a.** Inside, Above, and Outside.
   - **b.** Architecture, Landscape, and Interior Architecture.
   - **c.** Above, Below, and Intermediate.
4. What would be the easiest corner to start from, once we are ready to trace our photograph?

Apply Your Knowledge Using the photo-matching technique is rewarding experience once we gain some experience with the set up process. Walk around your neighborhood and locate a building that has four simple walls and that you can access easily from all four corners. Take one photograph from each corner, giving yourself some distance (to avoid unnecessary distortion and an unwelcoming fish-eye effect). Plan to build in 3D this structure using at least two of those pictures taken from opposite corners and remember to jump from one scene to the next, just to make sure that you are building the structure accurately.

8.3 MODELING ON TOP OF PHOTO TEXTURES

GOOGLE SKETCHUP IN ACTION

Understand the importance of texture being projected.

After you place a photo texture on the right face and in the correct place on that face, you may want to use the information in your photograph to help you add geometry to your model. It's a great way to be more or less accurate without having to measure much, and the combination of photo textures and a few simple push/pull operations can be very convincing.

Making a Texture Projected

Modeling with photo-textured faces isn't hard, but you must know one critical step before you can do it: you have to make sure that your texture is projected.

Figure 8-18

Pushing/pulling an opening in a textured face when the texture *isn't* projected (left), and when it *is* projected (middle).

Figure 8-18 shows what happens when you try to push/pull an opening in a photo-textured face: On the left, when the texture isn't projected, the inside faces are painted with random parts of the texture. On the right, when it is projected, note how the "inside" faces that are produced by the push/pull operation are a plain, easy-to-discern gray. This is known as painting with "stretched" pixels, and the result is typically more appropriate for what you're doing.

Consequently, it's a good idea to make sure that your face's texture is projected *before* you start drawing on top of it. Happily, telling SketchUp to make a photo texture projected is just a matter of flipping a switch. Simply right-click the face with the photo texture and choose Texture ⇨ Projected from the context menu. If you see a check mark next to Projected, your texture is already projected; don't choose anything.

Modeling with Projected Textures: A Basic Workflow

Follow these steps to get the hang of working with projected textures (see Figure 8-19):

1. **Make a basic rectangular box and then apply a photo texture to one of the side faces.** Check out the section "Adding Photos to Faces," earlier in this chapter.
2. **Right-click the textured face and choose Texture ⇨ Projected from the context menu.** Make sure that Projected has a check mark next to it.
3. **Draw a rectangle on the textured face and push/pull it inward.** Notice the "stretched pixels" effect?
4. **(Optional) Add other angles or features to your model, if you like.** Figure 8-19 illustrates an angled face.

Figure 8-19

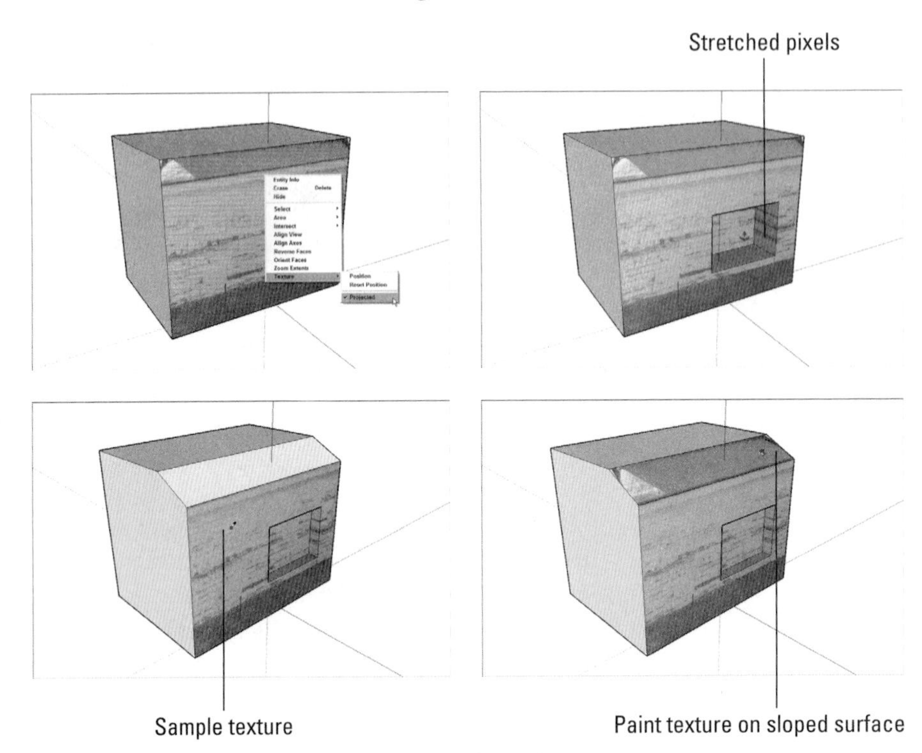

Working with projected textures.

5. **Switch to the Paint Bucket tool, hold down Alt (⌘ Command on a Mac) and click somewhere on the textured face to sample the texture.** (Your cursor should look like an eyedropper when you do this.) This "loads" your Paint Bucket with the projected texture.
6. **Release the Alt (⌘ Command) key to switch back to the Paint Bucket cursor, and click the angled face once to paint it with the projected texture.** You should see the "stretched pixels" effect here, too.

IN THE REAL WORLD

Serafim Alexiev works in Diva Art Design Studio in Sofia, Bulgaria. He has used Google SketchUp successfully to create profiles, 3D picture frames, and moldings. He likes to power up SketchUp with lots of plugins. He composes his presentation boards using Google Layout, and in those boards he makes sure to include dimensioned sectional views, rendered perspective views, orthogonal projections, or detail views. He often uses photo textures to enhance the look of his projects.

CAREER CONNECTION

Nicolas Rateau does architecture and engineering projects in Belgium and France, and he considers himself a proficient user of Google SketchUp. He likes it because it provides fast results and has a short learning curve, without having to compromise precision. He also likes to explore various plugins, such as the Follow and Keep plugin, that he finds more reliable than the original Follow Me tool, especially for landscape architecture and modeling of 3D roads and paths. He also finds being connected with a large community of users an enriching experience, through various blogs and specialized sites.

He has used most of the modeling tools and likes to combine his 3D creations with Google Earth in his captured scenes to create the right look. Also, he uses the Styles window to modify further the captured scenes, which are rendered in an external renderer—TheaRender most of the times for final texturing, lighting, and animation—and presented in Google Layout.

Nicolas recently created a lot of stills for a photovoltaic farm, a car closed driving range, and some urban planning projects.

SELF-CHECK

1. How do we call painting surfaces with pictures using 3D software?
2. The Move/Scale/Rotate/Shear/Distort Texture mode lets you edit your texture by stretching it to fit the face it's painted on. True or false?
3. When modeling with photo-textured faces, you have to make sure that your texture is:
   a. Projected
   b. Protected
   c. Protracted
   d. Stretched

Apply Your Knowledge Take a picture of a structure that has some interesting volumetric play, such as a building with many balconies on the front facade. Recreate in Google SketchUp the basic shape of that building and project that image as a photo texture onto one of the front. After you do that, model the volume of the balcony by drawing rectangles on the façade and using Push/Pull to create your volumes.

SUMMARY

Section 8.1

- You can apply photos to faces and then use the information in the pictures to help you model.
- Painting surfaces with pictures using 3D software is called *mapping*.
- You can map photos to two different kinds of faces: flat ones and curved ones.
- Use the Tape Measure tool to create guides that you can use to accurately stretch your face.
- For painting an image onto a complex curved surface, use the Projected Texture method.

Section 8.2

- From one or many photographs of the structure you want to model in SketchUp, photo-matching can help you set things up so that building your model is much easier.
- Your level of success with photo-matching depends to some extent on the photograph you start out with:
  - Make sure that the edges of two perpendicular surfaces are visible in the shot.
  - Shoot at a 45-degree angle if you can.
  - Watch out for lens distortion.
  - Modeling by photo-matching is not a linear process.

Section 8.3

- Modeling with photo-textured faces is not hard, but you *must* make sure that your texture is *projected*.

ASSESS YOUR UNDERSTANDING

SUMMARY QUESTIONS

1. What tool do we use to stick artwork to 3D prototypes or packaging designs?
2. Models to which you apply photo textures can be submitted to the 3D Warehouse. True or false?
3. To map an image to a face, you need at least:
   a. One face
   b. Two faces
   c. Three faces
   d. Four faces
4. Free Pin mode is also referred to as the Stretch Texture mode. True or false?
5. When you create a new model using photo-matching, you create a new ______.
6. The perspective bars come in two pairs: yellow and blue. True or false?
7. Modeling with photo-matching is a linear process. True or false?
8. The key to mapping photo textures to curved surfaces is to:
   a. Sample a flat surface.
   b. Line up a flat surface with the curved surface to which you want to apply the photo texture.
   c. Paint the curved surface with a texture.
   d. Paint the flat surface with a texture.

APPLY: WHAT WOULD YOU DO?

1. You have a photo of a home that you want to add to your model of a house. List the steps you would take to add this photo to the faces of your model.
2. Why would you want to have the option of modeling on top of photo textures? Describe a specific scenario where this technique would be useful.
3. You want to create a new model using photo-matching in your SketchUp file. What steps do you take to do this and why?

BE AN ARCHITECT

Adding a Photo to a Model of a House

Create in SketchUp an exterior model of a real-life house in your neighborhood using a photograph that you would have taken. Next, use the Texture Tweaker to map this photo to the appropriate faces in your model.

Building a Model From Scratch

Take a photograph of a coffee table or other simple piece of furniture that is blocky, such as a dresser or armoire. Use Photo Match to build a model of that object from scratch.

KEY TERMS

Geo-location
Mapping
Move/Scale/Rotate/Shear/Distort
Texture mode
Shearing
Stretch Texture mode

CHAPTER

CHANGING YOUR MODEL'S APPEARANCE

Applying Styles and Shadows

Do You Already Know?

- How to apply styles to your model?
- When and where to use styles?
- Ways to make changes to existing styles?
- How to evaluate the way shadows make models look more realistic?
- How to create and display accurate shadows in your model?

For the answers to these questions, go to **www.wiley.com/go/chopra/googlesketchup2e**

What You Will Find Out	What You Will Be Able To Do
9.1 How to change your model's appearance.	Understand how, when, and where to apply styles. Learn how to change the way faces look. Apply watermarks to your models.
9.2 How to create new styles.	Learn how to save and share the styles you make. Apply your styles in other models.
9.3 How to work with shadows.	Learn to use shadows to add depth and realism to your model.

INTRODUCTION

It's all fine and well to build elegant and efficient models, but that's only part of what this software's all about. SketchUp is also a very capable tool for presenting the models you build. Deciding how your models should look—loose and sketchy, quasi-photorealistic, or anything in between—can be lots of fun, and making the right decisions can help your models communicate what they are supposed to.

The first half of this chapter is about Styles. If you are the sort of person who likes to draw, you are in for a treat. If you can't draw a straight line with a ruler, you are in for an even bigger treat. SketchUp Styles is all about deciding how your geometry—all your faces and edges—will actually look. Reference Color Plate 21 for an idea of what styles can do.

The second half of this chapter is dedicated to SketchUp's Shadows feature. Displaying shadows is an easy operation; it's a matter of clicking a button. Adding shadows to your model views offers lots of ways to make them look more realistic, more accurate, and more readable.

9.1 CHANGING YOUR MODEL'S APPEARANCE WITH STYLES

GOOGLE SKETCHUP IN ACTION

Understand how, when, and where to apply styles.

This section provides a complete rundown of how to use styles in SketchUp 8.

To begin with, think about why you'd want to use styles in the first place. With so many options, it's important not to get stuck in applying too many styles.

9.1.1 Choosing How and Where to Apply Styles

Styles
A collection of settings that determines how the geometry appears in a given SketchUp model.

One important thing to remember about **styles** is that they're endless. With a million permutations of dozens of settings, you could quite literally spend all day altering the way your model looks. Thus, keeping one guiding question in mind—"Does this setting help my model say what I want it to say?"—will help you focus on what's important. There's no doubt styles are fun to use, but making them *useful* is the key to keeping them under control.

To make smart decisions about SketchUp styles, you should consider at least two factors when you're "styling" your model:

- **The subject of your model's "level of completeness":** For instance, you might consider reserving "sketchy" styles for models that are still evolving. The message this sends is "this isn't permanent/I'm open to suggestions/all of this can change if it has to." As your design gets closer to its final form, the appearance of your model should generally get less rough and more polished, and your styles should reflect this change. In this way, you can use styles to communicate how much input your audience can have and what decisions still need to be made.

- **How much your audience knows about design:** When it comes to how styles are perceived, there's a big difference between an architecture-school jury and a nondesigner client who's building a house for the first time. Design professionals are more experienced at understanding 3D objects from 2D representations, so they don't need as many visual "clues" to help them along. The essence of styles is really to provide these clues, so here's a rule of thumb: the more your audience knows about design, the simpler you should keep your styles.

Before you dive in to styles, also remember that a little style goes a long way. No matter how tempting it is to go crazy with the styles settings, please resist the urge. Remember that the purpose of styles is to help your model communicate and *not* to make it look "pretty" or "cool." If the *style* of your work is getting noticed more than the content of your work, you should tone down your uses of styles. Figure 9-1 shows an example of going overboard with styles and then reining things in.

9.1.2 Applying Styles to Your Models

The easiest way to get started with styles is to apply the pre-made styles that come with SketchUp. You will find plenty of them, which is great, because seeing what's been done is the best way to see what's possible. In addition, as you read through this section of the chapter, you'll no doubt get ideas for your own styles, and that's where the fun begins.

Applying an existing SketchUp style to your model is a four-step process:

1. Choose Window ⇨ Styles to open the Styles dialog box.
2. Click the Select tab and then choose a Styles collection from the Styles Collections drop-down list. You will be introduced to the collections that come pre-installed with SketchUp 8 further down.
3. Click a style in the Styles window to apply it to your model.

Figure 9-1

Abusing styles is easy. Remember, a simpler model has often a greater effect.

Figure 9-2

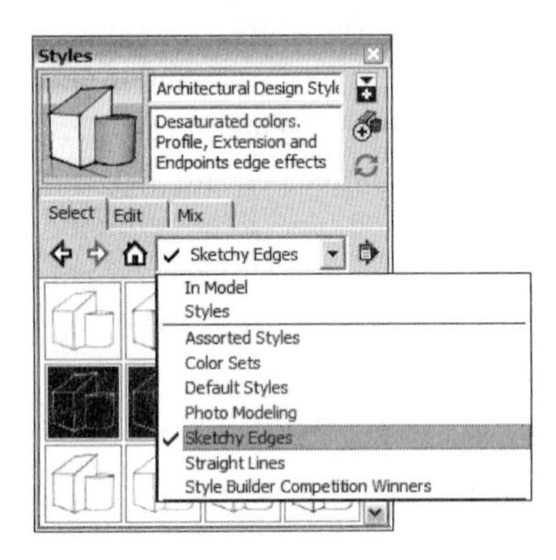

The Styles Libraries drop-down list is where you'll find all of your styles.

This may come as a surprise, but it's not possible to view your model without any style at all. This is because styles are really just combinations of display settings. Some styles are fancier than others, but no matter what you do, you always must have a style applied. If you want to get a relatively neutral view of your model, try choosing a style from among the selection provided in the Default Styles library.

One of the best things about SketchUp is that it offers so much premade content. Whether that content is styles, components, or materials, SketchUp comes with numerous examples to get you started. Figure 9-2 is a shot of the Styles Collections drop-down list you'll see when SketchUp 8 is "new out of the box." As the figure shows, a variety of different styles are built right in to the software.

Here's a quick introduction to the most interesting options in the Styles Collections drop-down list:

- **In Model:** The In Model collection shows you all of the styles you've applied to your model. It keeps track of every style you've *ever* applied to the model, whether that style is still applied or not. Thus, to see a current list of styles in your SketchUp file, do the following:
  1. Choose the In Model styles collection to show a list of all styles you've applied to your model.
  2. Click the Details flyout menu and choose Purge Unused to get rid of any styles that you aren't currently using.
- **Default Styles:** With the exception of the first style in this library (which is the default style for all new SketchUp files you create), these styles are as minimal as it gets: white background, black edges, white-and-gray front-and-back faces, and no fancy edge effects. Consider using these

styles to get back to a clean starting point for your models; it's usually easier to start simple and build from there.

- **Photo Modeling:** These styles make it easier to work when you are building models that are *photo-textured* (completely covered in photographs.) Use them when you model for Google Earth. Chapter 8 covers modeling with photos in detail.
- **Sketchy Edges:** The Sketchy Edges styles in SketchUp 8 are the result of more than a year's work on non-photorealistic rendering. (See the sidebar, below "Running from Realism: NPR Styles," for the whole story.)

Basically, these styles involve using real hand-drawn lines instead of digital ones to render edges. The result is that you can make your models look more like manual sketches than ever before. As mentioned earlier in

FOR EXAMPLE

RUNNING FROM REALISM: NPR STYLES

In the world of 3D modeling software, the trend in recent years has been toward *photorealism*. Today, the standard of perfection is usually how close a model comes to looking like a photograph, and in many cases, that standard has been met.

But what about models of buildings or other things that aren't completely finished? Perhaps you're an architect who's designing a house for a client. If you aren't sure what kind of tile you'll end up using on your roof, how are you supposed to make a photorealistic rendering of it? You *could* just go ahead and throw any old tile up there as a placeholder, but that could backfire. Your client could hate the tile and decide not to hire you without ever telling you why—and all because of something you didn't even choose.

What you need is a way to show only the decisions you've made so far, and that is exactly why architects, interior designers, and other such professionals make sketches and doodles instead of photorealistic renderings. When you are designing, decisions don't all happen at once, so you need to be able to add detail as your design evolves. Sketching allows you do that because it offers a continuum from "cartoony" to photographic, with everything in between. The following figure is an illustration of this.

Programs like SketchUp offer what's called *NPR*, or *non-photorealistic rendering*, as a way to solve this problem for people who design in 3D. Instead of dedicating processor power to making representations that look like photographs, the people who created SketchUp went in the opposite direction—they made a tool that lets you make drawings that are useful throughout the design process. And because SketchUp's NPR engine works in real time, you can make changes on the fly, in front of your audience.

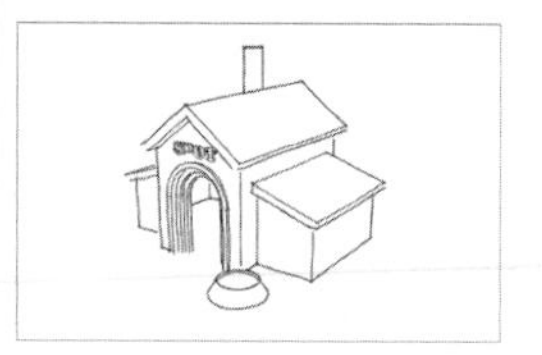

the chapter, you can use Sketchy Edges styles to convey any of the following ideas:

- that your design is in process
- that your model is a proposal and not a finished product
- that you welcome feedback in any and all forms

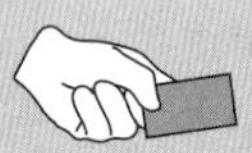

CAREER CONNECTION

Tony is a theatrical and lighting designer and constantly uses Google SketchUp. He likes it because it has excellent presentation tools built in; especially the styles and the scene manager.

He also "likes how well it interacts with other software, so that the work done in SketchUp is not wasted when further documenting the model." When preparing his work, Tony says, "I create a series of scenes, usually with simple non-realistic styles, and walk through it or use sections, etc. as appropriate. Sometimes I will export to AVI, and using a video editor I intersperse rendered scenes, or other info. There is no question that a 3D model which can be walked through, rotated, and explored in real time is the way to present concepts both to lay people and other professionals—I now never need to create a physical model."

He usually renders his scenes with Rhino "which has a fantastic SketchUp export." He recently designed a makeover of a large existing wharf shed into a temporary venue for the New Zealand Arts Festival, including a 600-seat theater and a bar/restaurant/club. Although the design was documented in Rhino and all the working drawings were done in that program, SketchUp was used to work up the initial concept and to present it to the Board.

9.1.3 Making Changes to Styles

If you're handy in the kitchen, you've probably heard the saying that cooking is an art and baking is a science. Cooking allows you to experiment—adding a little of this and a dash of that won't destroy your final result. When it comes to baking, however, taking liberties with a recipe can be a train wreck.

Luckily, in SketchUp, making your own styles has a lot more in common with cooking than it does with baking. Go ahead and experiment; you can't do any irreversible harm. Playing with styles doesn't affect the geometry in your model in any way, and because styles are just combinations of settings, you can always go back to the way things were before you started.

Of the three tabs in the Styles dialog box, Edit is definitely the most used of the group. Because you find so many controls and settings here, SketchUp's designers broke the Edit tab into the following five sections:

- **Edge:** The Edge section contains all the controls that affect the appearance of edges in your model. This includes their visibility, their color, and other special effects that you can apply.

- **Face:** This section controls the appearance of faces in your model, including their default colors, their visibility, and their transparency.
- **Background:** The Background section has controls for setting the color and visibility of the background, the sky, and the ground plane in your model.
- **Watermark:** These are images (think of a faint orthographic plan view or a logo) that you can use as backgrounds or as overlays. The Watermark tab gives you control over these.
- **Modeling:** The Modeling section provides controls for setting the color and visibility of numerous elements in your model, including section planes and guides.

Watermark
A graphic element that can be applied either behind or in front of a model to produce certain effects such as a faint orthographic plan view or a logo.

The following sections explain each tab in detail and provide suggestions for using some of the settings.

Tweaking Edge Settings

The Edge tab is tricky because it changes ever so slightly depending on what kind of style you currently have applied to your model. Specifically, NPR styles have different settings than regular, non-NPR styles. Figure 9-3 shows both the regular and the NPR versions of the Edge tab, which you open by choosing Window ⇨ Styles, selecting the Edit tab, and then clicking the box icon on the far left.

Figure 9-3

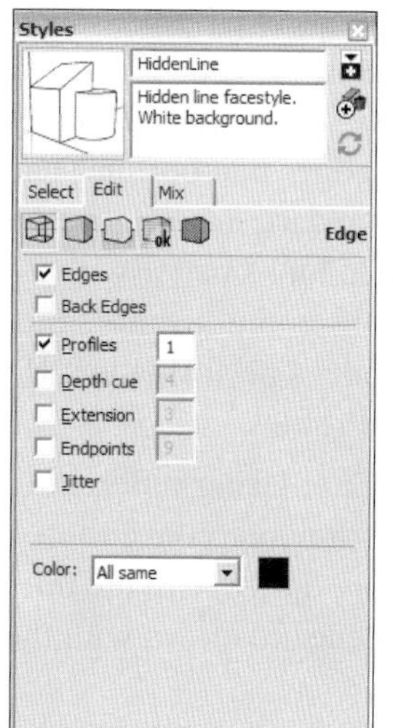

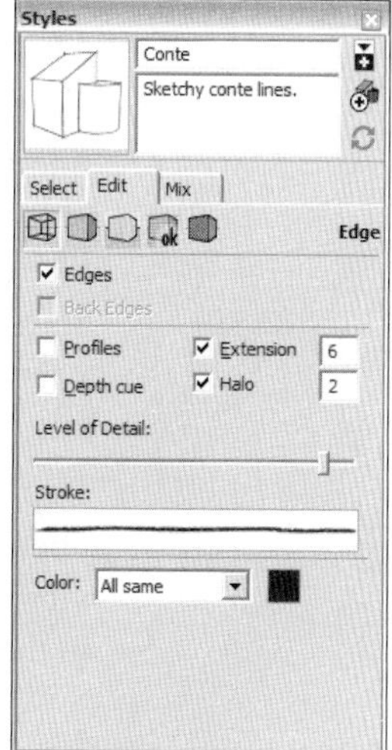

The appearance of the Edge tab is slightly different when using regular styles as opposed to NPR styles.

FOR EXAMPLE

INTRODUCING STYLE BUILDER

If you're using the Pro version of SketchUp 8, you have access to a relatively new Style Builder tool. It's a completely separate application (just like LayOut) that's put on your computer when you install SketchUp Pro 8.

Style Builder lets you create NPR styles based on edges you draw. Consequently, you can make your SketchUp models look like you drew them by hand with your medium of choice. All you need is a scanner and a piece of software like Photoshop, and you are good to go. The best thing about the styles you create with Style Builder is that they are completely unique. Unless you share them with someone else, no one can ever make SketchUp models that look like yours.

Because Style Builder is a whole other program and because it *is* only included in the Pro version of SketchUp, no more information is provided in this book. Reference this book's Web site for lots more information, though Style Builder is an incredible tool.

SketchUp 8 comes with two kinds of styles: regular and NPR. In NPR, SketchUp uses digitized, hand-drawn lines to render the edges in your model. All the styles in the Sketchy Edges collection, as well as all the ones in the Assorted Styles collection, are NPR styles. Because you can create your own styles based on existing ones, all the styles you create using edge settings from one of these NPR styles is an NPR style, too.

Below are some of the less-obvious settings in the Edge section. (See Figure 9-4 for a visual reference.)

- **Back Edges:** Switching this on tells SketchUp to draw all your model's *obscured* (hidden behind a face) edges as dashed lines. This comes in surprisingly handy. For instance, there is the ability to inference to edges and points that couldn't be seen before. Also, myriad dashed lines make a technical drawing look impressive and complex.
- **Profiles:** Selecting the Profiles check box tells SketchUp to use a thicker line for edges that outline shapes in your model. Using profile lines is a pretty standard drawing convention that's been around for a long time. SketchUp looks better with Profiles on, but it comes at a price: Profiles take more computer horsepower to draw, which can seriously affect your model's performance. If your model's big, think twice before you turn on Profiles.
- **Depth Cue:** Using different line thicknesses to convey depth is another popular drawing convention. With this method, objects closest to the viewer are drawn with the thickest lines, whereas the most distant things in the scene are drawn with the thinnest lines.

Depth Cue is SketchUp's automatic way of letting you apply this effect to your models. When its check box is selected, Depth Cue dynamically assigns line thicknesses (draftspersons call them "line weights") according

Figure 9-4

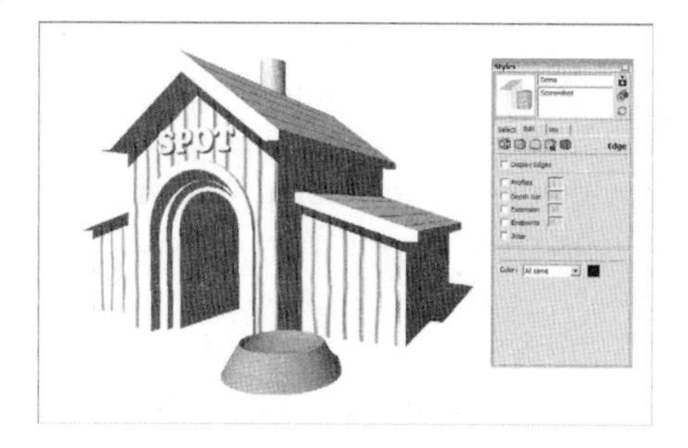

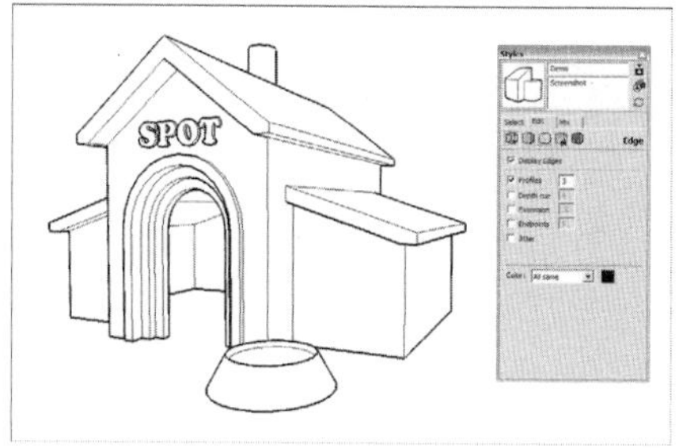

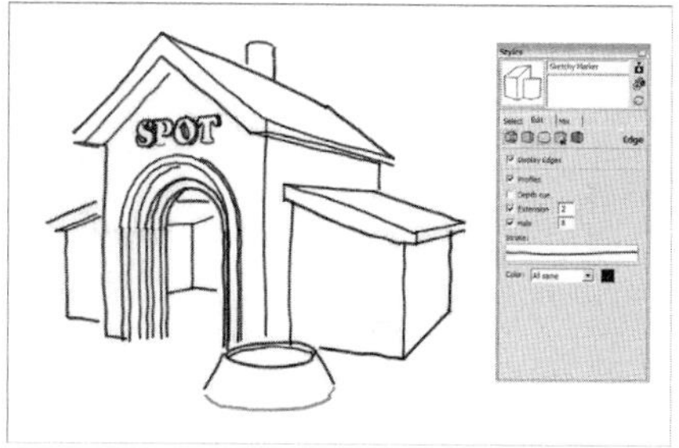

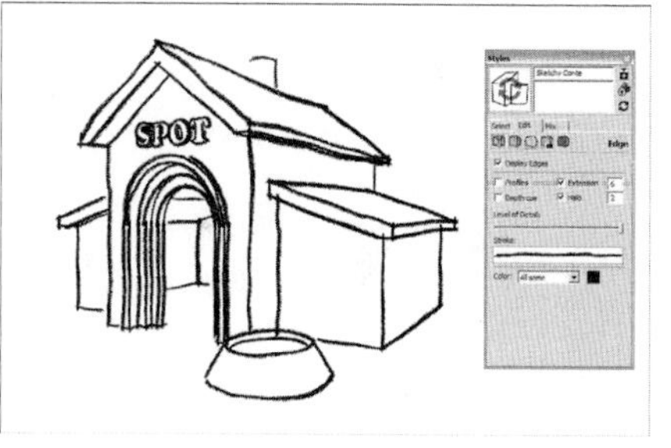

Choose among the edge settings to give your model the desired look, from realistic to sketchy.

to how far away from you things are in your model. The number you type in is both your desired number of line weights *and* the thickness of the fattest line SketchUp will use. Consider using a maximum line weight of 5 or 6 pixels. One last item: When using Depth Cue, turn off Profiles. These two drawing conventions don't work well together, so always choose to use one or the other.

- **Halo:** Halo simply ends certain lines before they run into other ones, creating a "halo" of empty space around objects in the foreground.

This keeps your model looking neat and easy to read. In fact, this is a drawing trick that pencil-and-paper users have been using for years to convey depth; look closely at most cartoons and you'll see what this means. The number you type into the Halo box represents the amount of "breathing room" SketchUp gives your edges. The unit of measure is pixels, but there's no real science to it; just play with the number until things look right to you. Take a look at Figure 8-7 to see Halo in full effect.

- **Level of Detail:** When you slide the Level of Detail controller (which only appears when you've applied an NPR style) back and forth, you're effectively telling SketchUp how busy you want your model to look. The farther to the right you slide it, the more of your edges SketchUp displays. You should experiment with this setting to see what looks best for your model. Figure 9-4 shows what happens when you slide the Level of Detail controller from left to right.
- **Color:** You use the Color drop-down list to tell SketchUp what color to use for all the edges in your model. Here's what each of the options means:
  - **All Same:** This tells SketchUp to use the same color for all the edges in your model. You tell it what color to use by clicking the color well on the right and choosing a color.
  - **By Material:** Choosing this causes your model's edges to take on the color of whatever material they're painted with. Because most people don't know that they can paint edges different colors, this feature doesn't get used very often.
  - **By Axis:** This is a useful but hidden gem. Choosing to color your edges "by axis" tells SketchUp to make everything that's parallel to one of the colored axes the color of that axis. Edges that aren't parallel to any of the axes stay black. Why is this so important? When something is not right with your model—for example, faces won't extrude, or lines won't "sink in"—switching your edge colors to by axis is the first thing you should do. You'll be surprised to see how many of your edges aren't what they seem.

Figure 9-5

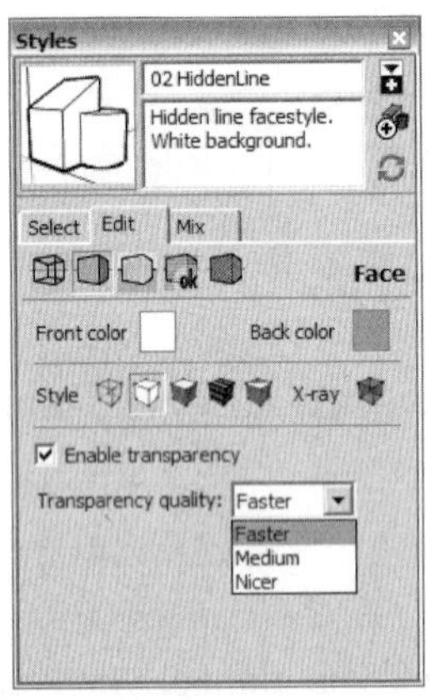

The Face section controls the appearance of your model's faces.

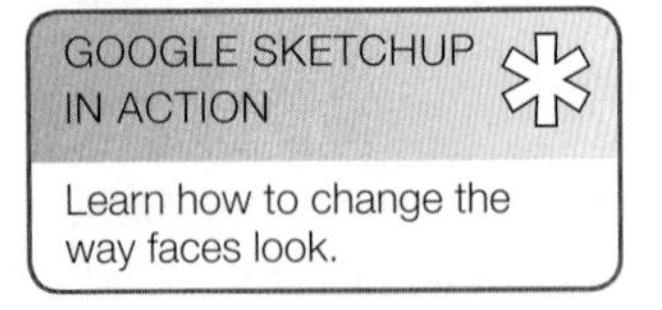

9.1.4 Changing the Way Faces Look

The Face section of the Styles dialog box is very simple—at least compared to the Edge section. This area of the SketchUp user interface controls the appearance of faces, or surfaces, in your model. From here, you can affect their color, visibility, and translucency. Figure 9-5 shows the Face section, which you can open by choosing Window ⇨ Styles, selecting the Edit tab, and clicking the box icon that's second from the left. The following sections describe each of the elements of this tab in detail.

FOR EXAMPLE

IN A FOG?

If you're looking for something else to provide a sense of depth in your model views, look no further than the Fog feature. Fog does exactly what it says—it makes your model look like it's enshrouded in fog (see figure below). Follow these three steps to let the fog roll into your model:

1. Choose Window ⇨ Fog to open the Fog dialog box.
2. Select the Display Fog check box to turn on the fog effect.
3. Experiment with the controls until you like what you see.

Unfortunately, the process of controlling how Fog looks is not particularly scientific. You just play around with the sliders until you have the amount of fog you want. However, just in case you absolutely need to know, here's what the sliders do:

- **Top slider (0%):** This controls the point in space at which Fog begins to appear in your model. When it's all the way to the right (toward infinity), you can't see any fog.
- **Bottom slider (100%):** This controls the point in space at which the fog is completely opaque. As you move the slider from left to right, you're moving the "completely invisible" point farther away.

Choosing Default Colors for Front and Back Faces

In SketchUp, every face you create has a back and a front. You can choose what colors to use by default for all new faces you create by clicking the Front and Back color wells and picking a color. Try sticking with neutral tones for your defaults; you can always paint individual faces later on.

Sometimes when you're modeling in SketchUp, a face will be turned "inside out." Follow these steps to flip a face around so that the right side is showing:

1. Select the face you want to flip.
2. Right-click and choose Reverse Faces.

Knowing which face is the front and which is the back is especially important if you plan to export your model to another program. Some of these, like 3D Studio Max, use the distinction between front and back to make important distinctions about what to display. In these cases, showing the wrong side of a face can end up producing unexpected results.

Choosing a Face Style

Even though these are called Face styles, they have nothing to do with styles. Face styles might as well be called Face modes because that what they are: different modes for viewing the faces in your model. You can flip between them as much as you like without affecting your geometry. All they do is change the way SketchUp draws your model on the screen. Each one has its purpose, and all are shown in Figure 9-6.

- **Wireframe:** In Wireframe mode, your faces are invisible. Because you can't see them, you can't affect them. Only your edges are visible, which makes this mode handy for doing two things:
  - When you select edges, switch to Wireframe mode to make sure that you've selected what you meant to select. Because no faces block your

Figure 9-6

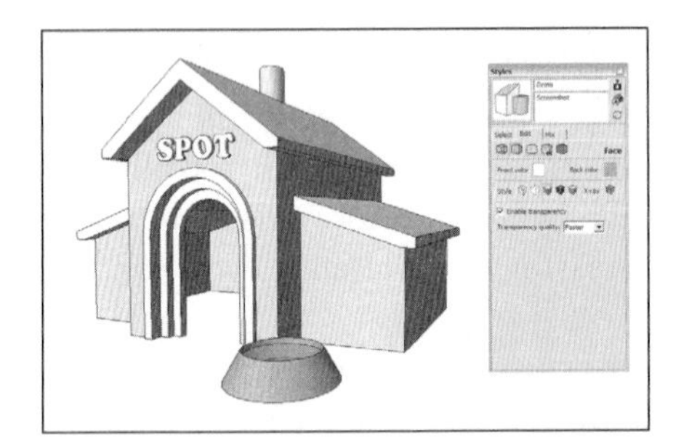

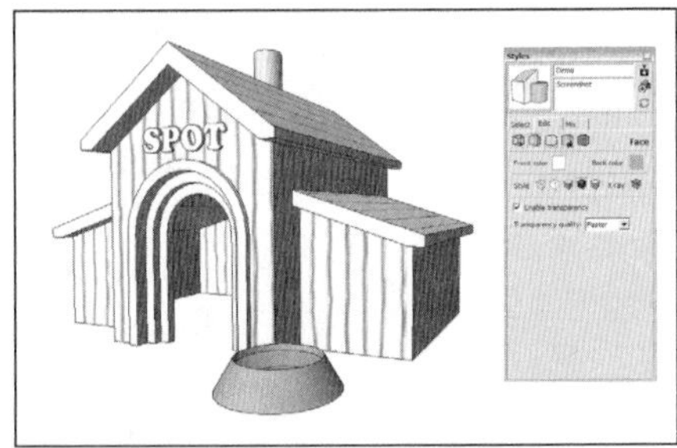

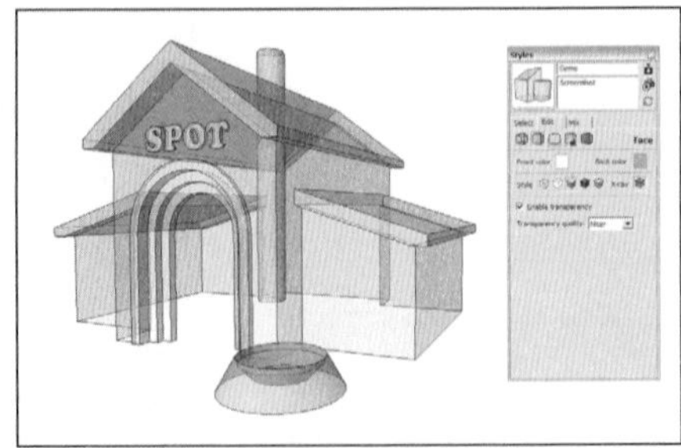

Use Face styles to change the way your faces appear.

view, this is the best way to make sure that you're getting only what you want. (The new Back Edges setting is handy for this, too.)

 – After you've used Intersect Faces, you'll usually have stray edges in your model. Wireframe is the quickest way to erase them because you can see what you're doing. See Chapter 4 for details on Intersect Faces.

- **Hidden Line:** Hidden Line mode displays all your faces using whatever color you're using for the background; it's as simple as that. If you're trying to make a clean, black-and-white line drawing that looks like a technical illustration, make your background white. This is discussed later in this chapter.
- **Shaded:** This Face style displays your faces with colors on them. Faces painted with a solid color appear that color. Faces to which you've added textures are shown with a color that best approximates their overall color. For example, if your texture has a lot of brown in it, SketchUp will pick a brown and use that. For models with a lot of textures, switching to Shaded mode can speed up orbiting, zooming, and otherwise navigating around. Consequently, unless you absolutely need to see the textures you've applied to your faces, you might consider staying in Shaded mode as you're working on your model.
- **Shaded Textures:** Use Shaded with Textures when you want to see your model with textures visible. Because this mode puts a lot of strain on your computer, it can also be the slowest mode to work in. Thus, you might think about turning it on only when you're working on a small model, or when you need to see the textures you've applied to your faces. Obviously, if you're going for a photorealistic effect, this is the mode to choose. It's also the mode that best approximates what your model will look like when (and if) you export it to Google Earth.
- **Display Shaded Using All Same:** When you want to quickly give your model a simplified color scheme, use this Face style; it uses your default front and back face colors to paint your model. You can also use this setting to check the orientation of your faces if you are exporting your model to another piece of 3D software.
- **X-Ray:** Unlike using translucent materials on only some of your faces (such as glass and water), turning on X-Ray lets you see through all your faces. Use it when you want to see through a wall or a floor to show what's behind it. If you are in a plan (overhead) view, it is a great way to demonstrate how a floor level relates to the one below it.

Adjusting Transparency

Displaying transparency (as in translucent materials) is an especially taxing operation for SketchUp and your computer to handle, therefore, you can decide how to display translucent materials:

- **Enable transparency:** Deselect this check box to display translucent materials as opaque. You should turn off transparency to speed SketchUp's performance if you find that it's slowed down.

Figure 9-7

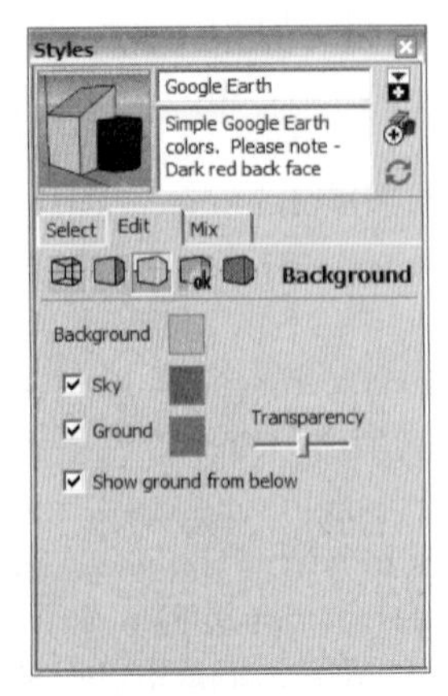

Use the Background section to turn on the sky and the ground and to choose colors for these elements.

- **Transparency quality:** If you decide to display transparency, you can further fine-tune your system's performance by telling SketchUp how to render that transparency.

Thus, you have the choice of better performance, nicer graphics, or an average of the two. Which one you choose depends on the size and complexity of your model, the speed of your computer, and the nature of your audience.

9.1.5 Setting Up the Background

In the Background section of the Styles dialog box, you choose colors and decide whether you want to be able to see a sky and a ground plane. Check out Figure 9-7 to get a view of the Background section, along with an idea of how this works. To open these options in your own copy of SketchUp, choose Window ⇨ Styles, select the Edit tab, and click the middle icon at the top of the tab.

You have the following options on the Background tab:

- **Background**: For most traditional models, the background is set to white.
- **Sky:** Displaying a sky in your modeling window makes things slightly more realistic, but the real purpose of this feature is to provide a point of reference for your model. In 3D views of big things like architecture, it's nice to be able to see the horizon. Another reason for turning on the sky is to set the mood—keep in mind that the sky isn't always blue. There are some beautiful SketchUp renderings wherein the sky was sunset or maybe orange.
- **Ground:** Just like the Sky tool, you can choose to display a ground plane in your model. You can pick the color by clicking the color well, and you can even choose to have the ground be translucent. Before turning on this feature, note that it can be very hard to find a ground color that looks good, no matter what you're building. Also note that you can't dig into the earth to make sunken spaces (like courtyards) with the ground turned on. Instead of turning on this feature, consider making your own ground planes with faces and edges. It's more flexible, and it often looks better.

9.1.6 Working with Watermarks

Watermarks are much easier to understand if you don't think about them as actual watermarks. They're not anything like watermarks; in fact, they're much more useful. Think of them this way: Watermarks are graphics that you can apply either *behind* or *in front* of your model to produce certain effects. Here are a few of the things you can do with SketchUp watermarks:

- Simulate a paper texture, just like the styles in the Paper Watermarks library.

Apply watermarks to your models.

- Apply a permanent logo or other graphic to your model view.
- Layer a translucent or cutout image in the foreground to simulate looking through a frosted window or binoculars.
- Add a photographic background to create a unique model setting.

Figure 9-8

Edit Watermark button

The Watermark section.

Understanding the Watermark Controls

Figure 9-8 shows the Watermark tab of the Styles dialog box. Here's a brief introduction to what the controls do:

- **Add, Remove, and Edit Watermark buttons:** The +, –, and gears icons allow you to add, remove, and edit (respectively) watermarks in the style you're editing.
- **Watermark list:** This list shows all your watermarks in relation to *Model Space*, which is the space your model occupies. All watermarks are either in front of or behind your model, making them overlays or underlays, respectively.
- **Move Up or Down arrows:** Use these buttons to change the stacking order of the watermarks in your model view. Select the watermark you want to move in the list and then click one of these buttons to move it up or down in the order.

Adding a Watermark

Watermarks are by no means simple, but working with them is. Follow these steps to add a watermark to your model view:

1. **Click the Add Watermark button to begin the process of adding a watermark.** The Open dialog box appears.
2. **Find the image you want to use as a watermark and then click the Open button to open the first Create Watermark dialog box (see Figure 9-9).** You can use any of these graphics file formats: TIFF, JPEG, PNG, and GIF.
3. **Type in a name for your watermark in the Name box.**
4. **Choose whether you want your new watermark to be in the background or in the foreground as an overlay, and then click the Next button.**
5. **Decide whether to use brightness for transparency.** Selecting this check box tells SketchUp to make your watermark transparent, which kind of simulates a real watermark. *How* transparent each part becomes is based on how bright it is. White is the brightest color, so anything white in your watermark becomes completely transparent. Things that are black turn your background color, and everything in between turns a shade of your background color.

Figure 9-9

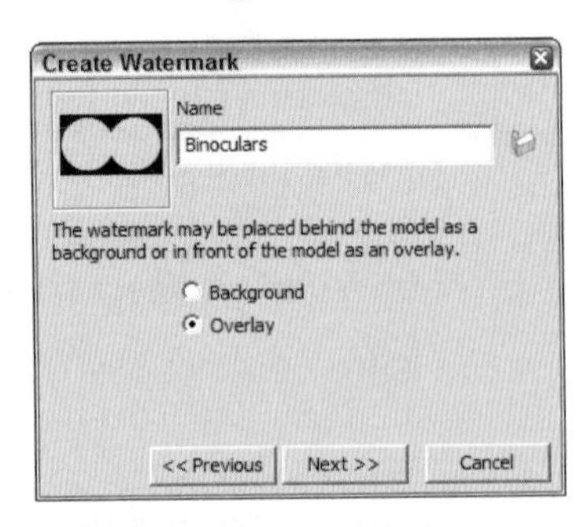

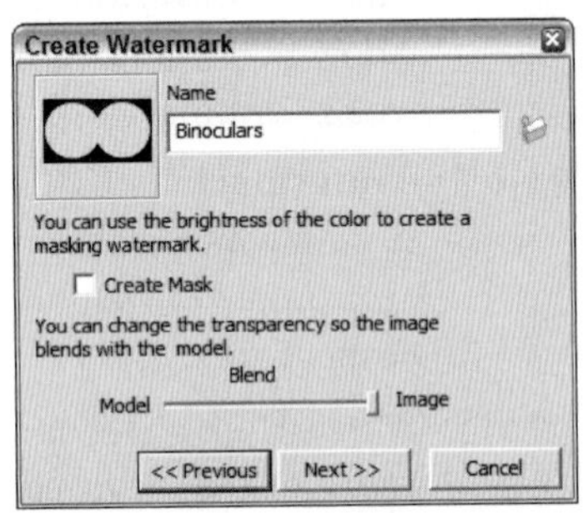

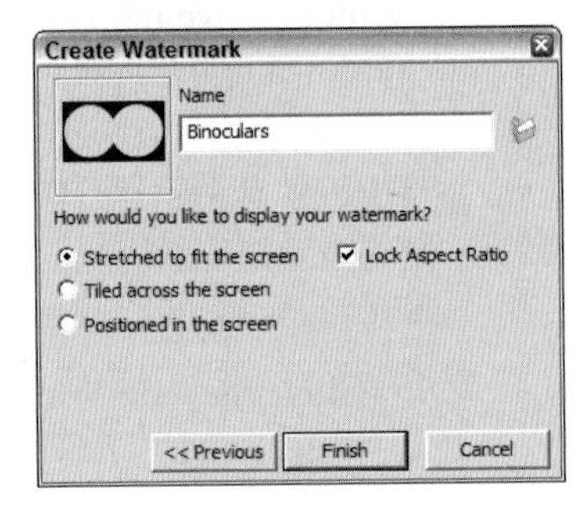

The Create Watermark series of dialog boxes.

6. **Adjust the amount that your watermark blends with what's behind it, and click the Next button.** In this case, Blend is really just a synonym for Transparency. By sliding the Blend slider back and forth, you can adjust the transparency of your watermark. Blend comes in handy for making paper textures because that process involves using the same watermark twice: once as an overlay and once as an underlay. The overlay version gets "blended" in so that your model appears to be drawn on top of it. To see how this works, apply one of the Paper Texture styles to your model, and then edit each of the watermarks to check out its settings.
7. **Decide how you want your watermark to be displayed and then click the Finish button.** You have three choices for how SketchUp

can display your watermark: stretched to fit the entire window, tiled across the window, and positioned in the window. If you select Stretched to Fit the Screen, be sure to select the Locked Aspect Ratio check box if your watermark is a logo that you don't want to appear distorted.

FOR EXAMPLE

WATERMARKS AND ALPHA CHANNELS

If you want to make a watermark out of an image that isn't a solid rectangle (like a logo), you need to use a graphics file format, like PNG or GIF, which supports alpha channels. An *alpha channel* is an extra layer of information in a graphics file that describes which areas of your image are supposed to be transparent. It sounds complicated, but it's really a straightforward concept. To make an image with an alpha channel, you need a piece of software like Photoshop. Try searching for "alpha channels" on Google for more information.

Editing a Watermark

You can edit any watermark in your SketchUp file at any time. Follow these simple steps to edit a watermark:

1. **Select the watermark you want to edit in the Watermark list.** You can find the Watermark list on the Watermark tab, in the Edit pane of the Styles dialog box.
2. **Click the Edit Watermark button to open the Edit Watermark dialog box.** The Edit Watermark button looks like a couple of small gears.
3. **Use the controls in the Edit Watermark dialog box and click the OK button when you're done.** For a complete description of the controls in this dialog box, see the previous section in this chapter.
4. **Consider the following tips about your watermarks**. You need to right-click on the name of one watermark if you want to export it as an image file. Watermarks only can be used either as a background or as an overlay but we can control the relative position of those in the page, their scale, whether they can be used as masks or how much they will blend with the model.

Figure 9-10

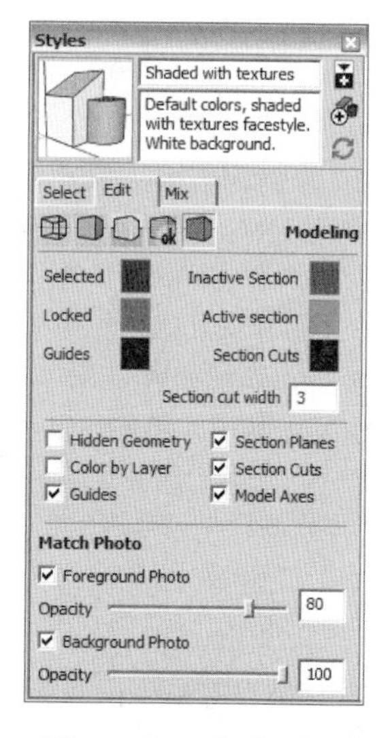

The controls in the Modeling section are every bit as simple as they look.

Tweaking Modeling Settings

All you need to know about the controls in the Modeling tab (see Figure 9-10) of the Styles dialog box is that there's not much to know. You use these controls to adjust the color and visibility of all the elements of your model that aren't geometry, such as the color of a selected entity. To open these options, choose Window ⇨ Styles, select the Edit tab, and click the box icon on the far right, at the top of the tab.

The controls found are work as follows:

- **Controls followed by color wells:** Click the wells to change the color of that type of element.
- **Section cut width:** This refers to the thickness of the lines, in pixels, that make up the section cut when you're using a section plane. For more about this, have a look at the information on cutting sections in Chapter 10.
- **Controls preceded by check boxes:** Use these to control the visibility of that type of element in your model. Three of these controls bear special mention because they can be a bit confusing:
- **Color by Layer:** This tells SketchUp to color your geometry according to the colors you've set up in the Layers dialog box. In order to work successfully with this option, you would have to first select Window ⇨ Layers and assign a specific color for each layer you have. Then, you can go ahead and choose Color by Layer under Window ⇨ Styles ⇨ Modeling.
- **Section Cut Width:** This refers to the thickness of the lines, in pixels, that make up the section cut when you use a section plane. For more about this, have a look at the information on cutting sections in Chapter 10.
- **Controls with check boxes:** Use these to control the visibility of that type of element in your model. Three of them are a little confusing:
  - *Color by Layer:* Tells SketchUp to color your geometry according to the colors you've set up in the Layers dialog box. Check out Chapter 7 for more on this.
  - *Section Planes:* This refers to the section plane objects that you use to cut sections. They're gray with four arrows on their corners.
  - *Section Cuts:* Unlike section planes, this setting controls the visibility of the section cut effect itself. With this deselected, your section planes don't appear to cut anything.
- **Match Photo settings:** When you photo-match (which you can read all about in Chapter 8), adjusting the visibility of your photograph is sometimes helpful. Use these controls to hide, show, and adjust the photo's opacity in both the background and the foreground.

Mixing Styles to Create New Ones

You can use the Mix tab to combine features of multiple styles to make new ones. Instead of working through the sections of the Edit tab, flipping controls on and off, sliding sliders, and picking colors, the Mix tab lets you build new styles by dropping existing ones onto special "category" wells. In addition to being a nifty way to work, this is the only way you can switch a style's edge settings between NPR and non-NPR lines.

Figure 9-11

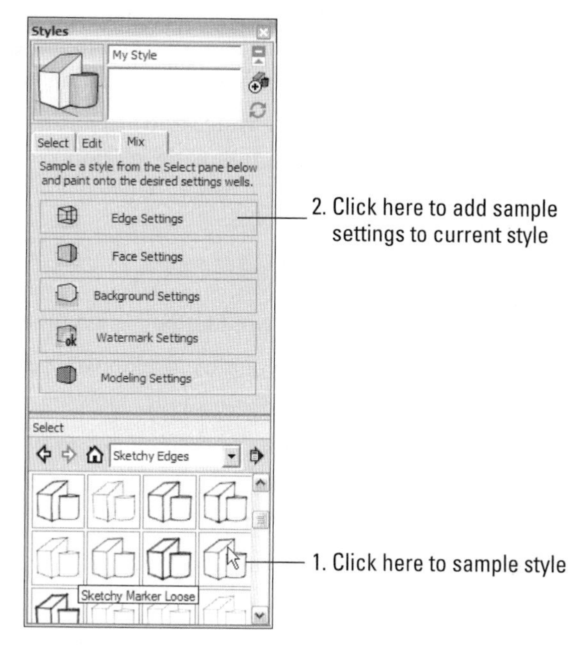

Sample from different styles to update the style you are working on.

NPR refers to the styles in the Assorted Styles, Sketchy Edges, and Competition Winners collections. These non-photorealistic rendering styles use scanned, hand-drawn lines to draw the edges in your model. If you have SketchUp Pro, you can use Style Builder to make your own NPR styles from lines you draw and scan in. Read the sidebar "Introducing Style Builder," earlier in this chapter, for more information.

Follow these steps to change a style using the Mix tab (see Figure 9-11):

1. **Choose Window ⇨ Styles and click the Mix tab in the Styles dialog box.** As part of the Mix tab, the secondary selection pane opens at the bottom of the dialog box. This provides you with a way to view your styles without having to switch from the Mix to the Select sections.
2. **Find the style you want to sample from the Select section. You can call this your *source* style.** Say that you're working on a new style, and you want your edges to look just like those in the Marker Loose style that came with SketchUp. In this example, choose the Sketchy Edges collection from the Styles Collections drop-down list, where you'll find the Marker Loose style.
3. **If you're using Windows, click the source style from the Styles list in the Select section to sample it and then click the category well that corresponds to the style setting you want to apply. If you're using a Mac, drag your source style from the Styles list in the Select section**

to the category well that corresponds to the style setting you want to apply. In this case, sample the Marker Loose style from the Select section and drop it on the Edge Settings Category well because you want the edge settings from that style to be applied to the style you're working on.

4. **To save your style after you're done adding all the various elements, see the following section.**

SELF-CHECK

1. When would it be a good time to apply a "sketchy" style to our model?
2. What edge feature do we use in Style Builder to instill a sense of depth to our model views?
3. Choose the correct answer:
   a. A watermark allows you to treat realistically large masses of water in our model, such as lakes and rivers.
   b. A watermark is a graphic that you apply either *behind* or *in front* of our model to produce certain effects.
   c. A watermark is synonymous with a logo added behind or in front of our model.

Apply Your Knowledge Working with styles is a very important skill to master when working in Google SketchUp. Find a project that you would have done already, such as a model of a house or a piece of furniture, and capture three different scenes applying one style to each. When you search for your styles, choose them based on these three categories:

- A style that would be suitable if you would be presenting a preliminary idea to your client.
- A style that would show your client how you are working with a volume, but would not show yet color or texture. That would approximately where you would want to be in the middle of the concept generation phase.
- A third style that you would share with your client when the concept is complete, with texture, shadows, and color.

9.2 CREATING A NEW STYLE

Creating a new style adds it to your In Model collection of styles, which means that you can come back and apply it to your model anytime you like. Follow these steps to create a new style:

1. **Click the Create New Style button in the Styles dialog box.** This duplicates the style that was applied to your model before you clicked

the Create New Style button. Your new style appears in your In Model collection as [*name of the original style*]1.

2. **Use the controls in the Edit tab to set up your style the way you want.** Frequently, you'll want to make a new style after you've already made changes to an existing one. If you want to create a new style that reflects modifications you've already made, just switch Steps 1 and 2.
3. **Use the Name box (at the top of the Styles dialog box) to give your new style a name and press Enter.** If you want, you can also give your new style a description in the Description box, though you might want to wait until later.
4. **Click the Update button.** This updates your new style with all the changes you made in Steps 2 and 3.
5. **Check the In Model collection in the Select tab to make sure that your new style is there.** Click the In Model button (which looks like a little house) to see your In Model Styles collection. Your new style should appear alphabetically in the list.

If a bunch of styles exist in your In Model collection that you aren't using anymore and you want to clean things up, right-click the Details flyout menu and choose Purge Unused. This gets rid of any styles that aren't currently applied to any scenes in your model. Have a look at Chapter 10 to find out more about scenes.

Remember, creating a new style doesn't automatically make it available for use in other SketchUp files. To find out how to do this, keep reading.

9.2.1 Saving and Sharing Styles You Make

GOOGLE SKETCHUP IN ACTION

Learn how to save and share the styles you make.

As you're working along in SketchUp, you'll want to create your own styles. You'll also want to save those styles so that you can use them in other models. In addition, if you're part of a team, it's likely that everyone will want to have access to the same styles so that all your models look consistent.

Saving the Styles You've Made

When it comes to creating your own styles, you can approach things in two different ways. Each of these ways gets its own button (see Figure 9-12):

- **Create new style:** Clicking this button creates a new style with the settings you currently have active. When you create a new style, it shows up in your In Model collection of styles and is saved with your model. The Create New Style button can be found in the upper-right corner of the dialog box, and looks like a couple of objects with a "+" sign on it.
- **Update style with changes:** This button updates the current style with any settings changes you've made in the Edit or Mix tabs. If you want to

Figure 9-12

The Update and Create buttons in the Styles dialog box.

modify an existing style without creating a new one, this is the way to go. You can find the Update button right below the Create button in the upper-right corner of the dialog box; it looks like two arrows chasing each other in a circle. Finally, proceed with caution when you update your styles. Sometimes you might want to go back and "undo" your changes, but that sometimes can prove difficult and you might end up losing the original style or the changes you made might affect other parts of your drawing that you did not intend to.

Updating an Existing Style

To make adjustments to a style in your model, you need to update it. Follow these steps to update a style:

1. **Apply the style you want to update to your model.** If you need help with this, follow the steps in the section, "Applying Styles to Your Models," earlier in this chapter.
2. **Use the controls in the Edit tab to make changes to the style.**
3. **Click the Update Style with Changes button in the Styles dialog box to update the style with your changes.**

You use the Update Style with Changes button to rename existing styles, too. Just type the new name into the Name box (at the top of the Styles dialog box), press Enter, and then click the Update button.

When you update a style, only the copy of the style that's saved with your model is updated. You aren't altering the copy of the style that shows up in every new SketchUp file you create.

9.2.2 Using Your Styles in Other Models

Using consistently a style or a combination of styles in SketchUp is part of what you would do to customize the look of your company's drawings and models. That way the material that is presented will be recognizable and differentiated from competitors with a branded look, much like an artist's signature on a painting, regardless of who was the designer or designers involved in the production.

After you update or create a style, you probably want to make that style available in other SketchUp models. To make this happen, you need to be able to create your own styles collections. *Collections* are folders on your computer that contain the styles that appear in the Styles dialog box. You can create your own collections to keep the styles you invent neat and tidy.

Follow these steps to create a collection to contain your styles:

1. Choose Window ⇨ Styles to open the Styles dialog box.
2. Click the Select tab, click the Details flyout menu, and choose Open or create a collection. The Browse For Folder dialog box opens.
3. **Navigate to the folder on your computer or network where you want to create your collection.** You can locate your new collection anywhere you like, but it is recommended to put it in the same folder as the other styles collections on your computer. Unless you choose to install SketchUp somewhere else in a different drive, the suggested path most likely would be:
   - *Windows:* C:/Program Files/Google/Google SketchUp 8/Styles
   - *Mac:* Hard Drive/Library/Application Support/Google SketchUp 8/SketchUp/Styles
4. Click the Make New Folder button (Windows) or the New Folder button (Mac). The new folder you create becomes your new collection.
5. Type a name for your new collection.
6. (Mac) Make sure that the Add to Favorites check box is selected.
7. Click the OK button. The Browse For folder dialog box closes, and your collection is added to the Favorites section of the Collections drop-down list. It will be there in every SketchUp model you open on this computer.

GOOGLE SKETCHUP IN ACTION

Apply your styles in other models.

After you create a new collection, you can add styles to it to make them available from any model you work on.

Follow these steps to make a style available for use in other SketchUp files:

1. **Choose Window ⇨ Styles.** The Styles dialog box appears.
2. **Click the Select tab and then click the In Model button to display your In Model collection.** The In Model button looks like a little

Figure 9-13

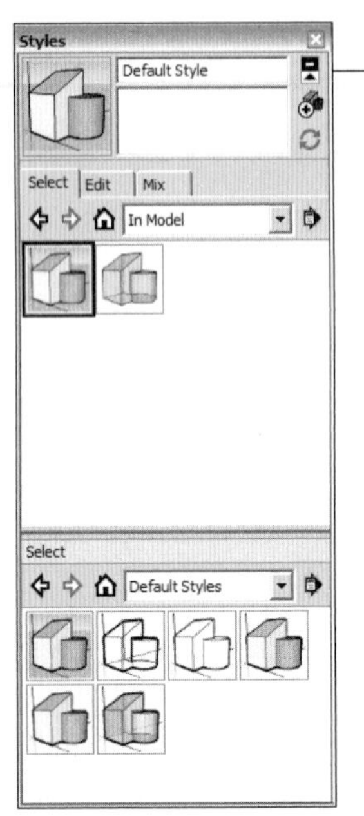

Use the Select section to manage your styles without leaving SketchUp.

house. The In Model collection contains all the styles you've used in your model, including the ones you've created.

3. **Click the Show Secondary Selection Pane button.** When you click this button, which looks like a black-and-white rectangle, and is in the upper-right corner of the Styles dialog box, a second copy of the Select section pops out of the bottom of the Styles dialog box (see Figure 9-13). Use this section to drag and drop styles between folders on your computer, which makes it easier to keep them organized.
4. **In the Select section, choose the collection to which you want to add your style.** If you've created a collection specifically for the styles you make, choose that one or you can pick any collection in the Collections drop-down list.
5. **Drag your style *from* the In Model styles list *to* the Styles list in the Select section.** By dragging and dropping your style from the upper list to the lower one, you make the style available to anyone who has access to that collection. This means that you can use the style in other SketchUp models you build on your computer. To share it with other members of your team, copy your style to a collection somewhere where other people can get to it, such as on a network.

IN THE REAL WORLD

Robert Pearce, AIA, from Pearce Design Studio uses SketchUp regularly in his studio. He likes how easy it is to use and how you can use the majority of the program before you make a financial commitment with the Pro version, which he says, is still sold at a reasonable price.

He usually sends high-resolution images to clients and occasionally walks them through the model, using the option of animating the scenes. He also likes SketchUp's ability to change materials quickly and how shadows can be shown at different times of the day or the year, as that is an important aspect to consider during presentations.

In his routine of work, he usually starts with an imported CAD floor plan and builds a full exterior 3D model, which is later exported as 2D CAD roof plans, elevations, and preliminary building sections. He usually uses SU's native image output with some retouching done in Photoshop.

SELF-CHECK

1. What button would you click to create a style with the settings you currently have active?
2. When you update a style, only the copy of the style that's saved with your model is updated. True or false?
3. A new style can be made available in other models. True or false?
4. In the context of a Style, explain what *Collections* are and where they can be found.
5. What steps do we have to take to share our styles with other members of our team?

Apply Your Knowledge Build a simple, one-story house and apply a new style to it, using an existing style as a starting point. Then, create a new collection and insert the newly created style on it. After you are done, create a new file and build a small tool shed that will carry that same new style.

9.3 WORKING WITH SHADOWS

Typically, you add shadows to a SketchUp drawing for two key reasons:

- **To display or print a model in a more realistic way:** Turning on shadows adds depth and realism, and gives your model an added level of complexity that makes it look like you worked harder than you really did.

- **To study the effect of the sun on what you've built (or plan to build) in a specific geographic location:** Shadow studies are an integral part of the design of any built object. If you are making a sunroom, you need to know that the sun is actually going to hit it. You can use SketchUp to show exactly how the sun will affect your creation, at every time of day, on every day of the year.

In this section is a brief, nuts-and-bolts description of how all the controls work, without diving too much into why you would want to pick one setting instead of another. The second part of this section is devoted to running through each of the preceding scenarios and using the controls to make SketchUp do exactly what you want it to.

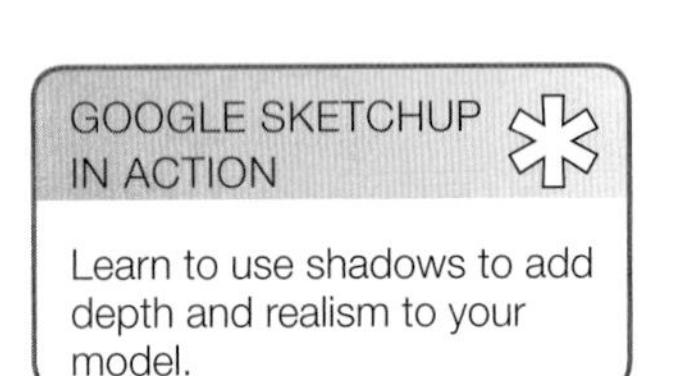

9.3.1 Discovering SketchUp's Shadow Settings

The basic thing to understand about shadows in SketchUp is that, just like in real life, they're controlled by changing the position of the sun. Because the sun moves predictably in the same way every year, you just pick a date and time, and SketchUp automatically displays the correct shadows by figuring out where the sun should be.

You do all these simple maneuvers in the Shadow Settings dialog box, as shown in Figure 9-14. The sections that follow introduce how the controls work so you can apply to them to your model.

Figure 9-14

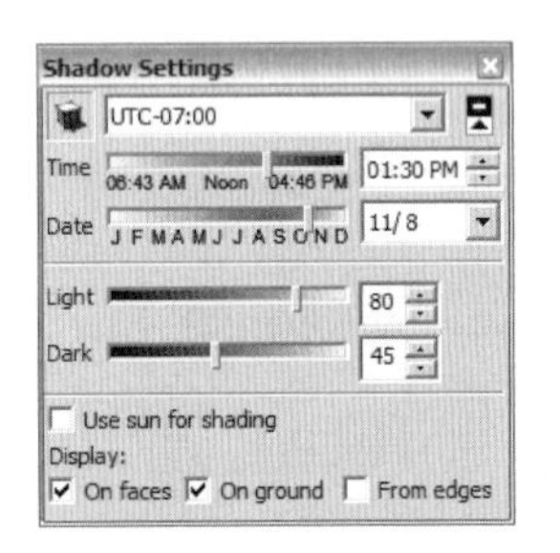

Dial up the sun in the Shadow Settings dialog box.

Turning On the Sun

Shadows are not on by default, so the first thing you need to know about applying shadows is how to turn them on. Follow these simple steps:

1. **Choose Window ⇨ Shadows to display the Shadow Settings dialog box.**
2. **At the top left corner of the dialog box, select the Show/Hide Shadows button.** Clicking it "turns on" the sun in SketchUp, casting shadows throughout your model.

Setting a Shadow's Time and Date

The Shadow Settings dialog box has time and date controls, which you use to change the position of the SketchUp "sun." The time and date you choose, in turn, controls the appearance of shadows in your model:

- **Setting the time:** Move the Time slider back and forth, type a time into the box on the right, or click on the up or down arrows next to the time. Notice the small times below each end of the slider? These represent sunrise and sunset for the day of the year you've set in the Date control, described in the next point.
- **Setting the date:** Just like the time of day, you set the day of the year by moving the Date slider back and forth, by typing in a date in the box on

the right or by selecting the date using the drop-down arrow to access an actual calendar. If you slide the Date control back and forth, notice that the sunrise and sunset times change in the Time control.

To toggle the extra shadow controls open or closed, click the triangular Expand button in the upper-right corner of the Shadow Settings dialog box.

Choosing Where Shadows Are Displayed

The Display check boxes in the Shadow Settings dialog box enable you to control *where* shadows are cast. Depending on your model, you may want to toggle these on or off.

- **On Faces:** Deselecting the On Faces check box means that shadows will not be cast on faces in your model. This is on by default, and should probably be left on, unless you only want to cast shadows on the ground.
- **On Ground:** Deselecting the On Ground check box causes shadows not to be cast on the ground plane. Again, this is on by default, but sometimes you'll want to turn it off. A prime example of this is when something you're building extends underground.
- **From Edges:** Selecting the From Edges check box tells SketchUp to allow edges to cast shadows. This applies to single edges that are not associated with faces—things like ropes, poles, and sticks are often modeled with edges like these.

9.3.2 Using Shadows to Add Depth and Realism

Shadows in SketchUp are easy to apply and to adjust. The previous sections dealt with the basic controls in the Shadow Settings dialog box. The following sections show how to use those controls to add depth, realism, and nuance to your models.

There are several situations when you'll need to use shadows to make your drawings read better; most of them fit into one of the following categories:

- **Indoor scenes:** The sun is the only source of lighting that SketchUp has, so any shadows you use in interior views have to come from it.
- **Objects that aren't in any particular location:** For things like cars and furniture, it doesn't matter that the shadows are geographically accurate; all that matters it that they help make your model look good.
- **2D views:** Without shadows, it's next to impossible to read depth in 2D views of 3D space.

Lighting Indoor Spaces

Adding shadows to interior views presents an interesting problem: There are no lights besides the sun in SketchUp, so can you make anything look realistic? With a ceiling in your room, everything's dark. If you leave off the ceiling, however, your model doesn't look right. A few techniques can solve this problem:

- **Decrease the darkness of the shadows.** Sliding the Dark slider to the right brightens your view considerably. You'll still be able to see the shadows cast by the sun coming through windows and other openings. Figure 9-15 illustrates this.

Brighten the room by decreasing the Dark setting.

Figure 9-16

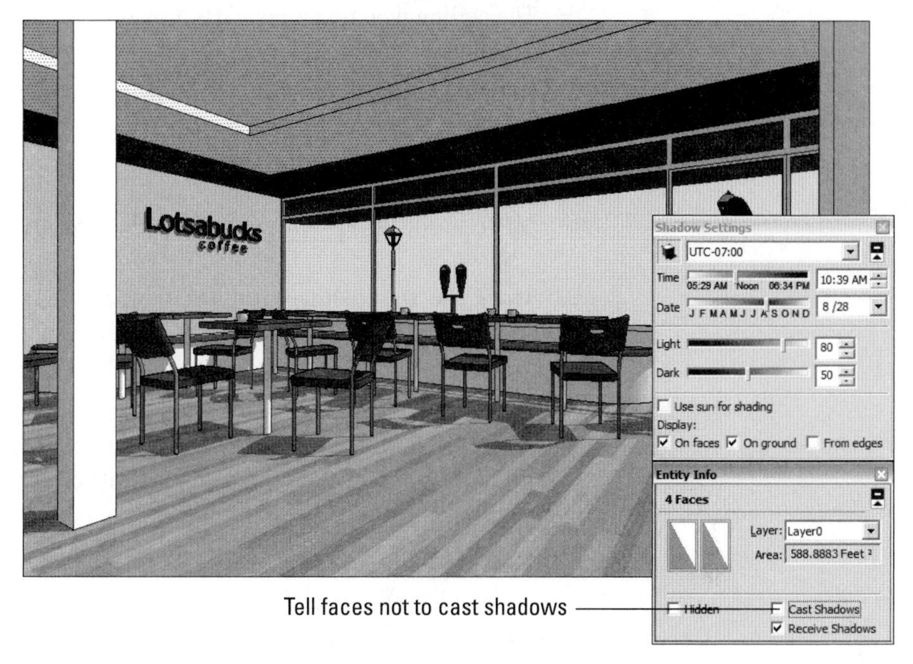

Tell the ceiling not to cast a shadow.

- **Make an impossible ceiling.** As long as you haven't modeled anything on top of the interior you're planning to show, you can tell the ceiling not to cast a shadow. That way, sunlight will shine directly onto your furniture, casting complex shadows.

Figure 9-16 shows the ceiling method in action; follow these steps to do it yourself:

1. **Adjust the settings in the Shadow Settings dialog box until the sun is shining through one or more windows in your view.** This ensures that shadows cast by objects in your room look like they're caused by light from the windows. To make it seem like overhead lighting is in your space, set the time of day to about noon. The shadows cast by furniture and similar objects will be directly below the objects themselves. One more thing: If you have lighting fixtures on the ceiling, remember to set them not to cast shadows in the Entity Info dialog box (described below).
2. **Choose Window ⇨ Entity Info.** This opens the Entity Info dialog box.
3. **Select any faces that make up the ceiling.** Hold down Shift to select more than one thing at a time.

4. **In the Entity Info dialog box, deselect the Cast Shadows check box.** The ceiling now no longer casts a shadow, brightening your space considerably.
5. **Repeat steps 3 and 4 for the following faces and objects:**
   - The wall with the windows in it.
   - The windows themselves.
   - Any walls in your view that are casting shadows on the floor of your space.
6. **Move the Dark slider over to about 50.** This brightens things even more and makes your shadows more believable.

Making 3D Objects "Pop"

Adding shadows to freestanding things like tables and lamps is a mostly aesthetic undertaking; just adjust the controls until things look good. Here are some things to keep in mind (which are illustrated in Figure 9-17):

- **Take it easy on the contrast.** This is especially true when it comes to very complex shapes or faces with photos mapped to them. When your model has too much contrast, it can be hard to read. To decrease the contrast, do the following:
  1. Move the Dark slider over to about 55.
  2. Move the Light slider down to 60 or 70.

Figure 9-17

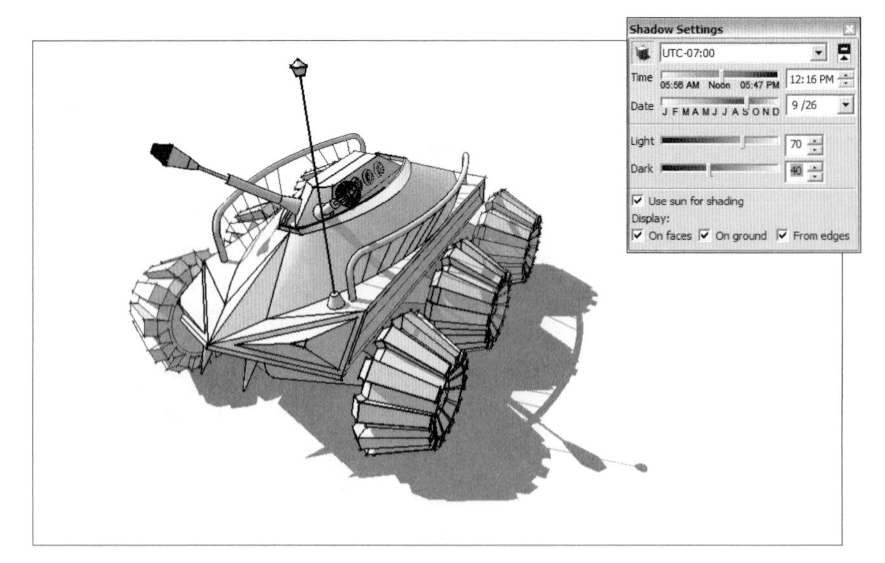

Tips for making objects stand out with shadows.

- **Shorten your shadows.** It's strange to see objects lit as though the light source is very far away; overhead lighting looks more natural. To make your shadows look better, follow these steps:

  1. Set the Date slider to a day in the early autumn.
  2. Set the Time slider to a time between 10 a.m. and 2 p.m.

- **Rotate your model around.** Remember that you can't get every possible shadow position by using only the controls in the Shadow Settings dialog box. To get the effect you want, you might have to rotate your model by selecting it and using the Rotate tool.
- **Select the From Edges check box.** Lots of times, modelers use free edges to add fine detail to models (think of a harp or a loom). Selecting the From Edges check box tells SketchUp to allow those edges to cast shadows.
- **Pay attention to the transparency of faces.** When you have a face painted with a transparent material, you can decide whether that face should cast a shadow—chances are that it shouldn't. In SketchUp, the rule is that materials that are more than 50 percent transparent cast shadows. So, if you don't want one of your transparent-looking faces to cast a shadow, do one of the following:

  – Select the face, and then deselect the Cast Shadows check box in the Entity Info dialog box.

  – Adjust the opacity of the face's material to be less than 50 percent in the Materials dialog box. For more information on this, reference Chapter 2.

IN THE REAL WORLD

Again, Tony the theatrical and lighting designer has more to say about the use of shadows and styles, "I use shadows sparingly in my theatrical projects; since they are locked to the geo location time and date, I usually turn them off or move the time around with the roof or ceiling hidden until they give a pleasing effect." He continues adding that "some kind of shadow usually helps give the scene a sense of depth, but in these cases I usually change the shadow color to something much lighter, sometimes just touched in, as opposed to an exterior in the sun."

Regarding styles, Tony says, "I will almost always run a project through a style before presenting it, either in print, on screen or exported as an AVI." He usually confides that he does not generally use the Style

(Continued)

IN THE REAL WORLD *(continued)*

Builder, but rather modifies the existing ones from the wide range of options available. He goes on to say that "I will generally use styles for one of two reasons: first, when the project is in a concept stage, I will find a style that softens the detail, makes the image looser, so those who are viewing the presentation are clear that details are not locked down, and what they are seeing is open to change. I personally find this feature to be one of SketchUp's greatest strengths, and something no other 3D program can match in real time. Secondly, I will sometimes use a different style when a project is further down the track, to in effect sharpen and 'pop' the visuals to emphasize particular details in a project. Either way, they are a great feature, and allow a more personal mode of presentation than just straight 'computer' visuals, and in real time. Finally, I make sure there are a couple of scenes left where all the styles are turned off, which I use for actual modeling."

9.3.3 Creating Accurate Shadow Studies

One of the most useful features in SketchUp is the ability to display accurate shadows. To do this, three pieces of information are necessary:

- the time of day
- the day of the year
- the latitude of the building site

The sun's position (and thus the position of shadows) depends on geographic location—that is to say, **latitude**. The shadow cast by a building at 3:00 on March 5 in Minsk is very different from that cast by a similar building, at the same time of day, on the same date in Nairobi.

Latitude
Geographic location measured by the angular distance north or south from the Earth's equator measured through 90 degrees.

If you're displaying shadows on a model of a piece of furniture, geographic location probably isn't important; the shadows are for effect only. But if you're trying to see how much time your pool deck will spend in the sun during the summer months, you need to give SketchUp your geographic location.

Telling SketchUp Your Geographic Location

Few people know the precise latitude of where they live. SketchUp makes it easy to tell it where your model is supposed to be. You can *geo-reference* your model (give it a geographic location) in two ways; which one you choose probably depends on whether you have an Internet connection available.

- **Using a geo-location snapshot:** This is by far the simplest approach, but it requires that you have a precise idea of where your model is supposed

Figure 9-18

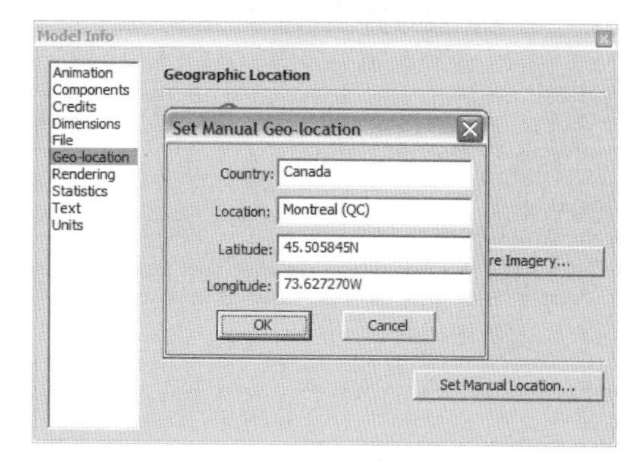

Giving your model a geographic location when you are online.

to be on the globe. It also requires that you be connected to the Internet for the operation. If you know exactly where your model is supposed to go, and you are online, use this method. Take a look at Chapter 11 for a complete set of instructions.

- **Using the Model Info dialog box:** This method is a little more complicated, but it's your only option if you're not online.

To give your model a geographic location (even if you think it might already have one); follow these steps (see Figure 9-18):

PATHWAYS TO... GIVING YOUR MODEL A GEOGRAPHIC LOCATION WHEN ONLINE

1. Somewhere on the ground in your model, draw a short line to indicate the direction of north as shown in Figure 9-19.
2. Starting at the southern endpoint of the edge you just drew; draw another edge that's parallel to the green axis. You have a V shape.
3. Select everything in your model *except* the edge you drew in Step 2 (to select everything, choose Edit-Select All and then hold Shift down to deselect the edge drawn on Step 2) . Your geo-location snapshot (if you have one) should have a red border around it; that's because it's locked. If for some reason it isn't, right-click it and choose Lock—you don't want to rotate it accidentally.
4. Choose the Rotate tool.

(continued)

PATHWAYS TO...
GIVING YOUR MODEL A GEOGRAPHIC LOCATION WHEN ONLINE *(continued)*

Figure 9-19

1. Draw two lines

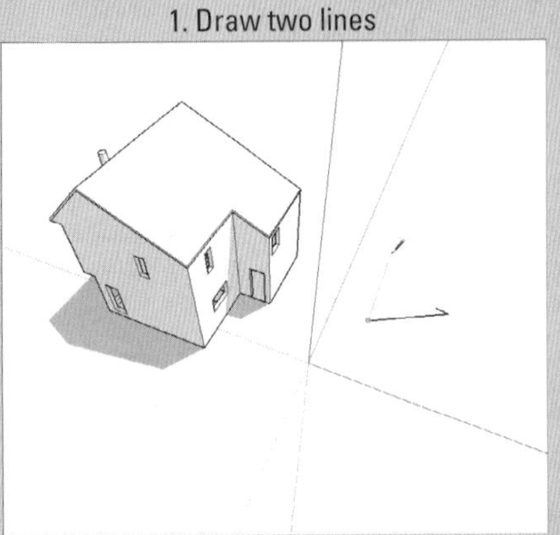

2. Select all except line parallel to green axis

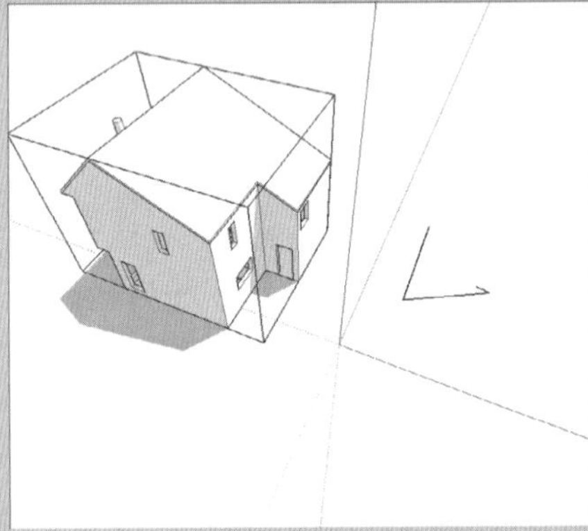

3. Rotate everything about point where lines meet

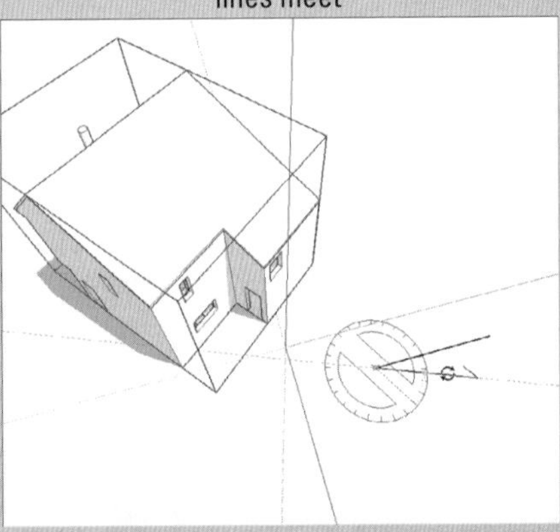

4. Rotate until lines are parallel

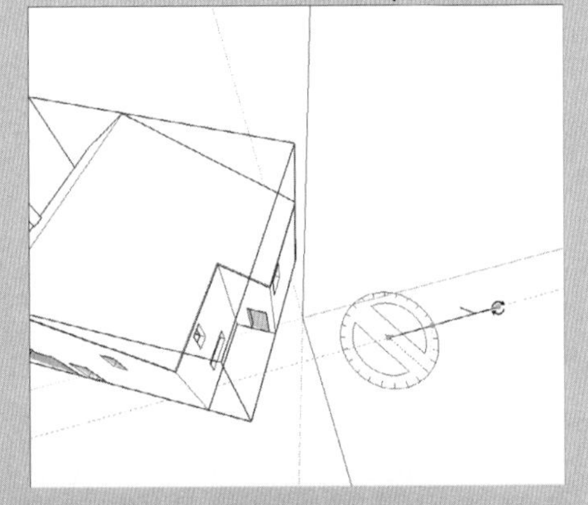

Make Sure Your Model is Correctly Oriented Relative to North.

5. Click the *vertex* (pointy end) of the V to establish your center of rotation.
6. Click the north end of the edge you drew in Step 1.
7. Click the north end of the edge you drew in Step 2.

Displaying Accurate Shadows for a Given Time and Place

Now that you've told SketchUp where your model is, it's a fairly simple process to study how the sun will affect your project, as shown in Figure 9-20.

Figure 9-20

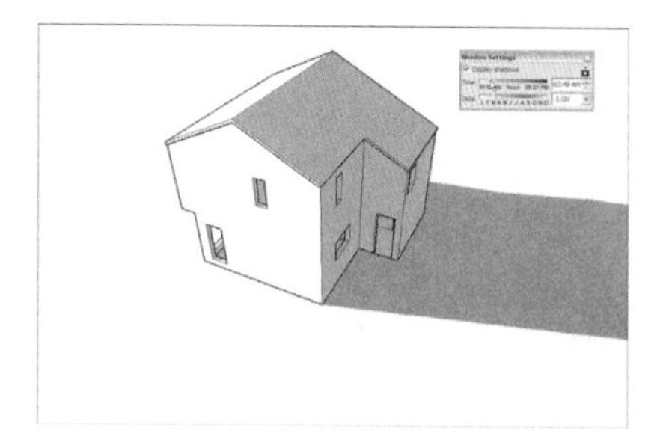

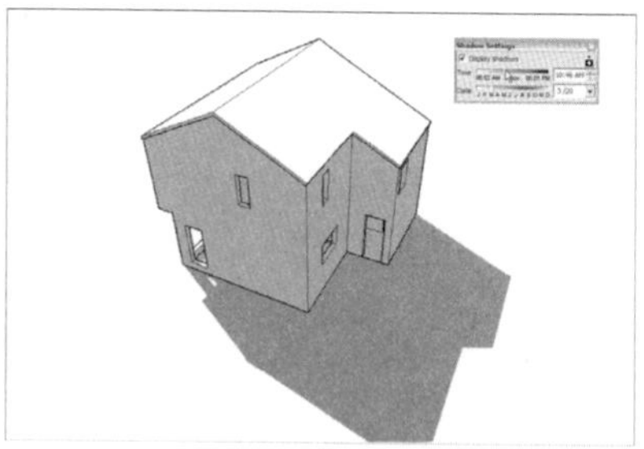

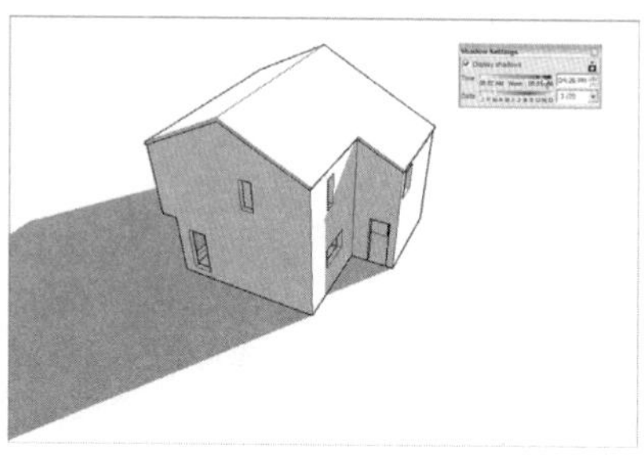

Studying the effect of the sun on your model.

PATHWAYS TO... HOW THE SUN WILL AFFECT YOUR PROJECT

1. Orbit, zoom, and pan around until you have a good view of the part of your project you want to study.
2. Choose Window ⇨ Shadows to open the Shadow Settings dialog box.
3. Select the Show/Hide Shadows button to turn on SketchUp's sun.
4. Make sure the time zone setting is correct for your location. SketchUp doesn't always get the time zone right for every location in the world; time zones don't always map directly to coordinates. If the time zone you see in the Time Zone drop-down list (at the top of the Shadow Settings dialog box) isn't correct, choose another one. Wondering what your time zone is in UTC? Try searching Google for "UTC time zones" to find a list to which you can refer.
5. Type a month and day into the box to the right of the Date slider, and press Enter.
6. Move the Time slider back and forth to see how the shadows will move over the course of that day.
7. Pick a time of day using the Time controls.
8. Move the Date slider back and forth to see how the sun will affect your project at that time of day over the course of the year.

FOR EXAMPLE

USES FOR SHADOWS

Even if you're not an architect, you might want to study shadows accurately for these reasons:

- To figure out where in your garden the plants that need the most light (or the most shade).
- To see when sunlight will be coming straight through the skylight you're thinking of installing.
- To make sure that the overhang behind your house will provide enough shade at lunchtime in the summer.

IN THE REAL WORLD

Rich O' Brien uses Google SketchUp because it is easy to use and intuitive. According to him, "shadows are extremely useful if you want to show clients shadow studies. In a past project, we constructed an Aircraft Mockup for a Training School and we wanted to create an area that was especially dark just to highlight the need to add the appropriate lighting unit when inspecting it. The result of sending technicians in with this visual aid was dramatic."

Rich continues, saying that he has used styles before to "wow" his clients during the final output. If time is against you and rendering isn't an option, Rich says that "Styles can give you an added appeal to your standard output." He mentions, too, that he has played with Style Builder before but he feels that it still requires a lot of pre-production to get good results. Finally, he mentions that the fog and watermark features are useful but not completely necessary but he would use them if the client is expecting to see high-quality images.

SELF-CHECK

1. For SketchUp to display accurate shadows, you have to tell SketchUp your precise _______ or latitude.
2. You can create a simple animation that shows shadow movement over the course of one day. True or false?
3. SketchUp shadows are controlled by the changing position of the sun. True or false?
4. SketchUp uses the following sources of light:
   a. Sun and artificial lights added in by the user.
   b. Artificial lights only.

c. Sun only.

d. The user can choose between using only the sun or using only artificial lights.

5. If you have a transparent-looking face and you don't want it to cast a shadow, what would you have to do to make it happen?

Apply Your Knowledge Build a simple, one-story house (you can use the house that you just built in the previous exercise) and add a sliding glass door on the back and a deck to be used by your client as a sunbathing area during the summer months. Now, follow the instructions to locate north in the model and do a shadow study to learn how the profile of the house will affect the hours of light that the deck will receive. Do you have to re-orient the whole house and deck because you are not getting enough light or on the contrary, are you getting enough sun during the central hours of the day?

SUMMARY

Section 9.1

- To make smart decisions about SketchUp styles, you should consider at least two factors when you're "styling" your model:
  - The subject of your model's "level of completeness."
  - How much your audience knows about design.
- Styles are really just combinations of display settings.
- NPR styles have different settings than regular, non-NPR styles.

Section 9.2

- Creating a new style adds it to your In Model collection of styles, which means that you can come back and apply it to your model anytime you like.
- If a bunch of styles exist in your In Model collection that you aren't using anymore and you want to clean things up, right-click the Details flyout menu and choose Purge Unused.

Section 9.3

- You add shadows to a SketchUp drawing for two key reasons:
  - To display or print a model in a more realistic way.
  - To study the effect of the sun on what you've built (or plan to build) in a specific geographic location.
- Shadows are controlled by the position of the sun.
- There are several situations when you'll need to use shadows to make your drawings read better; most of them fit into one of the following categories:
  - Indoor scenes.
  - Objects that aren't in any particular location.
  - 2D views.

ASSESS YOUR UNDERSTANDING

SUMMARY QUESTIONS

1. It's not possible to view a model without any style at all. True or false?
2. Typically, you should reserve "sketchy" styles for models that are:
   a. Closed to audience input
   b. Still evolving
   c. In their final form
   d. Required to be highly accurate
3. Which of the following is **not** one of the tabs on the Edit pane?
   a. Watermark
   b. Face
   c. Background
   d. Foreground
4. The ______ tab on the Edit pane is tricky because it changes slightly depending on whether you're using NPR or non-NPR styles.
5. When you create a new style, it shows up in your In Model library of styles and is saved with your model. True or false?
6. Creating a new style automatically makes it available for use in other SketchUp files. True or false?
7. Your Favorite Styles libraries are available:
   a. Only in models that you designate as "favorites."
   b. Only in models that you create from scratch.
   c. In every SketchUp model you're working on.
   d. To all users on your network.
8. You can add light bulbs in SketchUp and make them a source of light in your models. True or false?
9. To decrease the darkness of shadows in SketchUp, you should slide the Dark slider:
   a. Left
   b. Right
   c. Up
   d. Down
10. To create accurate shadows in SketchUp, which of the following information is needed?
    a. The time of day.
    b. The day of the year.
    c. The latitude of the building site.
    d. All of the above.

11. Under the Styles Collection drop down list, what tab would we use to show the styles that we have used already in our model?
12. Once we modify a style and apply it to a model, we cannot change its appearance any more. True or false?
13. NPR or *non-photorealistic rendering* allows us to make useful working drawings in real time. True or false?

APPLY: WHAT WOULD YOU DO?

1. In general, when should you consider toning down your use of styles within a model?
2. What are some reasons you might opt to use non-photorealistic rendering when creating a model?
3. What are some of the Styles settings you can select (or deselect) to make SketchUp run more efficiently on your computer?
4. What are the basic steps for making a style available for use in other SketchUp files?
5. What steps can you take to convey depth in 2D views?

BE A DESIGNER

Experiment with Your Model

Using the model of a house that you have already created, add the following effects and save each as a new model:

- Jitter
- Halo
- Transparency
- Change face styles

If you do not already have a model of a house, create a model of a simple doghouse and experiment with the effects listed above. Save each version of the doghouse as a different model.

Create a New Style

Experiment with combining elements of existing styles. Create a new style that you like and that looks good when applied to your model. Give your new style a name and save it for use with other models you create.

Adding Shadows to Your Model House

Collect the information you need to create accurate shadows for the geographic location where you reside. Then use this information to add shadows to the model house you built in previous chapters.

Building a Sunroom

Either create a freestanding model of a sunroom or add a sunroom to the model house you created in previous chapters. Input the necessary information into SketchUp to create shadows. Then create a simple shadow animation of your sunroom like the one that was presented in the last section of this chapter.

KEY TERMS

Latitude

Styles

Watermark

CHAPTER

PRESENTING YOUR MODEL INSIDE SKETCHUP

Showing Off Your Model

Do You Already Know?

- The reasons why you would want to walk inside your model?
- How you can capture particular views?
- How to animate scenes?
- When to cut slices through your model with section planes?
- The difference between plans and sections?

For the answers to these questions, go to **www.wiley.com/go/chopra/googlesketchup2e**

What You Will Find Out	What You Will Be Able To Do
10.1 How to walk through models.	Learn how and when to use work with scenes to speed the working process.
10.2 How to work with scenes.	Learn how to use scenes to make flyover animations, walk throughs, or shadow studies. Understand what scenes are and how to update and save them. Learn how to use scenes to showcase different design options.
10.3 How to work with sections.	Master cutting, controlling, and applying sections to create animations or dimensioned or orthogonal views or elevations.

INTRODUCTION

After you've made a model, you're probably going to want to show it to someone. How you present your work depends on the idea you're trying to convey. The challenging part about using SketchUp to present a model isn't actually using the tools; it's choosing the *right* tools to get your idea across without distracting your audience with too much extra information. Most 3D models have so much to look at that the real challenge is to figure out a presentation method that helps you focus on the features that you want to talk about.

In this chapter, you'll learn three different ways to show off your models without ever leaving SketchUp. Specifically, if you've made a building, you can walk around inside it. You can even walk up and down stairs and ramps, just like in a video game. You can also create animated slide shows by setting up scenes with different camera views, times of day, and even visual styles. Also, if you want to talk about what's *inside* your model, you can cut sections through it without taking it apart.

As you read this chapter, keep in mind what you want your model to communicate. Think about how you might use each method to make a different kind of point, and think about the order in which you'd like those points to be made. As with everything else in SketchUp, a little bit of planning goes a long way.

10.1 EXPLORING YOUR CREATION ON FOOT

GOOGLE SKETCHUP IN ACTION

Learn how and when to use work with scenes to speed the working process.

Few experiences in SketchUp are as satisfying as running around inside your model. After you've made a space, you can drop down into it and explore by walking around, going up and down stairs, bumping into walls, and even falling off ledges. You can check to make sure that the television is visible from the kitchen, for example, or experience what it would be like to wander down the hall. In a potentially confusing building (like an airport or a train station), you could even figure out where to put directional signs by allowing someone who's never seen your model to explore the space "on foot."

10.1.1 Using SketchUp's Tools

A couple of tools in SketchUp are dedicated to moving around your model as if you were actually inside it. The first step is to position yourself so that it seems like you're standing inside your model. This can be tricky with just the Orbit, Pan, and Zoom tools, so SketchUp provides a tool just for this purpose: Position Camera. Once you're standing in the right spot (and at the right height), then you can use the Walk tool to move around. It's as simple as that.

Just remember the Position Camera and Walk tools enable you to walk around inside your model.

Figure 10-1

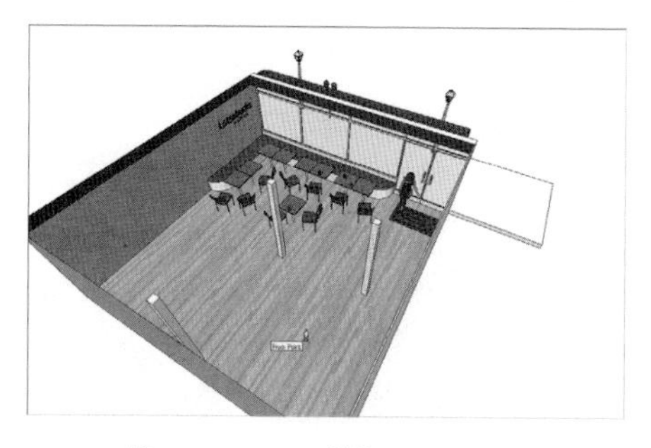

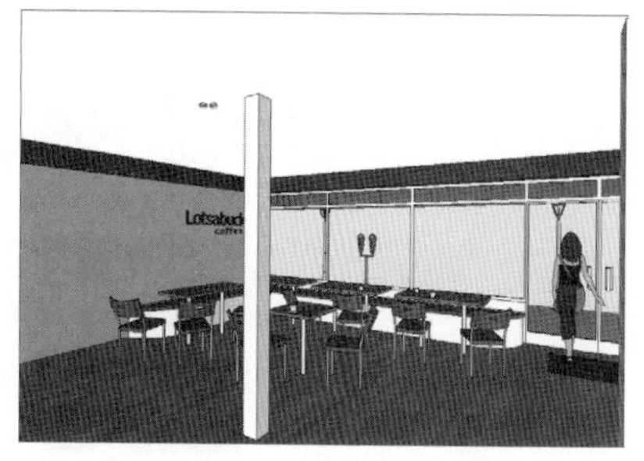

Drop yourself into your model with the Position Camera tool.

The Position Camera Tool

The essence of the Position Camera tool is its ability to precisely place your viewpoint in SketchUp in a particular spot. That's really all this tool does, but it works in two different ways:

- **If you want to be standing in a particular location:** Choose Camera ⇨ Position Camera from the menu bar and then single-click anywhere in the modeling window to automatically position your viewpoint 5 feet, 6 inches above wherever you clicked. Because this is the conventional *eye-height* of an adult human being, the result is that you are, for all intents and purposes, standing on the spot where you clicked (see Figure 10-1).

After using Position Camera, SketchUp automatically switches to the Look Around tool, assuming that, now that you're where you want to be, you might want to have a look around. (The Look Around tool is described in the next section of this chapter.) Note that you're not stuck being five-and-a-half-feet tall forever, though. After you use Position Camera, type in the height you'd rather "be" and press Enter. For example, type **18"** to see a golden retriever's view of the world, or type **7'** to pretend you play for a pro basketball team. Keep in mind that the VCB (the spot in the lower-right corner where numbers appear) displays your eye height as a distance from the ground, and not from whatever surface you're "standing on." Thus, to set your eye height to be 5 feet above a platform that's 10 feet high, you'd type in **15'**.

- **If you want your eyes to be in a specific location while you're looking in a particular direction:** Select Position Camera, click the mouse button while in the spot where you want your eyes to be, drag over to the thing you want to be looking at (you'll see a dashed line connecting the two points), and release the mouse button (as shown in Figure 10-2). Try

Figure 10-2

"Aim" your view by using Position Camera another way.

this technique a few times; it takes a bit of practice to master. You'd use Position Camera in this way if you wanted to be standing in a particular spot *and* looking in a particular direction. This technique works great with scenes, which are explained later in this chapter.

The Walk Tool

After you've used Position Camera to place yourself in your model, use the Walk tool to move through it. To walk around, click and drag the mouse in the direction you want to move, keeping the following rules in mind:

- Straight up is forward.
- Straight down is backward.
- Anything to the left or right causes you to turn while you're walking.

The farther you move your cursor, the faster you walk. Release the mouse button to stop walking. If you've ever played video games, you'll get used to this quickly.

You can also use the Walk tool to walk up and down stairs and ramps. Keep in mind that the highest step you can "climb" is 22 inches—anything higher and you get the "bump" cursor, just as if you walked into a wall. Also, if you walk off a high surface, you'll fall to the surface below.

Using modifier keys in combination with the Walk tool makes SketchUp even more like a video game:

- Hold down Ctrl (Option on a Mac) to run instead of walking.
- Hold down Shift to move straight up (like you're growing), straight down (like you're shrinking), or sideways (like a crab).
- Hold down Alt (⌘ command on a Mac) to disable collision detection, which allows you to walk through walls instead of bumping into them.

10.1.2 Stopping to Look Around

Look Around is the third tool in SketchUp that's dedicated to exploring your model from the inside. If using Position Camera is like swooping in to stand in a particular spot, and Walk is like moving around while maintaining a constant eye-height, Look Around is like turning your head while standing in one spot. It does exactly what it says. To use Look Around, follow these steps:

1. Choose Camera ⇨ Look Around from the menu bar to activate the Look Around tool.
2. Click and drag around in the modeling window to turn your virtual head.

While in any of the navigation tools, you can right-click to access any of the other navigation tools; this makes switching between them easier.

When you use Look Around with the field of view tool discussed in the next section, you get a fairly realistic simulation of what it would be like to be standing in your model.

10.1.3 Setting Your Field of View

Field of view
The amount of your model you're able to see in your modeling window at one time.

Field of view is the amount of your model you're able to see in your modeling window at one time. Imagine your eyesight like a cone, with the pointed end at your eyes and the cone getting bigger as it gets farther away from you. Everything that falls inside the cone is visible to you, and everything outside the cone isn't.

If you increase the angle of the cone at the pointed end, the cone gets wider and you see more of what's in front of you. If you decrease the angle, the cone gets narrower and you see less (see Figure 10-3).

Measured in degrees, a wide field of view means that you can see more of your model without having to move around. The bigger the angle, the more you can see. This comes in handy when you're inside a SketchUp model you're working on, because it's hard to work on things you can't see.

It's a good idea to adjust your field of view while walking around inside your model. Follow these steps to do so:

1. **Choose Camera ⇨ Field of View.** Notice the Measurements Box in the lower-right corner of your modeling window says "Field of View" and that the default value is "35 deg." This means that you currently have a 35-degree cone of vision, which is somewhat narrow.
2. **Type 60 and press Enter.** Your field of view is increased, and you now have a wider view of your model. The downside is that you see more distortion at the edges of your modeling window as more information is being displayed in the same amount of space.

Figure 10-3

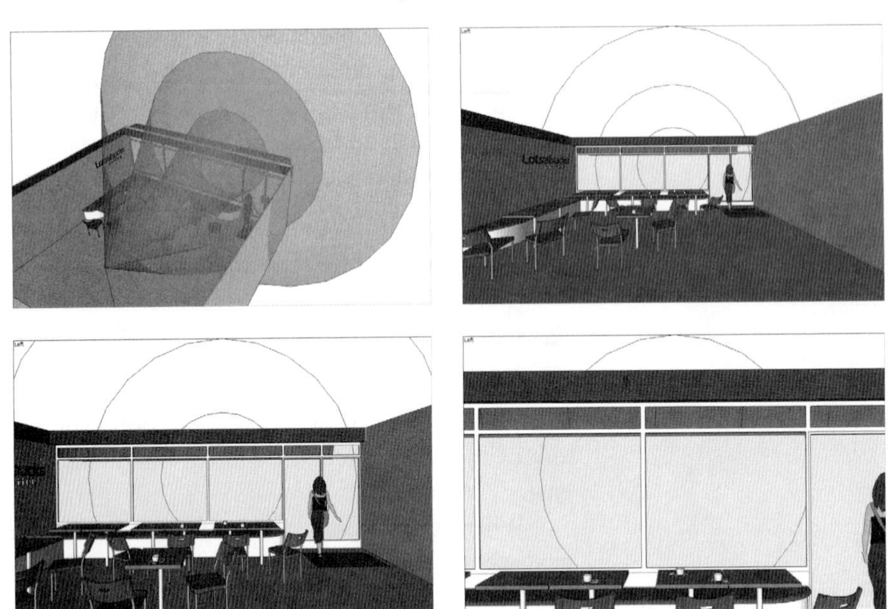

The wider your field of view, the more you can see.

A good rule of thumb for setting your field of view is to strike a balance between quantity and quality; a wider view always means more distortion. For views of the *outside* of something you've built, you may want to use a field of view of 35 to 45 degrees. For interior views, you might use 60 or 70 degrees.

If you are familiar with photography, you can also express field of view in millimeters, just as if you're using a camera lens. Typing in **28mm** gives you a wide-angle view, just like you're looking through a 28-mm lens. For people who think about field of view in these terms, this can be more intuitive than trying to imagine cones of vision.

IN THE REAL WORLD

Juliana Diehl is a Brazilian architect at Stúdio Arquitetura in Piracicaba, and she chose to work with Google SketchUp because it is an extremely fast and simple program to use. She utilizes SketchUp along with the V-Ray rendering plugin to create photo-realistic scenes to present to her clients. In some cases, she will also present the SketchUp model to the client so she can walk them through the entire design. Her clients always enjoy seeing the 3D images that she creates for them, as it is often the first time that they truly are able to understand the design's concept and how the completed project will look.

SELF-CHECK

1. What is the name of the tool that would allow you to control the angle of the cone of vision that you can obtain in the modeling windows?
2. The Walk tool allows you to look around and move through your model. True or false?
3. Which tool can precisely place your viewpoint in SketchUp in a particular spot?
   a. Pan
   b. Zoom
   c. Orbit
   d. Position camera

Apply Your Knowledge The ability of SketchUp to Pan and Orbit makes it easy to analyze models from all points of view, but often we forget to analyze them from specific viewing heights or field of view. For example, if we are building a model of a trade show exhibit that needs to be installed inside a big Convention Center, we must consider how the model would be seen from the entrance, from a determined viewing height and from a Field of View. Build a 10' x 20' exhibit booth that includes a reception counter, two small displays, a hanging banner, and mark the entrance of the trade show. Then, set up your viewing height from that spot, at a height of 5'-6" and an angle of 35 degrees. How would the view be different from a child's point of view of 3'-0"?

10.2 TAKING THE SCENIC ROUTE

With SketchUp, you can save a particular view of your model so that you can come back to that view whenever you want. That saved view can also save things like styles and shadow settings. Plus, you can come back to any saved view by clicking a button on your screen.

> GOOGLE SKETCHUP IN ACTION
>
> Learn how to use scenes to make flyover animations, walk throughs, or shadow studies.

SketchUp **scenes** are saved views of your model. Scenes can be thought of as cameras, except that scenes can save much more than just camera positions. Although they don't get much space in this book, scenes are one of the most important features in SketchUp, for three reasons:

- **Scenes can save you hours of time.** It's not always easy to get back to exactly the right view by using Orbit, Zoom, and Pan. Sometimes

Figure 10-4

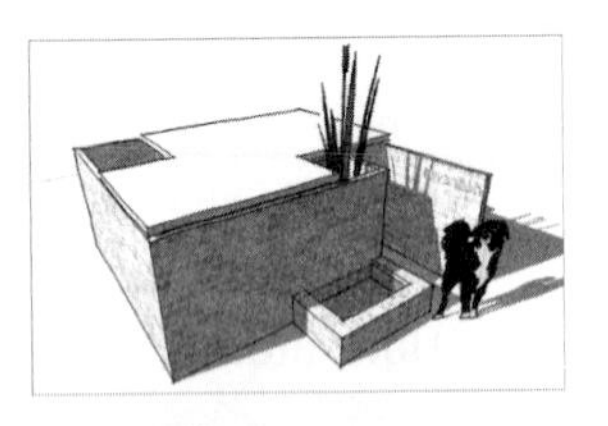
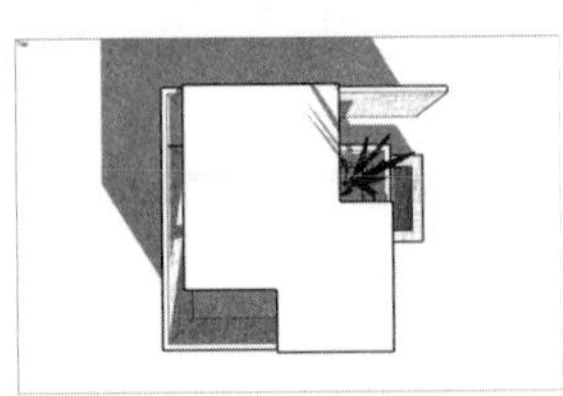
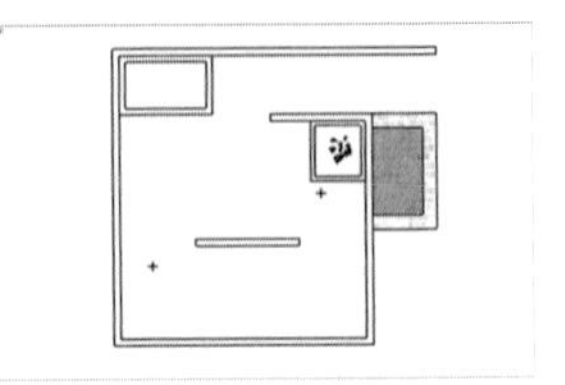
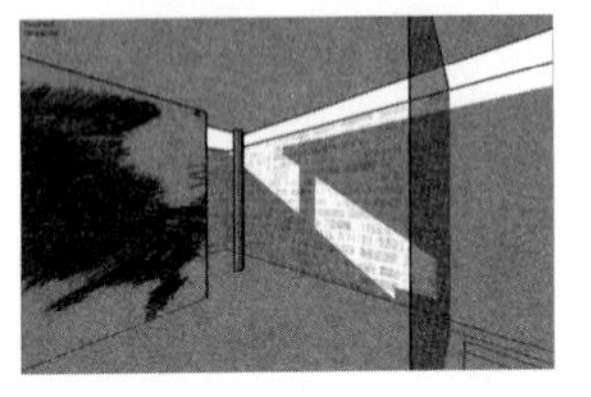
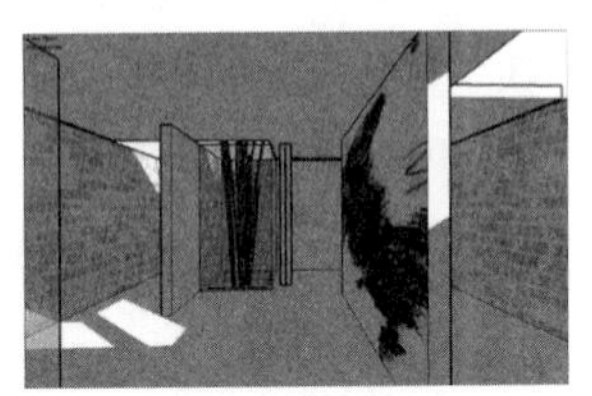

To show very specific views, create scenes.

a view involves shadows, styles, sections (you'll read about those later), and even hidden geometry. It can be frustrating to set up everything the way you need it, every time you need it. With SketchUp, you have a lot of different ways to view your model. Making a scene reduces the process of changing dozens of settings to a single click of your mouse.

- **Scenes are the most effective way of presenting your model.** Saving a scene for each point that you'd like to make in a presentation allows you to focus on what you're trying to say. Instead of fumbling around with the navigation tools, turning on shadows, and making the roof visible, you can click a button and SketchUp will automatically transition to the next scene (which you've already set up exactly the way you want it). Figure 10-4 illustrates a set of scenes to present a doghouse.
- **Scenes are the key to making animations.** You make animations by creating a series of scenes and telling SketchUp to figure out the transitions between them. The process, which is explained in later sections, is as simple as clicking a button.

After you get used to them, you'll find yourself using scenes all the time. Here are some of the most common uses for scenes:

- Showing shade conditions for the same area at different times of the day.
- Saving scenes for each floor plan, building section, and other important views of your model.
- Building a walk through or flyover animation of your design.

- Creating scenes to show several views of the same thing with different options.
- Demonstrating change over time by showing or hiding a succession of components.

10.2.1 Creating Scenes

GOOGLE SKETCHUP IN ACTION

Understand what scenes are and how to update and save them.

Making a scene in SketchUp is *not* like taking a snapshot of your model. If you create a scene to save a view, then do more modeling, and then return to that scene, your model will not go back to the way it was when you created the scene. The camera position will be the same and the settings will be the same, but your geometry won't be. This is an important concept, and one that makes using scenes so powerful.

A scene is simply a set of view settings, which means that they're automatically updated to reflect your changes every time you edit your model. You can make some scenes and use them all the way through your process, from when you start modeling to when you present your design.

Creating scenes is a simple process. The basic idea is that you add a scene to your SketchUp file whenever you have a view you want to return to later. You can always delete scenes, so there's no downside to using several of them. Follow these steps to make a new scene:

1. **Choose Window ⇨ Scenes to open the Scenes dialog box.** When it first opens, it doesn't look like there's much to the Scenes dialog box. Expanding it by clicking the Show Details button in the upper-right corner reveals more options, which are covered later in this chaper.
2. **Set up your view however you want.** Navigate around until you're satisfied with your point of view. If you want, use the Shadows and Styles dialog boxes to change the way your model looks.
3. **Click the Add button to make a new scene with your current view settings.** At this point, a new scene is added to your SketchUp file. If this is the first scene you've created, it will be called Scene 1 and will appear in two places (see Figure 10-5):
   - As a list item in the Scenes dialog box, right underneath the Add button.
   - As a tab at the top of your modeling window, labeled Scene 1.

A new feature has been added to SketchUp 8: thumbnail images of your scenes in the Scenes dialog box. To see them, just click the View Options button and choose either Small Thumbnails or Large Thumbnails.

Nothing is generated outside of SketchUp when you add a scene; it's not like exporting a JPEG or a TIFF. Scenes are just little bits of programming code that "remember" the view settings in effect when they were created. Scenes also don't add much to your file size, so you don't have to worry about using too many of them.

Figure 10-5

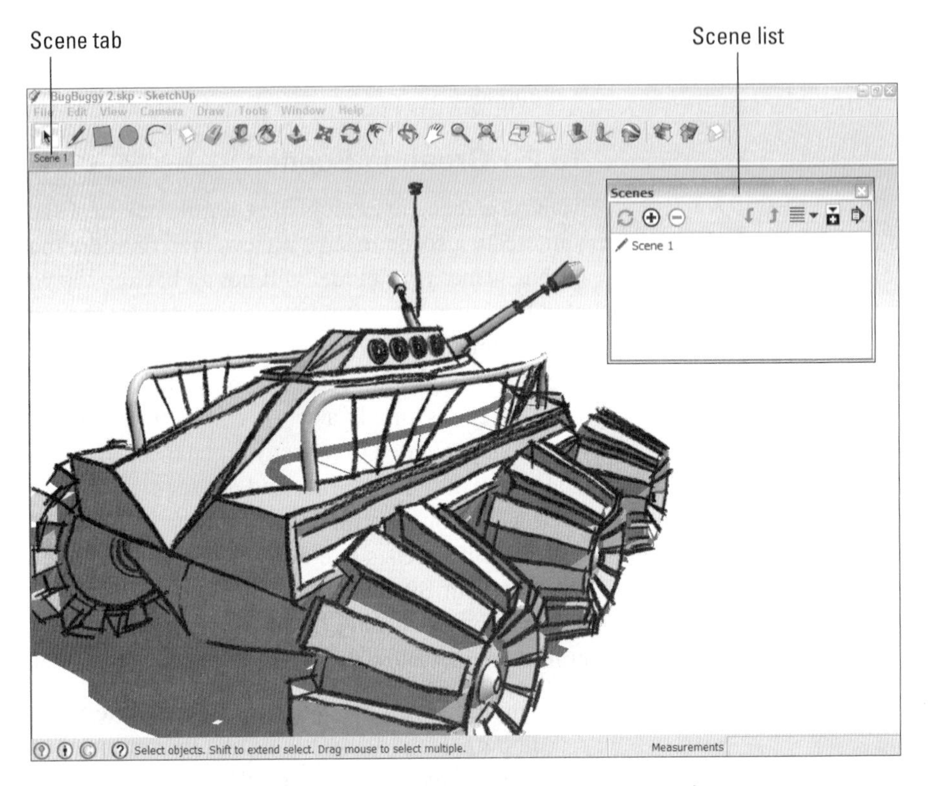

The scene you just added shows up in two places.

10.2.2 Moving from Scene to Scene

Activate a scene you've added earlier by doing one of three things:

1. Double-clicking the name of the scene in the Scenes dialog box.
2. Single-clicking the tab for that scene at the top of the modeling window.
3. Right-clicking any scene tab and choosing Play Animation to make SketchUp automatically flip through your scenes. (Choose Play Animation again to make the animation stop.)

Notice how the transition from one scene to the next is animated? You don't have to do anything special to make this happen; it's something SketchUp automatically does to make things look better.

You can adjust the way SketchUp transitions between scenes, which is handy for customizing your presentations. Follow these steps to access these settings:

1. **Choose Window ⇨ Model Info.**
2. **On the left side of the Model Info dialog box, choose Animation.** The Animation settings panel in the Model Info dialog box (see Figure 10-6)

Figure 10-6

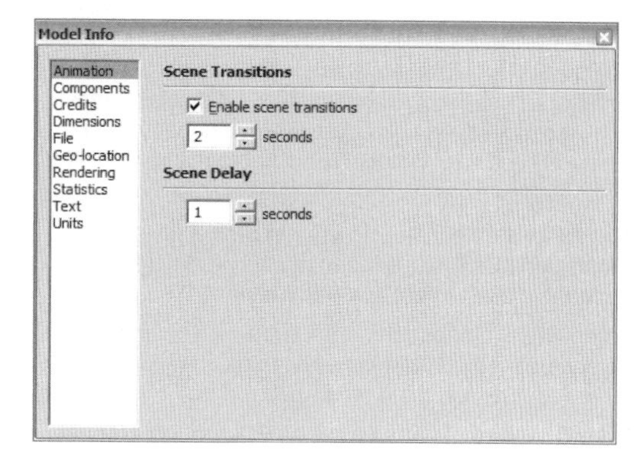

The Animation settings panel is helpful in customizing your presentations.

isn't very complicated, but it can make a great difference in the appearance of your scene-related presentations.

3. **In the Scene Transitions area, set how SketchUp transitions from one scene to another.** These settings apply to both manual (clicking on a page tab) and automatic (playing an animation) scene transitions:

   - **Enable Scene Transitions:** Deselect this check box to make SketchUp change scenes without animating the transitions between them. You'll probably want to do this if your model is so complex (or your computer is so slow) that animated transitions don't look good.
   - **Seconds:** If you've selected the Enable Scene Transitions check box, the number of seconds you enter here will be the amount of time it takes SketchUp to transition from one scene to the next. If you're "moving the camera" very far between scenes, it's a good idea to increase the transition time. Three seconds is generally a good transition time.

If you're presenting an incomplete model (perhaps you've thought about the garage and the living room, but nothing in between), it can be helpful to turn off scene transitions. That way, your audience won't see the things you haven't worked on when you click a tab to change scenes.

4. **In the Scene Delay area, set the length of time SketchUp pauses on each slide before it moves to the next one.** If you want it to seem like you're walking or flying, set this to 0. If you want time to talk about each scene in your presentation, increase this a few seconds.

FOR EXAMPLE

WHEN SCENES AND STYLES COLLIDE

Sooner or later, you'll be presented with the Warning—Scenes and Styles dialog box shown here. It pops up whenever you try to create a scene without first saving the changes you've made to the style applied to your model. In other words, SketchUp tries to help by reminding you to keep styles in mind while you work with scenes. (The first part of Chapter 9 is all about styles, just in case you need a refresher.) This warning dialog box gives you three options; here's some guidance on which one to choose:

- **Save as a New Style:** Adds a new style to your In Model styles library. When you come back to this scene, it looks exactly the way it did when you created it. Choosing this option is the safest way to proceed because it can't affect any other scene.
- **Update the Selected Style:** Choose this option only if you know what effect this will have on the other scenes in your model—if the style you're updating is applied to any of them, you'll affect the way they look. In models with lots of scenes and styles, this can have big implications.
- **Do Nothing to Save Changes:** Creates a scene with your current style applied, completely ignoring any changes you may have made to that style. When you come back to this scene, it looks different than it did when you created it. Only choose this option if you really know what you're doing, or if you enjoy doing the same thing more than once.

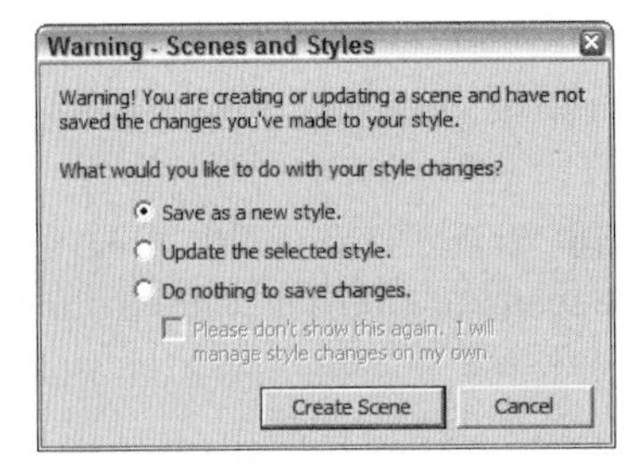

10.2.3 Modifying Scenes after You've Created Them

After you create several scenes, it's inevitable that you're going to need to adjust them in some way. After all, modifying something is almost always easier than making it all over again, and the same thing holds true for scenes. Because your SketchUp model will change many times, understanding how to make changes to your existing scenes can save you a lot of time.

Certain aspects of the scene-modification process can get a little complicated. Pay special attention to the section on updating scenes, and don't worry if it takes some time to figure things out.

Reordering, Renaming, Removing, and Viewing Scenes

Making simple modifications to scenes—such as reordering, renaming, and removing them—is easy. You can accomplish each of these in two ways: You either use the Scenes dialog box or you right-click the scene tabs at the top of your modeling window. Figure 10-7 is an illustration of this.

To access the modification controls in the Scenes dialog box, click the arrow-shaped expansion button in the upper-right corner.

Figure 10-7

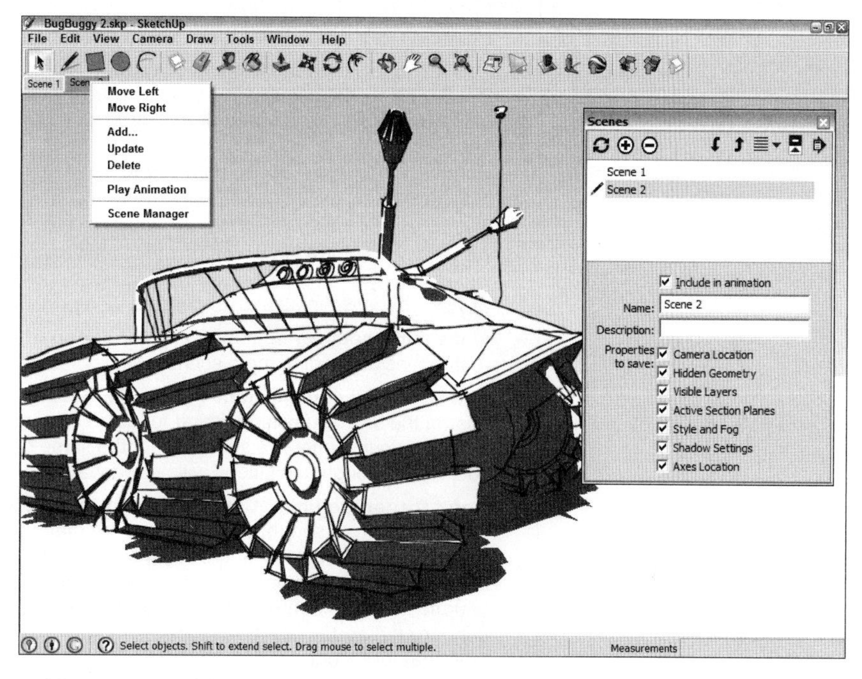

You can modify scenes by right-clicking scene tabs or by using the Scenes dialog box.

Here's how to reorder, rename, or remove scenes:

- **Reordering scenes:** You can change the order in which scenes play in a slide show. If you're using scenes, you'll need to do this often. Use one of the following methods:
  - Right-click the tab of the scene you want to move (in the modeling window) and choose Move Right or Move Left.
  - In the expanded Scenes dialog box, click the name of the scene you want to move to select it, and then click the up or down arrows to the right of the list to change the scene's position in the scene order.
- **Renaming scenes:** It's a good idea to give your scenes meaningful names: "Living Room," "Top View," and "Shadows at 5:00 p.m." are descriptive enough to be useful. "Scene 14" is not. Use one of the following methods to rename a scene:
  - Click on the scene name and over-write the name of that scene, or
  - Right-click on the thumbnail of the scene in the Scenes dialog box and choose Rename from the menu of options.
  - In the Scenes dialog box, select the scene you want to rename and type something into the Name field below the list. You may want to give it a description, too.

FOR EXAMPLE

MAKING WALK THROUGHS

A great way to use scenes is to pretend you're walking or flying through your model. By setting up your scenes sequentially, you can give a seamless tour without having to use the navigation tools. This is especially handy when you need to be able to "walk and talk" at the same time.

Here are some tips that can help you to simulate a person walking or flying through your model with scenes:

- **Adjust your field of view.** For interior animations, make your camera "see" a wider area by setting your field of view to 60 degrees. For exterior views, a field of view set between 30 and 45 degrees works well.
- **Make sure that your scenes aren't too far apart.** Instead of racing through a room, consider adding more scenes.
- **Add scenes at equal distance intervals.** Because SketchUp only lets you control the scene transition timing for all your scenes at once, it's best to make sure that your scenes are set up about the same distance apart. If you don't, your walk-through animations will be awkward.
- **Don't forget the animation settings in the Model Info dialog box.** Set the scene delay to 0 seconds so that your animation doesn't pause at every scene. For a normal walking speed, set your scene transitions so that you're moving about 5 feet per second. If your scenes are about 20 feet apart, set your scene transition time to 4 seconds. This gives your audience time to look around and notice things. For flying animations, pick a scene transition time that looks good.
- **Slide around corners.** When you're setting up a walking animation, you have an easy, reliable way to turn corners without seeming too robotic. Basically, the trick is to add a scene just short of where you want to turn—for example, a few feet ahead of a doorway. The key is to angle your view *into* the turn slightly. You should set up your next scene just past the turn, close to the inside, and facing the new view. This technique makes it seem like you're turning corners naturally.

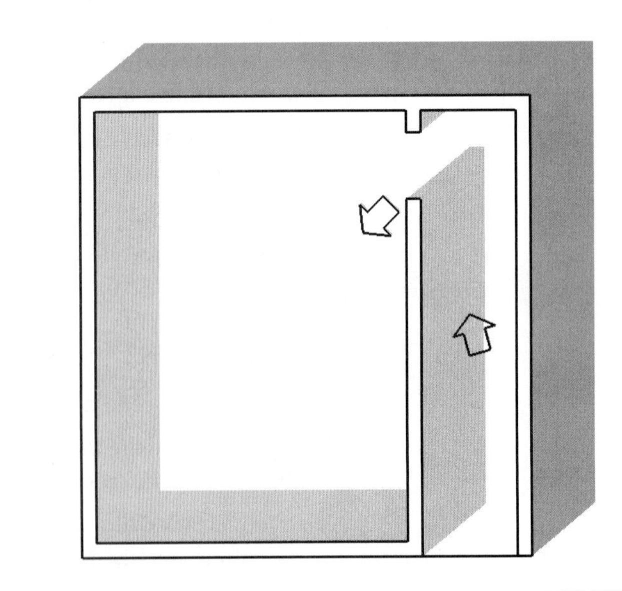

- **Removing scenes:** If you don't need a scene anymore, you can delete it. Use one of the following methods to remove a scene:
  - Right-click the scene tab and choose Delete to remove it permanently.
  - In the Scenes dialog box, select the scene you want to remove and choose Delete from the menu of options.

  However, if you have a scene that you don't want to appear in slide shows, you don't have to delete it. To exclude a scene from slide shows without getting rid of it, select its name in the list and deselect the Include in Animation check box. Once you are done with that step, the scene that is being excluded from the Animation will have a name shown in parenthesis in the scene dialog box.

  - Viewing scenes: If you have made a lot of scenes and the large thumbnails are crowding your Scenes dialog box, change how you see them to a list. Press the View Options icon (seen as five horizontal dashes stacked up) and choose from four choices: Small Thumbnails, Large Thumbnails, Details, and List. See Figure 10-8 for an illustration.

Updating Scenes

The process of updating scenes isn't altogether straightforward so make sure you have time and an environment to focus on this. Basically, a scene

Figure 10-8

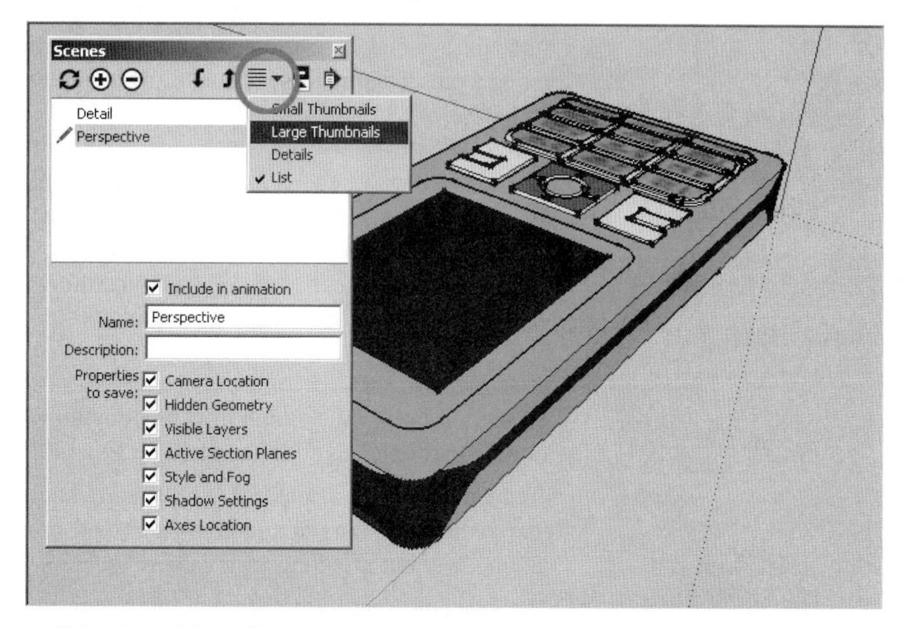

Selecting this pullout menu, we can choose the viewing mode of the scenes that we have captured in our model. In this example, we have two scenes, Detail view and Perspective, which are seen as a list.

Reproduced with the permission of Jorge Paricio, PhD.

is just a collection of saved viewing *properties*. Each of these properties has something to do with how your model looks:

- **Camera location:** Camera location properties include the position of the camera, or viewpoint, and the field of view. Field of view is discussed earlier in this chapter.
- **Hidden geometry:** Hidden geometry properties are simply what elements are hidden and what elements aren't. These properties keep track of the visibility of the lines, faces, groups, and components in your model.
- **Visible layers:** Visible layer properties keep track of the visibility of the layers in your model.
- **Active section planes:** Active **section plane** properties include the visibility of section planes and whether they are active. Sections are discussed in the last part of this chapter.
- **Style and fog:** Style and fog properties are all the settings in the Styles and Fog dialog boxes, and there are several.
- **Shadow settings:** Shadow settings properties include whether shadows are turned on and the time and date for which the shadows are set. They also include all the other settings in the Shadow Settings dialog box.
- **Axes locations:** Axes location properties are very specific. They keep track of the visibility of the main SketchUp red, green, and blue axes in your modeling window. Because you'll often want to hide the axes when you're giving a presentation, these elements get their own properties.

Section planes
Objects that let you cut away parts of your model to look inside.

Here's the tricky part: Scenes can *save* (remember) any combination of the preceding properties—it's not an all-or-nothing proposition. This means that scenes are *much* more powerful than they first appear.

By creating scenes that save only one or two properties (instead of all seven), you can use scenes to do some nice things. Here are three that are typically used:

- Create scenes that affect only your camera location, allowing you to return to any point of view without affecting anything else about the way your model looks (such as styles and hidden geometry).
- Create scenes that affect only styles and shadows, letting you quickly change between simple and complex (hard on your computer) display settings without affecting your camera location.
- Create scenes that have different combinations of Hidden Geometry to look at design alternatives without changing your model's style and camera location.

Figure 10-9

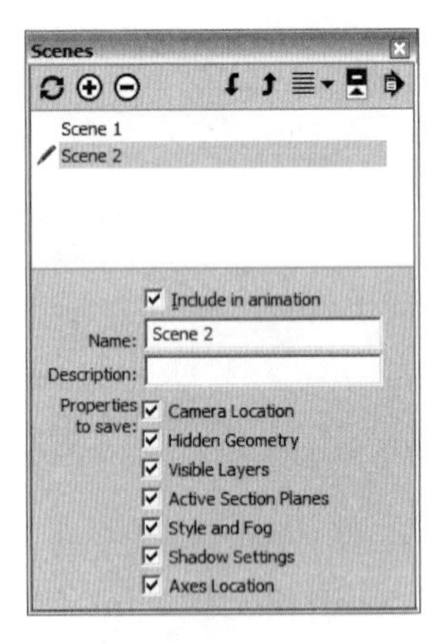

Choose which scene properties to save in the expanded Scenes dialog box.

The key to working with scene properties is the expanded Scenes dialog box, visible in Figure 10.9.

PATHWAYS TO... SETTING PROPERTIES A SCENE SAVES

1. **In the Scenes dialog box, select the scene whose properties you want to adjust.** You don't have to view this scene when you edit it; you can edit properties for any scene at any time.
2. **If not already expanded, click the Show Options button in the upper right corner of the Scenes dialog box.**
3. **Select the check boxes next to the properties you want to save.** That's it. You don't have to click Save anywhere to make your changes stick.

One terrific use of scene properties is to create scenes that help you show off different *iterations* (versions) of your design. See Chapter 7 for more information.

Updating All the Scene Properties at Once

The simplest way to modify a scene is to not worry about individual properties. If all you want to do is update a scene after you've made an adjustment to the appearance of your model, follow these steps:

1. **Go to the scene you want to update by clicking its tab at the top of the modeling window.**
2. **Make whatever styles, shadows, camera, or other display changes you want to make to your model.**
3. **Right-click the current scene tab and choose Update.** The old scene properties are replaced by the new ones. Be careful not to accidentally double-click the tab or you'll reactivate the scene and lose all the changes you made.

After you update a scene, you can't use Undo to revert things back to the way they were. You may want to save your SketchUp file right before updating a scene, and choosing File ⇨ Revert from the File menu if you don't like how things turn out.

Updating Scene Properties Selectively

Here's where things get complicated. At times, you'll want to update a scene without updating all its properties.

Updating scenes selectively involves making changes that you won't be able to see immediately; whenever you do this, it's a good idea to make a copy of your SketchUp file before updating more than one scene at a time, just in case a mistake occurs.

Maybe you've used scenes to create a tour of the sunroom you're designing for a client, and you want to change the shadow settings to make your model look brighter. You have 30 scenes in your presentation, and your meeting is in 5 minutes. You don't have time to change and update all 30 scenes one at a time. What to do? (This example deals with shadows, but the same method applies to any scene properties changes you'd like to make.) Follow these steps:

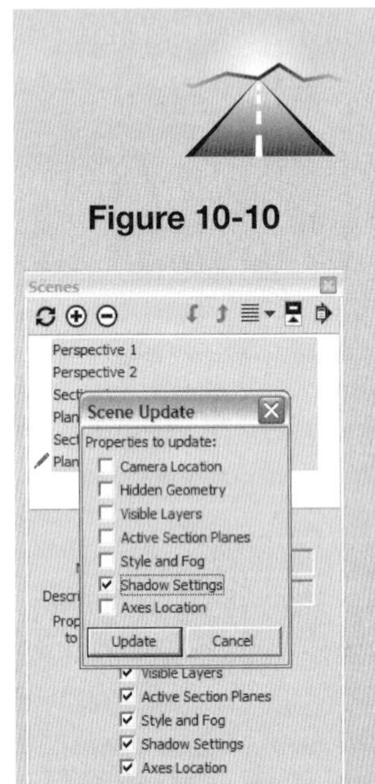

Updating only certain scene properties is a little more involved.

PATHWAYS TO... SELECTIVELY UPDATING SCENES

1. **Adjust the shadow settings to where you want them to be for all the scenes you'd like to update.**
2. **In the Scenes dialog box, select all the scenes you'd like to update.** Hold down Shift to select more than one scene at a time.
3. **Click the Update button in the Scenes dialog box.** A Properties to Update dialog box appears.
4. **Select the Shadow Settings check box and click the Update button** (Figure 10-10). If all you want to update are the shadow settings, make sure that only that check box is selected. More generally, you would select the check box next to each of the properties you want to update. All the selected scenes are updated with those new properties, and all the properties left deselected remain unchanged.

IN THE REAL WORLD

Henry Santos of Lionakis uses Google SketchUp because of its simple and intuitive workflow, which allows his ideas to be brought to light with unrivaled speed. According to him, SketchUp is the only program that allows designers to model at the "speed of thought."

A typical workflow for him is to create a model using an imported AutoCAD file or sketch as an underlay and then he builds his 3D model using components and staging scenes. These are later rendered using different plugins and software such as SU Podium, 3DS Max Design, or Maxwell Fire, and are finally retouched in Adobe Photoshop. These images are then shared with his clients to show how the final project will look. He goes on saying that in previous projects, he took pictures after the construction was completed and compared them to the original rendering and they were right on the mark.

SELF-CHECK

1. Scenes are a key element in making animations. True or false?
2. Every time we make a scene, a list item in the Scenes dialog box is updated and also a tab is created in the modeling window. True or false?
3. Name one way we can change a scene after it is created.
4. Name any two of the seven viewing properties we can control when we choose the option to update them.

Apply Your Knowledge We have learned that creating scenes can help us see our creation from different angles quickly. In this exercise we will use the previous model of your 10' x 20' exhibit and will collect a series of scenes. Before we start, choose the right viewing style and colors or textures and shadows and once you are done, gather a series of scenes from every corner. Once you are done, create two more scenes to show a front view and a top view, both in parallel projection mode.

10.3 MASTERING THE SECTIONAL APPROACH

Section
A from-the-side, two-dimensional, nonperspectival view of an object or space. Also referred to as a sectional view.

Sections are objects that let you cut away parts of your model to look inside. You place them wherever you need them, use them to create views you wouldn't otherwise be able to get, and then delete them when you're done. When you move a section plane, you get instant feedback; the "cut" view of your model moves, too. If you want, you can embed them in scenes and even use them in animations. Sections are easy to use, incredibly important, and impressive. People use sections in many ways:

- to create standard orthographic views (like plans and sections) of buildings and other objects
- to make cutaway views of complex models so these models are easier to understand
- to work on the interiors of buildings without having to move or hide geometry
- to generate sectional animations with scenes

10.3.1 Cutting Plans and Sections

GOOGLE SKETCHUP IN ACTION

Master cutting, controlling, and applying section animation.

The most common use for sections is to create straight-on, cut-through views of a model. These are some of the views that often include dimensions, and they are typical of the kinds of drawings that architects make to design and explain space.

Straight-on, cut-through views are useful because:

- they are easy to read
- you can take measurements from them if they are printed to scale
- they provide information that no other drawing type can

The following terms (which are illustrated in Figure 10-11) can help you more easily create different views of your model:

Plan
A top-down, two-dimensional, nonperspectival view of an object or space. Also referred to as a planimetric view.

- **Plan:** *A planimetric view*, or **Plan**, is a top-down, two-dimensional, non-perspectival view of an object or space. Put simply, it's every drawing of a house floor plan you've ever seen. You generate a plan by cutting an imaginary *horizontal* slice through your model. Everything below the slice is visible, and everything above it isn't.
- **Section:** Not to be confused with sections (the SketchUp feature about which this section of the book is written), a *sectional view*, or **section**, is a from-the-side, two-dimensional, nonperspectival view of an object or space. You make a section by cutting an imaginary *vertical* slice through your model. Just like in a plan view, everything on one side of the slice is visible, and everything on the other side is hidden.

You cut plans and sections by adding section planes to your model. These are a little abstract, because nothing like them exists in real life. In SketchUp, section planes are objects that affect the visibility of certain parts of your model. When a section plane is active, everything in front of it is visible and everything behind is hidden. Everywhere your model is "cut" by a section plane, a slightly thicker "section cut" line appears.

FOR EXAMPLE

CUTTING LIKE AN ARCHITECT

In architecture, the convention is to *cut* plans at a height of 48 inches, meaning that the imaginary horizontal slice is made 4 feet above the floor surface. This ensures that doors and most windows are shown cut through by the slice, whereas counters, tables, and other furniture are below it, and thus are fully visible. You can see this in Figure 10-11. These things are important when you try to explain a space to someone. After all, architectural drawings are two-dimensional abstractions of three-dimensional space, and every little bit of clarity helps.

When it comes to architectural sections (as opposed to sections, the SketchUp feature), there's no convention for where to cut them, but you should follow a couple rules:

- **Never cut through columns.** If you show a column in section, it looks like a wall. This is bad because sections are supposed to show the degree to which a space is open or closed. You can walk around a column, but you can't walk through a wall.
- **Try your best to cut through stairs, elevators, and other vertical circulation.** Showing how people move up and down through your building makes your drawings a lot more readable, not to mention interesting. See Figure 10-11 for an example.

Figure 10-11

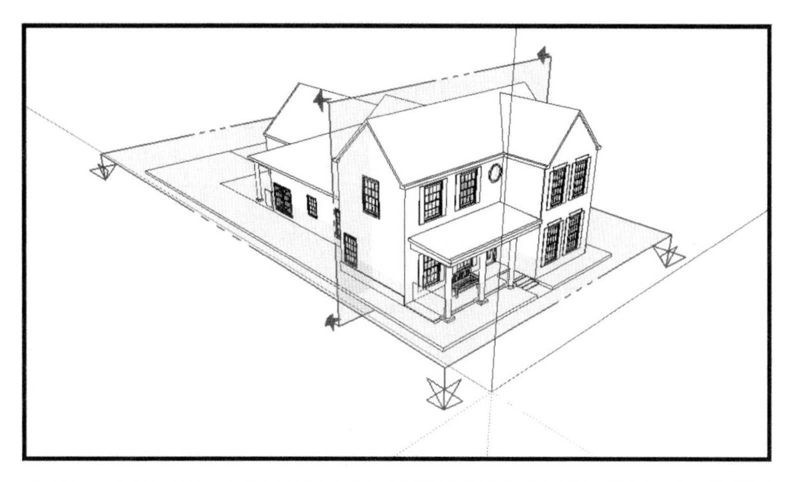

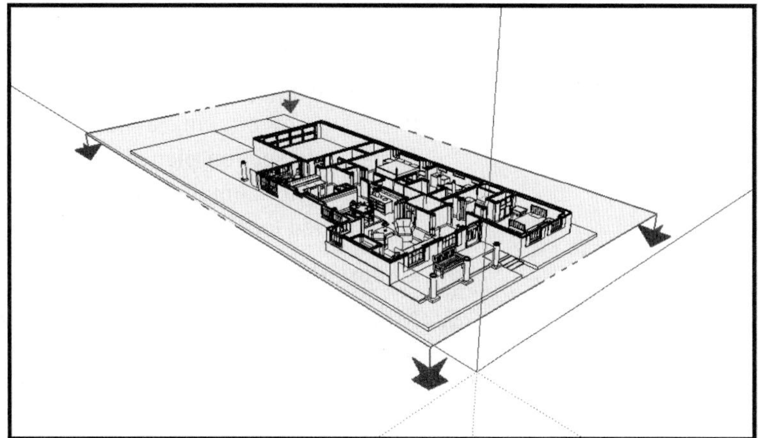

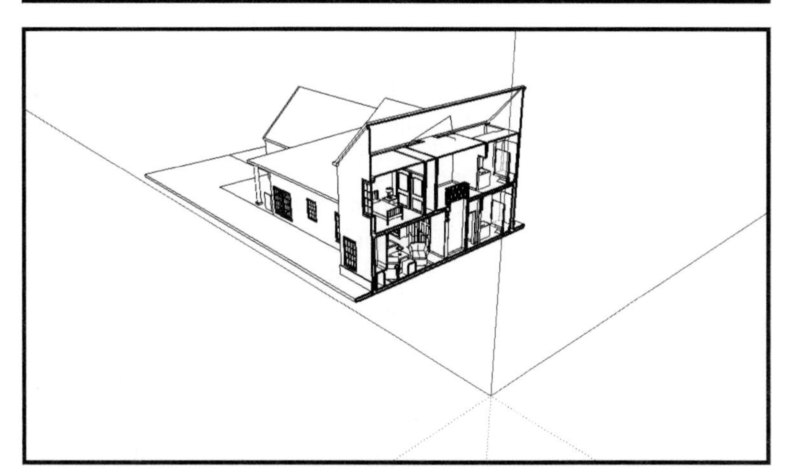

A plan is a horizontal cut (second image), whereas a section is a vertical one (third image).

Reproduced with the permission of Jorge Paricio, PhD.

If you are using Windows, now would be a good time to open the Sections toolbar by choosing View ⇨ Toolbars ⇨ Sections. If you're on a Mac, the Section Plane tool is in the Large Tool Set, which you can activate by choosing View ⇨ Tool Palettes ⇨ Large Tool Set. On both platforms, Section Plane looks like a white circle with letters and numbers in it.

To add a section plane, follow these steps:

1. **Choose Tools ⇨ Section Plane to activate the Section Plane tool.** You can also activate Section Plane by choosing its icon from the Large Tool Set (Mac) or the Sections toolbar (Windows), if you have it open.
2. **Move the Section Plane tool around your model.** Notice how the orientation of the Section Plane cursor (which is quite large) changes to be coplanar to whatever surface you're hovering over.
3. **When you've decided where you want it, click once to add a section plane.** To create a plan view, add a horizontal section plane by clicking a horizontal plane like a floor. For a sectional view, add a vertical section plane by clicking a wall or other vertical surface. You can, of course, add section planes wherever you want; they don't have to be aligned to horizontal or vertical planes. Figure 10-12 shows a section plane added to a model of a house.
4. **Choose the Move tool.**
5. **Move the section plane you just added by clicking it once to pick it up and again to drop it.** You can only slide your section plane back and forth in two directions; SketchUp only allows section planes to move perpendicular to their cutting planes. When you are deciding where to locate your cut, the example, "Cutting Like An Architect," offers helpful pointers. After you've added a section plane and moved it to the desired location, you can rotate and even copy it, just like any other object in your model. It will never affect your geometry—just the way you view it.
6. **If you need to rotate your section plane, select it and use the Rotate tool (which you can read more about in Chapter 6).** Why rotate a section plane? In certain circumstances, rotating a section plane (instead of creating a new one) can help explain a complex interior space. Showing a plan view *becoming* a sectional one is a powerful way to explain architectural drawings to an audience that doesn't understand them.
7. **To make a new section plane by copying an existing one, use the Move or Rotate tool to do it the same way you would make a copy of any other SketchUp object.** Chapter 2 explains these basic actions

Figure 10-12

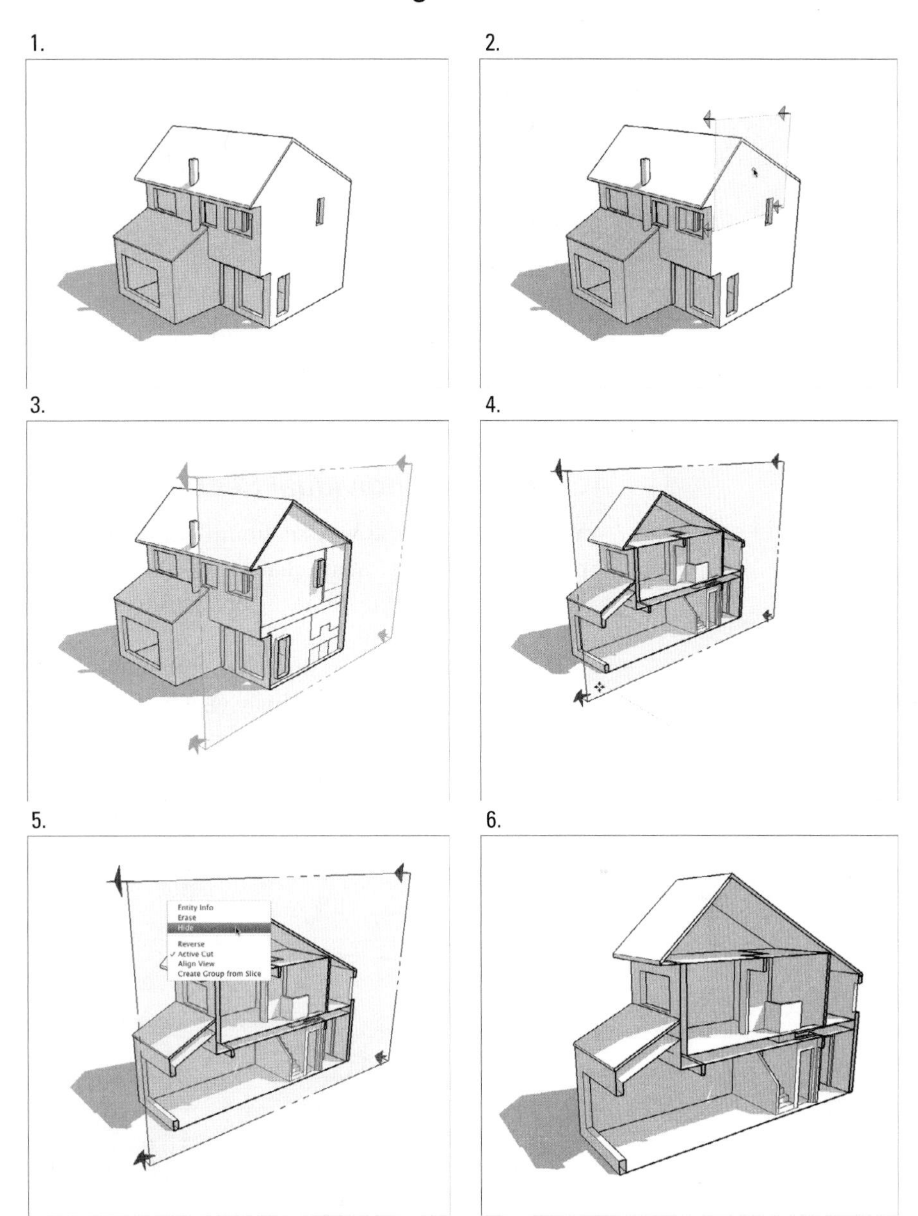

Add a section plane wherever you want one, and then move it into position.

in detail. Copying section planes is a great way to space them a known distance apart; this can be trickier if you use the Section Plane tool to keep adding new ones instead. Figure 10-13 shows moving, rotating, and copying a section plane.

Figure 10-13

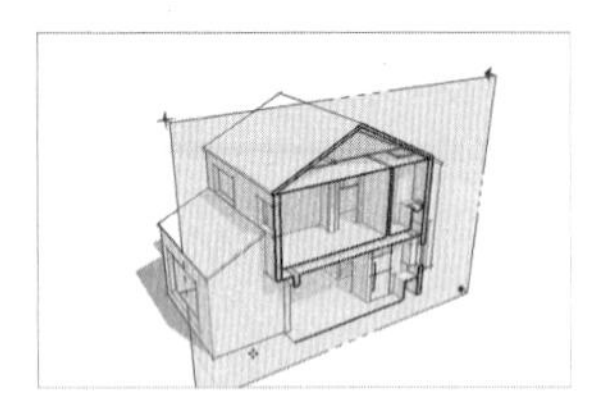
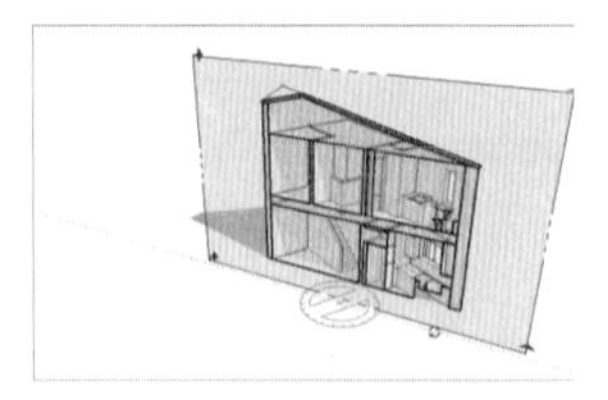
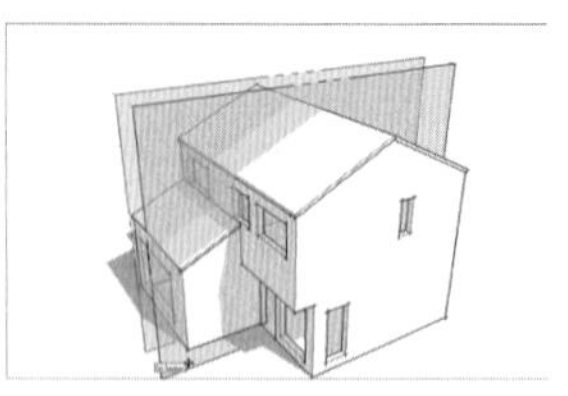

Moving, rotating, and copying a section plane.

When the section plane you've added is in position, you're ready to control how it impacts visibility in a number of other ways. See the following sections for details.

Controlling Individual Section Planes

You can control the way section planes behave by right-clicking them to bring up a context menu that looks like the one shown in Figure 10-14. Examples of what the following options do are shown in the same illustration:

- **Reverse:** This option flips the "direction" of the section plane, hiding everything that was previously visible, and revealing everything that used to be "behind" the cut. Use this when you need to see inside the rest of your model.

Figure 10-14

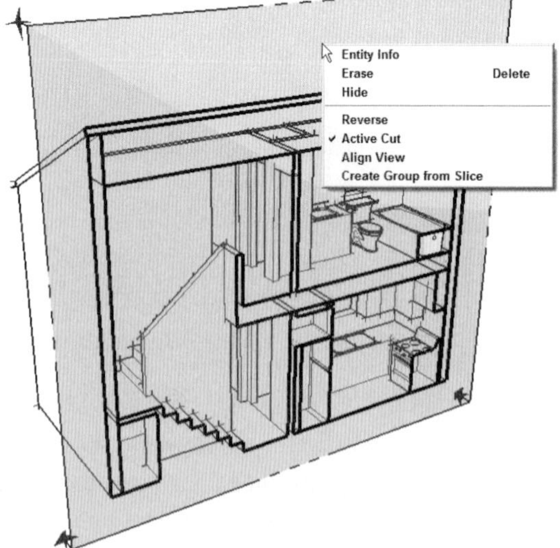

Right-clicking a section plane gives you some options.

Active cut
The section plane that is actually cutting through your model; others section planes are considered "inactive."

Nest
To embed an object in a separate group or component.

- **Active Cut:** Although you can have multiple section planes in your model, only one of them can be "active" at a time. The **Active cut** is the section plane that is actually cutting through your model; others are considered "inactive." If you have more than one section plane, use Active Cut to tell SketchUp which one should be active. You *can* have more than one active section plane in your model at a time, but doing so requires that you **Nest,** or embed, each section plane in a separate group or component. It's possible to achieve some fairly elaborate effects with this technique. You can read all about groups and components in Chapter 5.
- **Align View:** When you choose Align View, your view changes so that you're looking straight on at the section plane. You can use this option to produce views like the ones described in the section "Getting Different Sectional Views," later in this chapter.
- **Create Group from Slice:** This option doesn't have much to do with the other choices in this context menu; it's really a modeling tool. You can use this to do exactly what it says: create a group from the active slice, or section plane. This comes in handy for creating filled-in section cuts for final presentations. See Figure 10-15.

Setting Section-Plane Visibility

If you want to control the visibility of all your section planes at once, a couple of menu options can help. You use both of these toggles in combination to control how section cuts appear in your model. These two options, shown on the View menu, are illustrated in Figure 10-16.

- **Section Planes:** This choice toggles the visibility of section-plane objects without affecting the section cuts they produce. More simply, deselecting Section Planes hides all the section planes in your model, but doesn't turn off the section cut effect, as shown in the middle image in Figure 10-16. This is how you'll probably want to show most of your sectional views, so this is an important toggle.
- **Section Cut:** Deselecting this option toggles the section cut effect on and off without affecting the visibility of the section-plane objects in your model. This choice is the opposite of Section Planes, in the previous point, but it's every bit as important.

Getting Different Sectional Views

Using section planes, you can get a couple of useful and impressive views of your model without much trouble. The second builds on the first, and both are shown in Figure 10-17. A section perspective (left) is a special kind of way to view a three-dimensional space. The second type, an orthographic view (right), is straight on, doesn't use perspective.

Figure 10-15

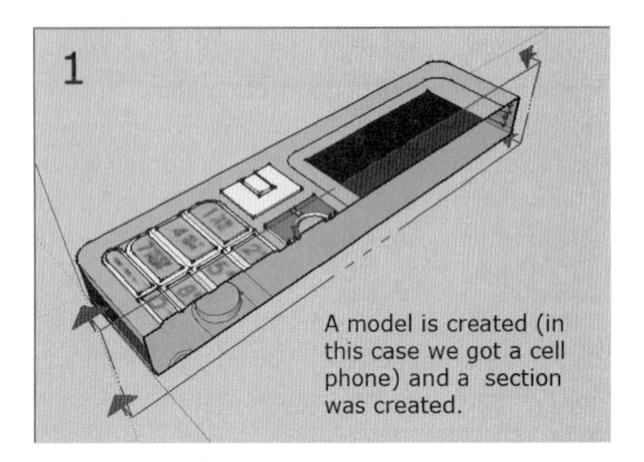

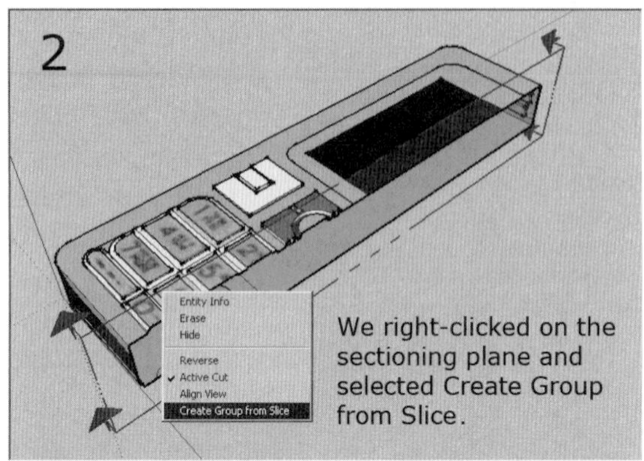

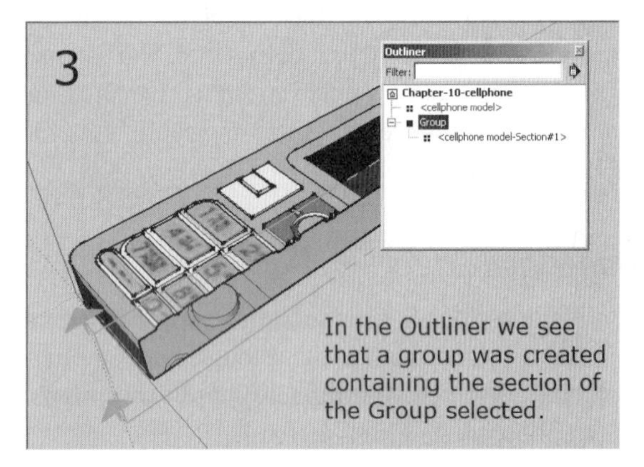

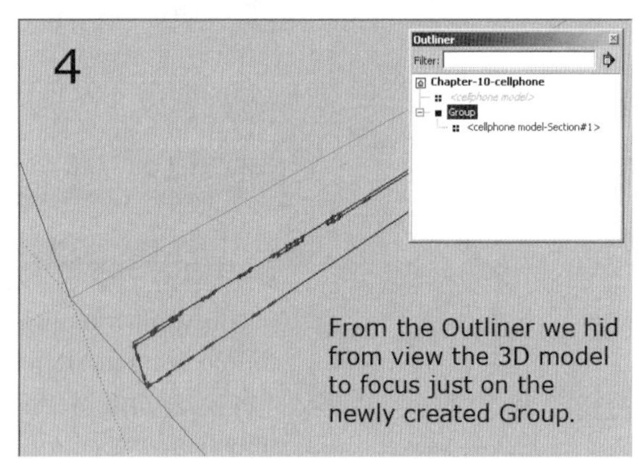

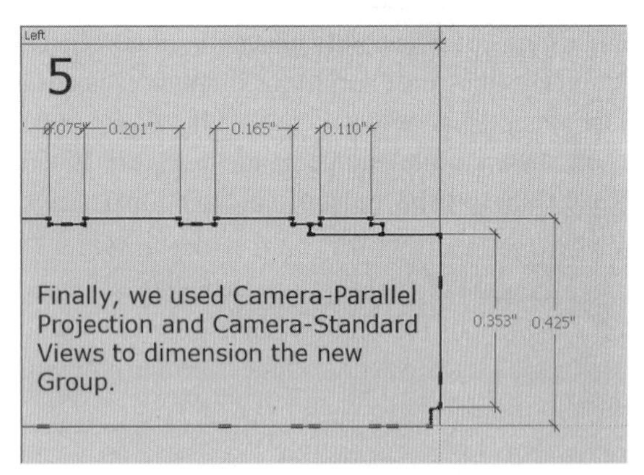

Create a group from the active slice or section plane.

Reproduced with the permission of Jorge Paricio, PhD.

Section perspective

The view of a cut building where objects seen inside the space appear to get smaller as they get farther away.

Making a Section Perspective

If you imagine cutting a building in half and then looking at the cut surface straight on while looking inside, you have a **Section perspective.** The *section* part of the term refers to the fact that the building has been cut

Figure 10-16

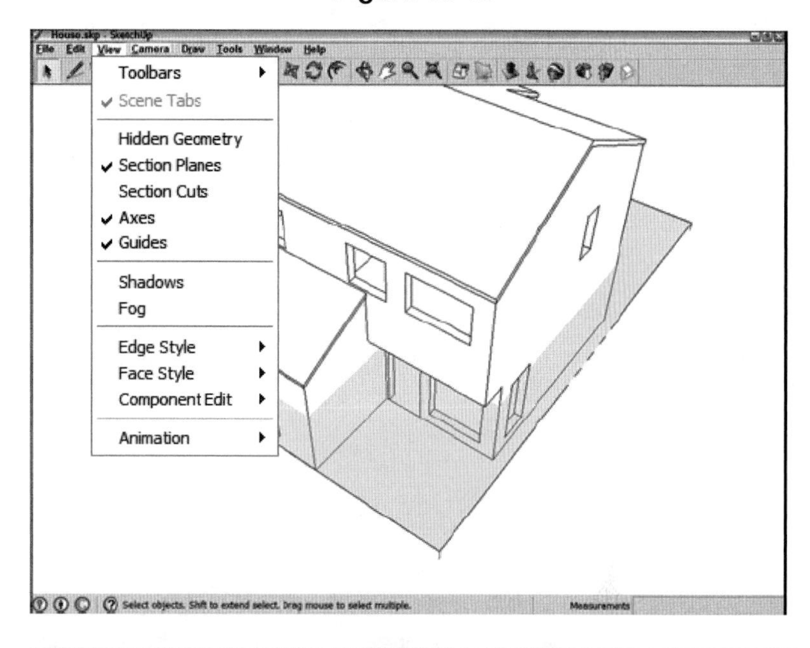

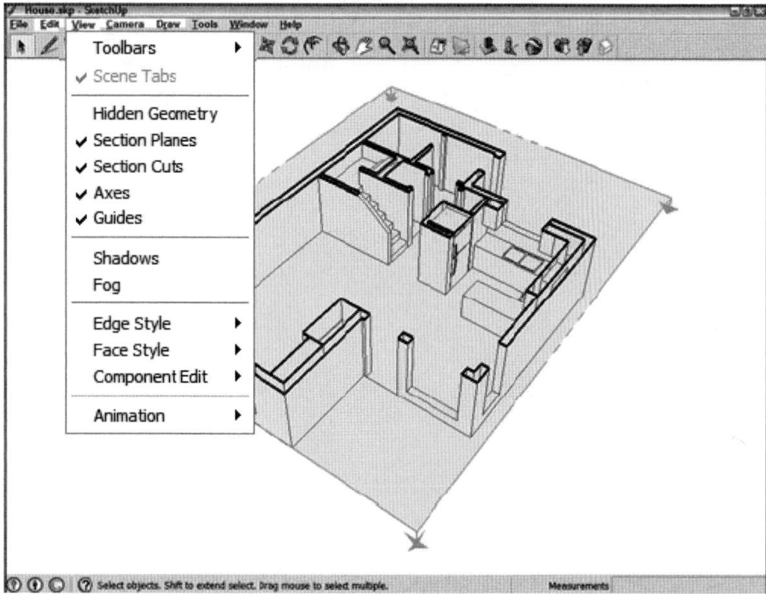

Control section plane visibility with Section Planes and Section Cut.

Reproduced with the permission of Jorge Paricio, PhD.

away. The *perspective* part indicates that objects seen inside the space appear to get smaller as they get farther away.

Section perspectives are a great way of showing interior space in a way that's understandable to most people. To create a section perspective using the Section Plane tool in SketchUp, follow these steps:

Figure 10-17

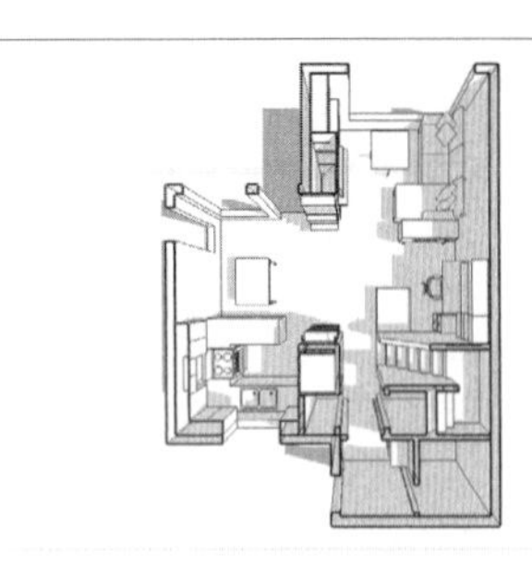
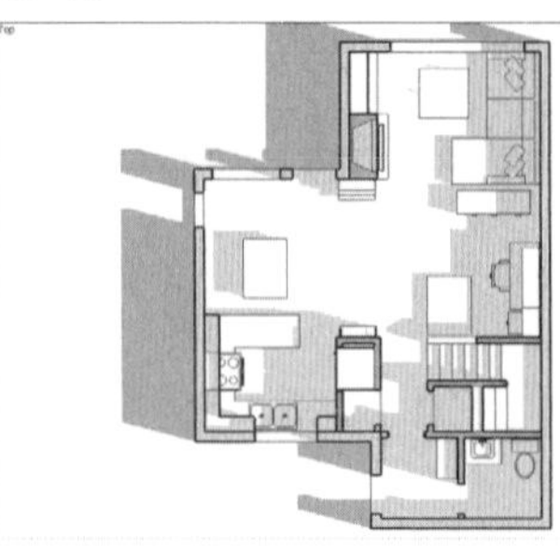

Turn Perspective on for a section perspective; choose Parallel Projection to produce an orthographic view.

PATHWAYS TO... TO MAKING A SECTION PERSPECTIVE

1. **Select the section plane you'd like to use to make a section perspective by clicking it with the Select tool.** When it's selected, your section plane turns blue, assuming that you haven't changed any of the default colors in the Styles dialog box.
2. **If the selected section plane isn't active, right-click and choose Active Cut from the context menu.** Active section planes cut through their surrounding geometry. If your section plane is visible but isn't cutting through anything, it isn't active.
3. **Right-click the selected section plane and choose Align View from the context menu.** This aligns your view so that it's straight on (perpendicular) to your section plane.
4. **If you can't see your model properly, choose Camera ⇨ Zoom Extents.** This zooms your view so that you can see your whole model in the modeling window.

Orthographic projection
A common way for three-dimensional objects to be drawn so that they can be built.

Generating an Orthographic Section

Ever seen a technical drawing that included top, front, rear, and side views of the same object? Chances are that was an **orthographic projection,** which is a common way for three-dimensional objects to be drawn so that they can be built.

Producing an orthographic section of your model is pretty easy; it's only one extra step beyond making a section perspective. Here's how to do it:

1. **Follow steps 1–3 in the preceding section, as if you're making a section perspective.**
2. **Choose Camera ⇨ Parallel Projection.** This switches off Perspective, turning your view into a true orthographic representation of your

model. If you printed it at a specific scale, you could take measurements from the printout.

To print a plan or section view of your model at a particular scale, have a look at Chapter 12, which explains the whole process.

10.3.2 Creating Section Animations with Scenes

This is probably one of the most useful and impressive things you can do with this software, but some people who have been using SketchUp for years don't know about it. The basic idea is that you can use scenes to create animations where your section planes move inside your model. Here are a few reasons you might want to use this technique:

- If you have a building with several levels, you can create an animated presentation that shows a cutaway plan view of each level.
- Using an animated section plane to "get inside" your model is a much classier transition than simply hiding certain parts of it.
- When you need to show the relationship between the plan and section views for a project, using an animated section plane helps to explain the concept of different architectural views to 3D beginners.

Follow these steps to create a basic section animation (a simple example is illustrated in Figure 10-18):

PATHWAYS TO... CREATING SECTION ANIMATIONS WITH SCENES

1. **Add a section plane to your model.** A complete explanation of how to create section planes can be found in the section "Cutting Plans and Sections" earlier in this chapter.
2. **Add a scene to your model.** Check out the section "Creating Scenes," earlier in this chapter, for a complete rundown on adding scenes.
3. **Add another section plane to your model.** You can add another section plane in one of two ways:
   - **Use the Section Plane tool to create a brand new one.** This is probably the easiest option, especially if you're just starting out.
   - **Use the Move tool to copy an existing section plane.** Copying section planes is discussed in the section "Cutting Plans and Sections," earlier in this chapter. Make sure that your new section plane is active; if it is, it'll be cutting through your model. If it isn't, right-click the section plane and choose Active Cut from the context menu.

(Continued)

PATHWAYS TO... CREATING SECTION ANIMATIONS WITH SCENES *(continued)*

Figure 10-18

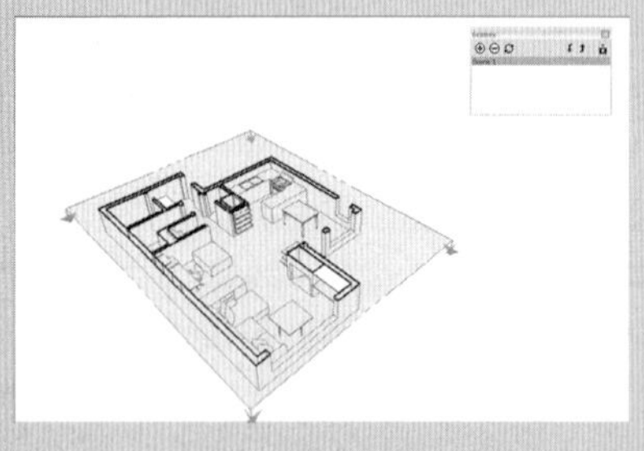

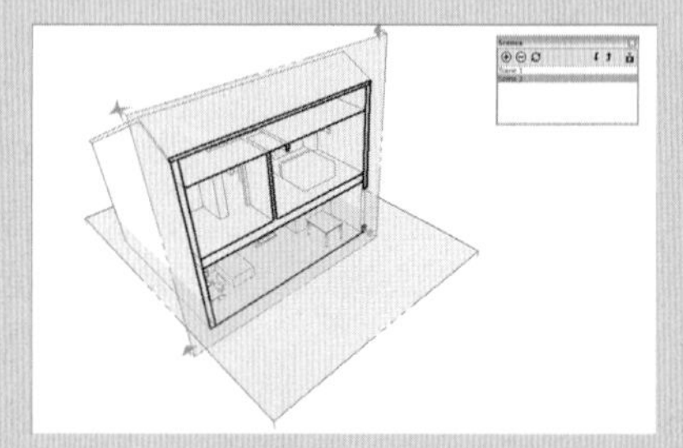

Making a section animation is a fairly straightforward process.

4. **Add another scene to your model.** This new scene "remembers" which is the active section plane.
5. **Click through the scenes you added to view your animation.** You should see an animated section cut as SketchUp transitions from one scene to the next. If you don't, make sure that you have scene transitions enabled. You can verify this by choosing Window ⇨ Model Info and then choosing the Animations panel in the Model Info dialog box. The Scene Transitions check box should be selected.

If you don't like being able to see the section-plane objects (the boxy things with arrows on their corners) in your animation, switch them off by deselecting Section Planes on the View menu. You should still be able to see your section cuts, but you won't see the gray rectangles.

The hardest thing to remember about using scenes and section planes to make section animations is this: You need a separate section plane for each scene that you create. That is to say, SketchUp animates the transition from one active section plane to another active section plane. If all you do is move the same section plane to another spot and add a scene, this technique won't work.

IN THE REAL WORLD

Grace Hwang is an interior designer who works at Gensler in San Francisco, and she uses Google SketchUp because the program is fast to work with and the visualization is immediate. She believes that one can use various readily available plugins to "jazz up" the final product and that SketchUp is also a great tool to be used when walking the client through the entire space. She goes on to say that clients love when she manipulates the model right in front of them during presentations and that she can adjust rooms and finishes very easily.

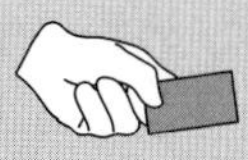

CAREER CONNECTION

Helen Keating, owner of Keating Design Studios, first began working with Google SketchUp at an architectural firm that was looking to replace the physical foam-core models they had been using. SketchUp enabled her work to be more detailed and she began to include interiors, as well as exterior and massing models.

Today, Helen uses SketchUp for creating conceptual exterior and interior renderings and presentation sheets. She then renders these scenes in Podium, V-Ray, or Shaderlight, and adds some final touches in Adobe Photoshop. She also creates elevations and section cuts, which she exports to AutoCAD when making construction documents.

SELF-CHECK

1. What tool do we use to make standard orthographic projections of objects and buildings?
2. Give one reason why we would use a sectional view (two-dimensional and nonperspectival) in a two-story building.
3. A section plane can be copied, moved, or rotated just like we would do with any other object in our model. True or false?
4. Explain one reason why we would want to create an animation to show different section views in a model.

Apply Your Knowledge For the beginner student involved in architecture, interior design, or even industrial design, understanding how orthographic sectional views work is very important, because it allows us to understand how an object or a building is on the inside, or how, for example, two floors relate to each other.

Build or use an existing two-story structure that either you would have used in a previous exercise or that you would have searched for in Google's 3D Warehouse. This is located under Windows-Components. Then type in the search area the words "two story house." A full explanation on how this tool works can be found in Chapter 11. Then, practice creating a plan view section at 4'-0" height for each floor, and with at least two orthographic sections, which are also called elevation views. Make one scene for each of the views you have created.

SUMMARY

Section 1

- The Position Camera tool gives you the ability to precisely place your viewpoint in SketchUp in a particular spot.
- Use the Walk tool to move through your model keeping in mind the following rules:
  - Straight up is forward.
  - Straight down is backward.
  - Anything to the left or right causes you to turn while you're walking.
- Measured in degrees, a wide field of view means that you can see more of your model without having to move around.

Section 2

- SketchUp **scenes** are saved views of your model.
- Scenes are one of the most important features in SketchUp, for three reasons:
  - Scenes can save you hours of time.
  - Scenes are the most effective way of presenting your model.
  - Scenes are the key to making animations.
- If you create a scene to save a view then do more modeling, and then return to that scene, your model will not go back to the way it was when you created the scene.
- A scene is simply a set of view settings, which means that they are automatically updated to reflect your changes every time you edit your model.

Section 3

- Sections are objects that let you cut away parts of your model to look inside.
- The most common use for sections is to create straight-on, cut-through views of a model.
- You cut plans and sections by adding section planes to your model.
- You can use scenes to create animations where your section planes move inside your model.

ASSESS YOUR UNDERSTANDING

SUMMARY QUESTIONS

1. With the Walk Tool, how do we control turning and moving forward or backwards?
2. What range of angles would we choose in our field of view if we wanted a perspective view of a building from the outside?
3. What type of views would be used to show interior spaces in a way that's understandable to most people?
4. Saving a scene for each point that you'd like to make in a presentation is tedious and consumes a lot of memory in our file. True or false?
5. If you have more than one section plane, use Passive Cut to tell SketchUp which one should be active. True or false?
6. How do we control the number of seconds that we give for each scene, before we make an animation?
7. What is the conventional height when we use Position Camera from the menu bar and click in the modeling window?
8. You can use scenes to create animations to show exactly where your section planes are situated inside your model. True or false?
9. Saved views of your model are called:
   a. Plans
   b. Sections
   c. Placemarks
   d. Scenes
10. Which term refers to the amount of your model that you're able to see in your modeling window at one time?
    a. Plan
    b. Section
    c. Orthographic projection
    d. Field of view

APPLY: WHAT WOULD YOU DO?

1. Describe two reasons why you would want to create a walk through of a model that you would have created.
2. You have created a model of a residence and you want to present it. How do you use SketchUp's scenes and sections to explore the model and present some views to your client?
3. Explain the differences between a section perspective and a orthographic section. What are the reasons to use one system versus another?
4. Write down a list of steps required to make a section animation with scenes.

BE AN ARCHITECT

Create Scenes

Take the model of your house and create four scenes:

1. From outside the house
2. From inside the living room
3. From inside the bathroom
4. From inside the bedroom

Modify Scenes

Take the scenes you have already created and put them in the opposite order that you originally had them. Also, add shadows to your scenes if you do not already have shadows present.

Create Animations

Taking the model of your house, create three section animations using scenes. Play your animations as a presentation for your class.

KEY TERMS

Active cut
Field of view
Nest
Orthographic projection
Plan
Scenes
Section
Section planes
Section perspective

WORKING WITH GOOGLE EARTH AND THE 3D WAREHOUSE

Tying Together All of Google's 3D Software

Do You Already Know?

- How to tie together all the Google 3D software?
- How to build a model in SketchUp and send it to Google Earth?
- That you can contribute your models to the 3D Warehouse?

For the answers to these questions, go to **www.wiley.com/go/chopra/googlesketchup2e**

What You Will Find Out	What You Will Be Able To Do
11.1 The definition of the various Google environments.	Understand how Google SketchUp, Google Building Maker, Google Earth, and Google 3D Warehouse work together.
11.2 How to geo-locate your model.	Work efficiently and effectively with Google Earth.
11.3 How to work with Google's 3D Warehouse.	Get to Google's 3D Warehouse and upload your models.

INTRODUCTION

If you've ever used Google Earth, you know what it's like to spend several hours traveling to Paris, Cairo, and the South Pole while checking out the peak of Mount Everest and looking at your old elementary school along the way.

What if you could see 3D models of buildings and other man-made structures in Google Earth the same way that you can see aerial images and 3D topography? You can. What if you could build your own models in SketchUp and see them in Google Earth? You can do that, too. What if you could allow everyone who uses Google Earth—there are hundreds of millions of them—to see your models in their copies of Google Earth, no matter where they are? That, too, is possible.

This chapter focuses on making SketchUp models that you and (if you'd like) anyone else can see on Google Earth. In this chapter, you will evaluate how to best use SketchUp, Google Earth, and the 3D Warehouse. You will learn how to navigate in Google Earth and how to build a model in SketchUp for Google Earth. Finally, you will learn how to contribute to the 3D Warehouse, a large online repository of free 3D models that anyone can contribute to or borrow from.

11.1 GETTING THE BIG (3D) PICTURE

GOOGLE SKETCHUP IN ACTION 

Understand how Google SketchUp, Google Building Maker, Google Earth, and Google 3D Warehouse are related.

SketchUp, Google Earth, the 3D Warehouse, and the new product Google Building Maker have a relationship that allows them to function together:

- **Google SketchUp:** Because SketchUp is especially good for architecture, landscape architecture, and urban planning, you can use it to make buildings that you can later view in Google Earth. If you want, you can also upload (send) what you make to the 3D Warehouse, where anyone who finds it can download (borrow) your model and use it in his or her own copy of SketchUp.
- **Google Building Maker:** Building Maker (www.google.com/buildingmaker) is the newest addition to Google's stable of 3D tools. This free, online application makes modeling existing buildings much, much easier; provided they are located in an area for which Google has special, multi-angle aerial imagery. You can use SketchUp to open Building Maker models; doing so lets you improve them by tweaking their geometry and textures.
- **Google Earth:** Google Earth (http://earth.google.com) is a software program that lets you explore the world by "flying" around, zooming in on things that interest you. The more you zoom, the better the detail is; in some places, you can see things as small as coffee cups. The imagery in Google Earth is anywhere from a couple of weeks to 4 years old, but it gets updated all the time. If you want, you can build models in SketchUp and view them in Google Earth. You can also see models that other people have made.
- **Google 3D Warehouse:** The Google 3D Warehouse (http://sketchup.google.com/3dwarehouse) is a collection of 3D models that is found on Google's servers. Anyone can contribute models, and anyone can use

them in their own SketchUp projects. Some of the best models in the 3D Warehouse are used in a special layer where anyone can see them while they are navigating in Google Earth.

Figure 11-1

1

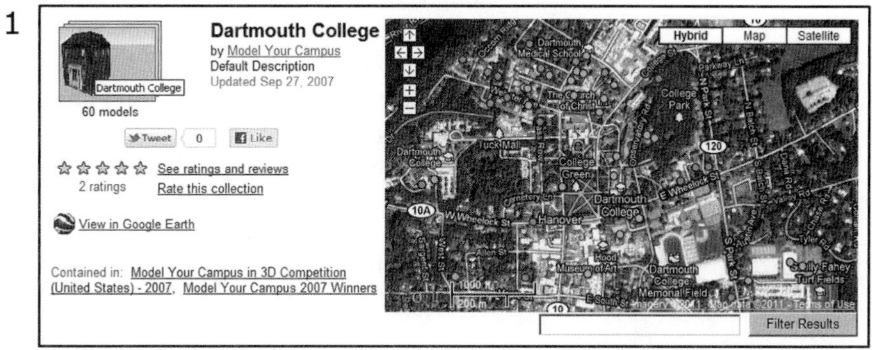

3D Warehouse can be accessed via Internet and you can easily make a group so that many people can collaborate on a single project.

2

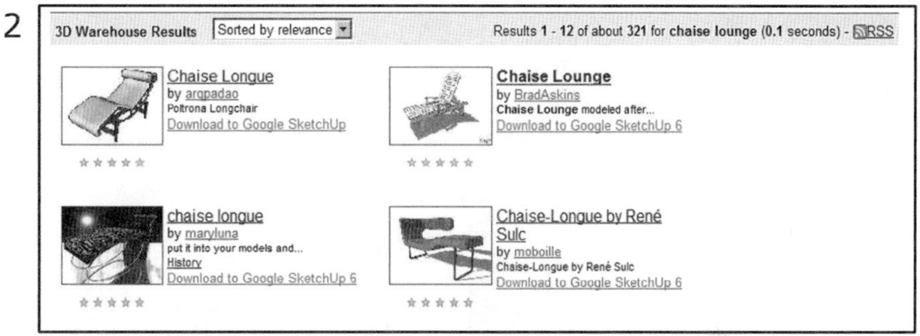

3D Warehouse can also help us find models via Internet that we do not have the energy or time to build.

3

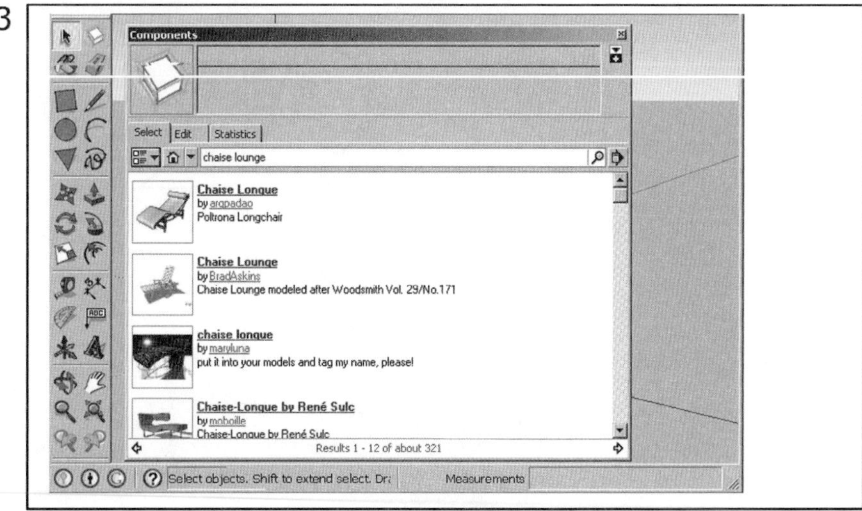

3D Warehouse can also be accessed via the Components dialog box within Google SketchUp.

Searching the 3D Warehouse.

There are two ways to access it, either via internet or through the Components dialog box and both methods render equal results. We need to remember that this is a powerful feature that we can utilize if we want many persons to collaborate on a single project. In Figure 11-1 you will see how in 2007 many students collaborated in building Dartmouth College's campus in 3D; a new work group was created and 61 models were added to the map. Also, if you are tight with time or do not have enough resources to build everything in your model, you can search for the models that you need in 3D Warehouse, and chances are that you will find what you are looking for, built by someone else, and free of use. In Figure 11-1 you will see how a search for a chaise lounge rendered 321 possibilities.

Figure 11-2 shows the SketchUp/Google Earth/Building Maker/3D Warehouse workflow in a diagram.

Figure 11-2

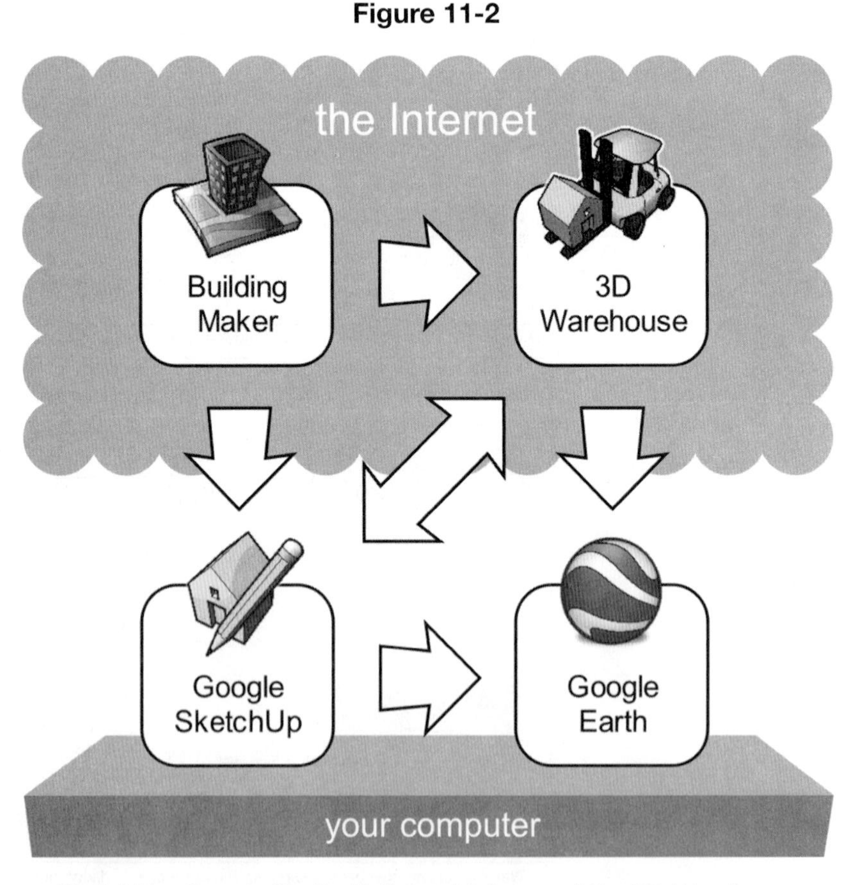

SketchUp, Google Earth, Building Maker, and the 3D Warehouse are all related.

IN THE REAL WORLD

Ann Barreca Young from ABY interior design in St. Louis started using Google SketchUp for numerous reasons. Among them, she cites that SketchUp is not only user-friendly, but also offers the possibility of importing her AutoCAD floor plans so she can easily build the space perspectives. She goes on saying that she really enjoys its 3D warehouse functions as well.

She has recently started to experiment using the SU Podium V2 plugin to create photo-realistic renderings, and has been able to easily train two other designers to use the tool. She also added that they picked up the program quickly and needed very little help to get started.

SELF-CHECK

1. Explain the differences between Google SketchUp and Google Building Maker.
2. Explain in your own words some of the most notable capabilities of Google Earth.
3. Google 3D Warehouse is a private repository of models that only subscribers can access. True or false?

Apply Your Knowledge Practice visiting the Google Earth, Building Maker, and 3D Warehouse Web sites listed in this section.

11.2 SENDING YOUR MODELS TO GOOGLE EARTH

The kinds of models you might like to view in Google Earth fall into two basic categories:

- **Existing buildings:** People make models of real-life buildings simply to provide context for a structure they are designing or to display their models on Google Earth's default 3D Buildings layer where everyone in the world can see them.
- **Unbuilt buildings:** If you are designing a new structure and you want to see (and show other people) how it might look in the context of the rest of the built environment, there is no better way than to view it in Google Earth.

If you're modeling existing buildings specifically so they will appear in everyone's Google Earth, you are geo-modeling and the next section is just for you. If, on the other hand, you already have a model of something you want previewed in Google Earth, please skip ahead to "Geo-Locating Your Model," later in this chapter.

11.2.1 Geo-Modeling for Google Earth

The folks at Google are on a mission: To make Google Earth into a hyper-realistic, fully-3D map of the planet, complete with everything from mountains to weather systems. The idea is to create a "mirror world" that can someday be the basis for experimenting, teaching, and shopping as well as tourism and just about anything else you can imagine. It's an exciting project, and plenty of 3D enthusiasts are pitching in to help make Google's mission a reality by modeling as many buildings as they can, one building at a time.

The thing is, geo-modeling (building models for Google Earth) is different from other kinds of SketchUp modeling. All Google Earth models are:

- **Geo-located:** They know where on earth they're supposed to sit.
- **Geometrically simple:** Having lots of edges and faces doesn't work; big models plus Google Earth equals sluggish performance and no one wants that.
- **Hollow:** Earth models are basically shells; they contain no interior walls or other details at all.
- **Completely photo-textured:** Photo-texturing your model means painting it with photographs of the building it's supposed to represent.

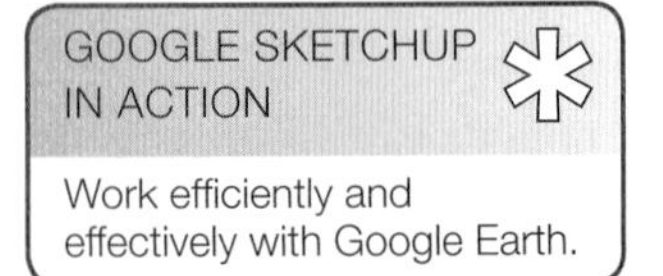

You can approach geo-modeling in a few ways; SketchUp is no longer the only tool at your disposal. The following sections are all about techniques for making models whose sole purpose is to reside in Google Earth.

IN THE REAL WORLD

Starting with Google Building Maker, Google Earth and Google SketchUp can work in together to create simulations geared to convince clients of the validity of a design proposal, through maximizing the use of the sun as it traverses the sky over one day, or over a whole year. For example, a company in Boulder, Colorado, Simple Solar Electric Systems uses this exact system to show their clients how their house will look after solar panels are installed, in addition to aiding designers in finding shadow spots. According to Drew Kundtz, "before committing to the final installation, customers can see in rich 3D how their array will look before the deployment occurs. Our customers typically love the final result, but SketchUp is like an insurance policy because everyone can accurately visualize the array on-site beforehand."

Building Maker is purpose-built for geo-modeling; making accurate, photo-textured models of existing buildings is literally all it does. Provided your modeling target is in an area where Building Maker data exists, this is definitely where you should start. Why? Because Building Maker gives you two things that are otherwise extremely hard to obtain:

- **Aerial imagery:** Being able to use photo-textures that have been taken from above (rather than from street level) automatically takes care of things like trees, cars, people, and anything else that would otherwise get in the way.

- **Building height:** Believe it or not, figuring out the height of an existing building you are trying to model is one of the hardest parts of the geo-modeling process. Because Building Maker makes you "trace" oblique photos to build models, you always end up with a building of acceptably accurate height.

To find out if you can use Building Maker for your model, search for 'building maker available locations' in your favorite search engine.

Here's another great thing about Building Maker: You can open the models it creates in SketchUp and refine them there. Here is what the workflow looks like; Figure 11-3 shows the steps along the way:

1. **Open a new SketchUp file and make sure you're online.** Building Maker is a Web application; it exists only on Google's servers. You need an Internet connection to use it.

Figure 11-3

1. Start in Building Maker

2. Export to SketchUp

3. Improve in SketchUp

4. Upload to 3D Warehouse

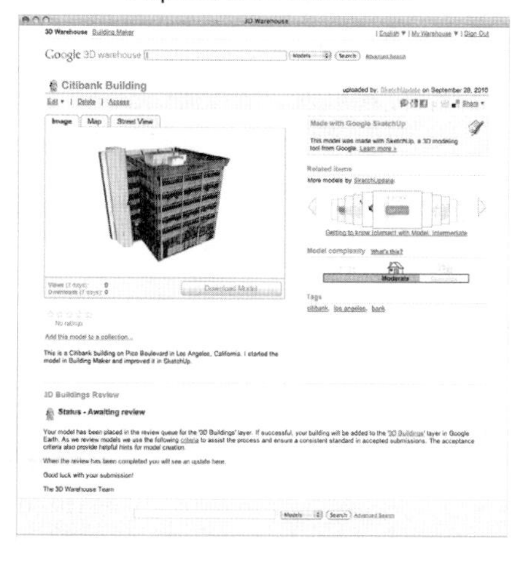

Start geo-modeling with Building Maker; then improve things with SketchUp.

2. Choose File ⇨ Building Maker ⇨ Add New Building. A Building Maker window appears.

3. **Model a building with Building Maker.** Building Maker is a completely separate program, with all the tools, widgets, and buttons that implies. What's more, its status as a Web application makes it easy to change with a fair amount of frequency. So anything written today could (and probably will) be obsolete next month. For both these reasons, how to use Building Maker in this book isn't depicted in detail. The following is a brief list of three things you should know about Building Maker:

   – Building Maker is nothing like SketchUp. You will quickly find out that there is no Orbit, no Line tool, no Push/Pull, and no faces. The whole modeling paradigm is different in Building Maker. All you are doing is matching basic 3D shapes to different views of the same building.

   – The Help link is in the upper-right corner of the screen. There is also plenty of contextual help that pops up while you model.

   – It saves everything to the 3D Warehouse. To save a Building Maker model, you need to be logged in to your 3D Warehouse account. Read more about this later in this chapter. If ever you want to see all the Building Maker models you have made, just check the My Models section of the 3D Warehouse.

4. **When you are ready, click the SketchUp Export button.** In time, your model pops open in SketchUp.

5. **Work on your model until it looks the way you want it to.** The point here is to do two things: Add geometric detail that you didn't (or couldn't) in Building Maker and improve photo-textures using all the tools in SketchUp. Here are some things to think about (see Figure 11.4):

   – Everything you brought in from Building Maker is just edges and faces. This means you can continue to model the way you always would.

   – Chapter 8 is a great reference. In Chapter 8, all of the methods are described for you to apply images to your models. You can use your own photos or even Google Street View imagery, if it's available.

   – Use the Match Photo scenes. Notice the little scene tabs at the top of your modeling window? They correspond to the aerial imagery from Building Maker. You can use the Match Photo dialog box to re-texture parts of your model as you work on it. Read Chapter 8 for more information about using Match Photo.

6. **When you're done, choose File ⇨ 3D Warehouse ⇨ Share Model.** Your revised SketchUp model automatically overwrites the original Building Maker model on the 3D Warehouse. Google presumes that it's better than its predecessor.

Figure 11-4

1

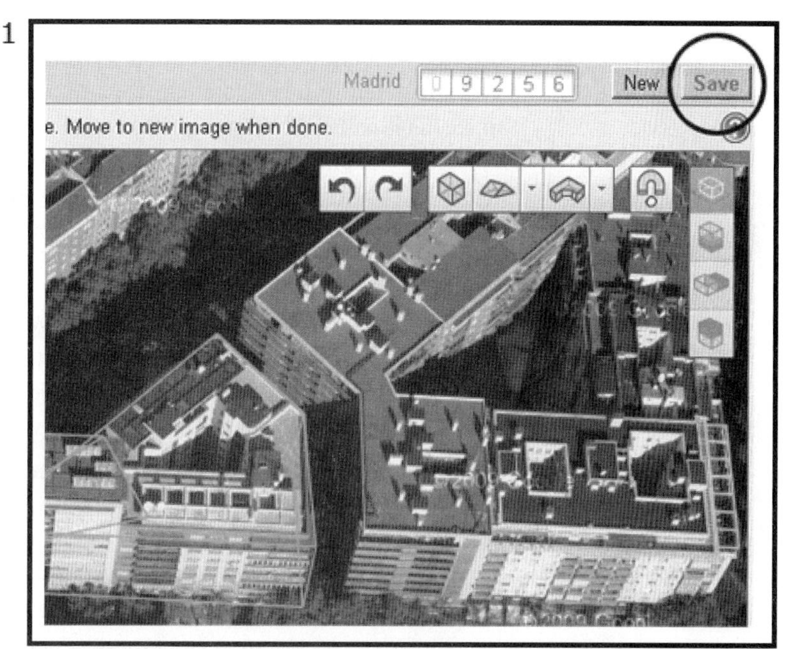

Once you are done building your model in Building Maker (this model has just been started), select Save on the top right corner of your screen.

2

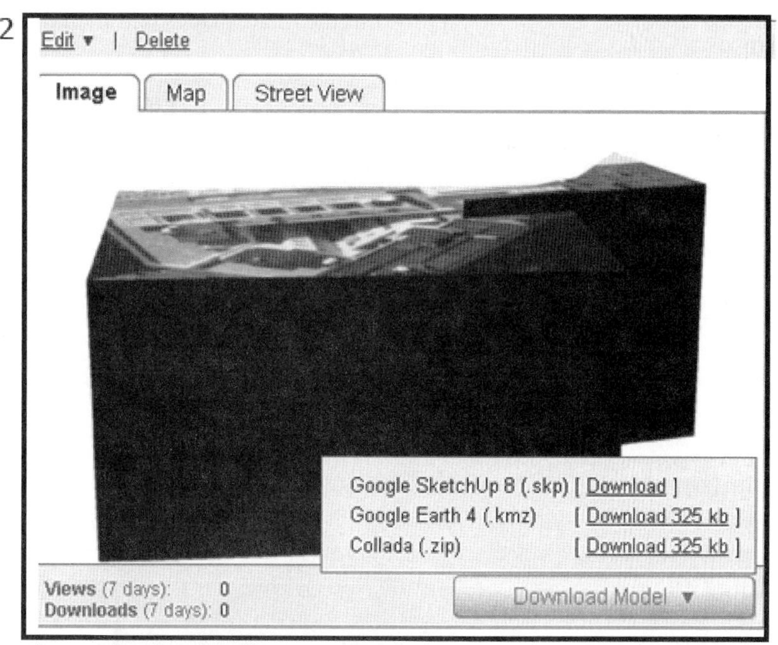

After saving it, you will see My Warehouse on the top right corner. From the list select My Models, and then choose the model you just worked on. Then, you will have an option to download it as a Google SketchUp file.

Building being created in Building Maker and shared in My Warehouse.

Doing It All in SketchUp

Of course, Building Maker is only an option for a miniscule percentage of the world's buildings. Google adds coverage all the time, but chances are you still need to use SketchUp for most of the geo-models you want to make, as it is the most appropriate tool.

Geo-modeling with SketchUp involves a several operations, most of which is covered elsewhere in this book. Given that, the steps that follow are mostly pointers to other sections.

1. **Open a new SketchUp file.**
2. **Import a geo-location snapshot by geo-locating your model.** Reference "Geo-Locating Your Model," later in this chapter, for complete instructions. Basically, you start modeling on top of a piece of color imagery and 3D terrain.
3. **Make sure that you look at the flat version of your terrain.** To choose a flat version you would have to go to Camera-Parallel Projection and again Camera-Standard Views-Top. Then, you would need to use Zoom Extents to make sure that you see all of the terrain. Then, choose File ⇨ Geo-location from the menu bar; make sure Show Terrain is deselected.
4. **Trace the footprint of the building you want to model directly on the color snapshot (see image A in Figure 11-5).** Use the Rectangle tool to block out the main shapes and then draw angles and other details with

Figure 11-5

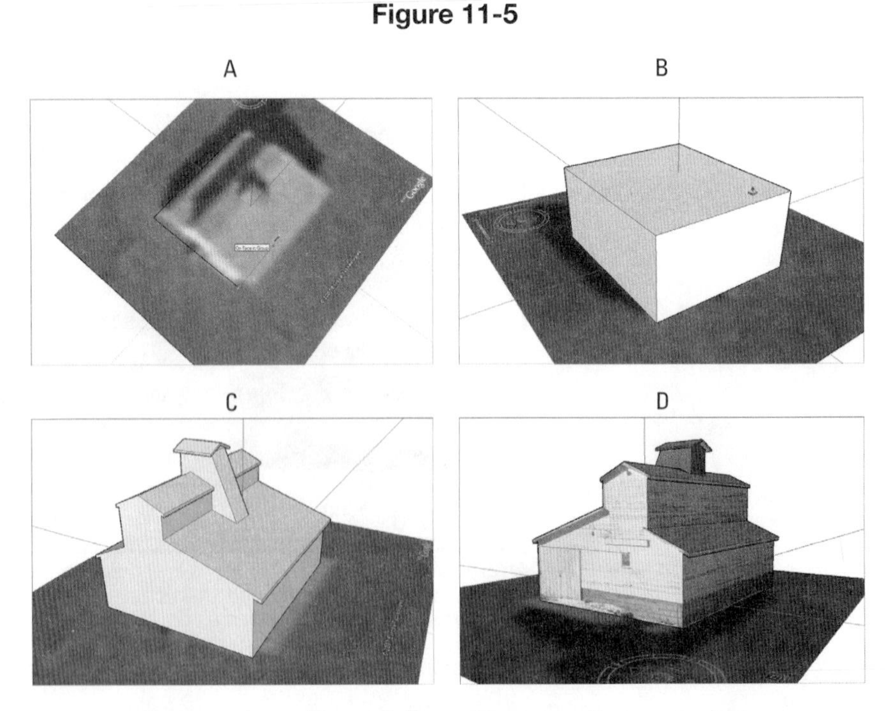

Draw on top of your flattened geo-location snapshot.

the Line tool. If the building you are trying to make doesn't line up perfectly with the colored axes, using the Line and Rectangle tools can be tricky. To fix this problem, reposition your main drawing axes by choosing Tool ⇨ Axes. Click once to set your origin. Then click again to establish the direction of your red axis (parallel to one of the edges in your photo), and a third time to establish your green axis. To make things easier, set origin at the corner of the building you are trying to make.

5. **Use Push/Pull to extrude the footprint to the correct height (see image B in Figure 11-5).** If you don't know how tall the building you are modeling is, some tricks help you make an educated guess. Search the Web for SketchUp estimating building height to find articles on the subject.
6. **Model until you are satisfied with what you have.** This is by far the most time-consuming part of the whole geo-modeling process. The key to building a great Google Earth model is to combine clean photo-textures with simple geometry to produce a model that is both detailed and lightweight. Chapter 8 is all about photo-texturing your models; definitely read those pages for more information. The section "Thinking Big by Thinking Small" offers tips on reducing the amount of geometry (faces, basically) in your work. Finally, the Tips from Professional section "Making Sure Your Models Get Accepted" enumerates Google's acceptance criteria for Google Earth.

GOOGLE SKETCHUP FACT

Be sure to make the building protrude far enough into the terrain so that we will not have any corner levitating in mid air, above ground. How much you will have to sink the building depends entirely on the pitch of the ground; the goal is to see all four corners from underneath, plus a few extra inches (6 inches might do).

7. **Flip to the 3D version of your snapshot.** Choose File ⇨ Geo- location ⇨ Show Terrain) and then move your building up or down until it sits properly, poking through the ground just a little bit, as shown in Figure 11-6. Select everything you want to move, and then use the Move tool to move it up or down. You can press the up- or down-arrow key to constrain your move to the blue axis if you want.
8. **Preview your model in Google Earth to make sure it looks right.** This is an important step, but you would be surprised by the number of people who skip it altogether. Read ahead to "Exporting from SketchUp to Google Earth," in this chapter, for a complete set of instructions.
9. **Upload your model to the Google 3D Warehouse.** The 3D Warehouse is the source for every user-generated, which means not modeled by Google, geo-model that is added to the 3D Buildings layer in Google Earth. By selecting the checkbox to indicate that it's Google Earth–ready and then uploading it, you're submitting your model for consideration.

Read the steps in "Uploading Your Models," near the end of this chapter, for everything else you need to know.

Thinking Big by Thinking Small

Lightness
The file size of a model, or the number of faces and textures used to build it.

When it comes to modeling for Google Earth, lightness is desired. **Lightness** refers to the file size of your model, or the number of faces and textures you use to build it. The more complicated your model is, the

Figure 11-6

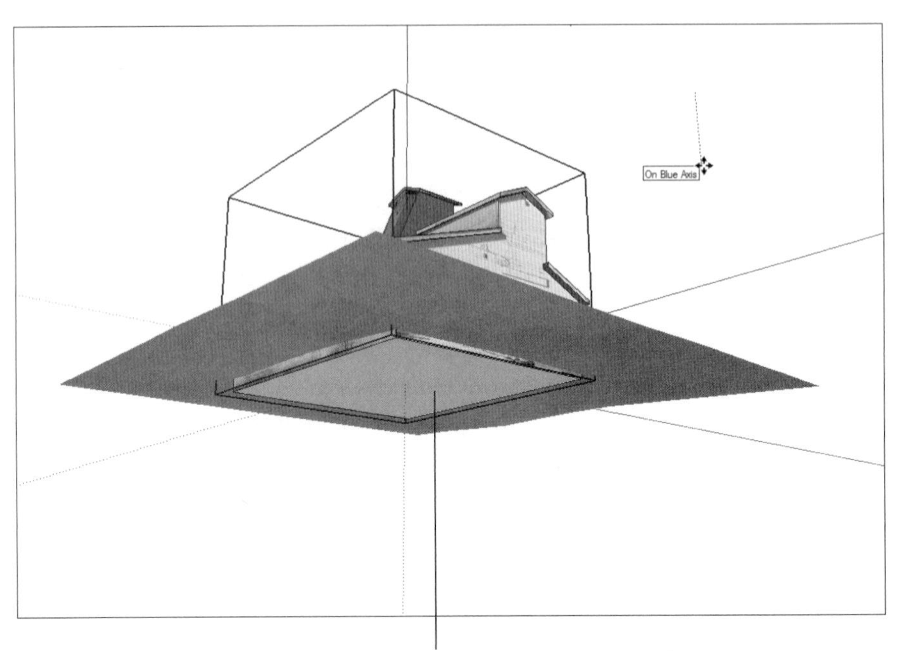

Make sure your building pokes through the ground

Move your model up or down until it sits properly on the terrain.

slower Google Earth will run. To do the most with the least geometry, follow these tips:

- **Get rid of extra geometry.** Often when you're modeling, you end up with edges (and even faces) that don't have a purpose.
- **Reduce the number of sides in your extruded arcs and circles.** SketchUp's default number of sides for circles is 24. This means that every time you use Push/Pull to extrude a circle into a cylinder, you end up with 25 faces: 24 around the sides, and the original face on top. As I suggested in earlier chapters, instead of using circles with 24 sides, reduce the number of sides by typing a number followed by the letter *s* and pressing Enter right after you draw a circle. The same thing goes for arcs; you change the number of sides in them in exactly the same way. For example, to draw a 10-sided circle, follow these steps:

  1. Draw a circle with the Circle tool.
  2. Type **10s** (this should appear in the lower-right corner of your modeling window), and then press Enter.

  The same thing goes for arcs; you change the number of sides in them in exactly the same way. I like to use 10-sided circles and 4-sided arcs when I model for Google Earth. The following image shows the same pipe

Figure 11-7

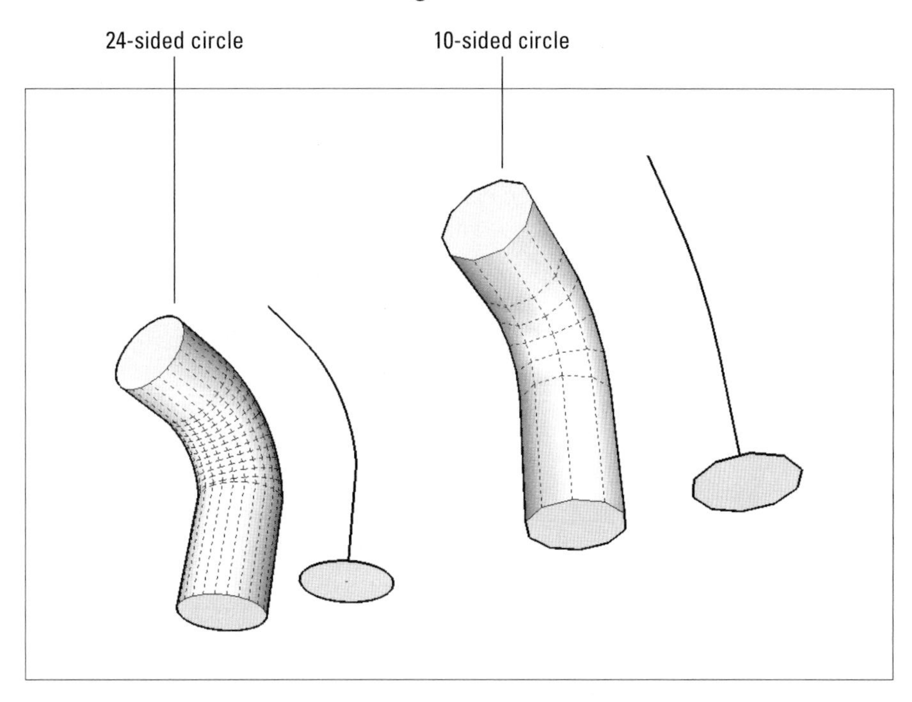

Pipe constructed by using Follow Me on two circles.

constructed by using Follow Me on two circles: one with 24 sides and one with only 10. Note the difference in the number of faces in each version.

- **When you can, use photo detail instead of geometry.** This really only applies if you're mapping photos (or using PhotoMatch) on the model you're making for Google Earth. If you are, it's a good idea to make as basic a model as you can, and let the detail in the photo do the work. Resist the temptation to model windows and doors. Consider using PNG images with transparency to represent columns, fences, and other thin building elements.

Geo-located objects
Objects that never move and have a fixed position, such as buildings and monuments.

Geo-location
The process where SketchUp sets the user's snapshot's latitude and longitude to match Google Earth, and it orients the snapshot in the right cardinal direction, That means, the model is placed and rotated according to the four cardinal points, North, South, East, and West.

11.2.2 Geo-Locating Your Model

No matter what kind of model you build, displaying it in Google Earth—for yourself, for a client, or for everyone in the world begins with giving it a **geographic location,** or **geo-location** for short. In SketchUp, geo-locating your model involves placing it where it belongs on a piece of terrain and aerial imagery (called a geo-location snapshot) that you download from the Web.

Adding a geo-location snapshot to your model in SketchUp 8 is much easier than it was in SketchUp 7. The 3D terrain data in SketchUp 8's geo-location snapshots is also more detailed and accurate than it used to be, and the aerial imagery is in full color.

PATHWAYS TO... ADDING A GEO-LOCATION SNAPSHOT TO YOUR SKETCHUP FILE

1. **Make sure you are online.** All Google's geo-data is stored on its servers; if you don't have an Internet connection, you can't use the geo-data.
2. **Open the SketchUp file you want to geo-locate.** You can add a geo-location snapshot to your model anytime as you work on it. If you haven't started modeling yet, it's perfectly okay to add a geo-location to an empty file.
3. **Choose File ⇨ Geo-Location ⇨ Add Location from the menu bar.** A new window that resembles a simplified version of Google Maps appears.
4. **Find the area where you want your model to be located.** You can type an address into the search bar in the upper-left corner if you like. You can also just use your mouse or the controls on the left side of the window to navigate around. Scroll your mouse wheel to zoom; click and drag to pan. When you are zoomed in close enough, you see a white, 1 km x 1 km square: This is the largest snapshot you can import all at once. That's still a very big area, so you probably want to keep zooming.
5. **Click the Select Region button to display a cropping rectangle.**
6. **Drag the blue pins to specify the precise corners of your geo-location snapshot, as shown in Figure 11-8.** Try to frame an area that's just big enough to provide a base for your model. Importing too much terrain data can slow down your computer. You can always bring in more terrain data later.
7. **Click the Grab button to add a geo-location to your SketchUp file.** The separate window closes, and a big, colorful rectangle appears in the middle of your model. That's your new geo-location snapshot.
8. **If you are geo-locating a model you have built already, move it into position on the snapshot.** Use the Move tool (and maybe the Rotate tool) to pick up your model and place it where it belongs. Make sure your model is vertically situated on the terrain. Follow these steps to do just that:
   a. Choose File ⇨ Geo-location ⇨ Show Terrain to switch to the 3D version of your geo-location snapshot.
   b. Select everything you want to move and use the Move tool to start moving; tap the up- or down-arrow key to constrain your move to the blue axis.
   c. Sink your model into the terrain until it sits properly. If you want to import another snapshot into SketchUp, you can. SketchUp

automatically tiles all the snapshots you “take” to form a patchwork in your model. This is useful if you find that you didn’t get everything you needed the first time.

Figure 11-8

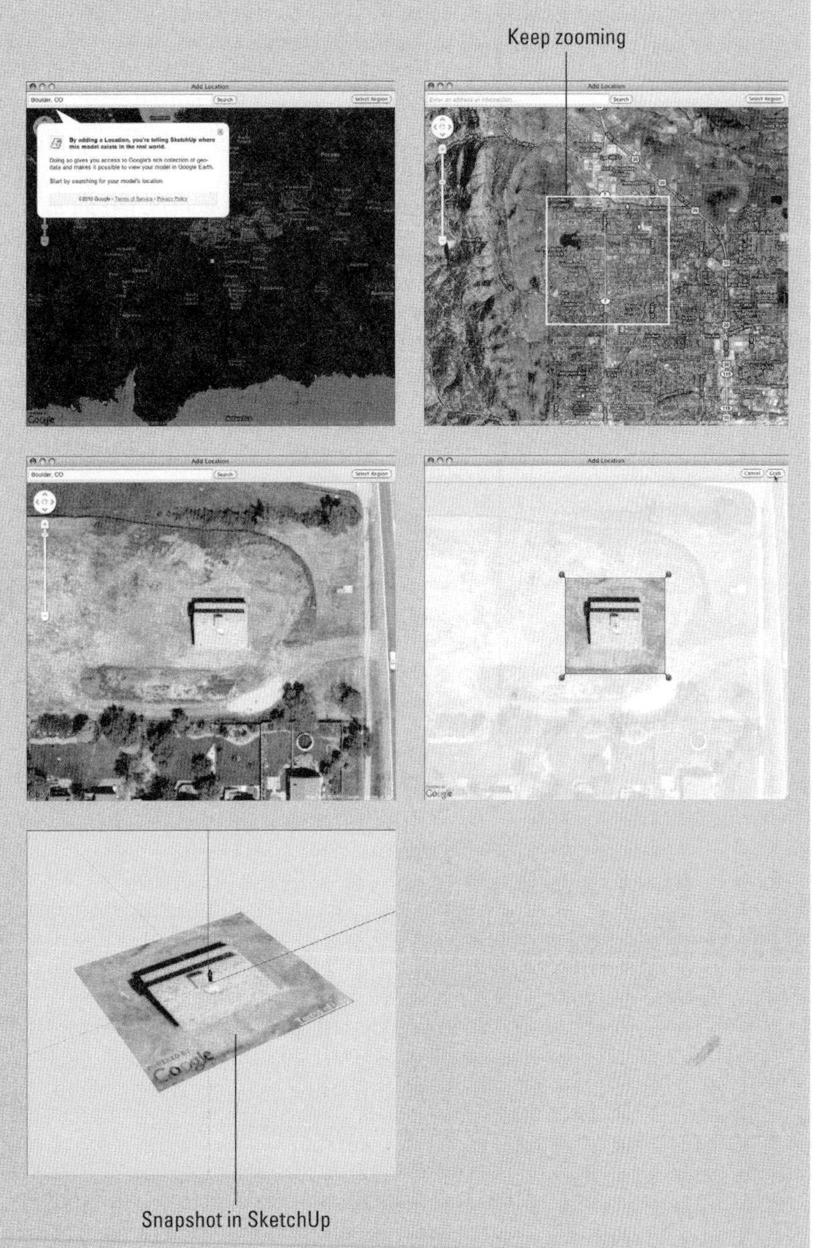

The area you frame with blue pins is imported into your model as a geo-location snapshot.

FOR EXAMPLE

ALL ABOUT GEO-LOCATION SNAPSHOTS

When you import a geo-location snapshot, you access Google's huge repository of geographic data; snapshots are much more than just pretty pictures:

- **A snapshot geo-locates your position automatically.** This means that it sets your model's latitude and longitude, and orients itself in the right cardinal direction. Any shadow studies you do with the Shadows feature are automatically accurate for your model's new geo-location. See Chapter 9 for more information.
- **Everything's already the right size.** Perhaps you take a snapshot of a football field; when you measure that football field in SketchUp, it is exactly 100 yards long. That's because SketchUp scales your snapshot to the correct size as part of the import process.
- **Snapshots look flat but contain terrain data, too.** The snapshot that SketchUp imports is more than just a color aerial photo—it also includes a chunk of topography—terrain. The terrain is flat when you first import it because it's easier to build on that way, but you can toggle between flat and 3D (not flat) views by choosing File ⇨ Geo-Location ⇨ Show Terrain. Don't worry if you don't see any difference when you flip between the views, you probably just chose a flat site. The image below shows the same snapshot with terrain toggled off (left) and on.

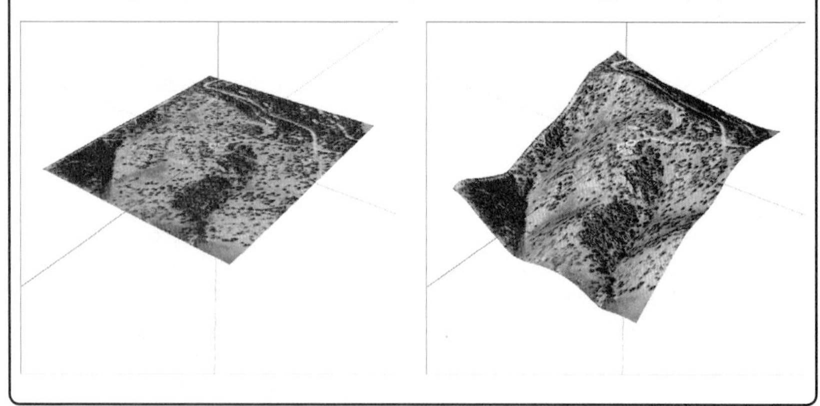

TIPS FROM THE PROFESSIONALS

Making Sure Your Models Get Accepted

When you take the time to build a model for Google Earth, you naturally want it accepted into Google Earth's 3D Buildings layer so that everyone in the world can see it. The following tips help you avoid common problems among rejected models:

1. **Match your model's terrain to the physical world:** When the 3D terrain you import isn't right, you may have to "build up" the terrain around your building by modeling it in SketchUp. Search the Web for "SketchUp create terrain skirt" to find an article that helps you figure out what to do.

2. **To replace an existing model, make your model better:** Before you submit your model for consideration, check Google Earth to make sure there isn't already a version of your building on the 3D Buildings layer. If there is, your model needs to be better than what's already there in order to replace it.
3. **Apply photo-textures that look accurate:** Use good photographs as textures when you model and be sure to position them accurately. See Chapter 8 for more information on this topic.
4. **Size and scale your model to match its photo in Google Earth:** This may seem obvious, but it is a common occurrence; if your model isn't the same size as it is in the aerial imagery in Google Earth, it won't be accepted. It also needs to appear in the correct spot; right on top of its own photo in Google Earth.
5. **Model existing buildings:** If the building you submit doesn't appear in Google Earth's aerial imagery, chances are good that it will not be accepted. If it really does exist, include Web links to images of the building in its surroundings in the model description when you upload it to the 3D Warehouse. If it's still under construction, include a link to somewhere on the Web that details when it's due to be completed.
6. **Sink your model into the ground just a bit so it doesn't float:** This one is simple; make sure your building doesn't float above the ground in Google Earth. It should be sunken into the ground just enough to appear realistic.
7. **Avoid ads:** You can't include ads or other visual "spam" on your models. If the building has a mural or a billboard on it, that's okay, but the ad has to be part of the original photo-texture on the model. In other words, it has to be a part of the structure in real life.
8. **Model only the building:** Don't include extras like cars, trees, streetlamps, people, or anything else that isn't a part of the building when you upload it.
9. **Check for and correct Z-fighting:** When two faces are in the plane, they flash when you orbit your model. This is Z-fighting, and it's very common. If you see this phenomenon, there's a good chance you accidentally chose Use as Image instead of Use as Texture when you imported the image into your model. This results in two co-planar faces. See Chapter 8 for the correct way to apply images when you model.
10. **Keep your model's file size lean:** Read "Thinking Big by Thinking Small," earlier in this chapter, for tips on minimizing the size of what you make. **Remember:** In Google Earth, smaller files are always better.

11.2.3 Viewing Your Model in Google Earth

After you make or position a model on top of a geo-location snapshot, sending it to your copy of Google Earth is a simple operation. And after you do that, you can save your model as a Google Earth KMZ file and it is ready to e-mail to clients.

Taking the Five-Minute Tour of Google Earth

Google Earth can do many things. This section will help get you started using this dynamic tool. Like Google SketchUp, the basic version of Google Earth is free. Here is some additional helpful information about the program:

- **You get Google Earth by downloading it.** Just go to http://earth.google.com, and click the Download button. While you are there, you can explore some of the other features of the site. You should be able to find answers to any questions you have, as well as links to online help, user communities, and more. You can also learn about other versions of Google Earth.
- **You need a fast Internet connection.** Google Earth is able to show you detailed imagery of the *whole world*—quite a bit of data that Google keeps on its servers until you "request" it by flying somewhere and zooming in. The faster your Internet connection, the faster you can stream imagery, 3D buildings, and topography into your copy of Google Earth. You must be online to use Google Earth effectively.
- **You need to be able to navigate**. On the upper-right corner of the screen, you'll find the navigation controls for Google Earth conveniently grouped together. Notice how they appear when you hover over them? Go ahead and experiment to figure out what they do. In the meantime, here's some help:
  - **Zoom:** Move this slider back and forth to zoom in and out on whatever's in the center of your screen. You can also use the scroll wheel on your mouse to zoom, just like in SketchUp.
  - **Pan:** You can move around by clicking the arrow buttons, but the easier way is to use your mouse. Just click and drag to "spin" the world in whatever direction you want.
  - **Rotate/Tile:** Turn the wheel to spin yourself around without moving. This works like the Look Around tool in SketchUp. Click the N button to reorient the world so that north is up. If you're looking at an area with mountains, they should resemble a 3D image (if they don't, make sure that the Terrain layer is enabled in the lower-left corner). You can also tilt by holding down your scroll wheel button, just as you do to Orbit in SketchUp. (See Chapter 2 for more on orbiting.)

Exporting from SketchUp to Google Earth

The process of exporting a model from SketchUp to Google Earth is really quite simple. Follow these steps to send your model from SketchUp to Google Earth on your computer:

1. **Choose File ⇨ Preview in Google Earth.** Doing this sends everything in your modeling window (with the exception of the Google Earth snapshot) over to Google Earth. Your computer should automatically switch you over to Google Earth and fly you in so that you're looking at your model (see Figure 11-9).
2. **If you decide you want to make changes to your model, go back to SketchUp, make your changes, and then choose Preview in Google Earth again.** Google Earth pops up a dialog box that asks you whether you want to overwrite the old version of the model you placed the first time.
3. **Click the Yes button if you are sure that's what you want to do.**
4. **Continue to go back and forth between SketchUp and Google Earth until your model looks exactly the way you want.**

When you preview your SketchUp model in Google Earth, it's visible only on your computer—no one else can see it. If you have modeled an

Figure 11-9

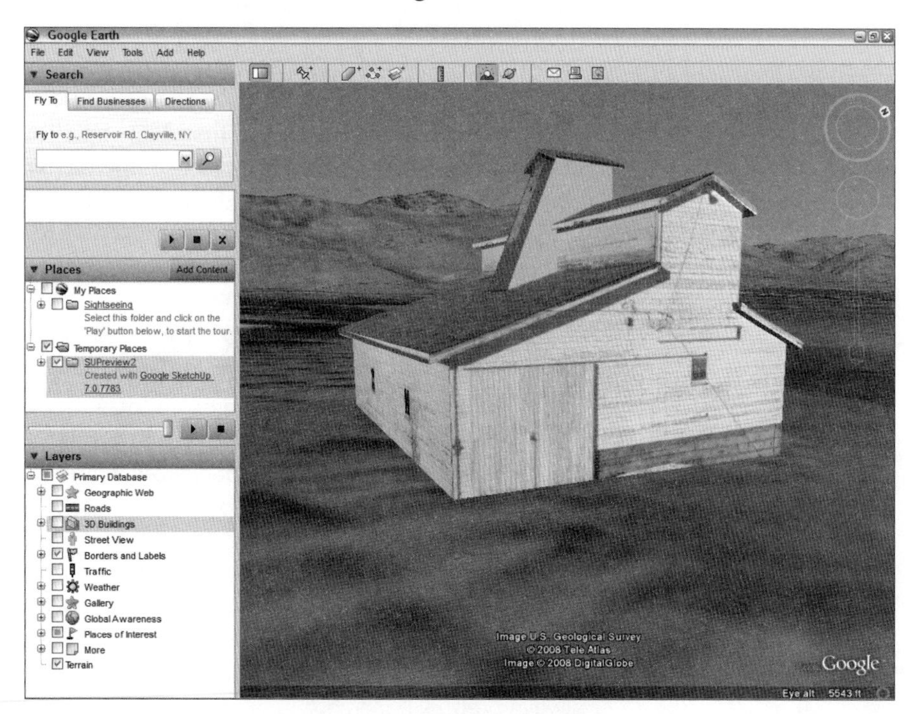

Your SketchUp model in Google Earth 2.

existing building and you want to contribute it to Earth's default 3D Buildings layer, you can; read "Geo-Modeling for Google Earth," near the beginning of this chapter, for more information.

Saving Your Model as a Google Earth KMZ File

You can save your SketchUp model as a Google Earth KMZ file that you can send to anyone. When someone opens the KMZ file, Google Earth opens on his or her computer (if he or she has Google Earth), and he or she is "flown in" to look at the model you made.

PATHWAYS TO... SAVE YOUR MODEL AS A KMZ FILE

1. In Google Earth, select your model by clicking it in the Temporary Places list on the left of the screen. Unless you've renamed it yourself, your model will be called *SUPreview1*.
2. Choose File ⇨ Save ⇨ Save Place As. The Save File dialog box opens.
3. Give your file a name and figure out where to put it on your hard drive.
4. Click the Save button to save your model as a KMZ file.

It's preferable to save KMZs from Google Earth because doing so forces a preview of the models first. You can skip the Google Earth step altogether if you are in a hurry. Choose File ⇨ Export ⇨ 3D Model while you are still in SketchUp, then choose Google Earth File (.kmz) from the Format drop-down list.

IN THE REAL WORLD

Charles Overy is an architect and owner of LGM Architectural Visualization in Minturn, Colorado. He uses Google SketchUp regularly and uses Google Earth as his company specializes in architectural 3D printing.

His working process starts with a drawing made in SketchUp and then he uses the export to Google Earth terrain tool. He has recently started to send .kmz files back and forth with clients during the proposal stage. According to him, it is becoming extremely important to combine design with geospatial data and collaborate. He goes on saying that when working with SketchUp and Google Earth together, the results are always great.

SELF-CHECK

1. Select the correct statement about the types of buildings that we can model in Google Earth:
   a. Buildings are always hollow but it is preferred that they are geometrically complex.
   b. Buildings are made hollow, photo-textured, geo-located, and simple in their geometry.
   c. Buildings are geometrically correct inside and outside and photo-textured.
2. Google's Building Maker allows the user to create buildings with accurate height with the help of aerial imagery. True or false?
3. Talking about Lightness in Google Earth, name one of the three options we can use to reduce the size our models.
4. Google Earth images cannot be used in our 3D models; we only have the option of exporting our SketchUp models into Google Earth. True or false?

Apply Your Knowledge Learning how to geo-locate any building in any part of the world is a great skill that will allow you to put your model in a larger context. Your clients will enjoy seeing a presentation that includes some perspective views with adjacent lots or with the terrain around the structure.

Locate in Google Earth the high school you attended when growing up. Now imagine that you have been contracted to build a brand-new gymnasium to either replace the existing one or to build a new one in the school's parking lot. Geo-locate your school first and place a marker in Google Earth. Then, build the gym in Google SketchUp and export it into Google Earth.

11.3 USING THE 3D WAREHOUSE

Google's 3D Warehouse is a huge collection of 3D models that is searchable and, most importantly, free for everyone to use. It's basically a Web site; it exists online, and you need an Internet connection to access it.

If you have a SketchUp model you want to share with the world, share with just a few people, or just store on Google's servers for safekeeping, the 3D Warehouse is where you store it.

11.3.1 Getting to the Google 3D Warehouse

Learn how to get to Google's 3D Warehouse and upload your models.

You can get to the 3D Warehouse in two ways:

- **From SketchUp:** Choose File ⇨ 3D Warehouse ⇨ Get Models. When you do this, a mini web browser opens right in front of your modeling window.
- **From the Web:** Browse to http://sketchup.google.com/3dwarehouse. This is a great way to search for the right 3D model without having to open SketchUp first.

Go ahead and explore the 3D Warehouse. It's amazing what you'll find; after all, thousands of people are adding new content every day. Most of this content isn't very useful, but you'll still find plenty of interesting things to download and look at. Taking apart models stored on the 3D Warehouse is a great way to see how they are built. Read Chapter 5 for details on how to download things you need from the warehouse.

FOR EXAMPLE

MINDING YOUR MODELING MANNERS

People sometimes wonder if anyone at Google is paying attention to what gets uploaded to the 3D Warehouse. The answer to that question is slightly complicated. For instance, nobody at Google minds if you refer to a website or enter tags that don't have anything to do with the model you upload.

On the other hand, it's frowned upon when people go anywhere near the usual taboo subjects for public, G-rated Web sites; pornography and/or foul language will get your model removed from the 3D Warehouse. Thousands of impressionable young people peruse Google's Web sites every day, so Google tries to keep its sites clean.

11.3.2 Uploading Your Models

You can break down the models in the 3D Warehouse into two categories:

- **Geo-located objects:** Things like buildings, monuments, bridges, and dams exist in a specific geographic location; they never move around. These are the kinds of models that show up on the 3D Buildings layer in Google Earth, and the 3D Warehouse is where they come from. Check out the section "Geomodeling for Google Earth," earlier in this chapter, for a full account of how to build geo-located models that you can upload to the 3D Warehouse.
- **Not geo-located:** Objects like toasters, SUVs, wheelchairs, and sofas aren't unique, and they don't exist in any one geographic location. For example, no physical address is associated with a model of a Honda Accord because millions of them exist, and because Honda Accords move around. Such objects never show up in Google Earth. That doesn't mean they don't belong in the 3D Warehouse, though; models of nongeolocated objects are incredibly valuable for people who are making their own SketchUp models.

PATHWAYS TO... UPLOADING YOUR OWN MODEL TO THE GOOGLE 3D WAREHOUSE

1. **Open the model you want to upload in SketchUp.**
2. **Adjust your view until you like what you see.** When you upload a model to the 3D Warehouse, SketchUp automatically creates a preview image that's a snapshot of your modeling window.
3. **Choose File ⇨ 3D Warehouse ⇨ Share Model.** A mini-browser window opens, and it shows the logon screen for the 3D Warehouse. If you want to upload models, you need a Google account. An account is free; you just need a valid e-mail address to get one. If you don't already have one, follow the on-screen instructions to sign up. When you're creating your Google account, be sure to type something in where the system asks for a "nickname." If you don't, everything you upload will be attributed to "Anonymous."
4. **Enter your Google account information and click the Sign In button.**
5. **Fill out the Upload to 3D Warehouse form as completely as you can:**

   - **Title:** Enter a title for your model. If it's a public building, you might enter its name. Something like "Royal West Academy" would do nicely.
   - **Description:** Models with complete descriptions are very popular with people who are searching the Warehouse. Try to use complete sentences here; the more you write the better.
   - **Tags:** Type in a string of words that describe the thing you modeled. Whatever you enter here will be used by the 3D Warehouse search engine to help people find your model. To increase the number of people who see what you made, add several tags. For example, to upload a modern coffee table, you might enter the following tags: coffee table, table, coffee, modern, living room, furniture, glass, chrome, metal, steel. Be exhaustive.
   - **Address:** This field only appears if your model is geo-located, meaning that you started with a Google Earth snapshot. If you know the physical address of the thing you made, type it in.

(continued)

PATHWAYS TO... UPLOADING YOUR OWN MODEL TO THE GOOGLE 3D WAREHOUSE *(continued)*

- **Google Earth Ready:** You only get this option if your model is geo-located. If your model is accurate, correctly sized, and in the right location, and if you want it to be considered for inclusion on the default 3D Buildings layer of Google Earth, select this check box. If you do, Google will consider adding it to Google Earth.
- **Web site:** If you have a Web address that you'd like people who view your model to visit, enter it here. For example, if your model is of a historic building, you might include the address of a Web site that provides more information about that building.
- **Viewing:** The check box next to the sentence Allow Anyone to View This Model and See It in Search Results lets you control access to your model on the Web. Checking it gives anyone the right to find, download, and use it however they like. If you deselect it, your model stays hidden (from everyone but you). Flip ahead to "Managing Your Models Online," in this chapter, for more about this topic.
- **Links for Containing Collections:** You (and everyone else) can create collections of models on the 3D Warehouse. Anyone can add any model to one of their collections; checking this box allows Google to display a list of collections that contain your model on your model's unique Web page.

6. **Click the Upload button to add your model to the 3D Warehouse.** If everything works properly, you should get a page with your model on it, along with all the information you just entered. The words "Model has been uploaded successfully" will be highlighted in yellow at the top of your browser window. You can't just upload any model to the 3D Warehouse, unfortunately. At the time this book was written, the maximum file size you could upload was 10MB, which is actually pretty big. You can check your model's file size in Explorer (on a Windows machine) or in Finder (on a Mac).

TIPS FROM THE PROFESSIONALS

We need to discuss one more issue with 3D Warehouse, and that is what constitutes proper etiquette when we are ready to upload one of our files into the database. We need to remember, many people might use our models, and if we did a lousy job, then they will be regretting that they used it, or it might slow their computers down. So, if we can answer yes to the following questions, then when we can upload them safely:

- Is our file small enough?
- Does it have hidden groups that we do not need any more?
- Is it overly-built with details that are not that necessary, and thus, will it slow down other computers?
- Can I substitute building complicated details on a surface for maybe a simple texture mapping?
- Did I use a name that is easy to find, and do my description and tags make sense?

FOR EXAMPLE

MANAGING YOUR MODELS ONLINE

The 3D Warehouse isn't a free-for-all of individual models floating around in cyberspace; it's actually a pretty organized place. Understanding collections and access settings makes your time as a member of the worldwide modeling community even more productive and enjoyable. If you want to group together a bunch of models so they're easier to find, make a collection. In the upper-right corner of your screen, choose My Warehouse ⇨ My Collections to bring up a page that lists all the collections you've made. Somewhere on the page, there's a link to Create a 3D Collection—click it and fill out the information on the form that opens up. Three more things you need to know about collections:

- **Collections can contain either models or other collections.** When you create a collection, you decide which kind of collection it should be.
- **You can share collections with other people.** On the main page of one of your collections, find the Share link. Clicking it lets you invite other people to collaborate with.
- **Your collections can contain other people's models.** And their collections can contain your models—provided you've set them to be publicly viewable. Which brings me to my next point . . . You actually have quite a lot of control over who gets to do what with the models you upload to the 3D Warehouse. Somewhere near the top of one of your model's pages, click the Access link.

A Privacy Settings page opens, where you set who can see, edit, download, or even delete your model. The settings themselves are pretty self-explanatory, but I thought I'd point them out—lots of folks don't even know they're there.

SELF-CHECK

1. Name the two ways we can access 3D Warehouse.
2. Out of the list below, mark those structures that are geo-located:
   a. Monuments
   b. Objects
   c. Cars
   d. A dining room set
3. To upload a 3D model into 3D Warehouse you need to enter a title but you do not need to sign in into your Google account or fill out a form about the model you just built. True or false?
4. Your uploaded model in 3D Warehouse can be part of a collection of models that other people might have contributed to. True or false?
5. Name two etiquette procedures that can be followed before uploading our models into 3D Warehouse.

Apply Your Knowledge Exporting 3D models into the 3D Warehouse requires that you observe some common sense rules so that other users can use your model. A good rule is to make your model full scale (build it with full dimensions) and with easy shapes; also, avoid any unnecessary details that would make it memory-heavy and be sure to group it. Inside that big group you can add smaller groups to organize it better. Finally, be sure that you did not select any loose line or a loose object far in the distance.

Build a medium-size tool shed in Google SketchUp following the requirements listed above. Be sure to build it with the right proportions, the right materials, or with proper photo-textured walls and upload it into 3D Warehouse. Enter an appropriate title and any other description that you might find adequate.

SUMMARY

Section 1:

- SketchUp, Google Earth, the 3D Warehouse and the new product Google Building Maker have a relationship that allows them to function together.

Section 2:

- The kinds of models you might like to view in Google Earth fall into two basic categories:
- Existing buildings
- Unbuilt buildings
- All Google Earth models are:
- Geo-located
- Geometrically simple
- Hollow
- Completely photo-textured
- Building Maker is purpose-built for geo-modeling; making accurate, photo-textured models of existing buildings and gives you two things that are otherwise extremely hard to obtain:
  - Aerial imagery
  - Building height

Section 3:

- Google's *3D Warehouse* is a huge collection of 3D models that is searchable and, free, is a Web site, exists online, and you need an Internet connection to access it.
- You can get to the 3D Warehouse in two ways:
- From SketchUp
- From the Web
- You can break down the models in the 3D Warehouse into two categories:
  - Geo-located objects
  - Not geo-located

ASSESS YOUR UNDERSTANDING

SUMMARY QUESTIONS

1. Google Earth, Google Building Maker, and the basic version of Google SketchUp can all be obtained free of charge. True or false?
2. In Google Earth can only bring real-life buildings as opposed to designs that are not built yet. True or false?
3. If we want to see a rugged terrain in Google Earth but the ground seems to be flat, what layer would we have to turn on to see it in 3D?
4. Name two of the most common problems that you could face when a SketchUp model is sent to be placed in Google Earth.
5. What is the most difficult dimension to guess when we are geo-modeling one or various buildings in Building Maker?
6. You do not need internet connection to see images of the whole world in Google Earth. True or false?
7. You can save your SketchUp model in a Google Earth XYZ file that you can send to anyone. True or false?
8. Lightness refers to the contrast of colors in your model. True or false?
9. A geographic location can be added to a SketchUp file any time, even if to an empty file. True or false?
10. SketchUp's default number of sides for circles is:
    - **a.** 6
    - **b.** 12
    - **c.** 24
    - **d.** 48
11. Is it possible to work back and forth from Google SketchUp and Google Earth even if we do not have our design fully finished in Google SketchUp?

APPLY: WHAT WOULD YOU DO?

1. You have a very complex model of a house that is causing Google Earth to run slowly. What steps can you take to simplify your model?
2. You want to build a model of your favorite downtown building and view it in Google Earth. What steps do you take to do this?
3. How does using the Google Earth import process save you time?

BE AN ARCHITECT

Visit America's Landmarks

Download Google Earth onto your computer and take some time to tour the program. Using the Fly To feature, tour Mount Rushmore, the White House, and the Golden Gate Bridge.

Model and Upload Your Home or School

Go to Google Earth and see if your home is visible. If it is, import a snapshot from Google Earth into SketchUp. If your house isn't visible, import a snapshot of your school instead. Build a simple model of the building in your snapshot and upload it to the 3D Warehouse. Next, e-mail a link to your model to a classmate.

KEY TERMS

Geo-located objects
Geo-location
Lightness
KMZ

CHAPTER 12

PRINTING YOUR WORK

With a Windows PC and a Mac

Do You Already Know?

- The various printing views of your model?
- How to figure out the printing dialog boxes?
- How to print to scale?

For the answers to these questions, go to **www.wiley.com/go/chopra/googlesketchup2e**

What You Will Find Out	What You Will Be Able To Do
12.1 How to print from a Windows based system.	Print using basic settings as well as configure specific settings.
12.2 How to print from a Mac.	Work with print dialog boxes.
12.3 How to print to scale.	Configure your system to print to scale.

INTRODUCTION

As much as it seems we live in an all-digital world, the truth is that we don't. People use more paper now than they ever have.

This chapter explains how to print views of your SketchUp model. Because the Windows and Mac versions of this procedure are different, a section is dedicated to each platform. The last part of this chapter is devoted to scaled printing—a topic that can sometimes intimidate even experienced architects. SketchUp makes printing to scale slightly more difficult than it could be, but it's still easier than drawing things by hand.

If you are using the Pro version of SketchUp, you can always use Layout to print views of your models. Making both scaled and non-scaled prints is easier in Layout than in SketchUp. Read Chapter 14 for more details.

12.1 PRINTING FROM A WINDOWS COMPUTER

It's very easy to print from SketchUp, as long as you're not trying to do anything too complicated, such as printing to a particular scale, which can be difficult the first time. (To print from a Mac, please visit Section 12.2.)

GOOGLE SKETCHUP IN ACTION

Learn to print using basic settings as well as configure specific settings.

12.1.1 Making a Basic Print (Windows)

Most of the time, all you need to do is to print exactly what you see on your screen. Follow these steps to do that on a Windows machine:

PATHWAYS TO...
BASIC PRINTING ON WINDOWS

1. **Make sure that you have the view you want to print in your modeling window.** Unless you are printing to scale, SketchUp prints exactly what you see in your modeling window.
2. **Choose File ⇨ Print Setup.** This opens the Print Setup dialog box, which is where you make choices about what printer and paper you want to print to.
3. **In the Print Setup dialog box (see Figure 12-1), do the following:**
   - Choose the printer you'd like to use.
   - Choose a paper size for your print.
   - Choose an orientation for your print; most of the time, you'll want to use Landscape, because your screen is usually wider than it is tall.
4. **Click the OK button to close the Print Setup dialog box.**

Figure 12-1

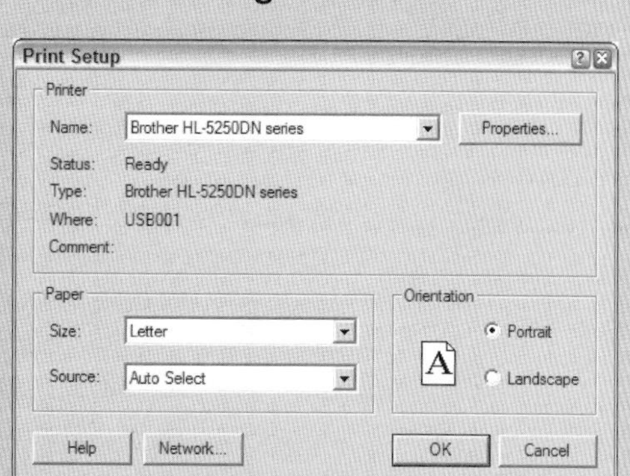

The Print Setup dialog box in Windows asks you to choose a printer, paper size, and orientation.

5. **Choose File ⇨ Print Preview.** This opens the Print Preview dialog box, which lets you see an image of what your print will look like before you send it to a printer.
6. **In the Print Preview dialog box, do the following:**
   - In the Tabbed Scene Print Range area, choose which scenes you'd like to print, if you have more than one. If you need to, you can read about scenes in Chapter 10.
   - Tell SketchUp how many copies of each scene you need.
   - Make sure that the Fit to Page check box is selected.
   - Make sure that the Use Model Extents check box *isn't* selected.
   - Choose a print quality for your printout (High Definition is best for most jobs).
7. **Click the OK button to close the Print Preview dialog box and generate an on-screen preview of what your print will look like.**
8. **If you are satisfied with what you see, click the Print button in the upper-left corner of the Print Preview window to open the Print dialog box.** If you *don't* like what you're about to print, click the Close button (at the top of the screen) and go back to step 1.
9. **In the Print dialog box (which should look exactly like the Print Preview dialog box), click the OK button to send your job to the printer.**

12.1.2 Decoding the Print Preview and Windows Print Dialog Box

The Print Preview and Print dialog boxes in SketchUp are exactly the same. Figure 12-2 shows the former, but the descriptions in this section apply to both.

Figure 12-2

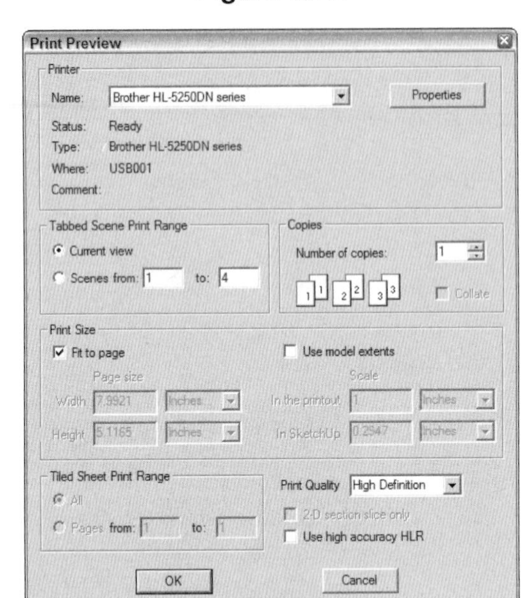

The Print Preview dialog box in Windows looks exactly the same as the Print dialog box.

Printer

If you used the Print Setup dialog box first, you shouldn't need to change any of the settings in this section. If you prefer, you can choose which printer to print to from the drop-down list. You can also click the Properties button to make adjustments to your printer settings. (Because these are different for every printer, you may need to consult your printer's user manual.)

Tabbed Scene Print Range

Use this area to tell SketchUp which of your scenes you'd like to print, if you have more than one. This is handy for quickly printing all your scenes. Select the Current View option to print only what is currently in your modeling window.

Copies

This step is fairly basic: Choose how many copies of each view you'd like to print. If you're printing multiple copies of multiple scenes, select the Collate check box to print "packets," which can save you from having to assemble them yourself. Here's what happens when you're printing three copies of four scenes:

- Selecting the Collate check box prints the pages in the following order: 123412341234.
- Deselecting the Collate check box prints the pages like this: 111222333444.

Figure 12-3

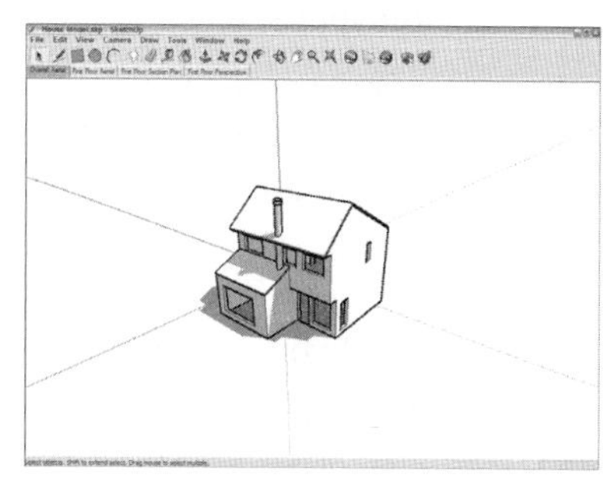

My SketchUp screen

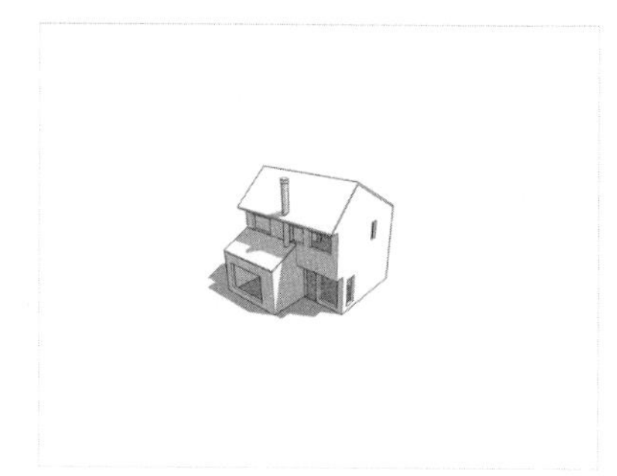

Fit to Page

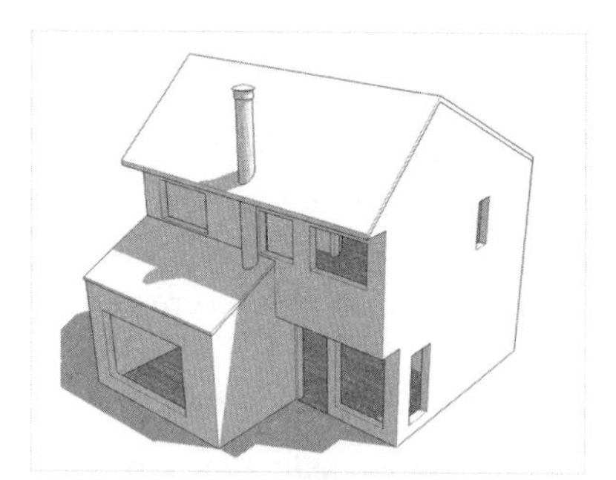

Fit to Page and Use Model Extents

Different Print Size settings have different results when applied to the same view in SketchUp.

Print Size

This is the most complicated part of this dialog box; you use the Print Size controls to determine how your model will look on the printed page. Figure 12-3 shows the effect of some of these settings on a final print.

The Print Size controls are as follows:

- **Fit to Page:** Selecting this check box tells SketchUp to make your printed page look like your Modeling Window. As long as the Use Model Extents check box isn't selected, you should be able to see exactly what you see on your screen.

- **Use Model Extents:** All this option does is instruct SketchUp to zoom in to make your model (excluding your sky, ground, watermark, and whatever else might be visible on your screen) fit the printed page. If you want this effect, you might instead use choose Camera ⇨ Zoom Extents from the menu bar before you print your model; it's easier, and you know exactly what you're getting.
- **Page Size:** As long as you don't have the Fit to Page check box selected, you can manually enter a page size using these controls. If you type in a width or height, SketchUp figures out the other dimension and pretends it's printing on a different-sized piece of paper. This option is especially useful if you want to make a big print by tiling together lots of smaller pages. See the next section in this chapter for more details.
- **Scale:** This step is slightly complicated. To print to scale, you have to do two things before you use the Print or Print Preview dialog boxes:
  - Switch to Parallel Projection mode.
  - Make sure that you're using one of the Standard views.

Take a look at the section "Printing to Scale (Windows and Mac)," later in this chapter, for a complete rundown on printing to scale in SketchUp.

Tiled Sheet Print Range

Perhaps you're printing at a scale that won't fit on a single page, or you've entered a print size that's bigger than the paper size you chose in the Print Setup dialog box. The Tiled Sheet Print Range area lets you print your image on multiple sheets and then attach them all together later. In this way, you are able to create posters using your small-format printer.

Print Quality

When selecting a print quality for your image, the result of each setting depends on your model, so you should probably try a few different settings if you have time:

- Draft and Standard are really only useful for making your model appear the way you want it to on the printed page.
- Use High Definition first, then increase to Ultra High Definition if your computer/printer setup can manage it.

Other Settings

You can control the following settings in the Print Preview dialog box, too:

- **2D Section Slice Only:** If you have a visible section cut in your model view, selecting this check box tells SketchUp to only print the section cut

Figure 12-4

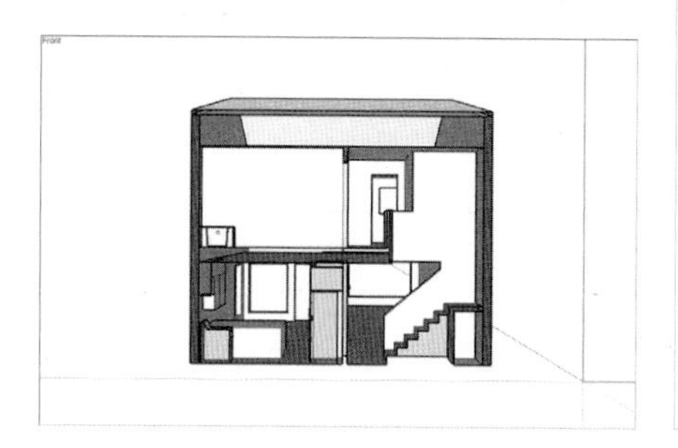

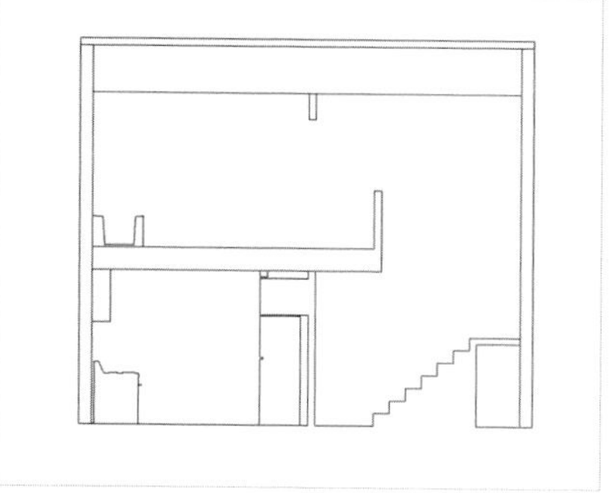

Printing only the 2D section slice yields a simple drawing that's easy to sketch over.

edges. Figure 12-4 shows what the same model view would look like without (on the left) and with (right) this option selected. This can be used to produce simple plan and section views that you can sketch on by hand.

- **Use High Accuracy HLR:** Selecting this check box tells SketchUp to send **vector** information to the printer instead of the usual **raster** data. These terms are further described in Chapter 13. Vector lines look much smoother and cleaner when printed, so your whole model will look better—with one exception: Gradients (especially the smooth shadows on rounded surfaces) don't print well as vectors. If you have a lot of rounded or curvy surfaces in your model view, you probably don't want to choose this option. Also, if your model view includes a sketchy edges style, don't use high accuracy HLR; you won't see any of the sketchy effects in your final print.

Vector
Term describing images that consist of instructions written in computer code.

Raster
Term describing images that are composed of pixels.

IN THE REAL WORLD

Ann Barreca Young mentions that she has used a variety of viewing styles, straight and sketchy, during her PowerPoint presentations to clients and also likes to print out straight-line drawings to color-render them by hand.

She points out that she uses SketchUp in conjunction with AutoCAD as she imports her floor plans into SketchUp to use as a base to build the 3D models from. She prints the models out in various modes to hand-render over the top in marker or she simply displays them in an architectural style. According to her, either way, the response has been great as clients usually like to see 3D perspectives because they offer a better visual of how the space will look like if compared to a floor plan made in AutoCAD.

Tabbed Scene Print Range

Use this area to tell SketchUp which of your scenes you'd like to print, if you have more than one. This is handy for quickly printing all your scenes. Select the Current View option to print only what is currently in your modeling window.

SELF-CHECK

1. If you want to print ordered packets of material in this fashion, 1234, 1234, what check box do you need to select?
2. Print Size is the most complicated section of the Print dialog box. True or false?
3. What option would be use if we want to print an image that is larger than the size of paper that we have available in our printer?
4. In the Print Dialog Box we can choose whichever scene we want from our model. True or false?

Apply Your Knowledge Quite often we do not give enough thought to our printing alternatives until we are almost done with our project, then, suddenly we realize that we have just enough time to print it quickly to show it to our client. It is important that we budget enough time to study our options so that the printing process will be painless. Also, we would have to make sure that we have our supplies so that we can print with the right papers and the right ink.

Use one of the projects you have built in the previous chapters, such as the gym for the high school done on Chapter 11. Capture some scenes. Then include those images on a board that is larger than your printer size and practice with the Tile Sheet Print option. How does that differ from the Tabbed Scene Print Range option?

12.2 PRINTING FROM A MAC

If you're using a Mac, printing is slightly simpler than it is for people who use Windows computers. The first part of the following sections explains the procedure for generating a simple, straightforward print of what you see in your Modeling Window. The second part goes into some detail about what each setting does.

12.2.1 Making a Basic Print (Mac)

Follow these steps to print exactly what you see in your Modeling Window on a Mac:

PATHWAYS TO... BASIC PRINTING ON THE MAC

1. **Make sure that your Modeling Window contains whatever you want to print.** SketchUp prints exactly what you see in your Modeling Window, unless you're printing to scale. This is considerably more complicated, and it is explained in a section at the end of this chapter.
2. **Choose File ⇨ Page Setup.** This opens the Page Setup dialog box, where you decide what printer and paper size to use.
3. **In the Page Attributes dialog box (see Figure 12-5), do the following:**
   - Choose the printer you'd like to use from the Format For drop-down list.
   - Choose a paper size for your print.
   - Choose an orientation for your print.
4. **Click the OK button to close the Page Setup dialog box.**
5. **Choose File ⇨ Document Setup.** This opens the Document Setup dialog box.
6. **In the Document Setup dialog box, make sure that the Fit View to Page check box is selected.** Read the next section in this chapter for a full description of what everything does.
7. **Click the OK button to close the Document Setup dialog box.**
8. **Choose File ⇨ Print to open the Print dialog box.**
9. **In the Print dialog box, click the Preview button.** This generates an on-screen preview of what your print will look like on paper.
10. **If you are satisfied with the preview, click the Print button to send your print job to the printer.** If you're not satisfied with the preview, click the Cancel button and start again at step 1.

Figure 12-5

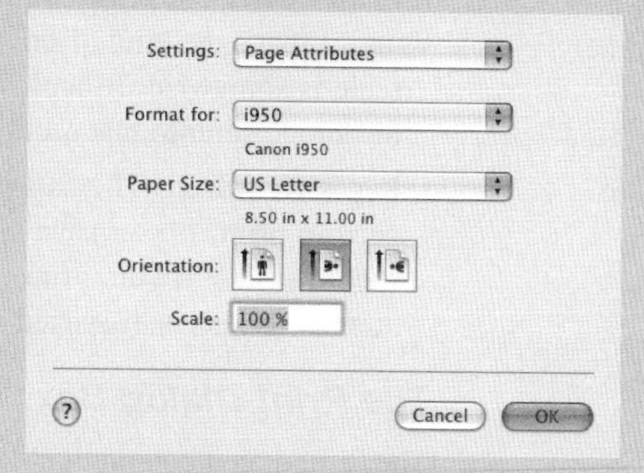

The Page Setup dialog box on a Mac lets you select a printer, a paper size, and a page orientation.

Figure 12-6

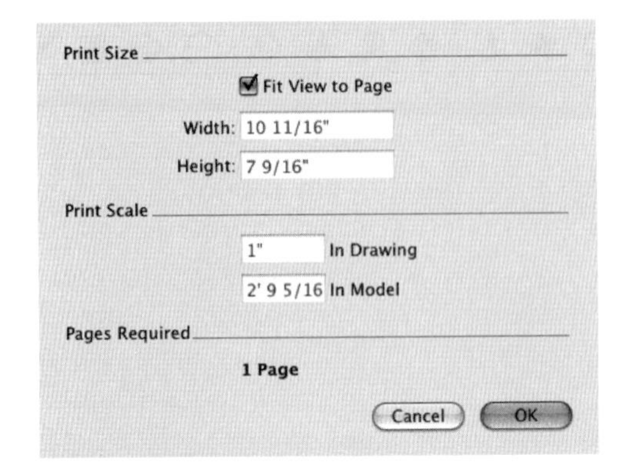

The Mac Document Setup dialog box.

Understand how to work with print dialog boxes.

12.2.2 Deciphering the Mac Printing Dialog Boxes

Because printing from SketchUp on a Mac involves two separate dialog boxes, both are described in the following sections.

The Document Setup Dialog Box

You use the settings in the Document Setup dialog box (see Figure 12-6) to control how big your model prints.

Here's what everything in the Document Setup dialog box does:

- **Print Size:** This one's fairly self-explanatory, but here are some details:
  - **Fit View to Page:** Selecting this check box tells SketchUp to make your printed page look just like your Modeling Window on-screen.
  - **Width and Height:** If the Fit View to Page check box is deselected, you can type in either a width or a height for your final print. This is what you should do if you want to print a tiled poster out of several sheets of paper; simply enter a final size.
- **Print Scale:** Use these settings to control the scale of your printed drawing, if that's the kind of print you're trying to make. Because printing to scale is somewhat difficult, the last section of this chapter is devoted to the topic. Refer to that section for a description of what these settings do.
- **Pages Required:** This is simply a read out of the number of pages you need to print. If you have selected the Fit View to Page check box, this should read 1. If your print won't fit on one sheet, it will be tiled onto the number of sheets displayed in this section of the dialog box.

The Print Dialog Box

The Print dialog box on the Mac contains several more panels in the Copies & Pages drop-down list. You only need to use two. Both are pictured in Figure 12-7 and described in the following list:

Figure 12-7

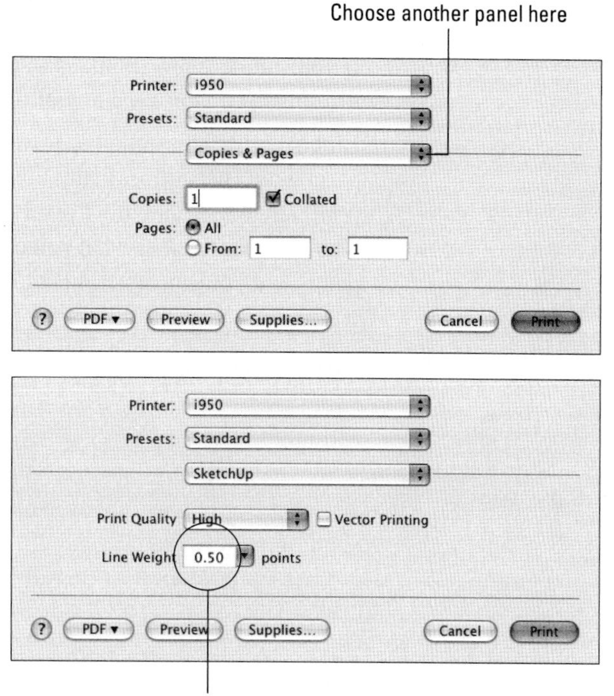

The Copies & Pages and SketchUp panels of the Print dialog box.

- **Copies & Pages panel:** The controls in this part of the Print dialog box are fairly straightforward; use them to tell SketchUp how many copies and pages you want to print:
  - **Copies:** If you're printing more than one copy of a print that includes multiple pages, select the Collated check box to tell SketchUp to print "packets," which can save you from having to collate them yourself.
  - **Pages:** If the Pages Required readout at the bottom of the Document Setup dialog box said that you need more than one sheet to print your image, you can choose to print all or some of those pages here.
- **SketchUp panel:** You use the settings in this panel to control the final appearance of your print:
  - **Print Quality:** The results you get with this setting depend on your printer model. In general, Draft or Standard shouldn't be used unless you're simply making sure your page will look the way you want it to. If you have time, try both High and Extra High and see which one looks better.
  - **Vector Printing:** When you select this option, SketchUp sends vector (instead of raster) information to the printer. Read Chapter 13 for more details on the description of these terms. Vector printing makes edges

look smoother and cleaner, but it does poorly with gradients. Use vector printing if your model view is made up of mostly flat faces, but try printing both ways (with vector printing on and off) to see which looks better. If your model view includes a sketchy edges style, don't choose Vector Printing; you won't see any of the sketchy effects in your final print.

– **Line Weight:** This option only works if you've selected the Vector Printing check box. The number in this box represents the thickness of edges in your print; any edges that are 1 pixel thick in your model view will be drawn with a line as thick as what you choose for this option. The default is 0.50 points, but you can experiment to see what looks best for your model.

IN THE REAL WORLD

Christopher Wulff loves the ease of use and speed that Google SketchUp offers to its users. He says that his learning process was so intuitive that he went from playing with SketchUp on his first week to creating a model to present to his client in the next.

His work preparation starts with Google Earth as he first grabs a shot of the site. The following step is to work with the architect and engineers to sketch out the basic layout. After the basic layout is completed, he recreates it in AutoCAD and then imports it into SketchUp to start working on the 3D model. According to him, once the 3D model is finalized, he can take still-shots and easily move through to video. In the end, he presents his clients not only well rendered still-shots but also a video file of the project. He points out that his clients have always been impressed with the final product due to the speed of creation and the quality of his work.

SELF-CHECK

1. The Line Weight option only works if you've selected the ______ _______ check box.
2. What option would we have to choose if we want to print a tiled poster using several sheets of paper?
3. What would be the value of using Draft or Standard print under Print Quality?
4. What option would allow you to print smoother edges?

Apply Your Knowledge Like described, proper printing requires time and it is never a good idea to rush printing if you are preparing work to be shown to a client. Use one of the projects you built in the recent past, such as the gym for the high school done in Chapter 11, and practice printing with the Vector Printing option turned on or off. Then, compare the results. Also, experiment with the option of Pages Required, to print your work in several sheets of paper.

12.3 PRINTING TO A PARTICULAR SCALE

GOOGLE SKETCHUP IN ACTION

Configure your system to print to scale.

Sometimes, instead of printing exactly what you see on your screen so that it fits on a sheet of paper, you might need to print a drawing to scale. Read the box, "Wrapping Your Head Around Scale" for more information about drawing to scale. Keep in mind that if you have SketchUp Pro, you can use Layout to generate scaled views of your model very easily. Read Chapter 14 for more information.

12.3.1 Preparing to Print to Scale (Windows and Mac)

Before you can print a view of your model to a particular scale, you have to set things up properly. Here are some things to keep in mind:

- **Perspective views can't be printed to scale.** In perspectival views, all lines appear to "go back" into the distance, which means that they look shorter than they really are. Because the point of a scaled drawing is to be able to take accurate measurements directly off your printout, views with perspective don't work.

FOR EXAMPLE

WRAPPING YOUR HEAD AROUND SCALE

When you print to scale, anyone with a special ruler (called a *scale*) can take measurements from your drawing, as long as he or she knows the scale at which it was printed. You can use three different kinds of drawing scales:

Architectural: In the United States, most people use feet and inches to measure objects. Most architectural scales substitute fractions of an inch for a foot. Three common examples of architectural scales are as follows:

- 1/2 inch = 1 foot (1 inch = 2 feet)
- 1/4 inch = 1 foot (1 inch = 4 feet)
- 1/8 inch = 1 foot (1 inch = 8 feet)

Engineering: When it comes to measuring large things like parcels of land and college campuses, U.S. architects, engineers, industrial designers, and surveyors still use feet, but they use engineering scales instead of architectural ones. Three common engineering scales include the following:

- 1 inch = 20 feet
- 1 inch = 50 feet
- 1 inch = 100 feet

Metric: Outside of the United States, most people use the metric system. Because all measurement is based on the number 10, metric scales can be applied to everything from very small things (such as blood cells) to very large things (such as countries). This is called the DIN system. Metric scales use ratios instead of units of measure; here are three examples:

- 1:10 (the objects in the drawing are 10 times bigger in real life)
- 1:100 (the objects in the drawing are 100 times bigger in real life)
- 10:1 (the objects in the drawing are 10 times smaller in real life)

The following figure shows the same drawing printed at two different scales.

Figure 12-8

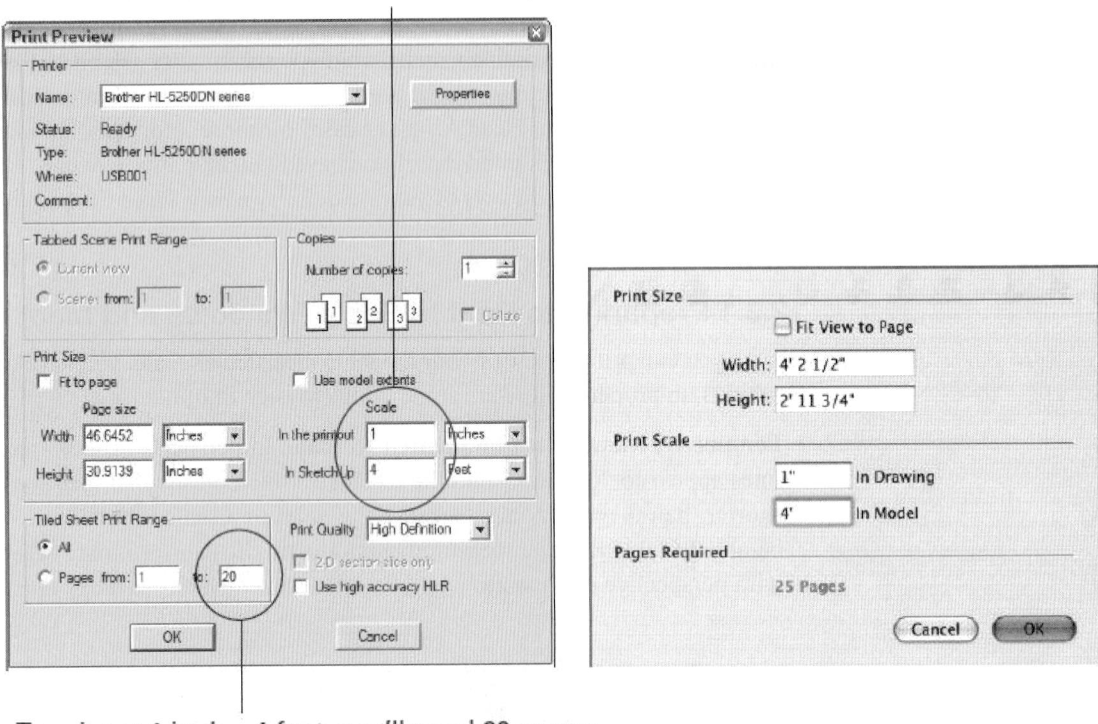

Setting up to print at 1 inch = 4 foot (1/4 inch = 1 foot) scale.

Reproduced with the permission of Jorge Paricio, PhD.

- **Switch to Parallel Projection if you want to print to scale.** To change your viewing mode from Perspective to Parallel Projection, choose Camera ⇨ Parallel Projection.
- **You must use the Standard views.** SketchUp allows you to quickly look at your model from the top, bottom, and sides by switching to one of the Standard views. Choose Camera ⇨ Standard and pick any of the views except Iso.

12.3.2 Printing to Scale (Windows and Mac)

The steps in this section allow you to produce a scaled print from SketchUp; Windows instructions are presented first, and then Mac instructions. When the user-interface elements are different for the two platforms, the ones for Mac are shown in parentheses. Figure 12-8 shows the relevant dialog boxes for printing to scale in Windows and on a Mac.

Before you begin, check that you've switched to Parallel Projection and that your view is lined up the right way. See the previous section of this chapter for information on what you need to do to prepare your model view for scaled printing. Follow these steps to produce a scaled print:

PATHWAYS TO... PRODUCING A SCALED PRINT

1. **Choose File ⇨ Print Setup (Page Setup).**
2. **Select a printer, paper size, and paper orientation, and then click the OK button.**
3. **Choose File ⇨ Print Preview (Document Setup).** You might be tempted to skip this step if you are pressed on time but try not to do that; this will allow you to see clearly what will be printed and if your lines will be visible. Also you can verify here your text size; especially at small page sizes, you might need to modify the correct text size (font size) a few times until your text appears legible.
4. **Deselect the Fit to Page (Fit View to Page) check box.**
5. **Make sure that the Use Model Extents check box is deselected.** Mac users don't have this option.
6. **Enter the scale at which you'd like to print your model view.** If you wanted to print a drawing at 1/4-inch scale, you would enter the following:
   - **1 Inches** into the In the Printout (In Drawing) box.
   - **4 Feet** into the In SketchUp (In Model) box.

     Similarly, if you wanted to produce a print at 1:100 scale, you would enter the following:
   - **1 m** into the In the Printout (In Drawing) box.
   - **100 m** into the In SketchUp (In Model) box.
7. **Take note of how many pages you'll need to print your drawing.** If you are using Windows, you can check this in the Tiled Sheet Print Range area of the dialog box. On a Mac, the number of pages you'll need appears in the Pages Required section of the Document Setup dialog box. If you want to print on a different-sized piece of paper, change the setting in the Print Setup (Page Setup) dialog box.
8. **If you want to print your drawing on a single sheet and it won't fit, try using a smaller scale.** Using the 1/4 inch = 1 foot example, try shrinking the drawing to 3/16 inch = 1 foot scale. To do this, you would enter the following:
   - **3 Inches** into the In the Printout (In Drawing) box.
   - **16 Feet** into the In SketchUp (In Model) box.
9. **When you are satisfied with how your drawing will print, click the OK button.**
10. **(Windows only) If you're satisfied with what you see in the Print Preview window, click the Print button (in the upper-left corner) to open the Print dialog box.**
11. **(Mac only) Choose File ⇨ Print.**
12. **In the Print dialog box, click the OK button to send your print job to the printer.**

Refer to the "Making a Basic Print" section for your operating system, earlier in the chapter, for a complete review of basic printing from SketchUp.

CAREER CONNECTION

Jeremy Fretts is an interior designer working for Niles Bolton Associates who enjoys using Google SketchUp. He considers this program a joy to use and the antithesis of other architectural software. He believes that the ease of use that SketchUp offers as a design and visualization tool changed his career and enabled him to unleash his potential as a designer. According to him, there is no other software so intuitive to use, with very minimal initial training.

He goes on to say, that among the many uses that SketchUp has, he likes to utilize crisp model images made with this program and water-color them in Photoshop using a pen tablet. As an example, he says that he turned a crisp 3D model of a seven-building site into a "fuzzy" public presentation in only two hours.

SELF-CHECK

1. Perspective views and plan views in parallel projection can both be printed to scale. True or false?
2. When printing in scale, which option do we need to choose under Camera, Parallel Projection, or Perspective view? Also, under View what option would we have to choose to obtain a top, side, or front view?
3. Scale printing allows us to take accurate measurements of our printouts. True or false?
4. If you wanted to print at 1:200 scale, you would enter which of the following numbers in the Drawing and Model box, respectively:
   a. 1:2
   b. 1:200
   c. 1:20
   d. 10:20

Apply Your Knowledge In many design disciplines we have to include view of our creations in a parallel projection mode so that we can obtain credible measurements off the different views. For example, if we have a hand-held object we might want to have a plan view, a front view and left and right views of the unit. Similarly, if we have a building we might want to show the plan view and also the front views (also called elevations) of the front, both sides, and back.

Using the model of the high school you worked on in previous exercises, practice printing the design of your gym on your home printer at different scales. Which scale fits better on your page?

SUMMARY

Section 1

- The Print Preview and Print dialog boxes in SketchUp are exactly the same.
- You can click the Properties button to make adjustments to your printer settings.

Section 2

- You use the settings in the Document Setup dialog box to control how big your model prints.
- The Print dialog box on the Mac contains several more panels in the Copies & Pages drop-down list than Windows, but only two are necessary: Copies and Pages.

Section 3

- Before you can print a view of your model to a particular scale, you have to set things up properly:
  - Perspective views can't be printed to scale.
  - Switch to Parallel Projection if you want to print to scale.
  - You must use the Standard views.

ASSESS YOUR UNDERSTANDING

SUMMARY QUESTIONS

1. Deselecting the Collate check box prints pages like this: 123412341234. True or false?
2. What type of images consists of instructions written in computer code?
   a. Raster
   b. Gradient
   c. Vector
   d. Scaled
3. To print at a 1:500 m scale, what number would you enter on the Drawing box?
4. To print at a 1:500 m scale, what number would you enter on the Model box?
5. If you want to print your model on a single sheet of paper and it won't fit, should we use a larger or a smaller number in our Drawing box?
6. You can use the Tabbed Scene Print Range to tell SketchUp which of your scenes you would like to print. True or false?
7. Vector printing is recommended when our model is done with a visual balance of lines and gradients. True or false?

APPLY: WHAT WOULD YOU DO?

1. What does printing to scale mean, and what is a major advantage of having scaled printouts?
2. List any three settings in SketchUp print panel and describe when you would use each one of them. What option works best with your own printer?

BE AN ARCHITECT

Print Your Model to Scale

Taking your model of a house, print it at the following scales:

- 1 inch = 2 feet
- 1 inch = 16 feet
- 1 inch = 100 feet

Which option is best and why?

KEY TERMS

Raster

Vector

CHAPTER 13

EXPORTING IMAGES AND ANIMATIONS

Making Image Files and Movies

Do You Already Know?

- How to 2D views of your models as TIFFs, JPEGs, and PNGs?
- Understand what are Pixels and Resolutions?
- How to choose which is the best image in which to export?
- How to export to animations?

For the answers to these questions, go to **www.wiley.com/go/chopra/googlesketchup2e**

What You Will Find Out	What You Will Be Able To Do
13.1 How to export 2D images of your model.	Export your models to TIFF, JPEG, and PNG images.
13.2 How to make animated movies.	Prepare and export your model using the Animation Export feature.

INTRODUCTION

Want to e-mail a JPEG of your new patio to your parents? How about a movie that shows what it's like to walk out onto that new patio? If you need an image or a movie of your model, forget about viewing or printing within SketchUp. Exporting is the way to go.

SketchUp can export both still images and animations in most of the major graphics and movie formats. Here's the part that's a bit confusing: Which file formats you can export depend on the version of SketchUp you have. If you have regular Google SketchUp (the free one), you can create *raster* image files as well as movies. If you have Google SketchUp Pro, you can also export *vector* files and a lot of 3D formats; all of them are discussed in Chapter 14.

This chapter focuses on the export file formats that are common to both versions of Google SketchUp. Thus, the 2D raster image formats that you can create with SketchUp are explained. More specifically, in this chapter, you will create 2D views of your models as TIFFS, JPEGS, and PNGs. You will also evaluate pixels and resolution. Then, in the final part of this chapter, you will learn how to export animations as movie files that anyone can open and view.

13.1 EXPORTING 2D IMAGES OF YOUR MODEL

Even though the free version of SketchUp can only export 2D views of your model as *raster* images, it s helpful to know about graphics file formats in general. Should you feel comfortable with this information, you can read ahead to the section titled, "Exporting a Raster Image from SketchUp."

If you have SketchUp Pro, you have a much easier way to export images through Layout. Layout is so comprehensive that you might be able to completely bypass the application you are trying to export to. Read Chapter 14 for more information on this.

Pictures on your computer are divided into two basic types: *raster* and *vector*. The difference between these two categories of file types has to do with how they store image information:

- **Raster:** Raster images are made up of dots. (Technically, these dots are called pixels, just like the pixels that make up the images you take with a digital camera.) Raster file formats consist of information about the location and color of each dot. When you export a raster, you decide how many dots (pixels) it should include, which directly affects how big it can be displayed. SketchUp exports TIFF, JPEG, and PNG raster images; the Windows version also exports BMPs. More on raster images is presented later under "Understanding Rasters."
- **Vector:** Vector images consist of instructions written in computer code. This code describes *how* to draw the image to whatever software is trying to open it. The major advantage of using vector imagery (as opposed to raster) lies in its scalability. Vectors can be resized larger or smaller without

affecting their image quality, while rasters lose quality if you enlarge them too much. The free version of SketchUp can only export raster images, but SketchUp Pro can export vectors in both PDF and EPS file formats; you can read about it in Chapter 14.

13.1.1 Exporting a Raster Image from SketchUp

The process of exporting a view of your SketchUp model is fairly straightforward. Depending on which format you choose, the export options are slightly different, but all of them are addressed in this section.

Follow these steps to export a raster image from SketchUp:

1. **Adjust your model view until you see exactly what you'd like to export as an image file.** With SketchUp's raster image export, your entire modeling window view is exported as an image, so use the navigation tools or click on a scene to set up your view. Use styles, shadows, and fog to make your model look exactly the way you want it to. To change the proportions of your image, resize your SketchUp window. Follow these steps to do so:

   - If your SketchUp window is full-screen (Windows only), click the Minimize button in its upper-right corner.
   - Drag the Resize tab in the lower-right corner of your SketchUp window until the modeling window is the right proportion.

   Figure 13-1 shows adjustment to the proportions of a modeling window in order to export a wide view of a modeled house. You might be

Figure 13-1

SketchUp modeling window

Exported image

Adjust your view and your modeling window until things look the way you want them to in your exported image.

wondering whether everything in your modeling window shows up in an exported raster image. The red, green, and blue axes don't, but guides do. If you don't want your guides to be visible in your exported image, deselect Guides in the View menu.

2. **Choose File ⇨ Export ⇨ 2D Graphic.** This opens the Export 2D Graphics dialog box.
3. **Choose the file format you'd like to use from the Format drop-down list.** Before you choose JPEG by default, you should know that this file type isn't always the best choice. For a complete description of each format (as well as recommendations for when to choose each), see the section "Looking at SketchUp's Raster Formats," later in this chapter.
4. **Choose a name and a location on your computer for your exported image.**
5. **Click the Options button.** This opens the Export Options dialog box, where you can control how your image is exported. Figure 13-2 shows what this dialog box looks like for each of SketchUp's raster file formats.
6. **Adjust the settings in the Export Options dialog box.** Here's a description of what the settings do:

   - **Use View Size:** Selecting this check box tells SketchUp to export an image file that contains the same number of pixels as are currently being used to display your model on-screen. If you're just planning to use your exported image in an e-mail or in an on-screen presentation (like PowerPoint), you select Use View Size, but it's still better to manually control the pixel size of your exported image. If you're planning to print your exported image, don't select this check box.
   - **Width and Height:** When you don't select the Use View Size check box, you can manually enter the size of your exported image.

Figure 13-2

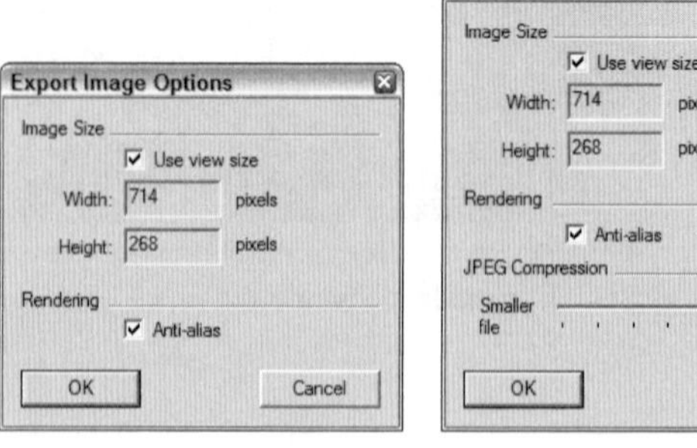

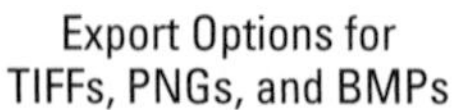

Export Options for TIFFs, PNGs, and BMPs

Export Options for JPEGs

The Export Options dialog boxes for TIFFs, PNGs, and BMPs (left) and JPEGs (right).

Figure 13-3

No anti-aliasing

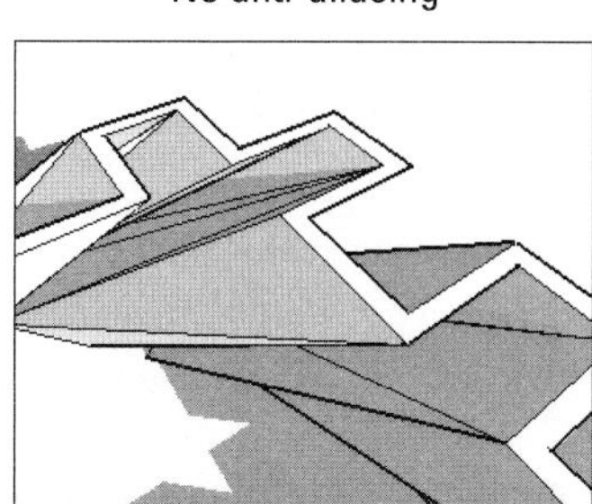

With anti-aliasing

A view of the same image with anti-aliasing off (left) and on (right).

Because this process requires a fair amount of figuring, a whole section is devoted to it; take a look at "Making Sure That You're Exporting Enough Pixels," later in this chapter, to find out what to type into the Width and Height boxes.

- **Anti-alias:** Because raster images use grids of colored squares to draw pictures, diagonal lines and edges can sometimes look jagged. **Anti-aliasing** is a process that fills in the gaps around pixels with similar-colored pixels so that things look smooth. Figure 13-3 illustrates the concept. In general, you want to leave anti-aliasing on.

Anti-aliasing
A process that fills in the gaps around pixels with similar-colored pixels so that things look smooth.

Resolution
An image's pixel density.

- **Resolution (Mac only):** This is where you tell SketchUp how big each pixel should be, and therefore how big (in inches or centimeters) your exported image should be. Pixel size is expressed in terms of pixels per inch/centimeter. This option is only available when the Use View Size check box isn't selected. Just like with the Width and Height boxes, the next section of this chapter goes into a lot of detail about image resolution.
- **Transparent Background (Mac only, not for JPEGs):** Mac users can choose to export TIFFs and PNGs with transparent backgrounds, which can make it easier to "cut out" your model in another piece of software. Exporting your image with a transparent background is also a nice way to use image-editing programs like Photoshop to drop in a sky and ground plane later on.
- **JPEG Compression (JPEG only):** This slider lets you decide two things at the same time: the file size of your exported image and how good the image will look. The two are, of course, inversely related; the farther to the left you move the slider, the smaller your file will be, but the worse it will look. You may not want to set JPEG compression any lower than 8.

7. **Click the OK button to close the Export Options dialog box.**

FOR EXAMPLE

UNDERSTANDING RASTERS

When you look at a photograph on your computer, you're really looking at a lot of tiny dots of color called pixels. These are arranged in a rectangular grid called a raster. Digital images that are composed of pixels arranged in a raster grid are called raster images, or rasters for short. Have a look at the first image in the figure below for a close-up view of a raster image. Here are some things to keep in mind about rasters:

- **Rasters are everywhere.** Almost every digital image you've ever seen is a raster. TIFF, JPEG, and PNG are the most common raster file formats, and SketchUp exports all three of them.
- **Rasters are flexible.** Every two-dimensional image can be displayed as a raster; a grid of colored squares is an incredibly effective way of saving and sharing picture information. As long as you have enough pixels, any image can look good as a raster.
- **Rasters take up a lot of space.** If you think about how raster images work, it takes a lot of information to describe a picture. Digital images are made up of anywhere from thousands to millions of pixels, and each pixel can be any one of millions of colors. To store a whole picture, a raster image file needs to include the location and color of *each* pixel; the bigger the picture, the more pixels it takes to describe it, and the bigger the file size gets.
- **Rasters are measured in pixels.** Because every raster image is made up of a specific number of pixels, you use a raster's *pixel dimensions* to describe its size. If you were told that you'd been e-mailed a photograph that was 800 × 600, you could expect to receive a picture that is 800 pixels wide by 600 pixels tall (see the figure below). Pixels don't have a physical size of their own—they're just dots of color. You determine a picture's physical size by deciding how big its pixels should be; this is referred to as *resolution*, and is generally expressed in terms of *pixels per inch* (ppi). Read the section, "Making sure that you are exporting through pixels," later in this chapter for more information.

Why use pixels instead of inches or centimeters to describe the size of a digital image? It all has to do with how computer screens work. Because not all screens display things at the same size, it's impossible to predict how *big* an image will look when it shows up on someone's computer. Depending on the person's display settings, an 800-×-600-pixel image might be a few inches across, or it might take up the whole screen. Giving a digital image's dimensions in pixels is the only accurate way of describing how "big" it is.

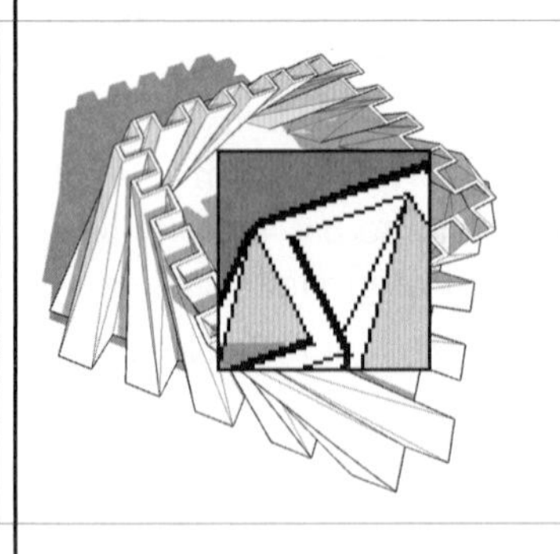

8. **Back in the Export 2D Graphic dialog box, click the Export button to export your raster image file.** You can find your exported file in whatever location on your computer you specified in step 4, above. What you do with it is entirely up to you—you can e-mail it, print it, or use it in another software program to create a presentation. The export process may take longer than you think it should. If you're exporting a large image (one with lots of pixels), the export will take a while.

13.1.2 Looking at SketchUp's Raster Formats

Export your models to TIFF, JPEG, and PNG images.

So you know you need to export a raster image from SketchUp, but which one do you choose? You have four choices in Windows; three of them are available on a Mac. The following sections give you the details.

When you export a raster image, you're saving your current view in SketchUp to a separate file somewhere on your computer. As a raster image, that file consists of tiny, colored dots called pixels. When you look at all the pixels together, they form an image.

Tagged Image File (TIFF or TIF)

TIFFs are the stalwarts of the raster image file format world; everyone can read them and just about everyone can create them. TIFF stands for Tagged Image File Format. Here's everything you need to know about TIFFs:

- **When image quality is important, choose TIFF.** Unless file size is a concern (because, for example, you need to send an image by e-mail), always export a TIFF if you need a raster image. For everything from working in Photoshop to creating a layout in InDesign or QuarkXPress, a TIFF can provide the image quality you need.
- **TIFFs do not compress your image data.** That means they don't introduce any unwanted elements like JPEGs do, but it also means that they're especially big files.
- **Pay attention to your pixel count.** If you're exporting a TIFF, you're probably looking for the best image quality you can get. And if that's the case, you need to make sure that your TIFF is "big" enough—that it includes enough pixels—to display at the size you need. See the next section in this chapter for more information.

JPEG (or JPG)

JPEG stands for Joint Photographic Experts Group. Almost every digital image you've ever seen was a JPEG (pronounced *JAY-peg*); it's the standard file format for images on the Web. Below are a few details about JPEGs:

- **When file size is a concern, choose JPEG.** The whole point of the JPEG file format is to compress raster images to manageable file sizes so that they can be e-mailed and put on Web sites. A JPEG is a fraction of the size of a TIFF file with the same number of pixels, so JPEG is a great choice if file size is more important to you than image quality.
- **JPEGs compress file size by degrading image quality.** This is known as **lossy** compression; JPEG technology basically works by removing a lot of the pixels in your image. JPEGs also introduce a fair amount of pixel garbage; these smudges are called **artifacts.**
- **JPEG and SketchUp are a dangerous combination.** Because of the way the JPEG file format works, JPEG exports from SketchUp are particularly susceptible to looking terrible. Images from SketchUp usually include straight lines and broad areas of color, both of which JPEGs have a hard time handling. If you're going to export a JPEG from SketchUp, make sure that the JPEG Compression slider is never set below 8. For more details, see the section "Exporting a Raster Image from SketchUp," earlier in this chapter.

Lossy
Type of compression that occurs with JPEGs. JPEGs compress file size by degrading image quality.

Artifacts
Smudges and other image degradation introduced by file compression; common in JPEGs.

Portable Network Graphics (PNG)

Pronounced *ping*, PNG is a graphics file format. Unfortunately, it isn't as widely used as many other formats. As far as SketchUp is concerned,

IN THE REAL WORLD

Geri Cruickshank Eaker, owner of Freespace Design in Charlotte, first started to work with Google SketchUp due to its free access, short learning curve, and ease of use.

When preparing presentations for her clients, she normally exports to JPEG or animation and inserts into Microsoft PowerPoint. According to her, the response is always great but it is important to point out that the overall quality of the presentation depends quite a bit on one's level of skill when using SketchUp. The more artistic and detail-oriented the designer is, the better the quality of the rendering that he created will be, which is true in any modeling software.

Her working process starts by importing the floor plan made in AutoCAD to SketchUp, where she builds the 3D model and then uses a plugin for rendering. She says that she is still experimenting with the many SU rendering plugins available in the market, which makes the options endless.

PNG combines all the best features of TIFF and JPEG. PNG details are as follows:

- **PNGs compress image data without affecting image quality.** As a lossless compression technology, PNGs are smaller files than TIFFs (just like JPEGs), but they don't mess up any pixels (unlike JPEGs). PNGs aren't as small as JPEGs, but the difference in image quality is worth a few extra bits.
- **If you're exporting an image for someone who knows about computers, choose PNG.** Some software doesn't know what to do with a PNG, so there's a risk in using it. If you plan to send your exported image to someone who knows what he or she is doing, go ahead and send a PNG. If the recipient of your export is less technologically sophisticated, stick with a JPEG or TIFF file; it's the safe choice.

The PNG file format wasn't developed to replace JPEG or TIFF; it was supposed to stand in for GIF (Graphics Interchange Format), which is a file type that SketchUp doesn't export. Without going into too much detail, people use JPEG for images like photographs and GIF for things like logos. Because exported SketchUp views usually have more in common with the latter, PNG (the replacement for GIF) is the better choice. So why can't PNG replace JPEG and TIFF? For most photographs (which are the majority of images on the Web), JPEG is better than PNG because it produces smaller files, which in turn yields faster load times when you're surfing the Internet. TIFF is more versatile than PNG because it supports different *color spaces*, which are important to people in the printing industry. For reasons that are beyond the scope of this book, that isn't relevant to exports from SketchUp; PNG is still the best—if not the safest—choice.

Windows Bitmap (BMP)

Windows Bitmap, or BMP, files can only be used on Windows, and they're big. BMPs should rarely be used for anything, with a couple of exceptions:

- **To send your exported file to someone with a late model Windows computer:** If the person to whom you're sending an exported image has a Windows computer that's more than about five years old, you may want to send a BMP.
- **To place an image in an old Windows version of layout software:** If your layout person is using a copy of Word or PageMaker that's a few years old, he or she might need a BMP file.

13.1.3 Making Sure You're Exporting Enough Pixels

When it comes to raster images, it's all about pixels. The more pixels your image has, the more detailed it is, and the bigger it can be displayed or

Figure 13-4

150 x 50 pixels

300 x 100 pixels

900 x 300 pixels

More pixels yield a much more detailed image.

printed. Figure 13-4 shows the same image three times. The first image is 150 × 50, meaning that it's 150 pixels wide by 50 pixels high. The second image is 300 × 100, and the third is 900 × 300. Notice that the more pixels an image has, the better it looks.

Why not always export as many pixels as possible, just in case you need them? There are two reasons:

- Image exports with lots of pixels take a long time to process.
- Raster images are very big files.

How many pixels you need to export depends on what you're going to use the image *for*. Very broadly, you can do two things with your image:

- Display or project it on a screen, digitally.
- Print it.

The next two sections discuss each of these possibilities in detail.

Exporting Enough Pixels for a Digital Presentation

If you plan to use your exported image as part of an on-screen presentation, it's helpful to know what computer monitors and digital projectors can display:

- The smallest, oldest devices currently in use have images that are 800 pixels wide by 600 pixels high.
- At the other end of the spectrum, high-end, 30-inch LCD monitors display 2560 × 1600 pixels.

Table 13-1: Suggested Image Sizes for On-Screen Use

How the Image Will Be Used	Image Width (Pixels)
E-mail	400 to 800
Web site, large image	600
Web site, small image	200
PowerPoint presentation (full-screen)	800 or 1024 (depends on projector)
PowerPoint presentation (floating image)	400

So it stands to reason that if you're exporting an image that will be viewed on-screen only, you need to create an image that's somewhere between 800 and 2500 pixels wide. Table 13-1 provides some guidelines on image sizes for different digital applications.

Understanding Resolution: Exporting Images for Print

Images that you want to print need to have more pixels than ones that are only going to be displayed on-screen. That's because printers—inkjet, laser, and offset—all operate very differently than computer monitors and digital projectors. When you print something, the pixels in your image turn into microscopic specks of ink or toner, and these specks are smaller than the pixels on your computer screen. To make a decent-sized print of your exported image, it needs to contain enough *pixels per inch* of image. An image's pixel density, expressed in pixels per inch (ppi), is its **resolution**.

What kind of resolution you need depends on three things:

- **The kind of device you'll be printing to:** For home inkjet printers, you can have a resolution of as little as 150 ppi. If your image will be appearing in a commercially produced book, you need a resolution of at least 300 ppi.
- **How far away the image will be from the audience:** There's a big difference between a magazine page and a trade-show banner. For close-up applications, a resolution of 200 to 300 ppi is appropriate. Large graphics that will be viewed from several feet away can be as low as 60 ppi.
- **The subject matter of the image itself:** Photographic images tend to consist of areas of color that blur together a bit; these kinds of images can tolerate being printed at lower resolutions than drawings with lots of intricate detail. For images with lots of lines (like SketchUp models), it's best to work with very high resolutions—300 to 600 ppi—especially if the image will be viewed close-up.

Table 13-2: Recommended Resolutions for Prints

How the Image Will Be Used	*Image Width (Pixels)*	*Image Resolution (pixels/cm)*
8.5 × 11 or 11 × 17 inkjet or laser print	200 to 300	80 to 120
Color brochure or pamphlet	300	120
Magazine or book (color and shadows)	300	120
Magazine or book (linework only)	450 to 600	180 to 240
Presentation board	150 to 200	60 to 80
Banner	60 to 100	24 to 40

Table 13-2 provides some guidelines for exporting images that will be printed.

Keep in mind that the biggest raster image that SketchUp can export is 10,000 pixels wide or tall (whichever is greater). This means that the largest banner image, printed at 100 ppi, that SketchUp can create is about 100 inches wide. To make larger images, you need to export a vector file.

Follow these steps to make sure that you're exporting enough pixels to be able to print your image properly:

1. **In the Export Options dialog box, make sure that the Use View Size check box is deselected.** To get to the Export Options dialog box, follow Steps 1 through 6 in the section "Exporting a Raster Image from SketchUp," earlier in this chapter.
2. **Decide on the resolution that you need for your exported image. (Refer to Table 13-2.)** Remember the resolution or record it.
3. **Decide how big your exported image will be printed, in inches or centimeters.** Note your desired physical image size, just like you did with the resolution in the previous step.
4. **Multiply your resolution from step 2 by your image size from Step 3 to get the number of pixels you need to export.** In other words, if you know what resolution you need to export, and you know how big your image will be printed, you can multiply the two numbers to get the number of pixels you need. Here's an example:

   300 pixels/inch × 8 inches wide = 2,400 pixels wide

   To export an image that can be printed 8 inches wide at 300 ppi, you need to export an image that's 2,400 pixels wide. Figure 13-5 illustrates this example.

Figure 13-5

8 inches wide x 300 ppi = 2400 pixels

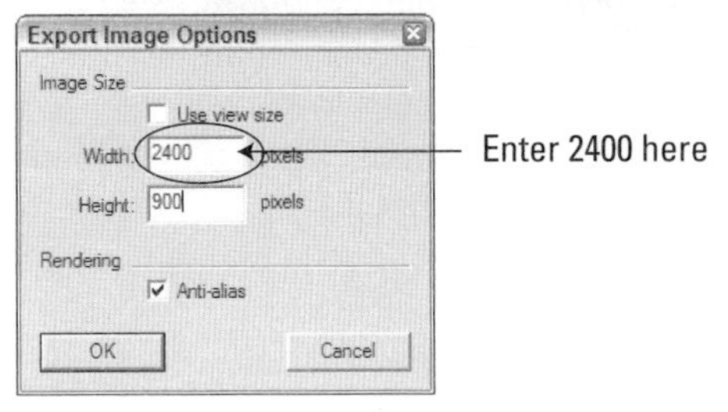

To figure out how many pixels you need to export, multiply the resolution by the physical size.

SketchUp's default setting is to make your exported image match the proportions of your modeling window; that is, you can only type in a width *or* a height, but not both. If you are on a MAC, to manually enter both dimensions, click the Unlink button (which looks like a chain). You can always click it again to relink the width and height dimensions later.

5. **Type in the width *or* height of the image you'd like to export, in pixels.** It's usually difficult to know *exactly* how big your image will be when it's printed, and even if you do, you probably want to leave some room for cropping. For these reasons, you may want to add 15–25 percent to the number of pixels you think you'll need. If your image calls for 2,400 pixels, export 3,000 pixels, just to be safe.

   If you're on a Mac, things are a little easier, because SketchUp's designers built a pixel calculator right into the Export Options dialog box. Just enter your desired resolution in the appropriate spot, change the width and height units from pixels to inches or centimeters, and type in your desired image size. SketchUp does the arithmetic for you.
6. **Click the OK button to close the Export Options dialog box.**
7. **Navigate to where you want to store the image, then click Export.**

IN THE REAL WORLD

Damon Leverett is an architect who has been using Google SketchUp for nearly 10 years. He first decided to use this program because of its great 3D capabilities and also because using it would create competitive advantages for both himself and his colleagues.

Currently, his primary model for presentation within this program is the exporting of JPEG images. According to him, "these images are then made available to a variety of staff for either publication or for viewing on MS PowerPoint, or [any] other presenting platforms." He believes that since many non-designers, such as marketing staff, utilize a variety of desktop software like the Adobe Suite platform, JPEG is a great common denominator for communicating with staff of varying backgrounds.

He also likes to use the animation tools in Google SketchUp to create small videos to present to his clients. There is no doubt in his mind that the ability to examine 3D form in real-time with shadows and perspective is extremely helpful in the architectural world for both understanding ones concept and in presenting to others.

SELF-CHECK

1. Raster images can be resized larger or smaller without affecting their image quality, while vector images lose quality if you enlarge them too much. True or false?
2. Why would it be a good idea to export an image file would be 15 to 25 percent bigger than what we think we will need it?
   a. So that if we lose the image can recover it easily.
   b. So that we have some room to crop it.
   c. So that we can modify the saturation and brightness.
3. When file size is a concern, you should choose:
   a. JPEG
   b. TIFF
   c. BMP
   d. PNG
4. When image quality is important, you should choose:
   a. JPEG
   b. TIFF
   c. BMP
   d. PNG

Apply Your Knowledge Choosing the correct resolution and image size is crucial in order to create presentations that can be viewed appropriately in the platform of your choice. One common mistake that novice designers often do is to capture, save, and retouch views for a project without opening them later in the program they were exported to.

Using one of the last projects you have worked on in past chapters and create a scene for each of the views you want to save. Then, save each scene as a raster image three times in PNG format, in three different image sizes:

- A small image for a Web site
- An image for a brochure or pamphlet
- An image for a presentation board

13.2 MAKING MOVIES WITH ANIMATION EXPORT

GOOGLE SKETCHUP IN ACTION

Learn how to prepare and export your model using the Animation Export feature.

What's so great about animation export is how *easy* it is to do. That's not to say that animation and digital video are simple topics—they're not. If you are primarily interested in 3D modeling, what you'll find in the following sections are instructions for doing what you need to do.

13.2.1 Getting Ready for Prime Time

The key to exporting animations of your SketchUp models is using scenes; see Chapter 10 for information on scenes. Scenes are saved views of your model that you can arrange in any order you want. When you export an animation, SketchUp strings together the scenes in your model to create a movie file that can be played on just about any computer made in the last several years.

Follow these steps to get your model ready to export as an animation:

PATHWAYS TO... PREPARING YOUR MODEL TO EXPORT AS AN ANIMATION

1. **Create scenes to build the "skeleton" of your animation.**
2. **To adjust the animation settings in the Model Info dialog box, choose Window ⇨ Model Info and then select the Animation panel.** All the controls are explained in the section about moving from scene to scene in Chapter 10.
3. **Click the Enable Scene Transitions check box to tell SketchUp to move smoothly from one scene to the next.**

(*Continued*)

PATHWAYS TO... PREPARING YOUR MODEL TO EXPORT AS AN ANIMATION *(Continued)*

4. **Enter a transition time to tell SketchUp how long to spend moving between scenes.** If your Scene Delay is 0 (see below), you can multiply your transition time by your number of scenes to figure out how long your exported animation will be.
5. **Enter a scene delay time to pause at each scene before moving on to the next one.** If you plan to talk about each scene, use this feature. If your animation is supposed to be a smooth walk through or flyover, set the scene delay time to 0.
6. **Adjust the proportions of your modeling window to approximate the proportions of your movie.** Unlike SketchUp's 2D export formats, the proportions of your exported movie don't depend on those of your modeling window; that is to say, making your modeling window long and narrow won't result in a long and narrow movie. You choose how many pixels wide and tall you want your movie to be, so to get an idea of how much you'll be able to see, make your modeling window match the proportions of your exported file (4:3 is common for video formats). See Step 1 in the section "Exporting a Raster Image from SketchUp," earlier in this chapter, for guidance on adjusting your modeling window.
7. **When your project is ready to go, move on to the next section to export your animation.**

13.2.2 Exporting a Movie

Fortunately, you have only one choice if you want to export a movie from SketchUp. If you're using Windows, you create an AVI file; Mac users create QuickTime MOVs.

If you're paying close attention to the available file formats for exporting movies, you'll probably notice three more choices in the drop-down menu: TIF, JPG, and PNG. You probably won't need these formats for animation (movie) export in SketchUp; choosing to export in any of these formats will give you a set of image files that each represents one frame in your animation. People who want to include their SketchUp animation in a Flash file should take advantage of this option, but explaining how to do so is beyond the scope of this book.

While exporting animations in SketchUp is a pretty simple operation, figuring out how to set all the animation export controls can be intimidating. What follows are step-by-step instructions for generating a movie file; settings recommendations are in the next section.

Figure 13-6

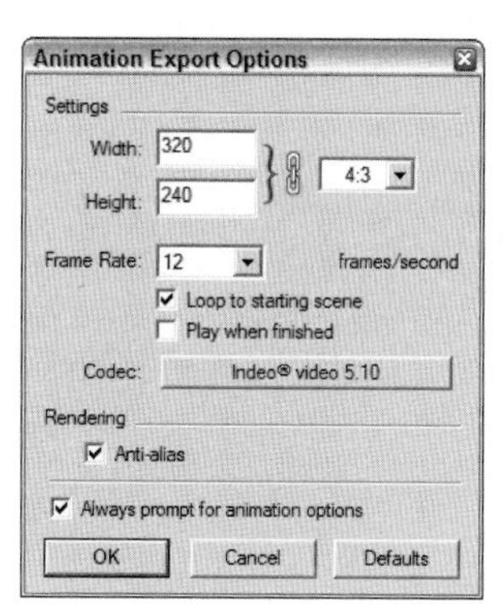

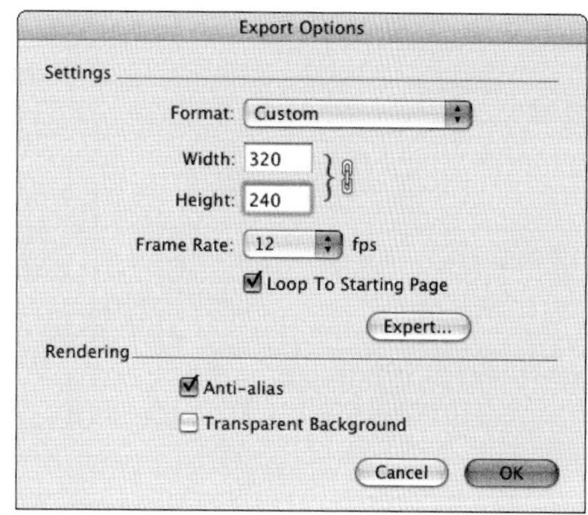

The Windows (left) and Mac (right) versions of the Animation Export Options dialog box.

Follow these steps to export a movie file from SketchUp:

1. **Prepare your model for export as an animation.** See the section "Getting Ready for Prime Time," earlier in this chapter, for a list of things you need to do before you export an animation.
2. **Choose File ⇨ Export ⇨ Animation.** This opens the Export Animation dialog box.
3. **Give your movie file a name, and choose where it should be saved on your computer system.**
4. **Make sure that the correct file format is selected.** In the Format drop-down menu, choose AVI if you're using Windows and QuickTime if you're on a Mac.
5. **Click the Options button to open the Animation Export Options dialog box (see Figure 13-6).**
6. **Adjust the settings for the type of animation you want to export.** How you set everything up in this dialog box depends on how you plan to use the animation you end up creating. See the next section in this chapter for recommended settings for different applications.

   If you're working on a Mac, there's an extra drop-down menu that you might find helpful: Format includes a short list of uses for your animation. Choosing one automatically sets most of the controls for you, though (as you'll see in the next section) you can improve things a bit by making some of your own selections.
7. **Select the Anti-alias check box, if it isn't already selected.** Choosing this doubles the amount of time it takes for your animation to export, but it makes your edges look much better in the final movie.
8. **Click the Codec button (Windows) or the Expert button (Mac).** This opens the Video Compression (Compression Settings on a Mac)

Figure 13-7

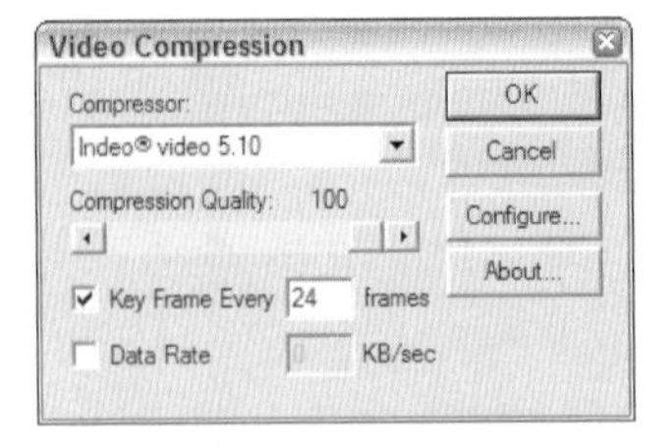

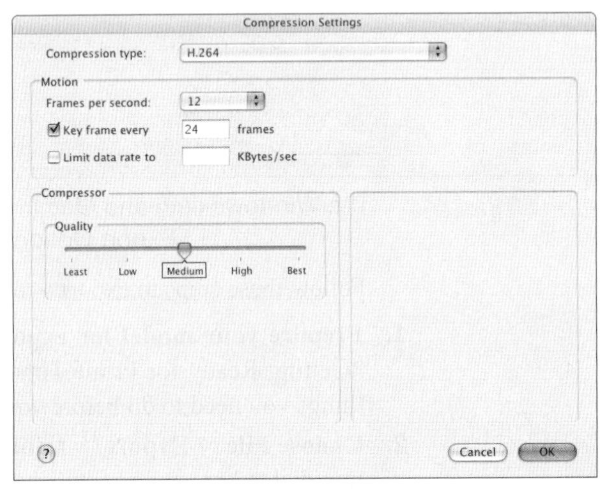

The Video Compression dialog box for Windows (top) and Mac (bottom).

dialog box (see Figure 13-7). Choose the correct settings for the type of animation you want to export, again referring to the next section of this chapter for details about what the options mean.

9. **Click the OK button in the compression dialog box, and then the OK button again in the Export Options dialog box.** This returns you to the Export Animation dialog box.

10. **Check to make sure that everything looks right, and then click the Export button.** Because exporting an animation takes a while, it pays to double-check your settings before you click the Export button. When the export is complete, you can find your animation file in the location you specified in step 3. Double-clicking it should cause it to open in whatever movie-playing software you have that can read it. On Windows computers, this is usually Windows Media Player; on Macs, it's QuickTime.

13.2.3 Figuring Out the Animation Export Options Settings

Digital video is complicated. Luckily, you don't have to know what everything means to export the right kind of movie; you just have to know how to set up everything.

What follows are a number of different things you might want to do with your animation, and recommended settings for getting good results. Feel free to experiment, but the following sections are a good place to start.

For Sending in an E-mail

If you're going to e-mail someone an animation file, you have to make the file as small as you can. These settings can help you do just that:

- **Width and Height:** 160 × 120
- **Frame Rate:** 10 fps
- **Codec (Windows):** Indeo Video 5.10
- **Compression Type (Mac):** H.264
- **Key Frame Every:** 24 frames
- **Compression Quality (Windows)**: 50
- **Quality (Mac):** Medium

For Uploading an Animation to YouTube

YouTube (www.youtube.com) is a video-sharing site that can host your animations for free. After your video is on YouTube, you can link to your video and even embed it on your own Web pages. You need to keep two things in mind when you create a video for YouTube: Videos need to be less than 1GB in file size, and they need to be less than 15 minutes in length. These settings yield a YouTube-ready video:

- **Width and Height:** 1280 × 720
- **Frame Rate:** 30 fps
- **Codec (Windows):** Indeo Video 5.10
- **Compression Type (Mac):** H.264
- **Key Frame Every:** 24 frames
- **Compression Quality (Windows):** 50
- **Quality (Mac):** Medium

For Viewing an Animation On-Screen (Computer or Projector)

If you plan to use your animation as part of an on-screen presentation (such as with PowerPoint or Keynote), you probably want it to look good full-screen. You'll probably be using a digital projector to present, and these days, most digital projectors come in two resolutions: 800 × 600 and 1024 × 768. You may know the resolution of the projector you'll be using, but if not, export at the lower pixel count, just to be safe:

- **Width and Height:** 800 × 600 or 1024 × 768
- **Frame Rate:** 15 fps
- **Codec (Windows):** Indeo Video 5.10

- **Compression Type (Mac):** H.264
- **Key Frame Every:** 24 frames
- **Compression Quality (Windows):** 100
- **Quality (Mac):** Best

You want your exported animations to look smooth—the transitions from one frame to the next shouldn't be jumpy or awkward. If your camera is covering a lot of ground (in other words, moving a large distance between scenes) in a very short time, you might want to experiment with increasing your frame rate to smooth things out. Doing so adds more frames between transitions, which means the camera isn't traveling as far between frames.

For Exporting to DV (To Be Viewed on a TV with a DVD Player)

If you need to export an animation that will be burned onto a DVD that will (in turn) be played in a DVD player, you should do everything you can for quality and file size. The export process will take a long time, but you'll get the best-looking movie you can get. Try these settings first:

- **Width and Height:** 720 × 480
- **Frame Rate:** 29.97 fps
- **Codec (Windows):** Full Frame
- **Compression Type (Mac):** DV/DVCPRO
- **Compression Quality (Windows):** 100
- **Quality (Mac):** Best
- **Scan Mode (Mac):** Interlaced

IN THE REAL WORLD

Ericka Violett is an interior designer who works at Benning Design Associates in Sacramento and started using Google SketchUp while still studying design at California State University. This allowed her to get comfortable with SketchUp even before entering the field.

When working on a large presentation for a client, she likes to use SketchUp primarily for walk throughs. In such cases, she enjoys creating walk-through animation videos of the 3D models and then works on photo-realistic renderings of focal spaces with SU Podium plugin.

On a daily basis though she normally uses exported 2D graphics to convey preliminary design intent or to update a client on design decisions. She believes that SketchUp is an extremely powerful tool to be used to get the clients to buy on the project as they often have a very positive reaction to it.

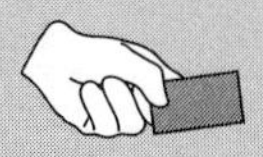

CAREER CONNECTION

Eric Schimelpfenig is an interior designer who works at Classic Kitchens in Springfield. He started using Google SketchUp because it is the easiest and best looking 3D tool that he could find in the market.

When preparing his presentations to clients, he normally exports the images to PDF and then presents his designs on a big screen. He goes on saying that the response is usually great.

SELF-CHECK

1. What is the key to exporting animations of your models?
2. What different saving extensions are used by Windows and Macs when you are creating a movie file?
3. The proportions of your exported movie depend on those of your modeling window. True or false?
4. If your camera is covering a lot of ground in a very short time, you might want to experiment with increasing your:
   a. Frame rate
   b. Movie size
   c. Pace of travel
5. What check box would we select if we want our edges to appear smoother in the final movie?

Apply Your Knowledge Frequently, creating an animation is the right way to go to convince a client that our design proposal is just what they needed. But planning ahead to have the right amount of scenes in our project is key to our success and that takes some practice and time to create the right pace in between transitions.

Build a very simple one-story hip-roof house with a front entrance and one window on each façade and a path leading to the front door. Then apply photo textures and/or colors to your design, shadows and other features, and finally zoom out. Capture a collection of scenes to create a small movie that would zoom into the house and then walk-through to see it from all four corners. Be particularly careful in making sure that turning the corners around your model will be done without jumps or sudden movements.

SUMMARY

Section 1

- Raster images are made up of dots.
- Vector images consist of instructions written in computer code.
- When image quality is important, choose the TIF/TIFF format.
- When file size is a concern, choose JPEG.
- The more pixels your image has, the more detailed it is, and the bigger it can be displayed or printed.

Section 2

- The key to exporting animations of your SketchUp models is using scenes.
- You have only one choice if you want to export a movie from SketchUp:
  - Windows users create an AVI file
  - Mac users create QuickTime MOVs
- You need to keep two things in mind when you create a video for YouTube:
  - Videos need to be less than 1GB in file size
  - Videos need to be less than 15 minutes in length

ASSESS YOUR UNDERSTANDING

SUMMARY QUESTIONS

1. Rasters and Vectors can both be scaled without losing image quality. True or false?
2. Which format compresses image data without affecting image quality?
   a. TIFF
   b. JPEG
   c. PNG
   d. BMP
3. Which files can only be used with Windows?
   a. TIFF
   b. JPEG
   c. PNG
   d. BMP
4. What type of movie file must Mac users export from SketchUp?
   a. AVI
   b. JPEG
   c. TIFF
5. When you export an animation, SketchUp strings together the scenes in your model to create a movie file. True or false?
6. What is the standard key frame rate for a movie?

APPLY: WHAT WOULD YOU DO?

1. You want to export your animation to a DVD so you can send it in to a design competition. How do you do this, and what settings do you use?
2. When should you use PNGs, TIFFs, and JPEGs?

BE AN ARCHITECT

Exporting a High-Resolution Image

Choose a view of your model house, and export it as a TIFF that can be printed at least 14 inches wide at a good resolution.

Posting Your Animation on the Web

Create an animation of your model house, and post it on the Web for your classmates and colleagues to view. Be sure to use settings that will optimize the quality of your animation for the web.

E-mailing Your Animation

Create an animation of your model house, and e-mail it to yourself and others. Be sure to use settings that will optimize the quality of your illustration for e-mail.

KEY TERMS

Anti-aliasing
Artifacts
Lossy
Resolution

CHAPTER

14 EXPORTING TO CAD, ILLUSTRATION, AND OTHER MODELING SOFTWARE

Using SketchUp Pro

Do You Already Know?

- About vector graphics?
- How to generate 2D files for CAD and illustration software?
- How to export your model to a variety of 3D software?

For the answers to these questions, go to **www.wiley.com/go/chopra/googlesketchup2e**

	What You Will Find Out	What You Will Be Able To Do
14.1	**How to export your model.**	Export your model to other software.
		Understand and identify different file types for export.
14.2	**How to export 3D data to of other software.**	Differentiate among various file formats and their use.

INTRODUCTION

This chapter and the following apply to SketchUp Pro only. However, you might find it useful to peruse it. However, if you're using Google SketchUp (free), and you're wondering what's in Pro, you might find it useful to see what you're missing.

SketchUp Pro users have access to a few file export formats that aren't available in the free version of SketchUp. These file formats let you share your work with other "pro-grade" software programs like Illustrator, AutoCAD, Revit, and 3ds Max. Most people who design things for a living use a number of different pieces of software to get their work done, and they need the ability to move their data between them. That's where SketchUp Pro's exporters come in.

This chapter is divided into two halves: The first part talks about SketchUp Pro's 2D export formats, and the second part deals with the 3D ones. For each file format that SketchUp Pro exports, this chapter includes a description of what the settings do, and in some cases, why you might need them. Which format (or formats) you choose depends entirely on what other software you are using.

14.1 EXPORTING DRAWINGS IN 2D

Most people who design 3D objects eventually need to create 2D views of their designs to be used in specific applications they own, such as Illustrator, AutoCAD, or Revit. Sometimes these views are for presentations, and sometimes you need to import a 2D view into other software programs, where you can continue to work on it. You can use SketchUp Pro's 2D export formats to do both.

Note: The case of re-using SketchUp Pro models in Revit is different. While it is possible to use your SketchUp Pro model in Revit as a starting point, it might make more sense in some cases to start your 3D model directly in Revit, especially if you are fast and accurate on that software.

GOOGLE SKETCHUP PRO IN ACTION

Learn how to do a basic export.

14.1.1 Sizing Up the Export Formats

You can find a lot of other software out there, and luckily, SketchUp Pro provides enough export formats that you can interact with most of it. On the 2D export side, here's a brief rundown on what SketchUp Pro has to offer.

- **PDF:** Lately, Portable Document Format (PDF) files have become more and more common. Notwithstanding that almost anyone can read them, PDF files are great for sending information to vector-illustration programs like Adobe Illustrator.
- **EPS:** Encapsulated Postscript (EPS) files are what people previously used to transfer vector information, but these days, increasingly more people are using PDFs.

- **DWG:** This is AutoCAD's native file format, and it's the best one to use for transferring information to that program and other pieces of CAD software. DWGs can also contain 3D information, so they are discussed in the second part of this chapter, too.
- **DXF:** Document Exchange Format (DXF) is another type of DWG. It was developed by Autodesk to be the file format that other pieces of software use to transfer data into AutoCAD, because they weren't supposed to be able to create DWGs. The trouble is, DWG has been *reverse-engineered* (taken apart, figured out, unlocked, and put back together again), such that most CAD programs can exchange both DXF and DWG files.
- **EPIX:** You may have never heard of *Piranesi*, but you should know about it. It's a piece of software that lets you "paint" on top of 2D views of your model. EPIX lets you open 2D views of your SketchUp Pro model in Piranesi, if you have it.

The sections that follow discuss each of the formats in more detail.

14.1.2 Exporting a 2D Drawing in Any Format

Regardless of which 2D format you choose to export, the procedure is always the same. Follow these steps to export a 2D image from SketchUp Pro:

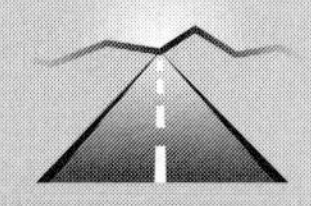

PATHWAYS TO...
EXPORTING A 2D DRAWING

1. **Adjust your model view until you have the view you want to export.**
2. **Choose File ⇨ Export ⇨ 2D Graphic.** This opens the File Export dialog box.
3. **Choose the file format you'd like to use from the Format drop-down list.** Each format is described in a fair amount of detail later in this chapter.
4. **Choose a name and a location on your computer for your exported image.**
5. **Click the Options button.** This opens the Export Options dialog box for the file format you chose in Step 3. The options are unique to each file format, so see the section on the one you're using for details on that particular format.
6. **Adjust the settings in the Export Options dialog box, and click the OK button.** This closes the Export Options dialog box.
7. **Click the Export button to export your 2D image file.**

14.1.3 Getting to Know PDF

Learn about different file types for export.

You've probably already heard of PDF files. In the past (years ago), it was hard to send someone a digital graphics file because so many different kinds were available, and because the person to whom you were sending the file had to have the right kind of software to be able to open it. For example, if you made a brochure in QuarkXPress and wanted to send it to someone to review, that person had to have QuarkXPress, too—and chances were, that person didn't.

Adobe developed PDF to solve this problem, and it's been quite successful. Because PDF reader software is already installed on tens of millions of computers, anyone can open and view a PDF file created by anyone else. Today PDF documents are universally accepted on computers and on any kind of small devices such as smartphones or electronic book readers but at the same time they are becoming more of a security risk and target for viruses and malware.

Lots of programs (including SketchUp Pro) can export PDFs, so now there's an easy way for people to share graphics files.

Here are some things to consider about PDF and SketchUp Pro:

- **PDF is universal.** Anyone with Adobe Reader (which is free from Adobe's website, www.adobe.com) can open and view a PDF created with SketchUp Pro. In fact, anyone with a Mac can open a PDF just by double-clicking it; the picture-viewing software that comes with the Mac operating system uses PDF as its native format.
- **PDF is consistent.** When you send someone a PDF file, you can be 99 percent sure that it will look just like it looks on your computer. The colors, line weights, and text will all stay the same, which is important if you're showing someone a design.
- **When you need to send a vector graphic file, use PDF unless someone specifically requests an EPS file.** Just about every piece of software handles PDFs with no problem. If you or someone else needs a vector export from SketchUp Pro, PDF works just about every time. Also, if you send a PDF or EPS of a plan or an elevation that has materials, the materials (i.e., tiles, bricks…) will not show up. If you want to show a plan or elevation with materials for a presentation for example, the PNG format will work better.

14.1.4 Getting to Know EPS

Postscript is a computer language that was developed to describe graphical objects. An EPS is an *encapsulated* Postscript file, meaning that it's a self-contained bundle of instructions for a how to draw an image. Back before PDFs existed, this was the best way to move vector information around. Today, you don't have much reason to use EPS when you could export a PDF file instead. In any case, here are some things you should know about the EPS file format:

- **EPS is complicated.** EPS files are different depending on what software and what operating system made them and some computer programs

FOR EXAMPLE

VECTORS: PICTURES MADE OF MATH

Remembering the location and color of millions of pixels isn't the only way that computers save images; it just happens to be the most common. The alternative to raster imaging is *vector graphics*, or *vectors* for short. In a vector, lines, shapes, and colors are described by mathematical functions. You don't need to do any math to create a vector. Your hardware and software do it for you. Here are some things to think about when it comes to vectors:

- **Vectors don't take up much space.** Imagine a straight line drawn using pixels. In order to draw the line, your computer has to be told where to put each and every pixel. In a vector image (which doesn't use pixels), a mathematical function is used to tell your computer how to draw that same line. Instead of having to provide details for every single pixel, only two instructions are needed: the function that defines the line and its color. Of course, actual vector graphics programming is more complicated than this, but you get the general idea. Thus, vector files are much smaller than raster files.
- **Vectors are scalable.** With a raster, you're limited by the number of pixels you have in your image. If you don't have enough, you might not be able to print or otherwise display your image very big. If you have a large number of pixels, your image might look terrific, but your file might be too large to work with. Because vectors are math-based instructions that tell your software how to draw an image, there's no size limit to how big—or small—your image can be. With the same vector file, you could print your company logo on your business card and on the side of a blimp, and they would both look great. The figure that follows shows what this means.

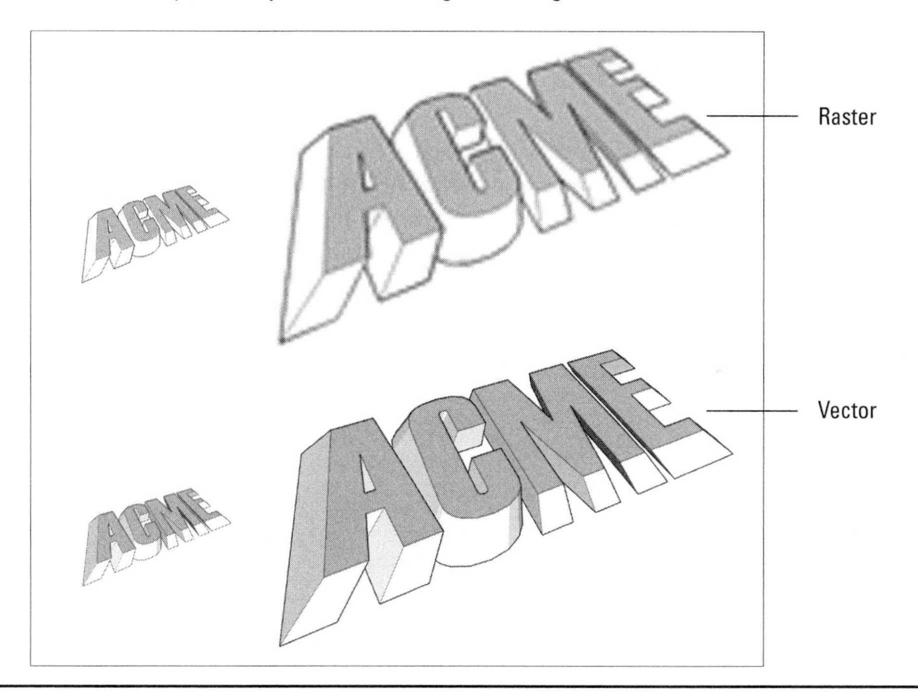

still accept them, but you can't edit or manipulate them. You will have better chances of working with them in programs such as Microsoft Word, Illustrator, or a layout program like InDesign, which most people don't have.

- **Use EPS if your other graphics software is more than a couple of versions old.** PDF support has only been widely incorporated in the last few years, so if you're using older image-editing, illustration, or layout software, you might need to use EPS.

- **If someone you're working with insists on EPS, go ahead and send an EPS file.** Some workflows have been designed with EPS files in mind, which is why SketchUp Pro exports EPS in the first place. It's good to know that EPS is there if you need it.
- EPS files can be opened in Photoshop but if you want to add materials, some touch-up will be necessary.

14.1.5 Navigating the PDF/EPS Options Dialog Box

Figure 14-1 shows what this dialog box looks like. It's the same for both PDF and EPS exports.

In Windows, the name of this dialog box, as written in its title bar, is PDF (or EPS) Hidden Line Options.

Drawing (Image) Size

You use these settings to control the physical dimensions of your exported image. If you want to produce a PDF that's a particular size, this is where you do it. You have these options:

- **Width and Height:** These controls are for telling SketchUp Pro how big you want your exported image to be. Because the proportions of your image will be the same as those of your modeling window, you can enter a width or a height, but not both. If you're on a Mac, you can click the Maintain Aspect Ratio button (it looks like a chain). Doing so allows you to enter both width and height, which is useful for exporting to common document sizes like letter and tabloid.

Figure 14-1

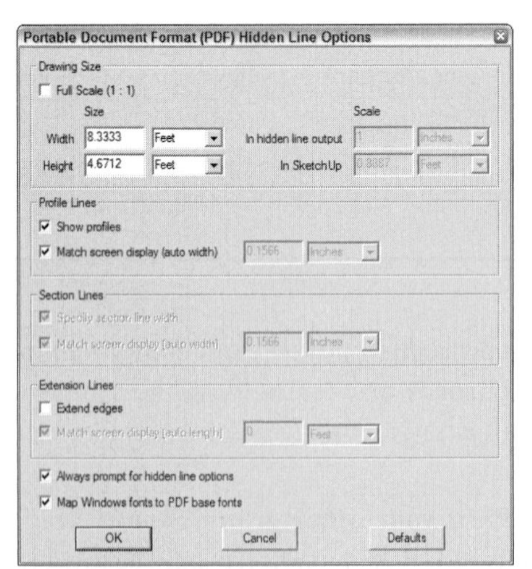

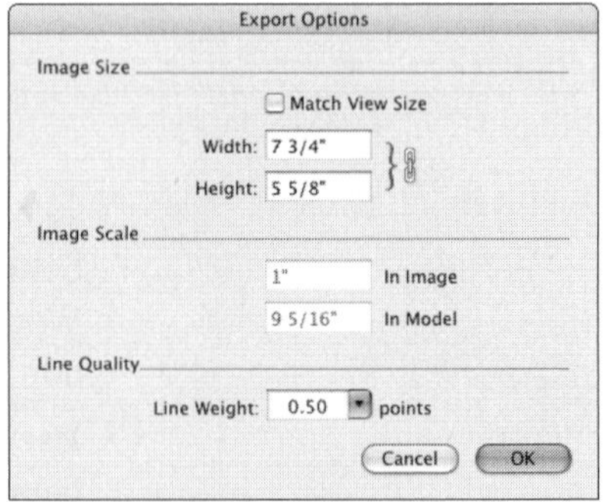

The PDF/EPS Options dialog box for Windows (left) and the Mac (right).

- **Match View Size:** (Mac only) Sometimes certain features just don't work; it's not even worth explaining what the Match View Size check box is supposed to do. *Always* leave Match View Size deselected.
- **Full Scale (1:1):** (Windows only) Selecting this check box tells SketchUp Pro to export a drawing that's the same size as your model. You can use this option if you want a full-scale image of something relatively small, like a teapot or a piggy bank. If you end up using Full Scale, remember to follow Steps 1 through 3 below, "Setting Up for Scaled Drawings," to make sure that your drawing exports the way you want it to.

PATHWAYS TO...
SETTING UP FOR SCALED DRAWINGS

1. Before you go anywhere near the 2D Graphic dialog box, switch from Perspective to Parallel Projection view. You do this by choosing Camera ⇨ Parallel Projection. It's impossible to make a scaled perspective view—drawings just don't work that way.
2. Switch to one of the standard views. Choose Camera ⇨ Standard and pick a view from the submenu to see a straight-on top, side, or other view of your model.
3. Use the Pan and Zoom tools to make sure that you can see everything you want to export. Whatever you do, don't orbit. If you do, you can always repeat Step 2.

Drawing (Image) Scale

Perhaps you're designing a deck for your neighbor's house, and you want to export a scaled PDF file so that the lumberyard can take measurements right off the printed drawing. You want your drawing scale to be 1/8 inch = 1 foot, and this is how you would specify that:

1. Enter 1" into the In Hidden Line Output (In Image) box.
2. Type 8' into the In SketchUp Pro (In Model) box.

If you want to export a scaled drawing from SketchUp Pro, you have to set up your model view a certain way. If you're not set up properly, the Scale controls are grayed out.

You can choose to set the Drawing (Image) Size or Scale, but not both. You can't, for instance, tell SketchUp Pro to export a 1/8-inch drawing as an 8 1/2-×-11-inch PDF file.

If you have SketchUp Pro 8, you also have a terrific companion app you can (and should) use: LayOut's main purpose is to make it easier to produce scaled PDFs from your 3D models. Before you go through the process of exporting a PDF from SketchUp Pro, spend a few minutes thumbing through Chapter 15 of this book to read LayOut's benefits.

Profile Lines (Windows only)

Profiles don't export very well to PDF and EPS, so you may only want to use this option if you're going to be tweaking your image in a vector-drawing program like Illustrator or Freehand. You have the following options:

- **Show Profiles:** Select this check box to export profile lines in your image, assuming that you're using them in your model.
- **Match Screen Display (Auto Width):** Selecting this check box tells SketchUp Pro to make the exported profiles look as thick, relative to other edges, as they do in your modeling window. If this check box isn't selected, you can type in a line width (thickness).

FOR EXAMPLE

WHAT YOU SEE ISN'T ALWAYS WHAT YOU GET

With their ability to scale without losing detail and their (relatively) tiny file sizes, vector images seem ideal, but unfortunately, they're not perfect. Whereas exported raster images look just like they do on-screen, vector images don't. Most of SketchUp Pro's graphic effects can't be exported as vector information. Here's a list of what you give up with PDF and EPS:

- Photo textures and transparency on faces
- Edge effects like depth cue, endpoints, and jitter
- NPR (Sketchy Edge) styles
- Shadows and fog
- Background, ground, and sky colors
- Watermarks

Section Lines (Windows Only)

If you have section cut lines in your model view, these options become available:

- **Specify Section Line Width:** *Not* selecting this option is the same as choosing Match Screen Display (see the next point).
- **Match Screen Display (Auto Width):** Select this check box to export section cut lines that look like they do on your screen. If you have another thickness in mind, don't select this check box; just enter a width right beside it.

Extension Lines (Windows Only)

As the only edge-rendering effect (besides profiles) that SketchUp Pro can export to PDF and EPS, extensions are important; this is where you control them. You have the following options:

- **Extend Edges:** Even if you have edge extensions turned on in your model, you can choose not to export them by deselecting this check box.
- **Match Screen Display (Auto Length):** Select this check box to let SketchUp Pro make your extensions look like they do on your computer screen. To put in a custom length, type one into the box on the right.

Line Quality (Mac Only)

Whereas Windows users get to be very specific if they want to, Mac users only have one point of control over exported line thickness, or weight. For very detailed models, turn the Line Weight setting down to 0.5 or 0.75 points; otherwise, leave it at 1.00 and see whether you like it. This setting depends on the complexity of your model and on your personal taste.

Other Windows-Only Controls

You have a couple of other options in this dialog box; both can be very helpful:

- **Always Prompt for Hidden Line Options:** Do yourself a favor and select this check box; it never hurts to look over your options before you do an export.
- **Map Windows Fonts to PDF Base Fonts:** Even though PDF almost always preserves the original appearance of your image, fonts are tricky. For this reason, the PDF file format comes with a set of "safe" fonts that will work on any computer, anywhere. Choosing this option tells SketchUp Pro to "map"—substitute—PDF-safe fonts for the ones in your SketchUp Pro file. Unless you're completely satisfied with the fonts you originally chose, it's a good idea to select this check box.

14.1.6 DXF and DWG (in 2D)

DXF and DWG are the only 2D CAD formats that SketchUp Pro can export. Just about every CAD program in existence can do *something* with either DXF or DWG, so your bases should be covered, no matter what you're using. Here are some things to keep in mind:

- **DWG is more capable than DXF.** Because the former is AutoCAD's *native* (private) file format, and DXF was developed by Autodesk to be an *exchange* (public) file format, DWG has more features. Thus, for exporting 2D drawings from SketchUp Pro to use in other CAD applications, using DWG usually yields better results.

- **Don't be afraid to experiment.** Data export from *any* program is a somewhat uncertain endeavor, and you never know what you're going to get until you try. Whenever you send your information from one piece of software to another, you may want to leave yourself an extra hour for troubleshooting. You can adjust settings until things work.
- **Don't get confused by all the version numbers.** Which version of DWG or DXF you decide to export depends on what CAD software you're using. In general, it's a good idea to use the most recent version available, which is DWG/DXF 2010 in SketchUp Pro 8. If your CAD program is older than that, try exporting an earlier version.

The most recent version of DXF or DWG that SketchUp Pro 8 can *import* is 2010. If you're working in AutoCAD 2011, you'll be able to open your DWG files just fine in SketchUp Pro, but should you have trouble, you can always save it as a version 2010 (or older) to bring it in.

Figure 14-2 shows what the DWG/DXF Hidden Line Options dialog box looks like, which contains your export options. The options are the same for both DWG and DXF, and the following explanation (the figure and the text), apply to both.

Figure 14-2

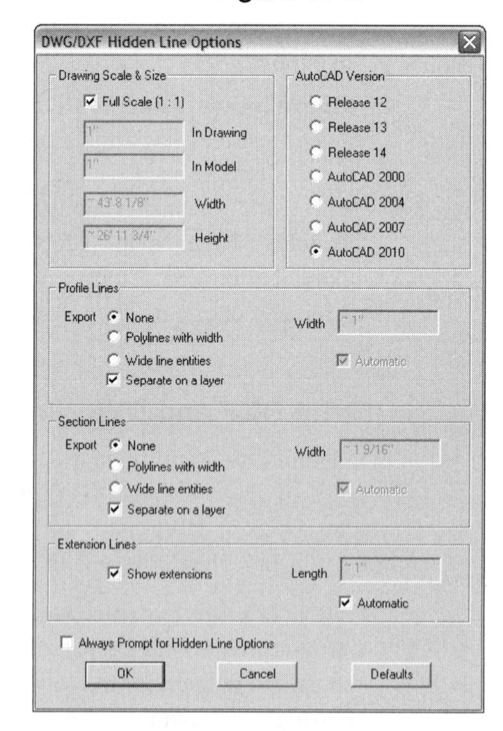

The DWG/DXF Hidden Line Options dialog box.

IN THE REAL WORLD

Again Juliana Diehl from Stúdio Arquitetura in Piracicaba, Brasil, says that in most cases importing AutoCAD files to Google SketchUp Pro helps quite a bit when building a 3D model in SketchUp Pro.

She begins by "cleaning" the entire floor plan in AutoCAD, keeping only the wall lines, the doors and windows dimensions, and the different types of flooring. The next step is to import this 2D drawing to SketchUp Pro, where she can easily elevate the floor plan to form the walls and openings. She goes on saying that after the 3D model is ready she exports back to AutoCAD as many elevations and section cuts needed to give a clear understanding of the project. According to her, by using SketchUp Pro to do part of the work, she speeds up the entire process while also making her job easier.

Drawing Scale & Size

These settings let you control the final physical size of your exported drawing. If you're in Parallel Projection view, you can assign a scale; if you're in Perspective view, scale doesn't apply. You have the following options:

- Full Scale (1:1): Most people use SketchUp Pro's DXF/DWG Export feature to produce nonperspectival, orthographic views of their models that they can use in their CAD programs. If that's what you are trying to do, you should select this check box; it will make opening your exported file in another program that much easier.

  To export a scaled view of your model, you need to set things up properly in your modeling window before you begin the export process. You need to be in Parallel Projection view, and you have to be using one of the standard views from the Camera menu. This is discussed in "Pathways to Setting Up for Scaled Drawings," earlier in this chapter. Keep in mind that unless you are set up properly, this option will be grayed out.
- **In Drawing and In Model:** If you're exporting a scaled drawing, and you haven't chosen the Full Scale option (see the previous point), you can set your drawing's scale using these controls. If you're not set up to export at scale, these settings won't be available. As an example, for a drawing at 1/16-inch scale, you would do the following:
  - Type **1"** in the In Drawing box
  - Type **16'** in the In Model box
- **Width and Height:** You can use these settings to determine the dimensions of your exported drawing, as long as you are not printing to scale.

Profile Lines

This is where you control how profile lines in your SketchUp Pro model view are exported. You have the following options:

- **None:** Exports profiles the same thickness as all your other edges.
- **Polylines with Width:** Exports profiles as polylines, which are a different kind of line object in CAD programs.
- **Wide Line Entities:** Exports profiles as thicker lines.
- **Width:** You can enter your own line thickness for exported profiles, or you can select the Automatic check box to tell SketchUp Pro to match what you see in your modeling window.
- **Separate on a Layer:** Puts your profiles on a separate layer in the exported file. This is handy for being able to quickly select all your profiles and give them a line weight when you open your exported file in a CAD program.

Section Lines

Section lines
Lines that occur where section planes create section cuts in a model.

Section lines occur where section planes create section cuts in your model. Traditionally, these lines are thick, which is why SketchUp Pro gives you control over how they export. The options in this section are identical to those for exporting section lines in the PDF/EPS Export Options dialog box. See the section "Section Lines (Windows only)," earlier in this chapter, for information on what everything in this part of the dialog box means.

Extension Lines

Extensions
The line overruns that you can choose to display in the Styles dialog box.

Extensions are the little line overruns you can choose to display in the Styles dialog box. If you want to include them in your exported file, you can. Just select the Show Extensions check box, and then either enter a length or select the Automatic check box to let SketchUp Pro try to match how they look on your screen. Extension lines are exported as several tiny, individual edge segments.

If you're using the Windows version of SketchUp Pro, you have an extra option in this dialog box: Always Prompt for Hidden Line Options. This simply means "Do you always want to see this dialog box when you're exporting to DXF or DWG?" Selecting this check box reminds you to look at these settings every time you export.

14.1.7 Peeking at EPIX

You only need to use the EPIX file format if you're using Piranesi, a great artistic rendering program that you can buy. Basically, EPIX is

FOR EXAMPLE

IT SLICES, IT DICES . . .

SketchUp Pro's Section Plane tool is quite helpful. See Chapter 10 for more information on this tool. If you have a section cut in your model that you'd like to export to a CAD program, follow these steps:

1. Make sure that the section cut you'd like to export is active.
2. Choose File ⇨ Export ⇨ Section Slice.
3. Click the Options button to open the Section Slice Export Options dialog box.
4. Set the export options the way you want them, and then click the OK button.
5. Click the Export button.

The most important part of the Export Options dialog box is right at the top: You need to choose either True Section or Screen Projection. See the following figure to see what happens when you choose each. Chances are, you'll want to pick True Section; it yields the most useful information. The top image is a screen shot of a modeling window in SketchUp Pro. The lower-left image shows what a file export as a true section looks like in AutoCAD. The lower-right image is an AutoCAD view of the same file exported as a screen projection. For a description of the controls in the rest of this dialog box, see the section "DXF and DWG (in 2D)," earlier in this chapter.

SketchUp modeling window

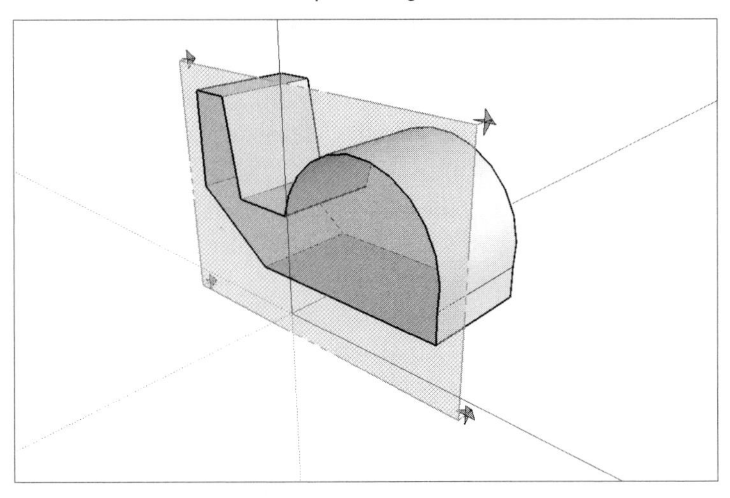

True section

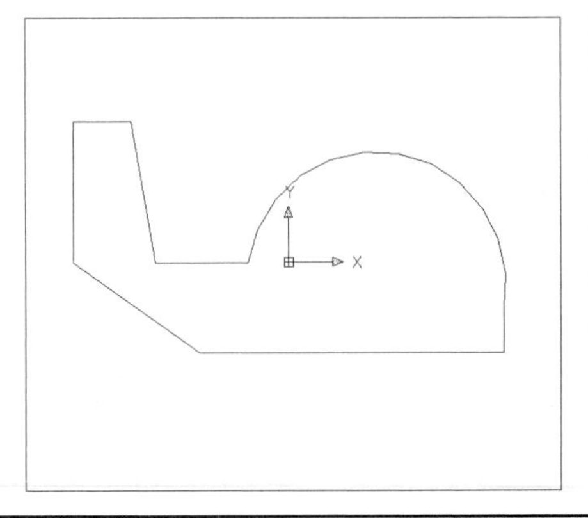

Screen projection

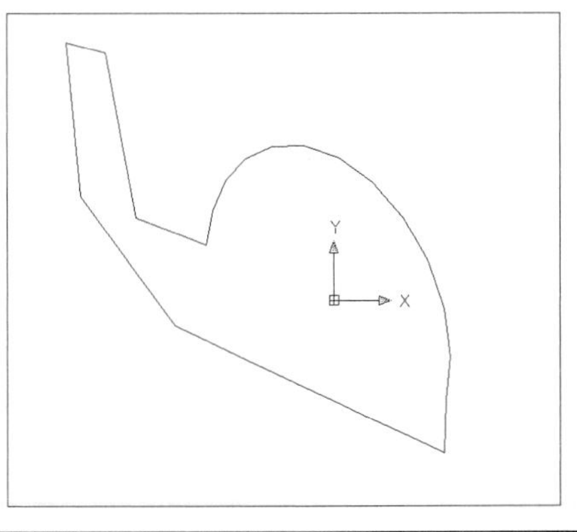

Figure 14-3

The Export Epx Options dialog box.

kind of a hybrid raster format that keeps track of pixels, just like other raster formats. But it also remembers another piece of information: the depth of each pixel in your scene. If you have a tree in the foreground, the pixels that make up that tree "know" they're, three feet away, for example.

Figure 14-3 shows the Export Epx Options dialog box. (Note that the file extension is pronounced Epix, but the Piranesi will save the files with the extension Epx). Here's what all the options do:

- **Image Size:** This part of the Export Epx Options dialog box is just like the Export Options dialog boxes you use for exporting 2D raster images (like JPEG, TIFF, and PNG). Chapter 13 includes an explanation of what everything means, and how to set things up properly.
- **Export Edges check box:** Choose this option to export the edge-rendering settings you currently have applied to your model.
- **Export Textures check box:** Piranesi can do some remarkable things with the textures you apply in SketchUp Pro; select this check box to export those textures as part of your EPIX file. Just a note: You need to have a style applied to your model that displays textures for those textures to be exported properly.
- **Export Ground Plane check box:** Exports a ground plane in your model view, regardless of whether a ground plane is in your currently applied style.

IN THE REAL WORLD

Chad Clary is an architect at Studio A Architecture in Louisville who works with Google SketchUp Pro because it is an easy to use software. According to him, he went from never using SketchUp Pro to being proficient in just one week.

When preparing presentations for his clients, he frequently creates animations circling his projects and providing context. He also exports still images, attempting to capture a project from a variety of angles so his clients can get a better feel of how the end result will be.

He goes on to say that most of his work is done in AutoCAD first and then imported to SketchUp Pro for a 3D model. He has just started experimenting with other rendering engines and says that his clients are always very receptive and happy with the final product.

SELF-CHECK

1. What is AutoCAD's native file format?
   a. PDF
   b. EPIX
   c. EPS
   d. DWG
2. The procedure for exporting a 2D file is always the same minus the fact of having to choose the format from a drop down list. True or false?
3. When you need to send a vector file, you should favor a PDF file over an EPS file because it is most widely used. True or false?
4. You only need to use the EPIX file format if you're using a rendering program called Piranesi. True or false?

Apply Your Knowledge Learning how to export your files correctly into other programs is a very important skill to master in Google SketchUp Pro. Quite often designers start their designs in Google SketchUp Pro, but later they refine their creations in other CAD programs, or finish their illustrations using software that will allow them to control precisely their color and graphic information.

Using Google's 3D Warehouse found under Window ⇨ Components, find a simple object such as a table or a chair and create one scene per view: Front, Side, and Top. Be sure to be in Camera ⇨ Parallel Projection and then go to Camera ⇨ Standard Views to choose each one.

After you have created your three scenes, export each one as a PDF file and import them into a single file in Adobe Illustrator. After you do that, export again the same three views as DWG files and import them into a single file in AutoCAD.

14.2 EXPORTING 3D DATA FOR OTHER SOFTWARE

The process of exporting 3D data from *any* program to *any* program is fraught with equal parts mystery and despair. There are so many pieces of software, with so many versions, and so many different file types, that finding a "step-by-step" recipe for success is practically impossible. The simple truth is that 3D file export from SketchUp Pro is about 50 percent knowing where to start and 90 percent trial and error after that.

14.2.1 Knowing Where to Start

Understand the various file formats and their uses.

This section focuses on the "knowing where to start" part. First, a list of 3D file types that you can export with SketchUp Pro is provided so you can make some decisions about which format you should use. Second, a general procedure for how to export a 3D model from SketchUp Pro is provided.

Examining Your 3D File Format Options

Here's a bit of information about the 3D file formats that SketchUp Pro 6 can export:

DAE (Collada)
A new 3D file format that is gaining acceptance with a large number of gaming and animation programs.

- **DAE (Collada):** Every couple of years or so, a new 3D file format comes out. More capable than 3DS, DAE (Collada) is being adopted everywhere by 3D software and gaming companies; Google has even chosen Collada as the file format for the 3D buildings in Google Earth. If you're using a recent version of any of the most popular 3D modeling programs, you might well be able to deal with Collada files; you should try it and find out.
- **DWG/DXF:** These are 2D formats. Ever since AutoCAD went 3D a few versions ago, its formats have gone 3D, too. These are a good choice for most of the CAD-type programs out there, but not as good a choice as 3DS. You need to know that, given the choice between the DXF and DWG, you should usually pick the latter.

3DS
One of the few standard 3D file formats.

OBJ
A 3D file format that can be used to send data to Maya.

XSI
An older 3D file format that is still used by some programs.

VRML
An older 3D file format that is still used by some programs.

FBX
A 3D file format used primarily by people in the entertainment industry who use Maya, 3ds Max, or Autodesk VIZ.

KMZ
The Google Earth native 3D file format.

- **3DS:** This has become one of the few standard 3D file formats in the industry. When in doubt, export a 3DS file and see whether your program can open it. SketchUp Pro's options for 3DS export are kind of complicated, so be sure to read the section about this later in the chapter.
- **OBJ:** OBJ is probably your best option for sending your data to Maya, which is now owned by Autodesk. While it doesn't offer some important things that 3DS does, it's still a pretty common 3D format.
- **XSI:** The folks at SketchUp Pro built XSI export into SketchUp Pro so that it would be more useful to people who use Softimage, a modeling/rendering/animation program from Canada that is now also owned by Autodesk.
- **VRML:** Though it's older, several people still use VRML to exchange 3D information.
- **FBX:** The FBX format is used primarily by people in the entertainment industry who use Maya, 3DS Max, or Autodesk VIZ (not many people will use VIZ now, as it got integrated into the newer versions of AutoCAD). Depending on what you're doing, you might want to use it instead of DAE (Collada), 3ds Max, or OBJ; try it and see whether you like it.
- **KMZ:** This is the Google Earth file format. Technically, the ability to export KMZ files isn't restricted to the Pro version of SketchUp Pro; anyone with regular, free SketchUp Pro can export them, too.

Exporting Your 3D Model

The process of exporting your 3D data from SketchUp Pro is the same no matter what file format you choose.

PATHWAYS TO...
EXPORTING A 3D MODEL FROM SKETCHUP PRO

1. **Choose File ⇨ Export ⇨ 3D Model.** This opens the File Export dialog box.
2. **Choose the file format you'd like to use from the Format drop-down list.**
3. **Choose a name and a location on your computer for your exported image.** Whenever you're exporting a 3D model from SketchUp Pro, it's a *very good* idea to create a new folder for your exported file. A lot of the file formats that SketchUp Pro exports save the textures in your model as separate files alongside your model. You'll want to have everything in a single folder.
4. **Click the Options button.** This opens the Export Options dialog box for the file format you chose in Step 2. Check out the sections in the rest of this chapter for more information about all the controls in the Export Options dialog box—they're anything but intuitive.
5. **Adjust the settings in the Options dialog box, and then click the OK button.** This closes the Export Options dialog box.
6. **Click the Export button to export your 3D model file.**

14.2.2 Getting a Handle on OBJ, FBX, XSI, and DAE (Collada)

The Export Options dialog boxes for each of the aforementioned file formats are almost identical, even if the file formats themselves aren't, which is why they are dealt with together in this section. Here are a few reminders about using OBJ, FBX, XSI, and DAE (Collada):

- **Try using Collada first.** Because it's relatively new, not all 3D programs support it yet, but your software might.
- **If Collada doesn't work, use OBJ for Maya, XSI for Softimage, and FBX if someone asks you to.** The sheer number of programs and versions out there means that it's impossible to give hard-and-fast rules.

- **Experiment.** Leave yourself some time to try exporting your model in more than one format. Then open each one in your other piece of software and see what works best.

Figure 14-4 illustrates the Export Options dialog boxes for all four file formats. Here's some help with what everything does:

- **Export Only Current Selection:** Tells SketchUp Pro to only export the geometry you currently have selected in your model.
- **Triangulate All Faces:** Some programs don't support faces with cutouts in them, so selecting this check box carves things into triangles so that you don't end up with any holes. Experimentation will tell you whether you need to use this; don't choose this option if you don't have to.
- **Export Two-Sided Faces:** SketchUp Pro's faces are two-sided, but not all 3D programs' faces are. Choose this option if you've spent a lot of time **texture-mapping** (painting with textures) your model in SketchUp Pro and you want it to look the same in the program you're sending it to. If

Texture-mapping
Painting with textures.

Figure 14-4

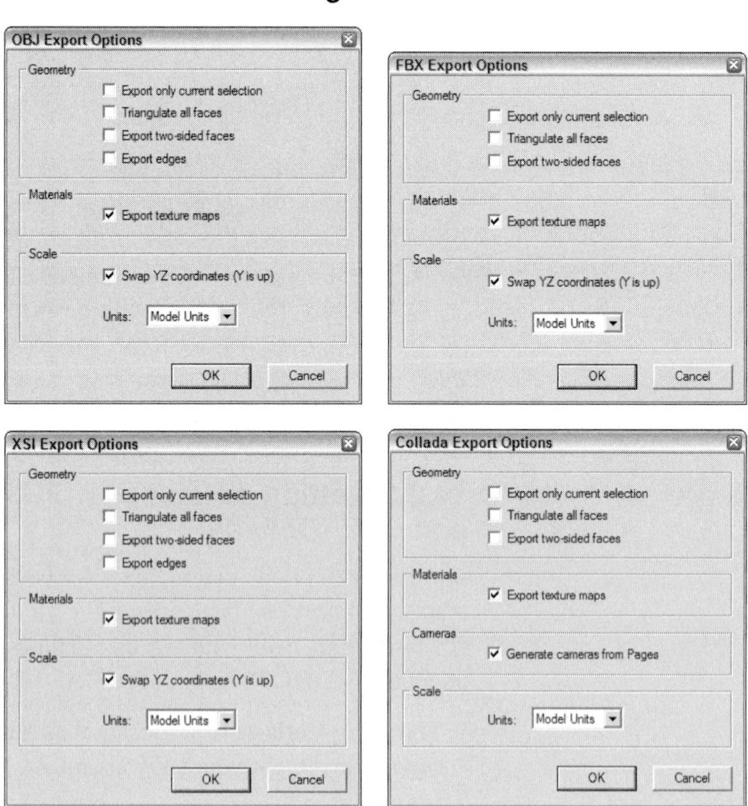

The Export Options dialog boxes for OBJ, FBX, XSI, and DAE (Collada).

you plan to use another piece of software to add textures to your model, don't select this check box.

- **Export Edges:** SketchUp Pro models *must* include edges and faces, but some programs' models only support the latter. Select this check box if you want to export the edges in your model; leave it deselected if you don't. Programs that don't support edges (and that includes most of them) will leave them out, anyway. The Export Edges option isn't available for FBX files.
- **Export Texture Maps:** This option includes the textures you used in your SketchUp Pro model in the exported file. If you plan to "paint" your model in another program, deselect this check box.
- **Generate Cameras from Scenes:** The Collada file format can store information about different views of your model. Selecting this check box exports each of your scenes (if you have any) as a separate camera object. Note that this option doesn't exist for OBJ, FBX, or XSI.
- **Swap YZ Coordinates (Y Is Up):** If your model ends up oriented the wrong way when you open it in another piece of software, try selecting this check box and exporting it again. Some programs set up their axes differently. This option isn't available for Collada.
- **Units:** Leave this option on Model Units unless something is wrong when you open your model in another program. If it is, make an adjustment and export again.

14.2.3 Wrapping Your Head around 3DS

3DS is almost guaranteed to work with just about any other piece of 3D software you're using, which is good. What's bad is that this flexibility comes at a price: a seemingly infinite number of options. Here are some things you should know about the 3DS file format:

- **You'll lose your layers.** 3DS doesn't support them, so you might be better off exporting a DWG file (if your other software can open it and if you aren't using textures). Another option is to use the Color by Layers option, which is described later.
- **You'll lose your edges, too.** You can always choose Export Stand Alone Edges, but few people recommend doing that.
- **Only visible faces get exported.** None of your hidden faces, or faces on hidden layers, will be exported, so make sure that you can see everything you want to export before you go to the File menu.
- **Make sure that you paint the correct side of your faces.** Faces (or the equivalent) in 3DS are one-sided, so only materials you apply to the front side of your faces in SketchUp Pro will end up getting exported. If you have materials on both sides of your faces, you might consider choosing the Export Two-Sided Faces option, described later in this section.

Figure 14-5

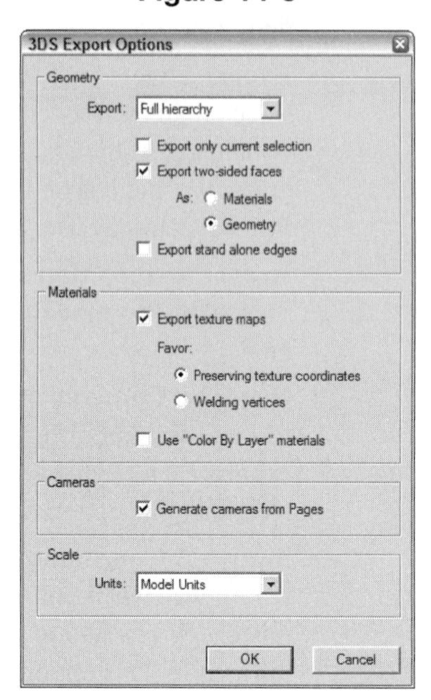

The 3DS Export Options dialog box.

Figure 14-5 shows the 3DS Export Options dialog box. Your options are as follows:

Meshes
Surfaces made out of triangles.

- **Export:** You have four choices here:
  - **Full Hierarchy:** This option tells the 3DS exporter to make separate **meshes** (surfaces made out of triangles) for each "chunk" of geometry in your SketchUp Pro model. Chunks are things like groups, components, and groupings of connected faces.
  - **By Layer:** This option tells the exporter to create separate meshes based on two things: chunks of geometry, and what layer things are on. If several faces are connected and they're on the same layer, they get exported as a single unit.
  - **By Material:** When you choose this option, you get a separate set for each grouping of connected geometry that shares the same material.
  - **Single Object:** Choosing this option exports all your geometry as one big 3DS unit.
- **Export Only Current Selection:** This option only exports the geometry you've selected in your SketchUp Pro model.
- **Export Two-Sided Faces:** Selecting this check box exports two faces (back to back) for every face you have in your SketchUp Pro model. Because

3DS supports only single-sided faces, this is necessary to preserve the appearance of your textures in your exported model. If you don't care about preserving your SketchUp Pro textures, or if you didn't apply any in the first place, don't bother choosing this option:

  - **As Materials:** Choose this option to export your back-side materials as 3DS materials without corresponding geometry.
  - **As Geometry:** Choose this option to export your extra set of faces as actual geometry. You should choose this option if you're wondering what to do.

- **Export Stand Alone Edges:** 3DS doesn't support edges the way SketchUp Pro does—you don't have a good way to export edges as "lines" to 3DS. When you select this check box, the exporter substitutes a long, thin rectangle for every edge in your model. The appearance is almost the same, but this can cause major problems in your file. You should not select this check box. If you really need to be able to see your edges, you should probably try another export format altogether.
- **Export Texture Maps:** If you have photo textures in your model (this includes the photo textures from the Materials dialog box), you might want to include them in your exported model file. 3DS handles these textures very differently than SketchUp Pro does, so you have to decide on an export method:
  - **Favor Preserving Texture Coordinates:** Choose this option if you've spent a lot of time getting the texture "maps" right in your SketchUp Pro model.
  - **Favor Welding Vertices:** Choose this option if it's more important that your geometry export as accurately as possible. In some cases, your textures won't look right, but your geometry will be correctly welded (stuck together) and smoothed.
- **Use "Color by Layer" Materials:** Because the 3DS file format doesn't support layers, you can choose to export your model with different colors assigned to the faces on each layer in SketchUp Pro.
- **Generate Cameras from Scenes:** Select this check box to export your file with a different camera position saved for each scene in your SketchUp Pro file. An extra scene (called Default Scene) is also exported to reflect your current model view.
- **Units:** If you leave this set to Model Units, most other 3D programs will understand what you mean—your geometry will appear the right size when you open your exported 3DS file. In some cases, it won't, and the best thing to do is to manually choose the units that you'll be using in the other program. Sometimes this doesn't work either, and you'll just have to experiment until something works.

Figure 14-6

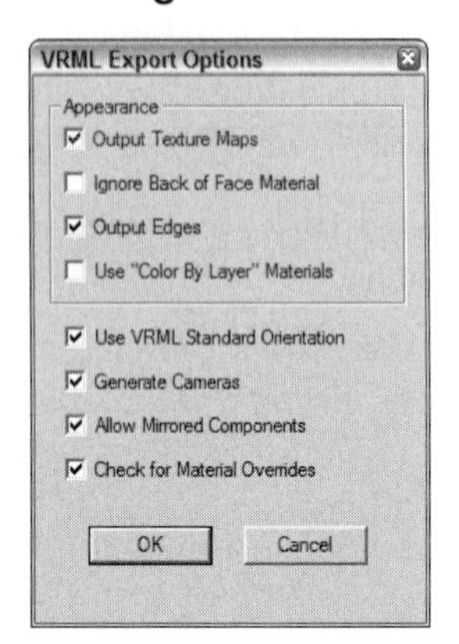

The VRML Export Options dialog box.

14.2.4 Dealing with VRML

The VRML file format is pronounced *vermal*. Virtual Reality Modeling Language is used by a large number of people around the world. There are newer, arguably better, formats out there, but VRML's been around long enough that it's tightly integrated into lots of professional workflows.

Figure 14-6 shows the VRML Export Options dialog box; what follows is some help with all the controls:

- **Output Texture Maps:** If you don't select this check box, you'll get colors instead of textures in your exported VRML file.
- **Ignore Back of Face Material:** Go ahead and select this check box unless your faces have different materials painted on either side of them.
- **Output Edges:** VRML supports edges, so select this check box if you want to export your edges (along with your faces) as part of your VRML file.
- **Use "Color by Layer" Materials:** VRML doesn't do layers, so if your layers are important to you, you should consider selecting this check box. All the faces in your exported model will be painted with the colors in the Layers dialog box (Chapter 7 has more information on assigning colors to layers).
- **Use VRML Standard Orientation:** You should select this check box to convert your model's "up" axis to match VRML's "up" axis.
- **Generate Cameras:** If you have scenes in your SketchUp Pro model, you might want to select this check box. It tells the exporter to create a separate camera view for every one of your scenes, and an extra one for your current view.
- **Allow Mirrored Components:** If you have components in your model whose instances you've mirrored (flipped over), you should select this check box. Because this is a standard technique for building symmetrical things like vehicles (as discussed in Chapter 5), this might apply to your model.
- **Check for Material Overrides:** This option makes sure that the materials in your exported model end up looking like they do in your SketchUp Pro model.

14.2.5 Handling DWG and DXF (in 3D)

You should use these two 3D file formats if you need to export your 3D data to AutoCAD or Revit, but for other programs, you're probably better off getting the right plugin. If you want to export the 3D model into any of these two programs, you will need to select File ⇨ Export ⇨ 3D Model and choose ACAD files in a .dwg format (reversely, if you wanted to import a Revit model into SketchUp Pro, you would have to save the file in Revit using the following directions: Export ⇨ CAD formats ⇨ under options, choose Solids ⇨ ACIS Solids).

Figure 14-7

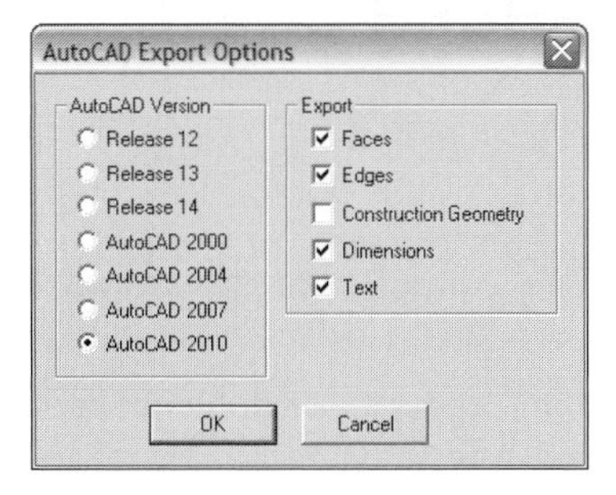

The AutoCAD Export Options dialog box.

Here is some additional information about DXF and DWG:

- **You'll lose your materials.** Your materials won't export to DXF/DWG, so pick another format if they're important to you.
- **You'll keep your layers.** The one great thing about exporting to DXF/DWG is that you get to keep your layers.
- **Go with DWG.** If you can, pick DWG instead of DXF; it's more robust, which means that it saves more of your data.

Figure 14-7 is a screen shot of the AutoCAD Export Options dialog box; it's quite simple compared to the one for 3DS. Just select the kinds of things you want to export, and then click the OK button.

IN THE REAL WORLD

Damon Leverett is an architect at EYP Architecture & Engineering in Washington who believes that the interoperability between 3D visualization and AutoCAD is extremely important in the design work flow. Google SketchUp Pro exports to both .dxf and .dwg formats, allowing designers to take their 3D model work and prepare it for production.

He says that while he often uses SketchUp Pro to communicate his ideas to other designers, the 2D export to CAD allows him to "jump-start" his production process. Once the design is conceptualized, he creates a set of parallel projections for the north, south, east, and west elevations and saves them as scenes. In addition, he also uses section planes from the top view to create plan drawings. Isometric views are frequently saved for export to help explain complex details and to reduce construction ambiguities.

Once all the views are prepared, the export 2D graphic is initiated and the views are transformed into CAD files that are ready for production. According to him, the beauty of SketchUp Pro is that it reduces the visual field into single line elements, resulting in a truly workable CAD file.

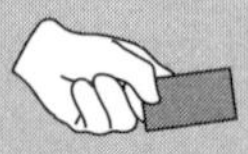

CAREER CONNECTION

Jeremy Wright is an architect at WFT Architects in Jackson, Mississippi. He remembers that the deciding factor when convincing a former employer to purchase Google SketchUp Pro was the cost of the software. He believes that its lower cost allows a professional to sway employers who would be unwilling to spend thousands of dollars on another piece of software that does the same job.

According to him, he typically does not render directly from SketchUp Pro for client presentations; instead, when photo-realism is not needed, he will render out three versions of the same image from the same camera (the first one with only the line work, the second one with only shadows and the third one with full textures) and composites them in Adobe Photoshop. He says that by doing this, he has better control over the final image, allowing him to employ filters on different layers to achieve a hand-drawn or painted look. He goes on saying that his clients usually like the "artistic" style but many prefer photo-realism.

He typically uses SketchUp Pro in conjunction with AutoCAD as he does the initial planning in AutoCAD and then exports the line work to SketchUp Pro to block out the shapes. From there, he imports it into 3DS Max, where he prepares it for the final rendering. He says that the reason why he uses SketchUp Pro prior to 3ds Max is because SketchUp Pro is a much faster program to work with, which is especially important during the schematic design phase.

SELF-CHECK

1. We would use SketchUp Pro's option of exporting two-sided faces when we want to make sure that our texture-mapping looks the same. True or false?
2. If our model ends up oriented the wrong way when we open it in another piece of software, what box would we have to select before we try exporting again?
3. 3DS is almost guaranteed to work with just about any other piece of 3D software you're using. True or false?
4. Some export options such as 3DS or VRML would not recognize layers but you can have the option of interpreting them as colors. True or false?

SUMMARY

Section 1

- Regardless of which 2D format you choose to export, the procedure is always the same.
- Vectors don't take up much space and are scalable.
- Profiles don't export very well to PDF and EPS, so you may only want to use this option if you're going to be tweaking your image in a vector drawing program like Illustrator or Freehand.
- DXF and DWG are the only 2D CAD formats that SketchUp Pro can export.
- You only need to use the EPIX file format if you're using Piranesi.

Section 2

- There are so many pieces of software, with so many versions, and so many different file types, that finding a "step-by-step" recipe for success is practically impossible.
- The process of exporting your 3D data from SketchUp Pro is the same no matter what file format you choose.
- 3DS is almost guaranteed to work with just about any other piece of 3D software you're using.

ASSESS YOUR UNDERSTANDING

SUMMARY QUESTIONS

1. What type of files should we favor when we want to transfer vector information?
   a. PDF
   b. EPS
   c. DXF
   d. EPIX
2. What file extension would be use if our other graphics software is more than a couple of versions old?
3. DXF is more capable than DWG, and for exporting 2D drawings from SketchUp Pro to use in other CAD applications, using DFX usually yields better results. True or false?
4. Data export from any program is a somewhat uncertain endeavor, so you should experiment and allow extra time for troubleshooting. True or false?
5. What data do we lose when we export to DXF or DWG?
6. DWG and DXF are 2D formats that are a good choice for most of the CAD-type programs out there, but not as good a choice as 3DS. True or false?
7. When in doubt, what would be the default program to export to, if we are unsure of what file extensions our program will open?
8. Name two characteristics of a PDF file that would make it the preferred choice to save a 2D image.
9. The process of exporting your 3D data from SketchUp Pro is the same no matter what file format you choose. True or false?
10. Most of the file formats that SketchUp Pro exports do not save the textures in your model as separate files alongside your model, so you needn't create a new folder for your exported 3D file. True or false?

APPLY: WHAT WOULD YOU DO?

1. Image that you have been commissioned to create a design of a new art institute in your area and that you need to e-mail five images to the institute's architecture committee. How do you go about deciding which file format to export out of SketchUp Pro?
2. What are the steps necessary to export a 2D image from SketchUp Pro as a PDF file?
3. List three 3D file formats we can export to from SketchUp Pro and some differences between them.
4. Why would we pick DWG over DXF when exporting our files?

BE AN ARCHITECT

Exporting a 2D SketchUp Pro Image

Choose a view of your model and export it both as a PDF and as a TIFF. Make note of the differences in the resulting images.

KEY TERMS

3DS
DAE (Collada)
DWG
DXF
EPIX
EPS
Extensions
FBX
KMZ
Meshes
OBJ
PDF
Section lines
Texture-mapping
VRML
XSI

CHAPTER 15

CREATING PRESENTATION DOCUMENTS WITH LAYOUT

Presenting 3D SketchUp Pro Models on Paper and On Screen: Working with Templates and Scrapbooks

Do You Already Know?

- How to use LayOut's tools?
- How to build a simple presentation from scratch?
- How to print and export your work?
- How to use layers and pages to streamline your work?
- How to create your own templates and scrapbooks?

For the answers to these questions, go to **www.wiley.com/go/chopra/googlesketchup2e**

What You Will Find Out	What You Will Be Able To Do
15.1 Getting acquainted with LayOut.	Work with the menu bar and panels.
15.2 How to build a quick LayOut Document.	Create templates and presentation pages, using model views, images, and annotation tools.
15.3 How to export and print your document.	Understand how to print. Export to a PDF, DWG, or DXF file. Make presentations.
15.4 How to stay organized with Layers and Pages.	Work with Layers and Pages.
15.5 About working with inserted model views.	Frame your model the way you want it.
15.6 How to work with dimensions in your document.	Edit your dimensions.
15.7 How to draw in LayOut.	Understand how and when to use LayOut's Vector tools.
15.8 How to customize LayOut.	Build your own Templates and Scrapbooks.

INTRODUCTION

People who design objects or structures in 3D have to present their ideas to other people, and most of the time, they have to present in a 2D format. Creating these presentations almost always involves the use of layout or illustration software like InDesign, Illustrator, or QuarkXPress; these programs are great, but they can be expensive and tricky to get the hang of, especially if you're not a graphic designer.

If you're lucky enough to have the Pro version of Google SketchUp Pro 8, you have access to a whole separate piece of software called Google SketchUp Pro LayOut, or LayOut for short.

LayOut is a program that lets you create documents for presenting your 3D SketchUp Pro models, both on paper and on-screen. LayOut was designed to be easy to use, quick to learn, and tightly integrated with SketchUp Pro. In fact, LayOut creates a file reference that keeps track of where it came from. The people who built it want you to use LayOut to create all your design presentations; here are some examples of what you can make with this software:

- Design packs, presentation boards, and posters
- Simple construction drawings with scaled views and dimensions
- Vector illustrations and diagrams
- Storyboards for planning camera shots

LayOut gives you the tools to create cover pages, title blocks, callouts, and symbols—whatever you need to accompany views of your model. You can create presentations that are any physical size, and you can export them as PDF files to send to other people. Best of all, when your design changes in SketchUp Pro, the views you've placed in LayOut update automatically to reflect the changes. If you make your living designing and presenting ideas in 3D, LayOut can save you valuable time.

This chapter gives a fairly high-level overview of what you can do with LayOut. First it discusses the different things you can use LayOut to accomplish, followed by a quick tour of the LayOut user interface, explaining where everything is and what it's supposed to do. Next, the chapter discusses the process of creating a simple presentation drawing set from one of your SketchUp models—not exhaustively, but it should be enough to see you through a tight deadline.

The last part of this chapter delves into more detail on LayOut's capabilities. It is filled with hints, recommended techniques, and details about the more powerful, but less obvious, aspects of LayOut.

There is information on how to organize your documents, inserting SketchUp Pro models, and designing them to look exactly the way you want. You will discover how to work with dimensions and learn how to create your own templates and scrapbooks.

15.1 GETTING YOUR BEARINGS

Even though LayOut comes with SketchUp Pro Pro, it's not just a SketchUp Pro feature—LayOut is its own program. As such, LayOut has its own menus, tools, dialog boxes, and Drawing Window.

Even though LayOut's user interface is pretty standard, this section provides a quick overview of the different elements. Knowing that LayOut

Figure 15-1

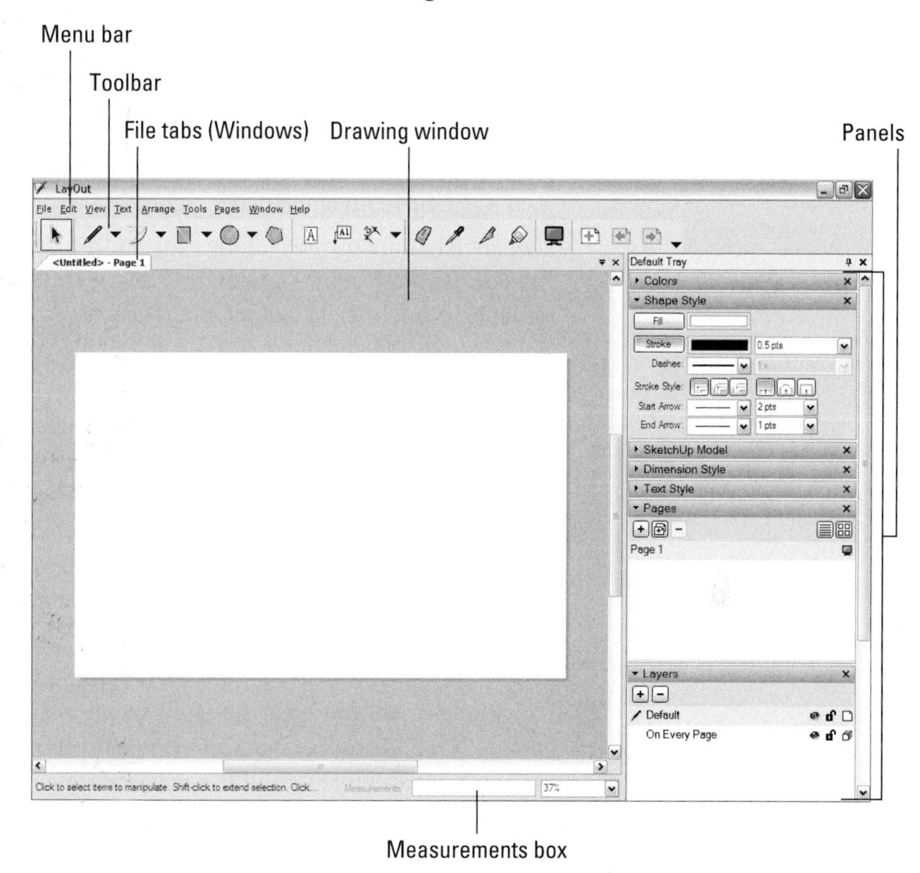

To get to see LayOut's interface, double-click on LayOut2 on the Desktop. Then, close the Tip of the Day dialog box and select a template (any type will do for now). Click Open.

is a lot like other software you've used (including SketchUp Pro) should help you come up to speed quickly. Figure 15-1 shows the LayOut user interface.

15.1.1 Some Menu Bar Minutiae

Learn to work with the menu bar and panels.

Just like almost every other piece of software, LayOut has a menu bar. And just like SketchUp Pro, you can use LayOut's menu bar to access the vast majority of its tools, commands, settings, and dialog boxes. Here's a brief description of each of the more important capabilities:

- **Preferences:** You find LayOut's application-wide Preferences dialog box on the Edit menu on Windows computers, and on the LayOut menu on Macs. The Preferences dialog box is where you do things like assign custom keyboard shortcuts and create new drawing scales.

- **Document Setup:** Located on the File menu, this is LayOut's version of SketchUp Pro's Model Info dialog box; it's full of settings you will frequently use:
  - *Grid:* Not only can LayOut display a helpful grid on your pages, it also lets you control the size and color of the gridlines or points. You can also choose to display the grid above your drawing elements, which some people really appreciate.
  - *Paper:* Right below the settings for paper size and margin width lies one of the most important controls in LayOut: Rendering Resolution. Both Edit Quality and Output quality are set to Medium by default, but you want to adjust them for almost every file you work on. This is covered in more detail later in this chapter.
  - *References:* When you insert a SketchUp Pro model or an image in your LayOut document, LayOut creates a file reference that keeps track of where it came from. If you edit the original file (which you probably will), this panel lets you know whether LayOut shows the most currently saved version. For people who go back and forth between design and presentation documents a lot, the References panel is a useful tool.
  - *Units:* When you insert a SketchUp Pro model or an image in your LayOutUse you can open the Units panel inside the Document Setup dialog box and you can determine the units used to measure in Layout, for both Measurements and Paper size settings.

Snap settings
Help you line up elements on your page with a grid or with other elements.

- **Snap settings:** These help you position elements on your page by making it easier to line up things with a grid or with other elements. Depending on what you want to do, you may choose to work with both types of snap settings, just one, or none at all. You can switch between Object Snap and Grid Snap while you work; rarely are both used at the same time. The snap settings are at the bottom of the Arrange menu.

In both the Windows and Mac versions of LayOut, you can have more than one document open at a time. On the Mac, separate files look just like they do for other programs; they are all in different windows. LayOut on Windows is a little different, though: Your open files display as tabs across the top of your drawing window, a little bit like scenes in SketchUp Pro. The tabbed files confuse some people who think that the tabs represent pages.

15.1.2 A Dialog Box Discourse

You can find most of LayOut's features and functions in its nine dialog boxes. In Windows, most of LayOut's dialog boxes are contained in a "tray" that appears on the right side of your screen by default. Those dialog boxes cannot be re-arranged in Windows; just expanded or contracted

On the Mac, your dialog boxes float around, but you can "snap" them together if you want. Choosing Windows ⇨ Arrange Panels cleans up your environment when you are having trouble finding things and they are not where they are supposed to be.

Here's a description of each:

- **Colors:** Just about all your LayOut documents will use color in some way, so you'll need this dialog box most of the time. The nice thing about Color is that it appears when you need it; clicking any color well in LayOut opens it (if it isn't already open). To hide a dialog box without closing it, click its title bar once to minimize it. Click it again to see it.
- **Shape Style:** A lot of the graphic elements in your presentation can have color fills and strokes (outlines). The Shape Style dialog box is where you control the appearance of those fills and strokes. Check out the options in the Start and End drop-down menus—you won't find callout styles like these in most other layout programs.
- **SketchUp Model:** The greatest thing about LayOut (at least with respect to other software like it) is its ability to include 2D views of your SketchUp Pro models. In the SketchUp ProModel panel, you can control all sorts of things about the way your "placed" SketchUp Pro model looks, including camera views, scenes, styles, shadows, and fog. For those who spend a lot of time laying out presentation drawings that include SketchUp ProSketchUp Pro Pro models, the SketchUp Pro Model dialog box is especially helpful.
- **Dimension Style:** LayOut in SketchUp Pro 8 lets you draw both linear and angular dimensions. This panel is where you control what they look like. More information on this is presented later in this chapter.
- **Text Style:** If you've ever used another piece of page layout or illustration software, you should be pretty familiar with what the Text Style panel lets you do. You use it to control the font, size, style, color, and alignment of text in your document.
- **Pages:** You use the Pages dialog box to manage the pages in your document. You can add, delete, and rearrange them as much as you like. The List and Icon buttons at the top let you toggle between views of your pages; you may prefer to use the former and give your pages meaningful names as you work. The little icons on the right control visibility for shared layers and full-screen presentations.
- **Layers:** You can have multiple layers of content in every LayOut document you create. Use the Layers panel to add, delete, and rearrange layers in your document. The icons on the right let you hide (and show), lock, and share individual layers. If we want to have a master layer that would show through different layers, that is called in LayOut, Shared Layers. To

reach it, select Layers and then add a new layer (the plus sign). This will be usually saved for elements that appear on every page in our presentation, such as a logo or a title block.

- **Shared layers:** Layers that let you automatically place elements on more than one page.
- **Scrapbooks:** This one's a little more difficult to explain; scrapbooks are unique to LayOut, so you probably haven't worked with anything like them before. Scrapbooks are LayOut files that exist in a special folder on your computer system. They contain colors, text styles, and graphic elements (like scale cars, trees, and people) that you might need to use in more than one of your LayOut documents. To use something in a scrapbook, just click it with the Select tool and then click again in your drawing window to stamp it in. You can also sample

TIPS FROM THE PROFESSIONALS

Switching to LayOut from Similar Software

If you are used to using other page-layout or illustration software, some things about LayOut are useful to know when you're just getting started. The people who designed LayOut did things a little differently on purpose, hoping to do for page layout in 2007 what they did for 3D design seven years earlier—make it easier for motivated people with no experience to produce good work, quickly.

Here are the five things you should keep in mind when you're exploring LayOut:

1. LayOut includes templates that help you get started in no time. See the nearby section "Starting Out with Templates" for details.
2. You can insert models from SketchUp Pro, skipping the process of exporting your model as an image file. Importing has the added benefit of helping you automatically update model views in your presentation.
3. The Layers feature in LayOut is a powerful tool for organizing your content. In particular, you can place content that appears on more than one of your pages on a master layer, so you only have to position it once.
4. In LayOut, you have enormous flexibility to crop images, including model views, with ease using clipping masks. Find out how later in this chapter.
5. When your presentation is ready to go, LayOut enables you to set up digital slide shows in full-screen mode, as well as to create printouts and PDF files.

things like colors, line weights, and text styles by clicking with any other tool.

- **Instructor:** The Instructor panel works just like it does in SketchUp Pro; it shows information on whichever tool you happen to be using. If you're just starting out with LayOut, make sure that this panel is open.

IN THE REAL WORLD

Joe Mesa Design Studio in Canada uses SketchUp Pro, extensively, for its ease of use and simple learning curve. Joe also likes it for its style and the wide range of available plugins that enhance its functionality. In the creation of his models he starts importing a CAD file or a photograph from scratch without any extra help. Once his images are done, he brings them into Podium to create high-quality renderings that are later retouched further in Photoshop. He often creates his own extrusions and details in his kitchen designs with Follow Me when he has to create a valance, a baseboard, or crown molding, and groups the modeled instances so that he can easily change them afterward. He also uses many building tools to model, entirely, his furniture pieces.

According to him, "presentation documents are the key to all of our design projects. As an artist and designer with a CAD background, I was looking for a tool that would provide us with a better workflow when presenting my models to clients. In the past, I would export multiple views of my model and then use two other programs to post process and template the images. The workflow was cumbersome and inefficient. Revisions become your worst nightmare! With LayOut, creating presentations is extremely easy and yields professional results. I love the fact that you can rotate the model on your layout to get just the right angle. No need to re-export or work back and forth between programs. This has been very useful when designing commercial sets as timing and presentation is important.

PDF files are exported from LayOut and are sent to our clients for approval via e-mail. LayOut allows us to make revisions to our model while being automatically updated in all the layout views. In LayOut 3, the new dimension tool is a godsend and a huge time saver. The need to export images is no more, which was a big plus for me. We have a collection of templates set up for different types of clients. Having your layers, styles, and templates setup simplifies the workflow. Having a collection of symbols is also definitely a time saver. The power of SketchUp Pro and LayOut combined can impress any client. It helps us to create beautiful, affordable, professional images for our clients."

SELF-CHECK

1. Name some of the diverse files found in the Scrapbook folder in LayOut.
2. LayOut has the ability to include 2D views of your SketchUp Pro models. True or false?
3. When you insert a SketchUp Pro model or an image in your LayOut document, LayOut creates a file reference that keeps track of where it came from. True or false?
4. Explain the difference between Layers and Shared Layers in LayOut.

Apply Your Knowledge Getting familiar with LayOut is important before we can even consider importing images. If you opened LayOut for the first time, you will notice that you have to choose a Default Template. Choose one that you think you will use most regularly (that can be changed later) and explore the different menu tabs. Once you have identified the different work areas in LayOut, go to File ⇨ Insert and insert the SketchUp Pro file of the chair or table that you worked on Apply your Knowledge at the end of Chapter 14, Section.1.

15.2 BUILDING A QUICK LAYOUT DOCUMENT

Imagine you have a major deadline coming up, and you have arrived at this page because you need to turn your SketchUp Pro model into a set of drawings.

GOOGLE SKETCHUP PRO IN ACTION

Learn to work with templates, model views, images and annotation tools.

There's little time and you need model views, page titles, basic dimensions, some annotations, and maybe a logo.

In the following sections are the steps to putting together a bare-bones LayOut document that includes all the elements listed in the preceding paragraph.

15.2.1 Starting Out with Templates

Generally, templates are the quickest way to get started with a new LayOut project. Follow these steps to load a LayOut template and customize it for your own purposes:

1. **Launch LayOut.** Keep in mind that LayOut and SketchUp Pro are separate software programs, so you need to launch them individually. If you've already launched LayOut, choose File ⇨ New to open

Figure 15-2

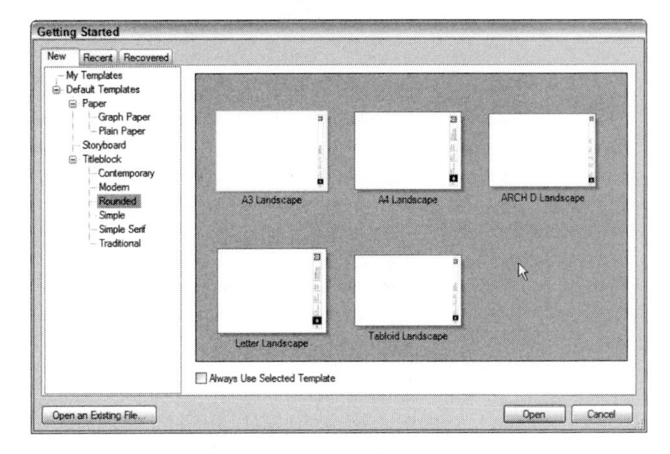

The Getting Started dialog box.

the Getting Started dialog box. (See Figure 15-2.) If you don't see it, you can switch it on in the Startup panel of the Preferences dialog box.

2. **In the Getting Started dialog box, click the New tab.** This shows a list of available templates on the left, with thumbnail previews of each template on the right. Nothing about these templates is special—they're just ready-made LayOut files you can use as a starting point for your document.
3. **Choose a template to use.** Expand the items in the list on the left to see the available templates by category. Browse the list, select the title block you want to use, and click the Open (Choose on a Mac) button to work with that template. If you change your mind about the template you picked, close the file you just created and choose File ⇨ New to pick another one.
4. **Unlock all your new document's layers.** Many templates that come with LayOut have multiple layers, and some layers are locked by default so that you can't accidentally move things. In this case, you want to unlock them all so you can customize the template with your own information. In the Layers panel, unlock all the locked layers by clicking their little lock icons one at a time. Figure 15-3 shows the Layers panel, among other things.
5. **Edit the default text on the page.** With the Select tool, double-click text to edit it. Click somewhere else on the page to stop editing. Roll your scroll wheel to zoom in and out on the page, just like in SketchUp

Figure 15-3

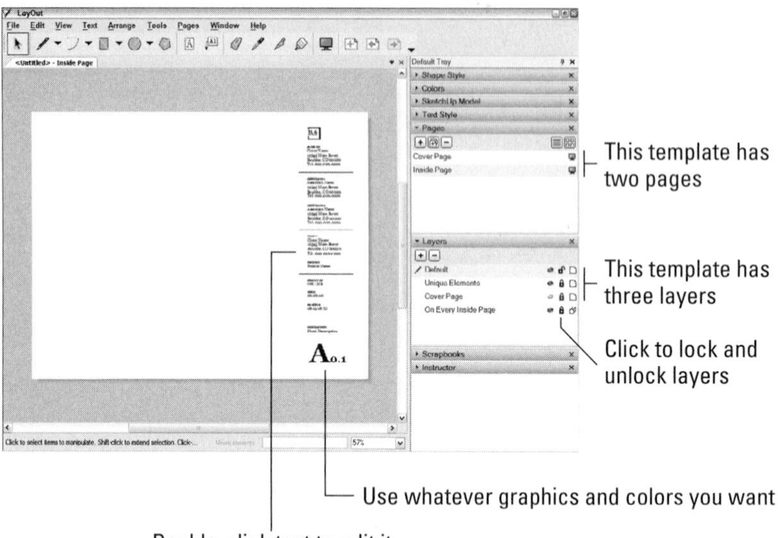

One of LayOut's nicer prebuilt templates.

Pro. Hold down the scroll wheel button to pan around. To fill your drawing window with the page you're viewing, choose Scale to Fit (Zoom to Fit on the Mac) from the Zoom drop-down list in your window's lower right corner.

6. **Edit the default text on all your other pages.** Most of the more interesting templates include at least two pages; many templates open on the second page. Use the Pages panel to switch between pages in your document. Repeat Step 5 for any default text that needs to change.

7. **(Optional) Change colors and line styles.** You can edit lines and other graphic entities you select in the Shape Style panel. Clicking a color well opens the Colors panel. To change the color of text, select it and click the color well in the Text Style panel (Fonts panel on a Mac). Sometimes the entity you are trying to edit is buried inside a group. Double-click a group to edit it. This works very similarly to SketchUp Pro.

8. **Swap out the generic logo for your own.** Delete the generic logo wherever it appears in your document by selecting it and hitting Delete on your keyboard. Follow these steps to bring in a logo of your own:

   a. Make sure you are not on your document's cover page.

b. In the Layers panel, click the On Every Inside Page layer to make it the active one and then choose File ⇨ Insert.
c. Find the logo image you want to use and click the Open button.
d. Activate the Select tool; then resize your logo by dragging its blue corner grips and pressing down the Shift key to keep from stretching your logo while you resize it.
e. Click and drag your logo to put it where you want on the page and then choose Edit ⇨ Copy to copy your logo to the clipboard.
f. Switch to your document's cover page and make the Cover Page layer active by clicking its name in the Layers panel.
g. Choose Edit ⇨ Paste to paste your logo on the page.
h. Repeat Steps d and e to place your logo where you want it, and then make the Default layer the active one before you forget.

15.2.2 Inserting SketchUp Pro Model Views

With every other page-layout program on the market today, the only way to include a view of a SketchUp Pro model is to export that view from Sketch Up as an image file and then place it in the layout program. Changing the SketchUp Pro file means going through the entire export-and-place process again, and if your presentation includes lots of SketchUp Pro model views, it can take a considerable amount of time.

This brings us to LayOut's *raison d'être:* Instead of exporting views from SketchUp Pro to get them into LayOut, all you need to do is insert a SketchUp Pro file. From within LayOut, you can pick the view you like best. You can also use as many views of the same model as you want. When your SketchUp Pro file is modified, LayOut knows about it and (using the References panel in the Document Setup dialog box) lets you update all your views at once by clicking a single button. Follow these steps to insert a SketchUp Pro *viewport* (model view) into your document:

Figure 15-4

To save time, save a scene in SketchUp Pro for every view you want to include in your LayOut document.

1. **In SketchUp Pro, create a scene for each view of your model that you want to show in your LayOut document (see Figure 15-4).** Read Chapter 10 for a refresher on using scenes. Be sure to give them meaningful names, and remember to save your SketchUp Pro file when you're done.
2. **In LayOut, navigate to the page where you want to insert a viewport.** Use the Pages panel to move between existing pages. The quickest way to add a new page is to duplicate an existing one: Just click the Duplicate Selected Page button (between Add and Delete) in the Pages panel.

Figure 15-5

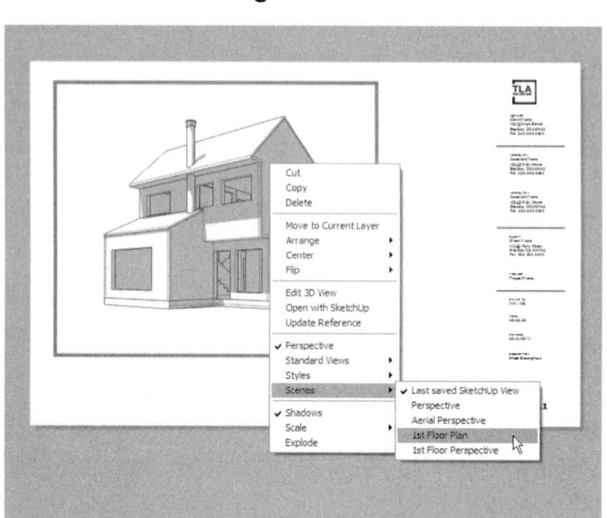

Associate a scene from your model with a viewport in LayOut.

3. **Insert a SketchUp Pro model viewport:**
   a. Choose File ⇨ Insert to open the Insert dialog box.
   b. Find the SketchUp Pro file on your computer that you want to insert and click the Open button.

   The Insert dialog box closes, and your SketchUp Pro model is placed on your current LayOut document page.

4. **Associate a scene with your model viewport (see Figure 15-5).** With the Select tool, right-click your viewport and choose Scenes, and then choose the name of the scene you want to associate. If you don't see a list of scenes, you probably forgot to save your SketchUp Pro file in Step 1. Save your SketchUp Pro file; then right-click your viewport (in LayOut) and choose Update Reference.

5. **Assign a drawing scale to your model view, if that's appropriate (see Figure 15-6).** If the scene you picked in Step 4 is an *orthographic* view (top, front, side) where perspective is turned off, it's very likely that you want to show your model at a particular drawing scale. With Select, right-click your viewport, choose Scale, and then choose one from the list that appears. LayOut doesn't prevent you from assigning a scale to any old view, but that doesn't matter. Drawing scales apply only to non-perspectival, straight on views of your model. If your view isn't orthographic, it isn't at scale. If a bright yellow exclamation mark icon appears in the lower-right corner of a viewport, you need to tell LayOut to render that viewport in order for it to reflect whatever changes you have made to it. Right-click the viewport and choose Render Model from the context menu.

Figure 15-6

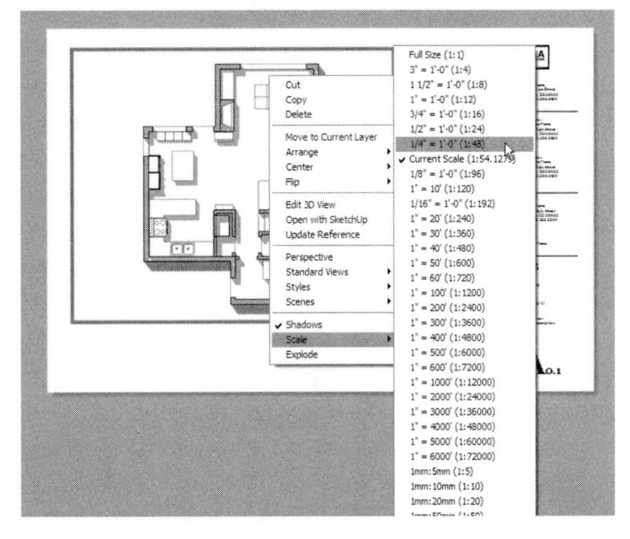

Assign a precise drawing scale to any orthographic viewport.

6. **Use the Select tool to position, rotate, or resize your model view.** Click and drag to move any element in your document on the page. Use the Rotation Grip (the little blue stick in the center of your image when it's selected) to rotate. You can resize anything by clicking and dragging any corner.

Repeat the preceding steps for all the additional viewports you want to add to your document. Read on in this chapter for more information about viewports.

15.2.3 Adding Images and Other Graphics

Inserting images into your LayOut document is straightforward; just choose File ⇨ Insert. A few more things to know about images you insert:

- **LayOut can insert raster images.** This means TIFFs, JPEGs, GIFs, BMPs, and PNGs are all graphics file formats that save pictures as a myriad of pixels.
- **The Mac version of LayOut can also insert PDF files.** This is indisputably the best way to bring in vector art, such as logos. You can use a program like Adobe Illustrator to save any AI (Illustrator) or EPS file as a PDF.
- **Images are a lot like viewports.** Moving, resizing, and rotating images works just like it does for SketchUp Pro model views; use the Select tool to do everything. Remember to hold down the Shift key when you resize to keep your images from stretching.

Unfortunately, LayOut offers no easy way to import editable vector (such as AI, EPS, and SVG) graphics. If you want to use vector graphics in your LayOut document, you have two choices:

- **Make your own.** LayOut is actually a great vector illustration tool. Read further in this chapter to discover all the nuances of LayOut's illustration toolset.
- **Borrow from the Scrapbooks panel**. One of the best things about LayOut is the hundreds of pre-drawn graphical elements you can find in the Scrapbooks panel (see Figure 15-7). You find things like:
  - Symbols: Arrows, section markers, north indicators, graphic scales, and column grids
  - Entourage elements: Trees, cars, and people at various scales and levels of detail
  - Color palettes: To help with producing attractive documents quickly

To use something you see in the Scrapbooks, just click it with the Select tool to "sample" it and then click again to stamp it onto your page. You can keep clicking to stamp more copies, or press the Esc key when you're ready to exit stamping mode.

Figure 15-7

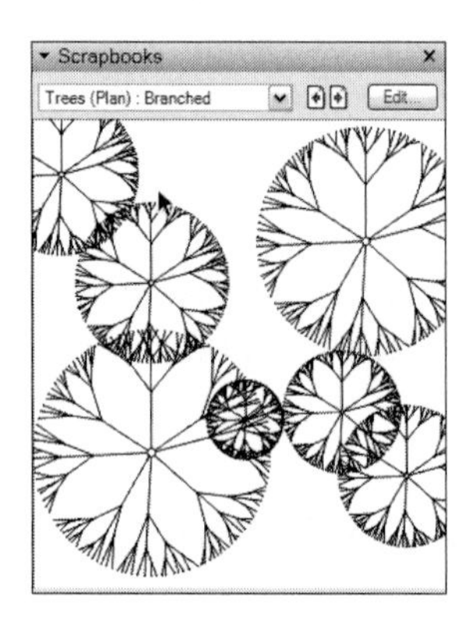

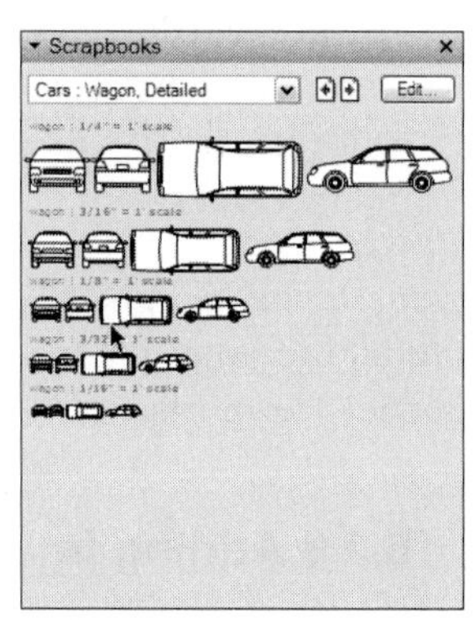

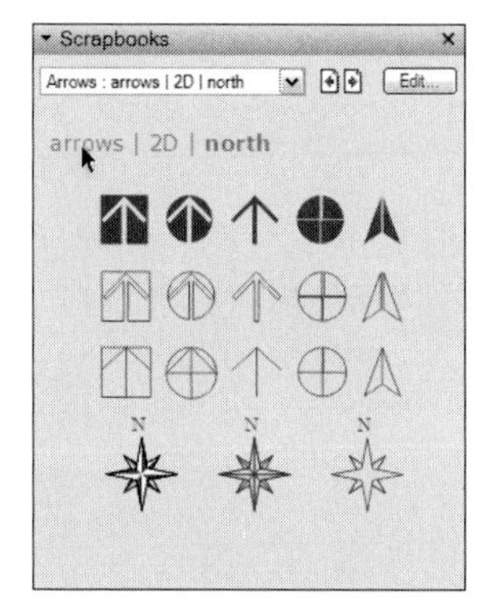

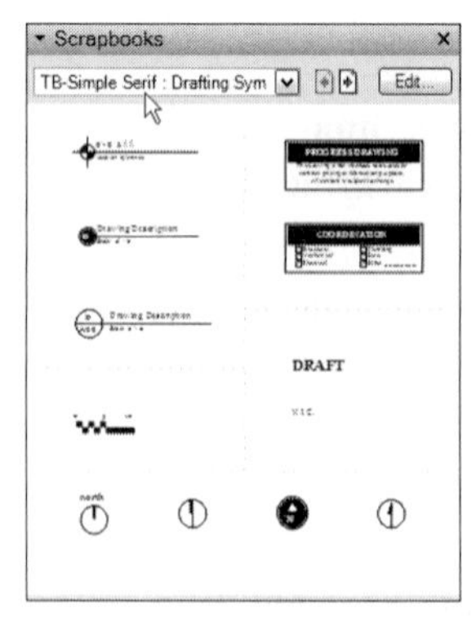

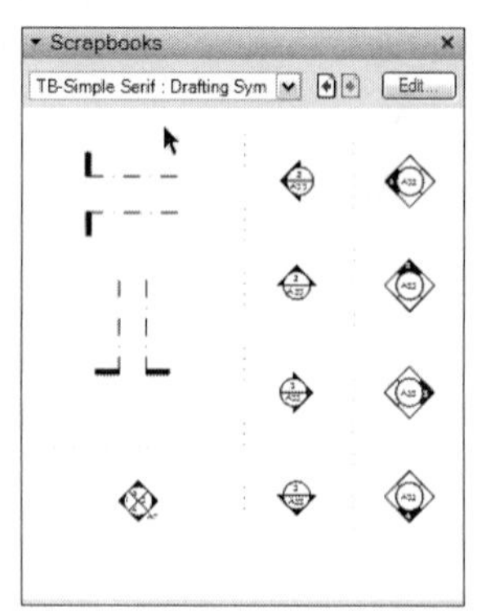

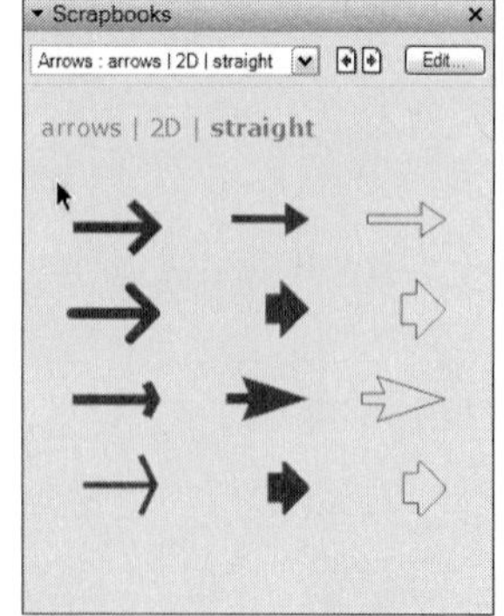

LayOut's Scrapbooks is the fastest way to add professional-looking graphics to your drawings.

15.2.4 Annotating with Text and Dimensions

LayOut has a few tools you can use to add blocks of text, titles, callouts, labels, and dimensions to your drawings. Luckily, none is terribly complicated to use. Here's a mini-section about each text tool; Figure 15-8 is a sampler of different drawing annotations:

The Text Tool

Text boxes in LayOut are classified into two broad types, depending on how you create them:

- **Bounded:** If you click and drag with the Text tool, the text box you create is *bounded*. Any text you enter into it that doesn't fit isn't visible, and you get a little red arrow at the bottom. That arrow tells you that there's more in your text box; you need to use the Select tool to make the box bigger to show everything that's inside. Use a bounded text box whenever your text needs to fit into a precise space in your design.
- **Unbounded:** If, instead of creating a text box with the Text tool, you simply click to place your cursor somewhere on your page, the text you create is *unbounded*. It stays inside a text box, but that text box automatically resizes to accommodate whatever text you put inside it. To turn an unbounded text box into a bounded one, just resize it with the Select tool or choose Text ⇨ Make Unbounded.

Naturally, you control things like text size, color, alignment, and font using the Text Style panel (Fonts panel on a Mac).

This only applies to Mac folks: Choosing Text ⇨ Show Rulers does more than just display ruled increments at the top of your drawing window.

Figure 15-8

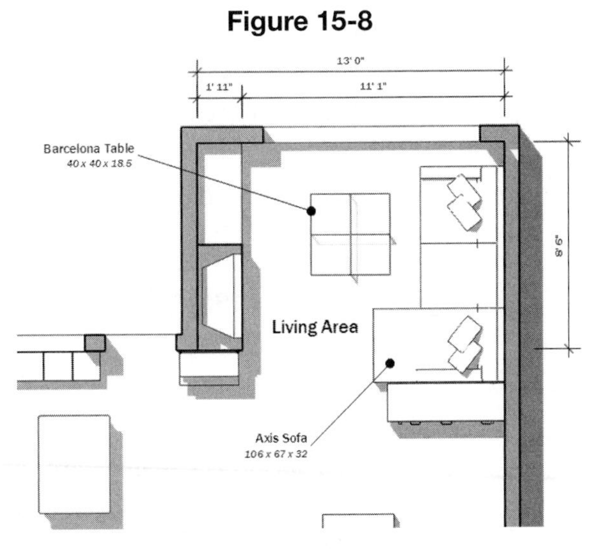

Text, callouts, and dimensions are all different forms of annotation.

It also enables extra controls for paragraph spacing and lists both bulleted and numbered. Just select text in your document to see them appear above the ruler.

The Label tool

Use the Label tool to add *callouts* (notes with leader lines) wherever you need them. Four important points about the most useful tool in LayOut:

- **Activate, click, click, type, and click.** Activate the Label tool, click once to "pin" the end of the leader line to an element in your drawing, click again to place your text cursor, type something, and click somewhere else to finish your label.
- **Leader lines stick to drawing elements.** When you move the thing your leader line is pinned to, the line moves with it.
- **Use the Shape Style panel to edit the look of your leader lines.** You can change the color, thickness, and endpoints (arrowheads, slashes, and dots) of any leader line very easily after you create it.
- **Save time by sampling.** After you edit a label you have made already, it's easy to set up things so every subsequent label you make looks the same:
    a. Activate the Label tool and then press the S key. Your cursor changes into an eyedropper.
    b. Click the text part of the label you sample and then click S again.
    c. Click the leader line of the label you sample.
       Now every label you create looks just like the one you sampled.
- **Use the Style tool to copy styles between labels you've created already.** If you have a bunch of labels and you want to make them all look the same, follow these steps:
    a. Activate the Style tool (it looks like an eyedropper) and then click the text part of the label whose style you want to copy.
    b. Apply (by clicking) that style to the text of every other label you want to change.
    c. Repeat Steps a and b, sampling the leader lines of your labels instead of the text.

The Dimension Tools

Both dimensioning tools (Linear Dimension and Angular Dimension) work very similarly to the Label tool; all three are made of lines and text. Here's a stripped-down version of what you find later in this chapter:

- **Turn on Object Snap before you start.** Choose Arrange ⇨ Object Snap to make sure your dimension leader lines can "see" the points they're supposed to be attached to.

- **Creating a new linear dimension is very simple.** Activate the Linear Dimension tool, click a start point, click an end point, click to define an offset, and you're done.
- **Double-click to create a string of linear dimensions.** After you create your first dimension and while the tool is still active, double-click the next point you want to dimension. The offset you set for the first is duplicated.
- **Angular dimensions are a little trickier.** Using the Angular Dimension tool is a five-click operation. Follow these steps to make it work (see Figure 15-9):
  a. Activate the Angular Dimension tool and then click once to establish the first "pin point" for your new dimension.
  b. Click again, somewhere along the same line as the point you clicked in Step a.
  c. Click once to establish the second pin point.
  d. Click again along the same line as your second point.
  e. Click one last time to position the text of your angular dimension.
- Use the Shape Style panel to edit your leader lines. Thickness, color, endpoints—it's all here.
- Use the Dimension Style panel to change formatting. By *formatting*, this means metric or imperial, decimal places, text position, and visibility.
- **Use the Style tool to copy formatting and other settings between dimensions.** Activate Style, click your "source" dimension, and then click each dimension you want to change. Creating separate layers for text, labels, and dimensions saves time in the long run. In the Layers panel, click the Add Layer button to make a new one. The layer with the little, red pencil next to it is your active layer.

IN THE REAL WORLD

Once again, Damon Leverett, an architect at EYP Architecture & Engineering in Washington, believes that Google SketchUp Pro Layout is an excellent tool to create development sketches throughout the design process.

He says that interior designers and architects spend quite a bit of time communicating ideas between themselves and the best form of communication is through sketching or drawing. According to him, with SketchUp Pro Layout, a brief sketch diagram can be created, allowing the circulation of the appropriate amount of information among team members. He goes on saying that he considers LayOut to be an intermediate tool between modeling and production as it acts to promote coordination between disciplines.

Figure 15-9

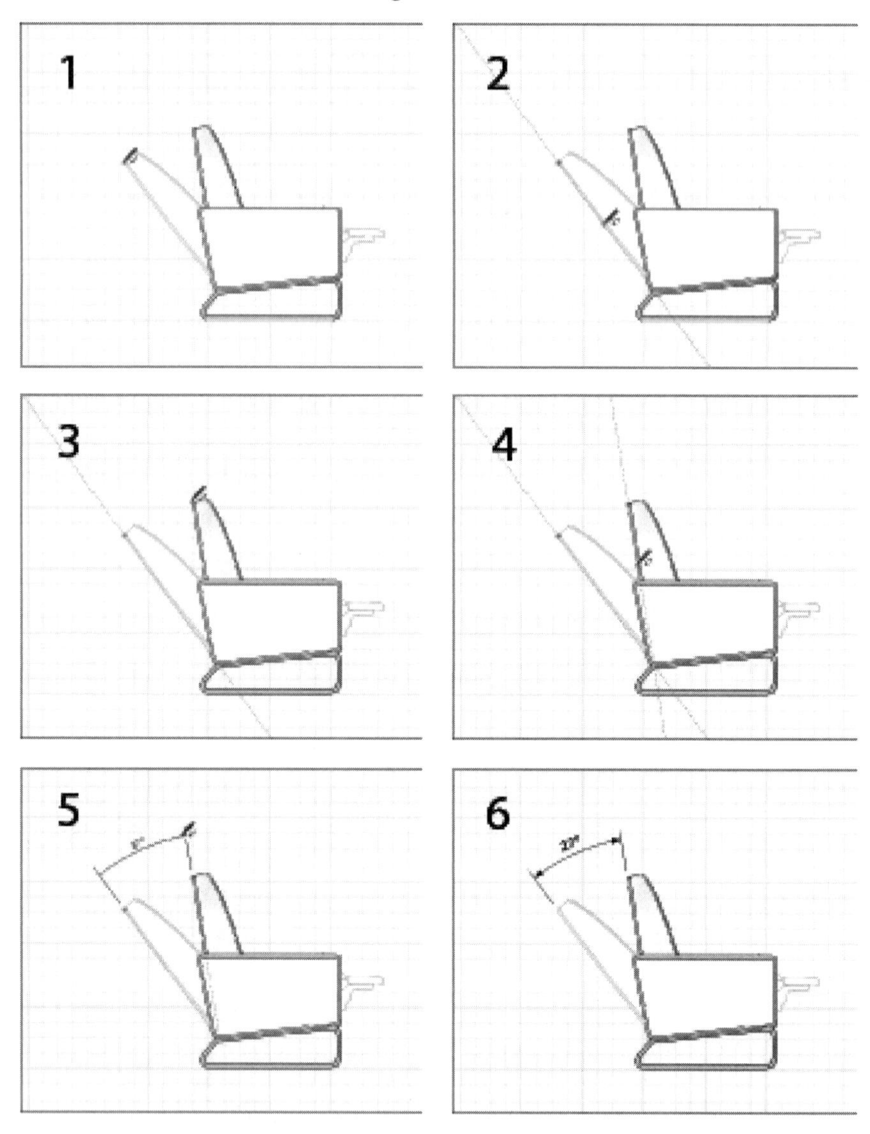

Creating an angular dimension takes some getting used to.

SELF-CHECK

1. LayOut and SketchUp Pro are separate software programs, so you need to launch them individually. True or False?
2. List the steps that are required to insert a SketchUp Pro model viewport into LayOut.

3. To insert TIFF, JPEG, GIF, BMP, or PNG images into your LayOut document choose the following path:
   a. File ⇨ Insert
   b. File ⇨ View
   c. File ⇨ Print
   d. File ⇨ Preferences
4. What tool do we use if we want to make a collection of labels to look all the same?

Apply Your Knowledge It is important to have a default template to work with in LayOut so we are always ready to put together a quick presentation for our clients. In addition, if we have already a series of scenes created in an original SketchUp Pro document, assembling a presentation in LayOut could be a straight-forward proposition.

Open LayOut and edit your default template. If you have not been prompted to do so at start up, go to Edit ⇨ Preferences and under New Document choose the option Use Default Template). Be sure to add your own logo or a business name (it would be a good time to come up with one if you do not have one yet) and unlock layers to edit the default text and change the graphic entities to make your own design template, such as background colors or line styles.

Once you have completed editing your default template, save it as a new template (go to Save ⇨ New Template). Then, locate the SketchUp Pro file that you created on the previous exercise—your table or chair—and create a new viewport and associate a view to it. It can be a plan view or an elevation. Then, select a drawing scale to the scene to fill your page nicely and practice adding some text (bound or unbounded), annotations (also known as callouts), and dimensions to your single page.

15.3 LIFE AFTER LAYOUT

After you have created your LayOut document, you can do the following five things:

1. Print it
2. Export it as a PDF file
3. Export it as an image file
4. Export it as DWG or DXF (CAD) file
5. View it as a full-screen presentation

The next five sections provide more detail on each of these options.

GOOGLE SKETCHUP PRO IN ACTION

Understand how to print.

15.3.1 Printing Your Work

Chapter 12 is devoted to printing from SketchUp Pro and presents many pages on the subject. The instructions for printing from LayOut are much shorter. Follow these steps to print your LayOut document:

1. **Choose File ⇨ Print.** In the Print dialog box, choose which pages to print, how many copies you want and if the right printer is selected.
2. **Click OK to send your document to the printer.**

This simple approach to printing is why you should always insert your SketchUp Pro models into a LayOut document if you need to print them. You can also export a PDF and use Adobe Acrobat (or Reader) to send the actual print job to the printer. The settings in Adobe's Print dialog box give you more control over the finished product.

GOOGLE SKETCHUP PRO IN ACTION

Learn how to export to a PDF, DWG, or DXF file.

15.3.2 Exporting a PDF File

Anyone with Adobe Acrobat Reader software (which is free and is already loaded on most computers, but do not confuse it with Adobe Acrobat Pro) can look at a PDF document you create; all you have to do is share it with the person that is expecting it.

PATHWAYS TO...
EXPORTING YOUR LAYOUT DOCUMENT AS A PDF FILE

1. **Choose File ⇨ Export ⇨ PDF.** If you are on a MAC, choose File ⇨ Export and then make sure PDF is selected in the Export dialog box. This opens the Export dialog box.
2. **Give your PDF file a name, and decide where to save it on your computer.**
3. **Click the Save button (in Windows) to open the PDF Export Options dialog box; click the Options button if you are on a Mac.**
4. **Set the PDF options the way you want them.** Here's what everything means:
   - **Page:** Choose which pages you want to export.
   - **Quality:** Read further in this chapter for a brief discussion about Output Quality. Here's a good rule of thumb: For documents that

are small enough to be hand-held, it is recommended to go with a setting of High. For anything bigger, go with Medium or Low.

- **Layers:** PDFs can have layers, just like LayOut documents do. If it makes sense to do so, you can export a layered PDF so that people who view it can turn the layers on and off.
- **Finish:** Select this check box to view your PDF after it's exported.

5. **Click the OK button to close the PDF Options dialog box (Mac only).**
6. **Click the Export button (Save button on a Mac) to export your document as a PDF file.**

15.3.3 Exporting an Image File

You can export the pages of your file as individual raster images in either JPEG or PNG format. Take a look at Chapter 13 for more information on the differences between JPEG and PNG if you need to. Follow these steps to export your LayOut document as one or more image files:

PATHWAYS TO...
EXPORTING YOUR LAYOUT DOCUMENT AS ONE OR MORE IMAGE FILES

1. **Choose File ⇨ Export ⇨ Images.** If you are on a Mac, choose File ⇨ Export, and make sure PNG or JPEG is selected in the Export dialog box. This opens the Export Image dialog box.
2. **Name your file and tell LayOut where to save it on your computer.**
3. **Click the Save button (Options on a Mac).** The Image Export Options dialog box opens.
4. **Set the Image Export options.** Here's what each option means:
   - *Pages:* Choose which pages you want to export. Each page in your LayOut document exports as a separate image file.
   - *Size:* See Chapter 13 for a complete rundown on pixel size and image resolution.
   - *Finish:* Select this check box to view your image after it's exported.
5. **Click OK to close the Image Export Options dialog box (Mac only).**
6. **Click the Export button (Save button on a Mac) to export your document as one or more image files.**

15.3.4 Exporting a DWG or DXF File

It is difficult to find a piece of professional computer-aided drawing (CAD) software that can't read the DWG and DXF formats, which are the industry standard for exchanging CAD files with people who use apps like AutoCAD and Vectorworks.

PATHWAYS TO... TURNING YOUR LAYOUT DOCUMENT INTO A CAD FILE

1. **Set all your SketchUp Pro viewports to vector rendering mode.** Here's the short version: LayOut treats your SketchUp Pro models' edges as either *raster* (dots) or *vector* (math) information. Viewports that are rendered as rasters export to DWG/DXF as raster images. That's usually not what you want to happen, especially if you're exporting a CAD file. Follow these steps to make sure your viewports are vector images:
   a. Select a model viewport and then click the View tab of the SketchUp Pro Model panel (choose Window ⇨ SketchUp Pro Model to open this panel).
   b. Change the Rendering style drop-down list from Raster to Vector. Depending on the complexity of your model, this might take a while.
   c. Repeat the preceding two steps for each viewport. If a viewport contains a view whose edges you don't want to manipulate in CAD (such as a glitzy rendering), leave it as a raster.
2. **Choose File ⇨ Export ⇨ DWG/DXF.** On a Mac, choose File ⇨ Export and make sure DWG/DXF is selected in the Export dialog box.
3. **Name your file, tell LayOut where to save it on your computer, and click the Save button (Options on a Mac).** The DWG/DXF Export dialog box opens.
4. **Set the DWG/DXF Export options.** Here's what all the various features do:
   - *Format:* Unless you know you need a DXF, export a DWG file. As for which version, choose the latest one: AutoCAD 2010.
   - *Pages:* Choose which pages you want to export. Keep in mind that each page in your LayOut document exports as a separate file.
   - *Output Space:* When you choose Paper Space, the lines in your resulting document file, when measured in the CAD program you use next, are exactly as long as they would be on a piece of paper printed from LayOut. Choosing Model Space tells CAD to draw the lines at a particular scale. Here's an example: You have a viewport

that shows a plan view of your building at 1/8 scale. Your building is 80 feet wide, so it looks 10 inches wide in the drawing in LayOut. If you export to Paper Space, open your drawing in AutoCAD and measure your building, it'll be 10 inches wide—probably not what you wanted. If you export to Model Space, choose 1/8" = 1'–0" as a scale, and then measure your building in AutoCAD, it'll be 80 feet wide. Most of the time, you want to choose Model Space.

- *Scale:* This setting is relevant only if you choose to export to Model Space. Pick the drawing scale that matches the scale of the viewports on the pages you're exporting. If your pages have viewports at different scales, you have to export them separately to make sure all the scaling is accurate. Have two viewports at different scales on the same page? Only one of them will be correct in your exported file.
- *Layers:* Compared to the preceding two settings, this one's mercifully clear. Decide which layers will appear in your exported files.
- *Ignore Fills: Fills* are shapes that are drawn in LayOut and filled with a color.

5. **Mac only: Click OK to close the DWG/DXF Export dialog box.**
6. **Click the Export button (Save button on a Mac) to export your document as one or more DWG/DXF files.**

If your LayOut file included any inserted raster images (such as JPEGs or PNGs,) you also end up with a folder that contains copies of those. They are necessary for the DWG/DXF files you produce.

15.3.5 Going Full-Screen

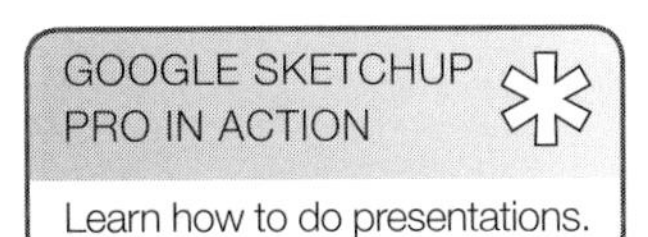

Many times, design presentations for clients go beyond printed boards and booklets; they include a digital slide show that usually involves a few hours of work in a program like PowerPoint or Keynote (if you're on a Mac).

LayOut was designed to help you skip the PowerPoint step by letting you display your presentation in a full-screen view. You can move back and forth between pages with the arrow keys on your computer, and you can even double-click SketchUp Pro model views to orbit them around. Follow these tips:

- **Switching to full-screen mode takes less than a second.** Choose View ⇨ Start Presentation to view your presentation full-screen. Press Esc to exit Presentation mode.
- Specify where you want your presentation to appear. Use the Presentation panel in the Preferences dialog box to tell SketchUp Pro which monitor (or projector) you want to use to show your presentation.

- **Move from page to page.** Use the left- and right-arrow keys to flip through pages.
- **Choose which pages to show full-screen.** You can decide not to show certain pages in full-screen mode by toggling the Show Page icon in the Presentations icon to the right of those page names in the Pages dialog box (make sure that you're in List view to be able to do this).
- **Double-click to change your view of a SketchUp Pro model.** When you're in full-screen mode, you can double-click any SketchUp Pro model view to orbit and zoom around inside it. Just use your mouse's scroll wheel button the same way you do in SketchUp Pro. Click anywhere outside the view to exit.
- **Draw while you're in full-screen mode.** Try clicking and dragging while you're in full-screen mode; doing so lets you make red annotations right on your presentation. For example, if a client doesn't like the porch you designed, you can scrawl a big, red *X* over it to let him know you understand. When you press ESC to exit Presentation mode, you can choose to save your annotations as a separate layer.
- **Play scene animations right in full-screen mode.** You can right-click on a model view with scenes that you've set up in SketchUp Pro and choose Play Animation. LayOut will transition from scene to scene, just like SketchUp Pro does. You can read more about scenes in Chapter 10.

IN THE REAL WORLD

Kristoff Rand, project manager at www.aboveallhouseplans.com, describes LayOut in the following terms: "although I don't use LayOut for construction documents, I do find the sheets made with LayOut superior for preliminary and presentation drawings. We modified one of the stock templates to work for our needs, and we feel it made our presentation drawings be more enticing and professional.

SELF-CHECK

1. After you've created your LayOut document, you can print it, export it as an image file, export it as a PDF, DWG, or DXF file, or view it as a full-screen presentation. True or false?
2. What steps do we take to save a LayOut presentation as a PDF document?
3. PDFs, unlike LayOut documents, cannot have layers. True or false?
4. When we are on full-screen mode, we can make annotations directly on each page, which can be later saved in a separate layer. True or false?

Apply Your Knowledge Experimenting with the different exporting options in LayOut helps us to become more confident with the possibilities of this software when we are ready to present it to our audience.

Create in LayOut a simple presentation of a small cottage that you would have previously designed in SketchUp Pro, and insert a few captured scenes in your customized template. Then, practice saving your entire document as a PDF file. Likewise, choose a single page in your document and save it as JPEG or PNG and open it in a digital image manipulation program such as Adobe Photoshop or Painter. Finally, choose again one single page and save it as a DWG or DWG and open it in a CAD program of your choice, such as AutoCAD, Vectorworks, or 3DS Max, and compare your results.

15.4 STAYING ORGANIZED WITH LAYERS AND PAGES

Here's something you already know: When you build a sophisticated document, cutting corners at the beginning of the process comes back to haunt you when you need to make a last-minute change. Knowing *exactly* how LayOut's layers work gives you the confidence to use them all the time. The frustration you avoid late in the schedule (when it really counts) is easily worth the extra few minutes it takes to "work clean."

15.4.1 Using Layers

LayOut has *layers* that (unlike SketchUp Pro's layers) act just like layers in InDesign, Illustrator, Photoshop, and every other graphics program available on the market today. Layers let you:

- Keep collections of similar elements separate and organized.
- Easily show or hide large numbers of elements at once.
- Lock elements so you can't accidentally change them.
- Stack one group of elements on top of another.
- Create design iterations by tweaking copies of the same elements.

Figure 15-10 shows a simple, one-page document with three layers. Different elements on the page are assigned to different layers. When you work with layers, keep these points in mind:

- To know which is the active layer, look for the little red pencil icon; anything you insert, draw, or paste is assigned to the active layer.
- To make another layer the active layer, click its name in the Layers panel.
- To show or hide a layer, click the "eye" icon next to that layer's name.
- To change the **stacking order** of layers, drag them around in the Layers panel.

Stacking order
The arrangement of all elements on the same layer in a document, determining which elements appear to be in front of which.

Figure 15-10

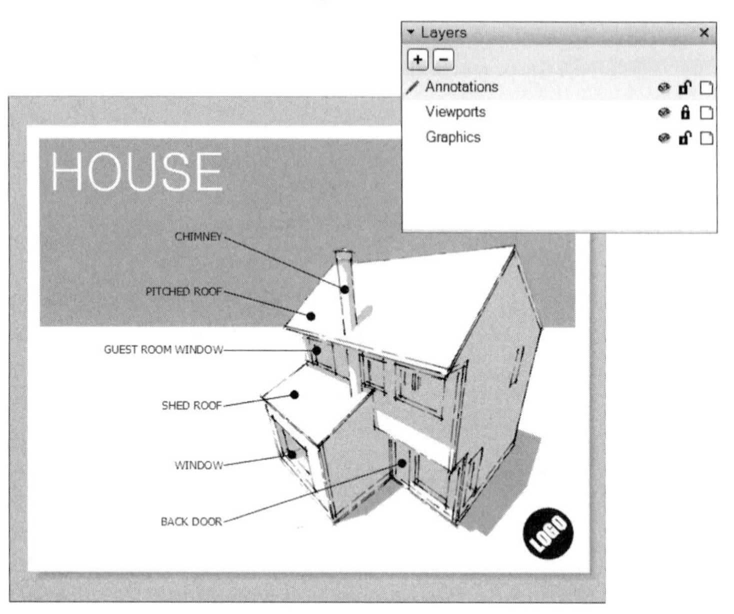

Layers let you organize the elements on your page.

- To see what layer an element is on currently, select the element and look for the tiny blue dot in the Layers panel. If you select two elements on two different layers, you see two blue dots.
- To change which layer something's on, select the destination layer in the Layers panel. Then right-click the element you want to move and choose Move to Current Layer. Selecting multiple elements, right-clicking one of them and choosing Move to Current Layer moves them all. Using layers is the absolute best way to work efficiently in LayOut. Check out the following tips for working with layers:
- Give your layers meaningful names. When you open your file sometime in the future, you want to be able to recognize it.
- Lock layers you're not using. Lock a layer by clicking the lock icon next to its name in the Layers panel.
- Improve performance by hiding layers. Make liberal use of the hide icon next to the name of each layer; hiding layers can really improve LayOut's performance, especially on slower computers. Hide any layers you're not working with, and you'll notice the difference.
- Duplicate a layer and its contents to save time. This is a quick way to iterate through several different versions. Here's how you can, too:

  a. Lock all your other layers and choose Edit ⇨ Select All.

  b. Create a new layer and name it. Make sure it's the active layer.

  c. Choose Edit ⇨ Paste.

- Move several elements from multiple layers to a single layer with Copy and Paste. Copying elements from multiple layers and pasting them pastes them all on the same layer—the active one.
- Group elements from different layers to create a group on the active layer.

15.4.2 Making Layers and Pages Work Together

Unshared layers
Layers containing elements that exist only on one page.

You can use layers to make certain elements appear on more than one page in your document. LayOut has two kinds of layers:

- **Unshared:** Any element (text, graphic, or otherwise) that you put on an unshared layer exists only on one page: the page you are on when you put the element on the layer.
- **Shared:** LayOut introduces the notion of *shared layers;* anything you put on a shared layer appears on every page of your document, as long as those pages are set up to show that layer. Think logos, title blocks, arrows, and other graphics that need to be exactly the same on most pages.

Shared layers can confuse new LayOut users, so here are a few quick tips about how you organize content on layers—including shared layers—as you create presentations in LayOut:

- You can make any layer a shared layer by clicking the sharing icon to the right of its name in the Layers panel. (See Figure 15-11.)
- You can make an element (such as a logo, a title block or a logo) appear in the same spot on more than one page by putting it on a shared layer. For example, the logo and the project title need to appear in the same spot on every page; these two elements on the shared On Every Page layer. In Figure 15-11, note how the logo and project title appear in exactly the same place on the second, third, and fourth pages.
- Put content that appears on only one page on an unshared layer. Again, on the last three pages of the document shown in Figure 15-11, the image boxes and page titles are different on each page, so put them on the unshared Default layer.
- You decide which pages should show which layers. For example, you don't want the logo and the project title to be on the cover (first) page. Toggle the show/hide icon beside the On Every Page layer to hide it on that page. Try working with at least four layers, organizing content on each as follows:
  - Elements that should appear in the same place on almost every page, such as logos and project titles.
  - Things that appear in the same place on most pages, but that change from page to page, such as numbers and page titles.

– Content (such as images and SketchUp Pro model views) that appears only on a single page.

– Unused elements that you are not sure what to do with but that you don't want to delete.

Figure 15-11

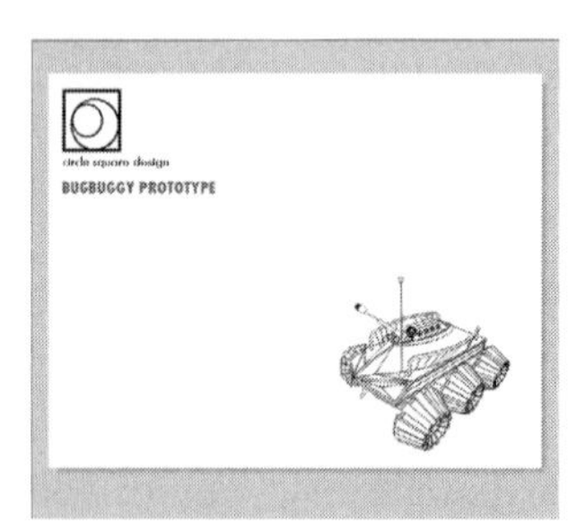

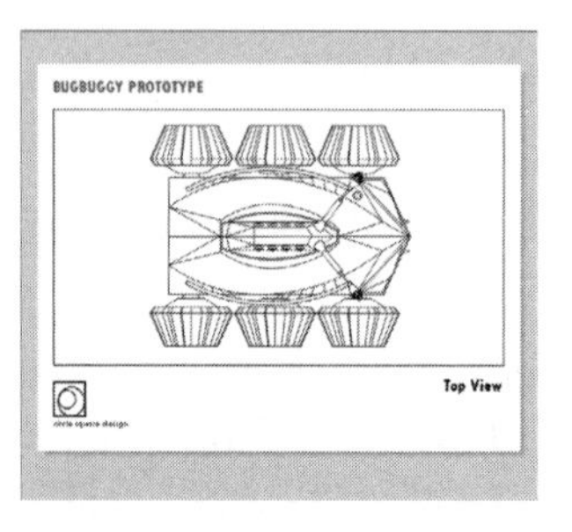

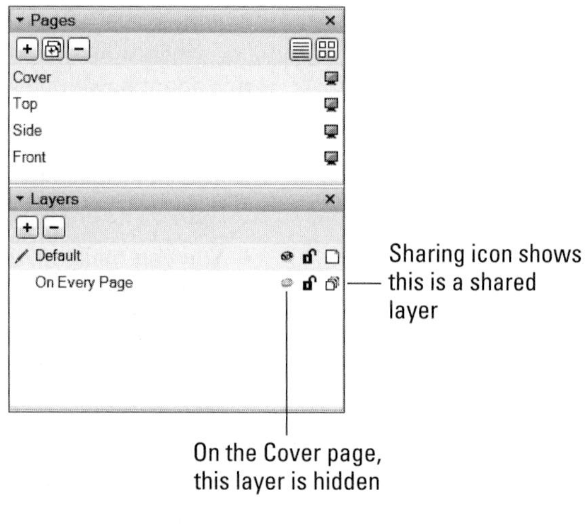

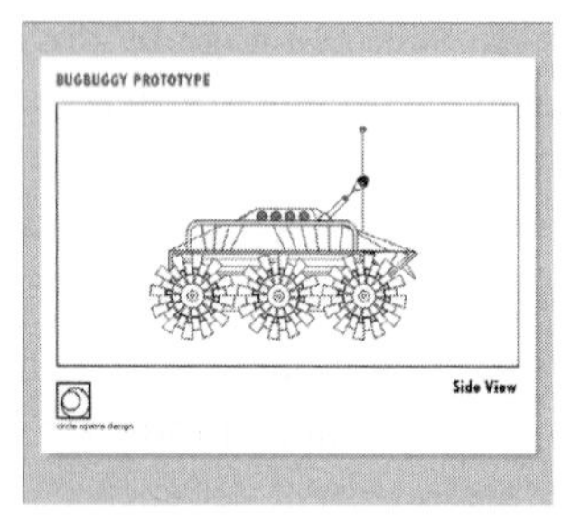

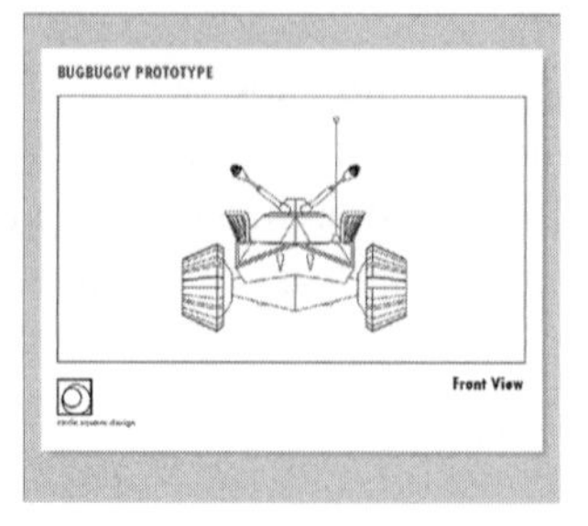

A simple document with two layers: One that's shared, and one that isn't.

SELF-CHECK

1. What are the two kinds of layers in LayOut?
   a. Unshared and Shared
   b. In and Out
   c. Top and Bottom
   d. Right and Left
2. We can make any layer a shared layer in LayOut. True or false?
3. If we are not sure of what to do with certain elements in our LayOut document, it is better to erase them than to put them in a hidden or invisible layer. True or false?
4. If we want to use a logo or another element that appears in the same place page in many pages, we would:
   a. Copy it on every single page.
   b. We would assign it to a shared layer.
   c. We would redraw it with the pencil tool.

Apply Your Knowledge Managing and shuffling information effectively throughout your layers in LayOut is one of the most effective ways to create compelling and visually engaging presentations. We need to remember that quite often our 3D creations need to be accompanied with a plethora of information (usually in the form of callouts and blocks of text) that describe our design with precision, so that it can be understood by other designers, vendors, or builders.

Using the same cottage design that we worked on in the previous exercise, experiment by adding on each page different elements, such as annotations, dimensions, text, and labels. Then practice including them in different layers. Is there any information that you would want to include on just one layer? What same graphic material or text would you add on more than one page, on the same spot? What information would you reject and would you compile your "leftover arsenal" on a hidden layer?

15.5 WORKING WITH INSERTED MODEL VIEWS

Being able to choose views of your SketchUp Pro models and put together documents to present them is what LayOut is all about. Text, vector drawing, raster images, and everything else aside, LayOut is a tool for presenting SketchUp Pro models.

Learn to frame your model the way you want it.

This section is about two things: managing the model views that you've inserted into your LayOut document and controlling how they look. You accomplish both by adjusting the controls in the SketchUp Pro Dialog Box, which you can open by choosing Window ⇨ SketchUp Pro Model from the menu bar. To insert a view of a SketchUp Pro model into your LayOut document, all you have to do is choose File ⇨ Insert and pick the model you want to work with. Creating the look you want, on the other hand, can be a challenge and that's where this section comes into play.

15.5.1 Framing Exactly the Right View

A SketchUp Pro model view that lives in your LayOut document is a *viewport*. You can have multiple viewports that show the same model. For example, you may have different viewports for a top view, a perspective, and a section through a building you're designing. They're all linked to the same model but show it in different ways.

The next few sections provide specific advice on setting up different model views in your documents' viewports. Of course, your LayOut document can also have viewports that correspond to more than one SketchUp Pro model.

Seeing Precisely what You Want to See

When you insert a SketchUp Pro model into a LayOut document, it shows up in a new viewport. Not only that—it shows up looking exactly the way it did in SketchUp Pro when you saved it (see Figure 15-12).

Figure 15-12

Saving while on this view in SketchUp...

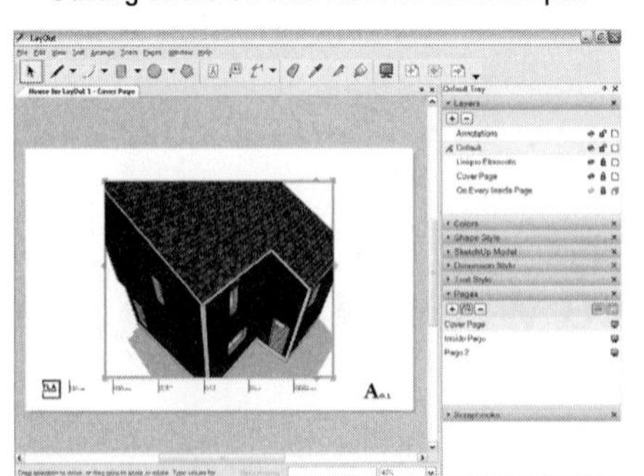

...yields this view when you insert the model into LayOut

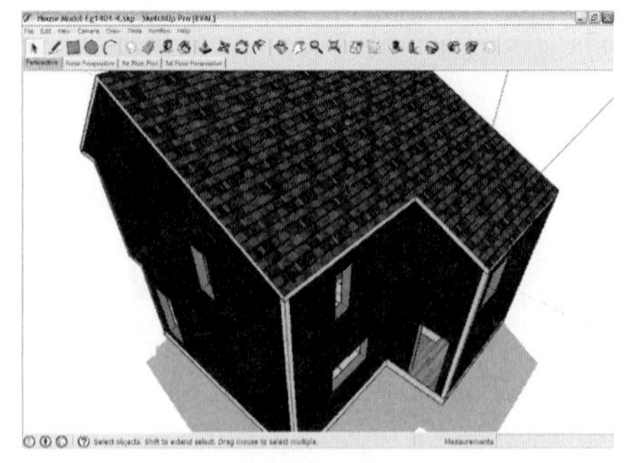

Newly inserted SketchUp Pro models look just like they did when you saved them.

You can do a couple things to change your viewport's point of view: Use the Camera tools or edit scenes with the Model panel.

Using the Camera Tools Directly

Double-clicking a viewport with the Select tool is a little bit like activating SketchUp Pro from inside LayOut. The model looks different and sometimes worse, than it did. That's because you're looking at the model itself, instead of the rendered image of it that LayOut made when you inserted it. To change your point of view, you can:

- Orbit, zoom, and pan around using your mouse, exactly the way you do in SketchUp Pro.
- Right-click the viewport and choose a specific Camera tool from the context menu.

When you're done repositioning your model, click somewhere else to stop editing the viewport. LayOut re-renders the view, and your model goes back to looking nice and crisp.

Using Scenes and the SketchUp Pro Model Panel

Using the SketchUp Pro Model panel is by far my preferred method for controlling what's visible in viewports. Instead of messing around with the Camera tools, choose a scene to display from the Scenes drop-down list in the panel's View tab; see Figure 15-13.

Of course, working this way requires that you first set up scenes in your SketchUp Pro model, but that's not hard at all; Chapter 10 describes the whole simple process of creating and working with scenes.

Figure 15-13

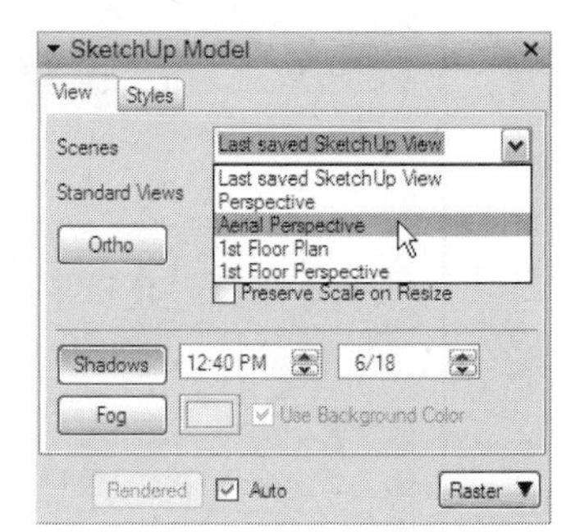

Use the View tab of the SketchUp Pro Model panel to control your viewport's point of view.

Clipping masks
Used in LayOut to hide the parts of images that you don't want to see.

FOR EXAMPLE

CROPPING WITH CLIPPING MASKS

Cropping an image means reframing it so that you can see only part of it; every page-layout program on the planet allows you to crop images, and each one insists that you do it a little differently. LayOut is no exception. In LayOut, use ***clipping masks*** to hide the parts of images, and viewports that you don't want to see. Follow these steps to use a shape as a clipping mask; the images below provide a visual reference:

1. **Draw the shape you want to use as a clipping mask and make sure that it's positioned properly over the image you want to crop.**
2. **Use the Select tool to select both the clipping mask object and the image you want to crop.**
3. **Right-click the selected elements and choose Create Clipping Mask from the context menu.** Sometimes, it works better to use Shift to select both the shape and the viewport (keep in mind that what is kept is what's inside the shape).

Here are some fun facts about clipping masks in LayOut:

- **Clipping masks work on inserted images.** This includes both raster images and SketchUp Pro model viewports.
- **Deleting clipping masks is easy.** To see a whole image again, select the image and choose Edit ⇨ Release Clipping Mask.
- **Edit clipping masks by double-clicking them.** When you double-click a clipping mask, you can see the whole image and the shape you used to create the mask. Now you can modify the shape, the image, or both. Clicking somewhere else on your page exits the edit mode.

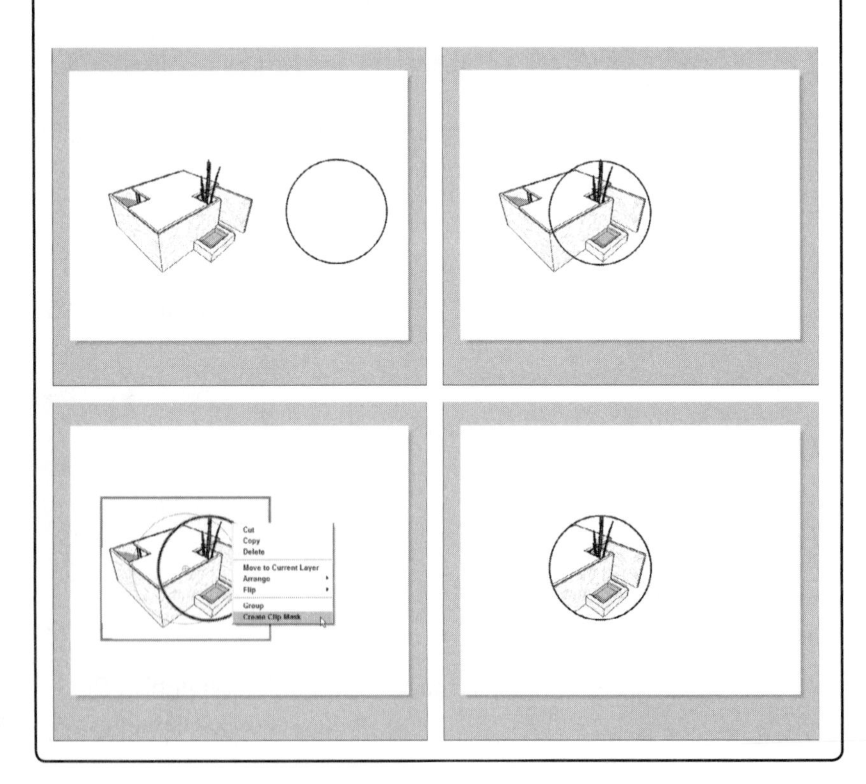

Setting up the views you want to use in LayOut by saving scenes in SketchUp Pro makes things easier for four main reasons:

- **You can see more.** Your SketchUp Pro modeling window is bigger than your viewport in LayOut, so it's easier to see what you're doing.
- **You can go back to a previous view.** Repositioning your model in a LayOut viewport is kind of a temporary thing; if you change things, there's no way to come back to the view you set up previously. SketchUp Pro scenes, on the other hand, are views that you can always return to.
- **You can show section cuts.** Scenes are the *only* way to save views of your model with different section planes active.
- **You have more control over shadows, fog, and styles.** LayOut provides basic tools for adjusting other aspects of your viewports' appearance, but they are nowhere near as easy to use as the ones in SketchUp Pro.

If the scenes you created in SketchUp Pro aren't visible in the SketchUp Pro Model panel, you probably forgot to save your model before you switched applications. You also need to make sure your viewport is current; right-click it and choose Update Reference to make sure everything's up to date.

Figure 15-14 shows a LayOut page with three viewports on it. All three show the same model. Before laying out this page, three scenes were created in SketchUp Pro:

- **Large plan view:** To get this point of view, lop off the temporarily top of the model by adding a section plane about 48 inches from the floor. Turn off Perspective view (Camera ⇨ Parallel Projection) and choose a top view (Camera ⇨ Standard Views ⇨ Top). Apply a Hidden Line style to make it black and white and then choose View ⇨ Section Planes to hide the section plane. Update the style and turn on Shadows to help the model read better on the page.
- **Smaller perspective views:** To create these two scenes, use a little trick: Create two section planes a couple inches apart. "Point" one down and the other up; right-clicking a section plane lets you reverse it (Edit ⇨ Section Plane ⇨ Reverse)—whatever was hidden becomes visible and vice versa.

Creating Scaled Orthographic Views

With the addition of honest-to-goodness dimensions (both linear and angular) to LayOut in SketchUp Pro 8, people have started to push the limits of what LayOut was intended to do. LayOut was never supposed to be a 2D drafting tool; its toolset has always been closer to Illustrator than to AutoCAD. That said, you absolutely can put together simple, scaled drawing sets, complete with title blocks, symbols, dimensions, and other forms of annotation.

The first step in creating a dimensioned drawing is to turn your viewport into a 2D orthographic view of your model. (See Chapter 10 for an

Figure 15-14

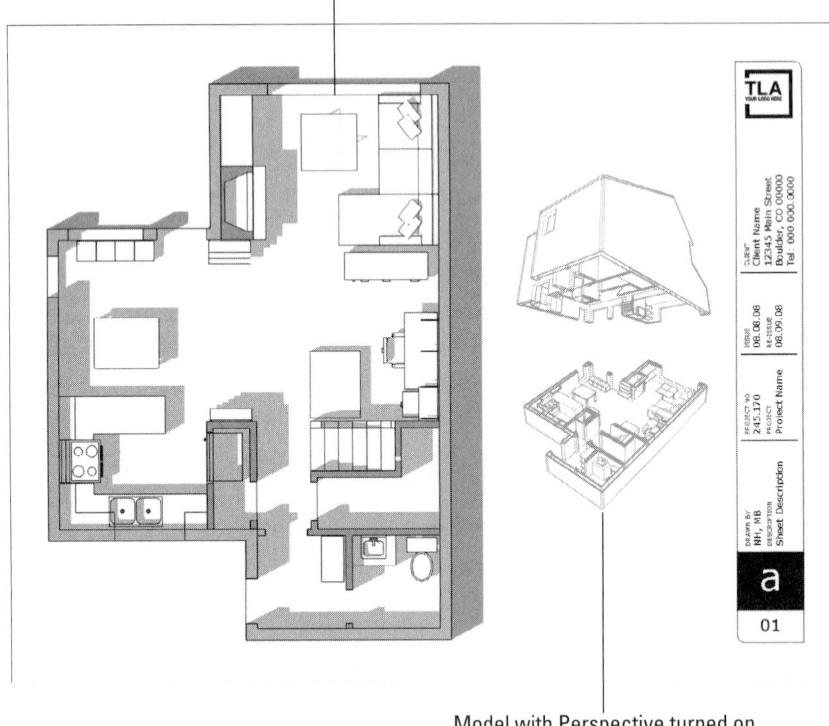

Three viewports inserted in a LayOut document.

introduction to orthographic views.) Although you can use the controls in the SketchUp Pro Model panel to accomplish this, my favorite way is to go back to the model and create a scene.

PATHWAYS TO... SAVING AN ORTHOGRAPHIC SCENE IN YOUR SKETCHUP PRO MODEL

1. In LayOut, right-click (with the Select tool) the viewport that contains your model and choose Open with SketchUp Pro.
2. If you plan to have an active section cut in your view, add it to your model (if you haven't already). Chapter 10 explains how to make section cuts.
3. In SketchUp Pro, choose Camera ⇨ Parallel Projection; then choose Camera ⇨ Standard Views ⇨ Top (or any other option from this list except Iso).

4. Zoom and pan (but don't orbit) until you have the view you want and then choose View ⇨ Animation ⇨ Add Scene.
5. Save your model and close it.
6. In LayOut, right-click the viewport and choose Update Reference.
7. In the View tab of the SketchUp Pro Model panel, choose your new scene to associate it with the viewport.

Now that you have an orthographic view of your model, you can assign a scale to it.

PATHWAYS TO... ASSIGNING A SCALE TO AN ORTHOGRAPHIC VIEW OF YOUR MODEL

1. Assign a scale using the scale drop-down list in the SketchUp Pro Model panel (see Figure 15-15). Don't forget to select the viewport you're working on first.
2. You can create your own scales if you want. Need a scale that doesn't appear in the default list? Choose Edit ⇨ Preferences (LayOut ⇨ Preferences on a Mac) to open the Preferences dialog box; then click Scales on the left. Click the plus sign to add a new scale to the list. Scales you add are available for any LayOut file you're working on.
3. Make sure Preserve Scale on Resize is selected. After you assign a scale to a viewport, you probably want to manually resize its boundaries with the Select tool. Before you do, make sure the Preserve Scale on Resize check box (in the SketchUp Pro Model panel) is selected. If it's not, you change the scale of your model view when you try to resize its viewport.

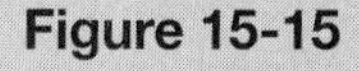

Figure 15-15

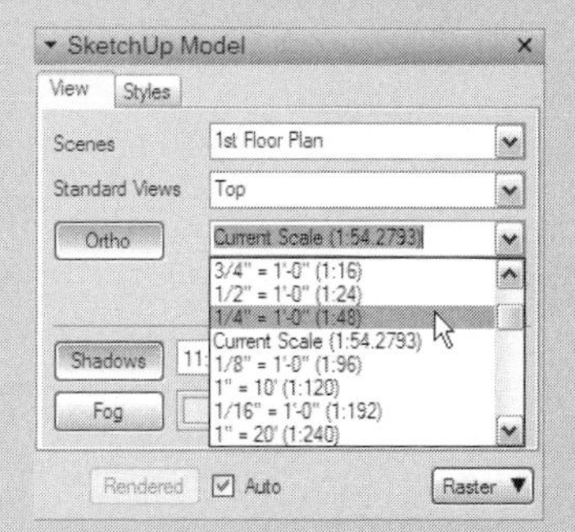

Use the SketchUp Pro Model panel to assign a scale to your viewport.

15.5.2 Making Your Models Look Their Best

Getting your models "posed" correctly on the page is only half the battle; they also need to look readable and compelling. That's what this section is about. You discover how LayOut *renders*, or draws, your models on the page and how adjusting line weights can make your drawings look their best.

Choosing Raster, Vector, or Hybrid

Every time you insert or edit a SketchUp Pro model view, LayOut renders an image of your model to display in the viewport. This rendering process is just like exporting an image from SketchUp Pro; it can produce either a raster or a vector, depending on the settings. Read Chapters 13 and 14 for more information about raster and vector images.

You control how your models look by choosing which method LayOut uses to render each viewport. Simply select an option from the rendering method drop-down list in the lower-right corner of the SketchUp Pro Model panel.

You have three choices, which I illustrate in Figure 15-16:

- **Raster:** Renders your viewport as an image comprising many, many little dots. If your model is rendered as a raster, it can display sketchy styles, shadows, and other effects that make it look like it does in SketchUp Pro. On the other hand, printing or exporting a raster image at larger sizes involves truckloads of pixels, and that can make LayOut choke. See the nearby Tips from the Professionals box "Balancing Performance and Quality" to find out more.

Figure 15-16

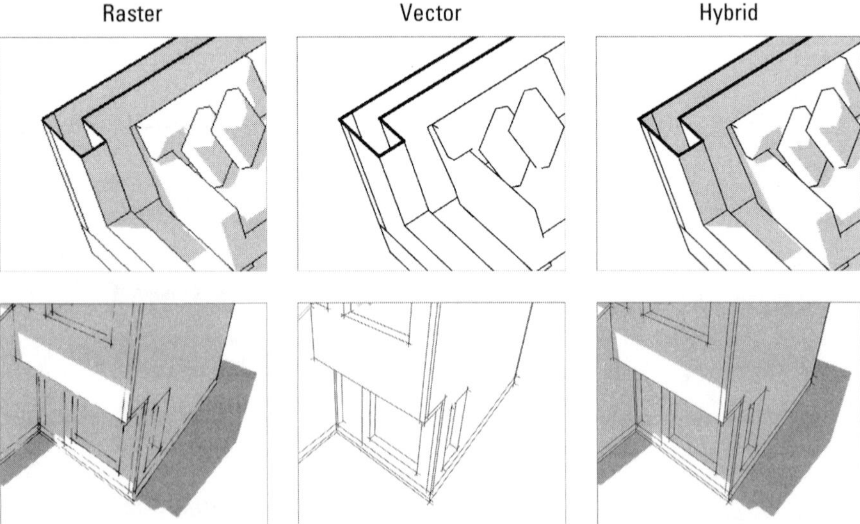

Choose a rendering method for each viewport in your LayOut document.

- **Vector:** Renders your selected model view as a vector image. Lines appear smooth and crisp, but things like shadows, textures, and sketchy styles don't appear. Also, choosing vector rendering for really complex models can take a long time to process.
- **Hybrid:** Combines clean vector lines with rich raster faces, shadows, and other goodies. Behind the scenes, LayOut actually renders twice—once as a vector and once as a raster. Hybrid rendering takes even longer than vector rendering, but it produces very nice results. If you have time, try hybrid rendering to see how it looks.

Consider using raster rendering for views of my models that involve Sketchy Edges styles and for any model with a lot of geometry. Use hybrid or vector rendering for any plans, sections, or other views that feature a lot of line work.

Line Weight

The SketchUp Pro Model panel's second tab—Styles—contains one of the most important settings in all of LayOut. In the lower-left corner, the Line Weight field lets you control how bold your models look; see Figure 15-17.

Figure 15-17

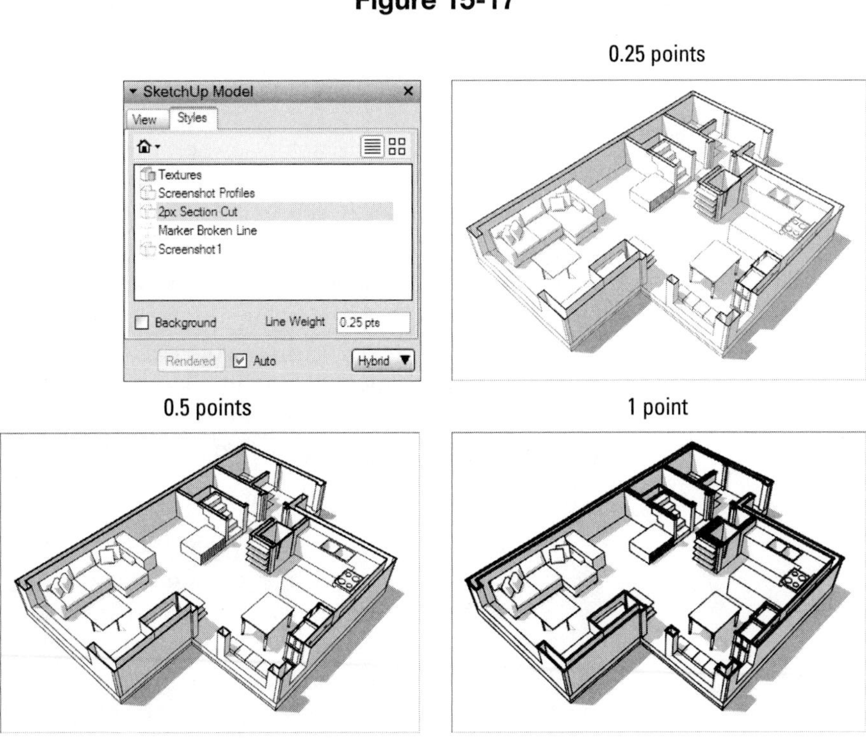

Use the Line Weight field to make your models look their very best.

TIPS FROM THE PROFESSIONALS

Balancing Performance and Quality

LayOut places rendered images of your SketchUp Pro models into viewports on the page; that's why your models look so much better in LayOut than they do in SketchUp Pro's modeling window. But that's also why LayOut can feel so slow at times; rendering is a very time-intensive activity. Luckily, you have a few ways to manage LayOut's speed.

Manage Your Rendering Resolution

At the bottom of the Dialog Box (File ⇨ Document Setup), you find two settings: Edit Quality and Output Quality. They both control the pixel resolution of raster-rendered viewports in your document. Low correlates to a resolution of 72 ppi (pixels per inch), Medium to 150 ppi, and High to 300 ppi. The higher the resolution, the more pixels LayOut has to figure out—and the longer it takes to render. By default, both settings are set to Medium for every new document you create.

When working with a big raster or hybrid-rendered viewports, dial down Edit Quality to Low. Doing so doesn't adversely affect the quality of my exports and prints; both of those are controlled by Output Quality. What it does, is ask LayOut to draw far fewer pixels every time you adjust a viewport.

Be Careful About Choosing Vector Rendering

When you set a viewport to be rendered as a vector, LayOut runs through every single edge and face in your SketchUp Pro model; even the ones you can't see. Really big models can take a very long time to vector-render. Another thing: Unlike rasters, vectors don't care how small your viewport is. A 1-millimeter viewport takes just as long to vector-render as a one meter one. Big viewports destroy rasters, whereas big models destroy vectors, which is worth remembering.

Here's a tip if you're using vector rendering: LayOut renders *everything* in your SketchUp Pro model when you use vector rendering on a viewport—even the geometry that isn't visible. To speed things up, put everything you don't need LayOut to render on a separate layer (in your SketchUp Pro model,) hide that layer, and save a scene to associate with your viewport in LayOut. Chapter 7 has everything you need to know about using layers in SketchUp Pro.

Switch Off Auto Rendering

Notice the controls in the lower-left corner of the SketchUp Pro Model panel? When the Auto check box is selected, LayOut automatically re-renders a viewport every time you edit it. If your model is big and heavy, you have to wait while LayOut works, and that can get old, fast. Deselecting the Auto check box lets *you* decide when LayOut should render your viewports.

Just select a viewport and click the Render button (also in the SketchUp Pro Model panel) to tell LayOut to start cranking.

The number you put into the Line Weight field tells LayOut how thick to draw the thinnest lines in your viewport. Entering **2** yields edges that are 2 points wide.* Typing **0.25** makes your edges a quarter point wide—much thinner and (in many cases) much nicer.

FOR EXAMPLE

MANAGING STYLES INSIDE LAYOUT

The Styles tab of the SketchUp Pro Model panel gives you access to all the styles included with SketchUp Pro, plus any that you've saved with your model. You can select any viewport in your document and apply a style to it at any time.

The tricky part lies in knowing which styles are actually saved with your model. When you add in elements like section plane visibility, background colors, and transparency, it becomes difficult to know what you have.

If you depend on styles to work in LayOut, make sure you do two things:

- **Create scenes.** The easiest way to make sure you can get the model views you want in LayOut is to create scenes in SketchUp Pro beforehand. As long as the style in each scene looks right, you should be just fine when you get to LayOut.
- **Update continuously.** Problems sometimes occur when people forget to update their current style before they save their model. This creates a disconnect between the styles in SketchUp Pro and the ones in LayOut. Before you save a SketchUp Pro model you're using in a LayOut viewport, make sure your current style doesn't need updating. If it does, its thumbnail (in the Styles dialog box) looks like the one in the following image. Click the icon to bring the style up to date.

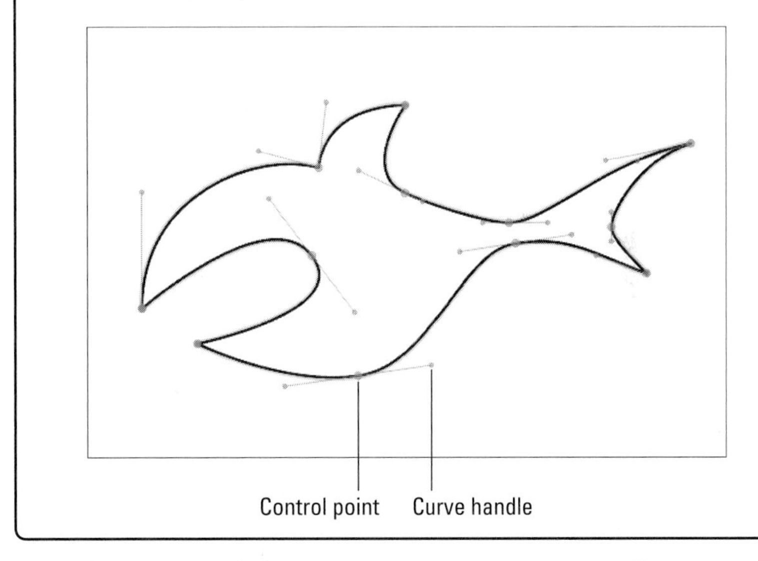

Changing the Line Weight number is the single best thing you can do for your models in LayOut. The line weights you use depend entirely on the size of your viewports and the complexity of your drawings. Try to avoid making anything look too wispy or too chunky; the key here is *readability*.

If the style that's applied to your viewport has Profiles enabled, some edges look thicker. To change the thickness of Profiles in a LayOut viewport, you need to edit the style that defines them in the SketchUp Pro model. Profile thickness

* In typography, a point is described as the smallest unit of measure, and its length is measured as 1/72nd of an inch.

is always a multiple: A setting of **4** produces Profiles that are four times as thick as regular edges. Read Chapter 9 for more about styles and how to edit them, and see the nearby For Example, "Managing Styles Inside LayOut" to find out how they relate to SketchUp Pro's companion application.

SELF-CHECK

1. What is the name of a SketchUp Pro model view that lives in your LayOut document?
2. When you insert a SketchUp Pro model into a LayOut document, it shows up in a new viewport. True or false?
3. Since LayOut places rendered images of our SketchUp Pro models, that can slow us down if the document is large. Name one of the three possible ways that we can utilize to make LayOut perform faster.
4. Name one thing we can do when we double-click on a viewport with the Select tool.

Apply Your Knowledge Design in SketchUp Pro a simple writing desk with some drawers on the side and some built-in stationary cubbies above the writing surface. Then, save three orthographic views (top, front, and one side view) and one perspective view. Then, using your LayOut template, create a two-page document.

On the first page insert the front view and after double-clicking on it, right-click on it and go to Camera Tools and practice with Orbit, Zoom, and Pan using your mouse. Once you are done, click outside of the viewport to exit and then right-click on the viewport again and choose a specific Camera tool (such as perspective or left view, for example) from the context menu.

On the second page insert the front view on a new viewport. Then draw a circle over the drawer unit on one of the viewports and, using the Select tool and with right-click, experiment with Create Clipping Mask and scaling the resulting detail view.

15.6 DISCOVERING MORE ABOUT DIMENSIONS

When Google added dimensions to LayOut in SketchUp Pro 7.1, designers, architects, engineers, woodworkers, and all kinds of people were pleased. With the addition of *angular* dimensions in SketchUp Pro Pro 8, LayOut became a full-fledged tool for creating scaled, annotated drawings from your models.

If you need a refresher, there is a set of instructions for using both of LayOut's dimension tools described earlier in this chapter.

Learn how to edit your dimensions.

If you have trouble getting your dimension tool to "see" any of the points in a model viewport, chances are you don't have Object Snaps (Arrange ⇨ Object Snaps) turned on.

15.6.1 Editing Your Dimensions

After you actually draw a dimension on your page, you can do many things to change what the dimension looks like, whether it is linear or angular. To begin with, take a look at the anatomy of a dimension. Figure 15-18 shows an example of each kind.

Now that you understand the nomenclature, here's some advice on making the kinds of changes you may want to make:

- **Use the Shape Style panel** to change colors, line styles, line weights, and arrow styles (the things at the ends of dimension lines). Basically, the Shape Style panel is for controlling everything about your dimensions except their text strings.
- **Use the Dimension Style panel** to change the format of text strings and their level of precision. Read ahead for a more in-depth look at the Dimension Style panel.
- **Double-click a dimension to get access to all its internals.** After you've double-clicked to start editing a dimension, you can move its connections points, offset points, or extent points all you like.
- **Click and drag to move a text string.** Need to reposition a text string to make it more legible? Just drag it someplace else after you've double-clicked the dimension to edit it.

Figure 15-18

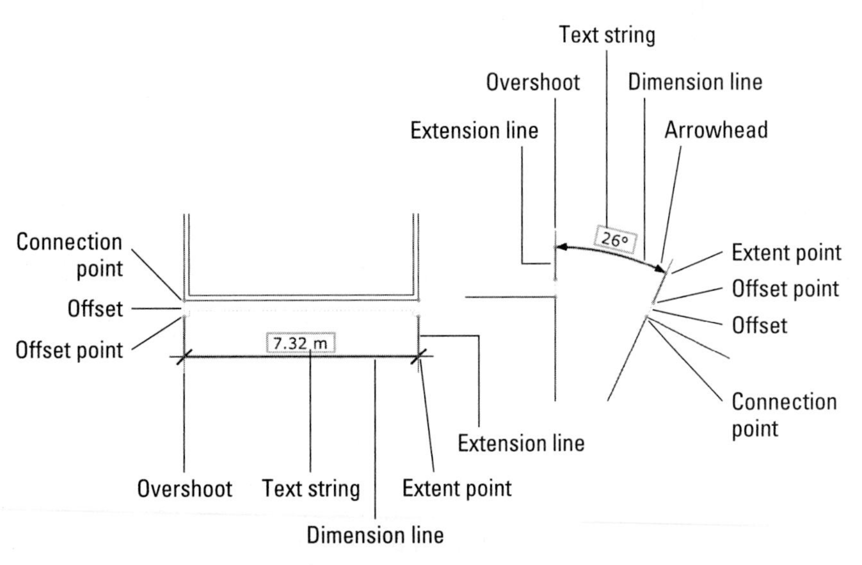

The anatomy of a LayOut dimension: linear on the left, angular on the right.

- **Double-click twice more to edit a text string.**
- **Select individual lines to edit them individually.** After you double-click a dimension, you can select its constituent lines one at a time. Draw the dimension lines slightly thicker than the extension lines; selecting the former individually allows this.
- **Overshoots can be challenging.** The *overshoot* (as shown in Figure 15-18) is the part of an extension line that extends beyond the dimension line. You can adjust your overshoots' length if you choose. Here's how:
  - **a.** Double-click a dimension to edit it.
  - **b.** Click to select the extension line whose overshoot you want to adjust.
  - **c.** Change the number beside the End Arrow setting in the Shape Style panel.

Unfortunately, there's no way to alter both extension lines' overshoots simultaneously. It is recommended to change one, choose Edit ⇨ Copy Style; then select the other and choose Edit ⇨ Paste Style.

Copying a dimension's style and applying it to your other dimensions is relatively simple. Just use the Style tool to transfer formatting from one dimension to the other. To sample a dimension's style so that every new dimension you draw matches it: Activate the tool, tap the S key, and click your "source" dimension before you draw the next one.

Look at the Dimension Style panel, as shown in Figure 15-19. Most of the controls here are obvious, but some definitely aren't.

- **Text position:** Choose to display a text string above, below, or in the middle of its corresponding dimension line.
- **Text alignment:** Force a text string to always be horizontal or vertical on the page, or aligned (parallel) or perpendicular to its dimension line.
- **Display units:** People who use the Imperial system of measurement tend to show the units on their dimensions. Metric folks tend not to. You have the choice.

Figure 15-19

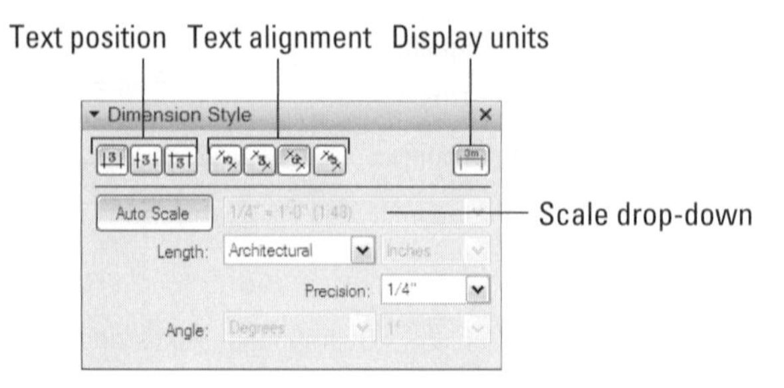

The Dimension Style panel is where you control your dimensions' text string format.

- **Auto Scale button:** Here's where dimensions start to get a little bit complicated. For a full discussion on this button's functions, see the next section.
- **Scale drop-down list:** This is available only when Auto Scale is deselected. Move ahead to read all about model space and paper space. It's advisable to spend some time on this topic in order to fully grasp the capabilities.
- **Length:** Different professions have different conventions for the dimensions they put on their drawings. Choose the one that suits you best.
- **Precision:** As a guideline, if you dimension the overall length of an airport runway, you probably don't need to be accurate the 1,000th of an inch. However, if you are designing an artificial heart valve, precision becomes much more critical.
- **Angle:** Degrees or radians that you can determine based on your need.

15.6.2 Keeping Track of Model Space and Paper Space

When you place a SketchUp Pro model viewport on your page, you end up with two different types of space in your LayOut document:

- **Paper space:** Distances that pertain to the physical sheet of paper you're working on are said to be in *paper space*. A 4" × 4" blue square in paper space is 4 inches long.
- **Model space:** Distances within a model viewport have nothing to do with the size of the sheet of paper the viewport's on. An 80' × 80' building shown at 1 inch = 8 feet scale is 80 feet long in model space. In paper space, it's 10 inches long.

  A dimension you draw in LayOut is either in paper space or in model space. Which one the dimension is in by default depends on what the dimension is connected to:

- **Viewports:** When you draw a dimension between two points in a model viewport, LayOut is smart enough to presume that you want to display the length between the points *in the model* (in model space).
- **Everything else:** When you create a dimension between two points that have nothing to do with a viewport, LayOut assumes that you want to see the *actual length on the page* (in paper space).

Figure 15-20 illustrates this point. Both dimensions are exactly the same physical length on the page: 3 inches. The difference is that the dimension on the left is attached to two points on a SketchUp Pro model that's shown at 1 inch = 8 feet scale. It displays the *model space* length of 24 feet (3 × 8), whereas the dimension on the right just shows its *paper space* length of 3 inches.

The Auto Scale button in the Dimension Style panel is automatically selected whenever you create a new dimension. If your dimension touches

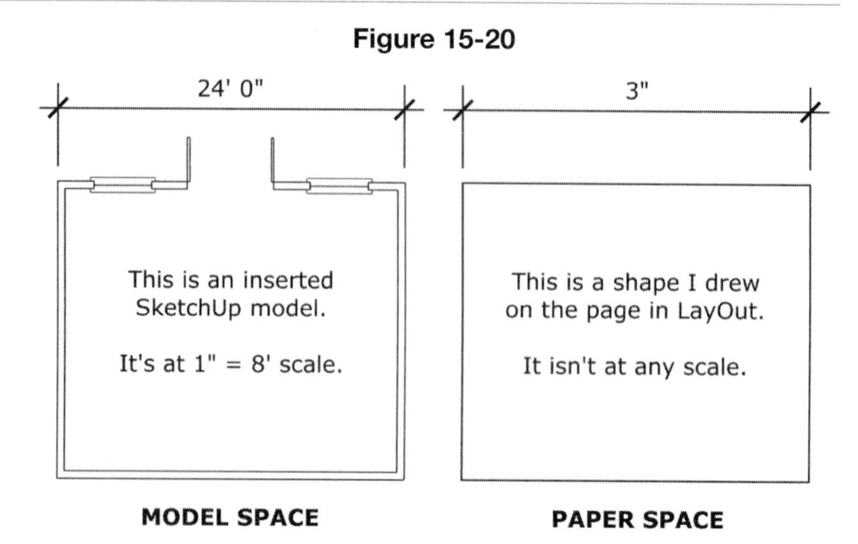

Dimensions can either show model space or paper space.

a point in a model viewport, the text string displays the length in model space. If it doesn't touch any model viewport at all, you get a length in paper space. Turning off Auto Scale lets you assign a scale to the dimension you select. Choosing 1" = 60 feet for a dimension that's physically 4 inches long makes its text string read 240 feet—no matter what it's attached to.

IN THE REAL WORLD

Interior designer Eric Schimelpfenig, Classic Kitchens in Springfield, goes on to say that Google SketchUp Pro is great because of its intuitive design and numerous plugins and resources, which makes it an excellent sales tool that is far more efficient than any other software used in kitchen design. He says that SketchUp Pro really helps him sell his designs quickly.

Once he sells a design to his clients, he begins preparing the construction documents in LayOut. The first step he follows is to include the images that he showed to his clients. By doing it this way, the builders will know exactly how he wants the final product to look. The next step is to create a series of 2D elevations and plan views, which are annotated and dimensioned to contain all the information needed for the construction of the project. He also likes to include an idea board at the end of the presentation with a collage of all the materials and finishes used in the design process. According to him, the beauty of creating everything in LayOut is that any changes made to the SketchUp Pro model will automatically be applied to all of the construction documents as well, making updating or modifying the design quite fast and easy.

SELF-CHECK

1. To change colors, line styles, line weights, and arrow styles use the:
   a. Dimension Style Panel
   b. Shape Style Panel
   c. Preferences Panel
2. What is the name of the small line that extends beyond the dimension line?
3. If you have trouble getting your dimension tool to "see" any of the points in a model viewport, what would you need to turn on?
4. What would we have to turn off to manually assign a scale to the dimension we select in the Dimension Style panel?

Apply Your Knowledge Create in SketchUp Pro a design for a storefront for your town and create a scene of the front elevation, in parallel projection either in white or in full color. Then, using your template in LayOut, insert a viewport that includes that scene. Then, choose a scale that will fit that viewport nicely in your page, such as 1" = 1'–0", ½" = 1'–0" or ¼" = 1'–0" depending on the size of the store front. Then, explore adding a collection of dimensions to that view, in Model space. Finally, add one or two dimensions in Paper space and notice the difference in the measurements.

As a second challenge, work with the controls inside the Dimension Style panel, such as text position, text alignment, auto scale, length, angle, or precision.

15.7 DRAWING WITH LAYOUT'S VECTOR TOOLS

GOOGLE SKETCHUP PRO IN ACTION

Understand how and when to use LayOut's Vector tools.

LayOut includes a full slate of drawing tools that you can use to create logos, title bars, north arrows, graphic scales—anything you want. The drawings you create are *vectors*, meaning that you can do the following:

- Scale the drawings without losing quality
- Change the fill and stroke (outline) colors
- Split lines and then rejoin them to make new shapes

Here are a few pointers when drawing:

- **Use the right kind of snaps.** Drawing exactly what you want is easier if you let the software help. Just like SketchUp Pro, LayOut includes an elaborate (but easy-to-use) *inference* system of red and green dots and

TIPS FROM THE PROFESSIONALS

Curves are Back

LayOut's Line tool is one of the most intuitive *Bézier* curve vector drawing tools on the market today. In vector software (such as LayOut and Illustrator), you can draw curved lines freehand, but they don't usually look very good. It's hard to draw smooth, flowing curves without assistance, and that's where the Line tool comes in. Figuring out how to draw Bézier curves usually takes a little getting used to, but after you do, you'll be pleased you took the time to learn it.

LayOut's Line tool excels because of two things:

- **The Line tool uses the inference system.** SketchUp Pro and LayOut share a similar "guidance" system of colorful points and lines that help guide you. By turning on Object Snap when you draw, you can align things and draw more accurately.
- **Draw and edit with two tools.** Use the Line tool to draw curves (and straight lines, of course) and then do all your editing with the Select tool. Compared to curve tools in other software, LayOut does a lot more with a lot less.

Here are pointers on creating and editing curves with the Line tool:

- **Click-click-click to draw straight-line segments.** The LayOut Line tool works just like the Line tool in SketchUp Pro.
- **Click-drag-release to draw curved-line segments.** You "shape" your curve while you drag.
- **Double-click with the Select tool to edit.** When you do, you see all your line's control points and *handles* (the antenna-looking elements coming out of your control points). Click and drag points and handles to edit your line, and then click somewhere else to stop editing.
- **Drag a control point on top of an adjacent point to delete it.** If you want to remove a control point while you edit a line, just drag it onto one of its neighbors.
- **Hold down the Ctrl key and click somewhere on your line to add a point.** Hold down the Option key on a Mac.
- **Hold down Ctrl and drag on a point to pull out curve handles.** (The Option key on a Mac.)
- **Hold down Ctrl (Option on a Mac) and drag on a handle to sharpen a curve.**

lines to help you line up things. LayOut also has a grid (that you define) to keep elements in your drawing aligned:

– *Snap to objects:* When you choose Arrange ⇨ Object Snap to turn on object snapping, colored hints to help you draw appear on-screen.

– *Snap to grid:* Choose Arrange ⇨ Grid Snap to turn on grid snapping. Now your cursor automatically *snaps* to (is attracted to) the intersection of grid lines in your document, whether your grid is visible or not. See earlier in this chapter for more information on setting up grids in LayOut.

You can use any combination of snapping systems (Object or Grid) while you work. To save time, assign a keyboard shortcut that toggles each system on and off. (To do that, use the Shortcuts panel in the Preferences dialog box, which you can find in the Edit menu; on a Mac, look on the LayOut menu.)

- **Type measurements and angles.** LayOut has a Measurements Box (in the lower-right corner of your screen), just like the one in SketchUp Pro. Read Chapter 2 for tips on working accurately with this box.
- **Build complex shapes out of simpler ones.** For example, Figure 15-21 shows how to draw a simple arrow. Follow these steps to build one just like it:
  a. Make sure that Grid Snap is turned off and Object Snap is turned on, and then draw a rectangle with the Rectangle tool.
  b. Draw a triangle with the Polygon tool. To do so, type **3s** and press Enter before you start drawing to make sure that you're drawing a triangle. Hold down the Shift key to make sure that the bottom of the triangle is a horizontal line.
  c. Shift-click to select both shapes and then choose Arrange ⇨ Align ⇨ Vertically to line up the rectangle and the triangle vertically. You can always use the Undo feature to go back a step; it's in the Edit menu whenever you need it.
  d. Deselect both shapes by clicking once somewhere else on your page, and then select a shape and move it up or down on the page (by pressing the up- and down-arrow keys) until the two shapes overlap.
  e. Use the Arc tool to draw a half-circle at the bottom of the rectangle.
  f. With the Split tool, click and hold down the mouse button over an intersection point, and don't release the mouse button until all the lines stop flashing blue. Do this for all four intersection points to split the shapes into a series of line segments.
  g. Use the Join tool (it looks like a bottle of glue) to connect all the line segments by clicking once on the arrowhead part, once on each half of the stem part, and once on the half-circle. You have one shape instead of three; verify this by clicking the shape once with the Select tool. You see one red "selected" rectangle around your new shape. If you don't have one shape, use the Join tool again.
  h. Move your new shape somewhere else and then delete the leftover line segments you don't need.
- **Open the Shape Style dialog box.** Use the Shape Style dialog box to change the fill and stroke characteristics of elements in your document. In plain English, this is where you pick colors for the things you draw. The controls are pretty straightforward. Try experimenting and see what happens.

Figure 15-21

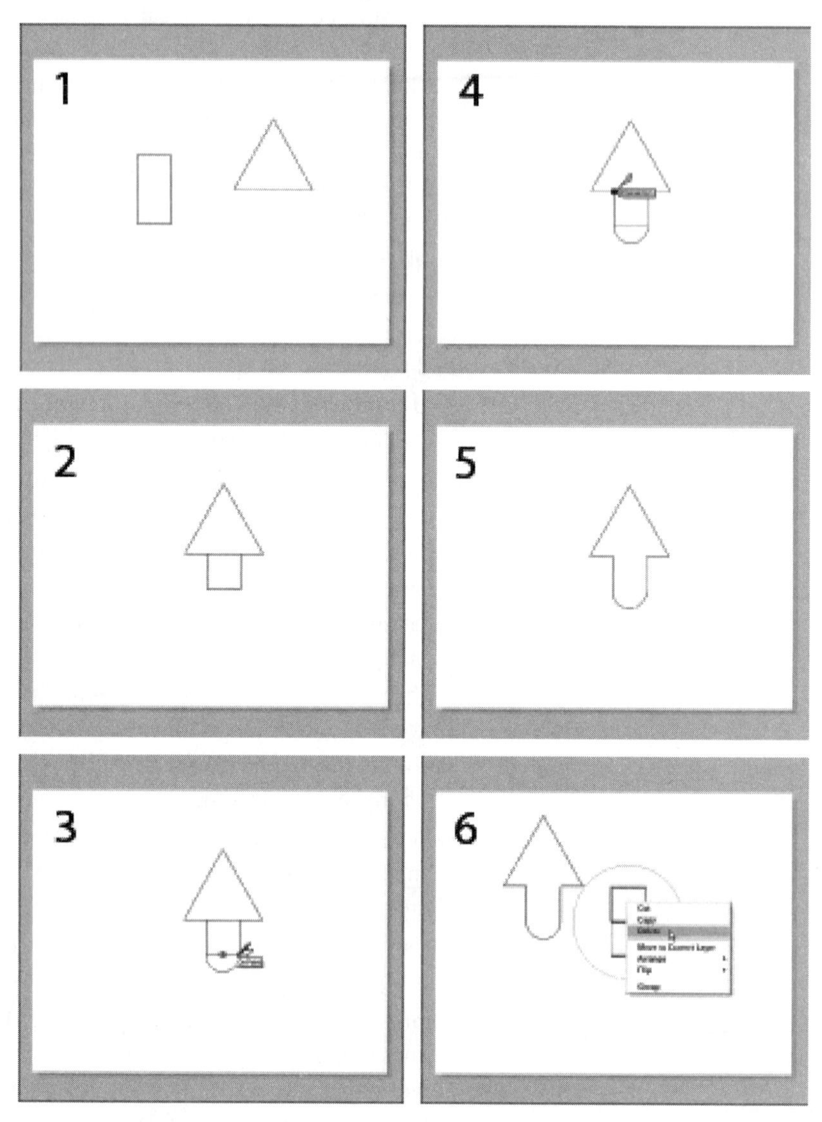

Drawing a simple arrow by combining a rectangle, a triangle, and an arc.

TIPS FROM THE PROFESSIONALS

Watch those Defaults

All LayOut's drawing and text tools come "out of the box" with a default setting, depending on the template you use. For example, the Line tool may be set to create a single-pixel black line with a white shape fill.

The Text tool may be configured to create 10-point Verdana. Tools automatically use the settings that were applied the last time they were used.

It's easy enough to change the default settings for your tools. Just follow these steps:

1. **Click the tool icon you want to use.**
2. **In the Shape Style or Text dialog box, choose the settings you want to use.**
3. **Draw or type to use the new settings.**

Keep in mind that changing the default settings in one LayOut file doesn't change them for other files. Tool defaults stick to particular documents. If you want to change the default settings for every new document you create, you need to create your own template. Discover how in "Creating Your Own Templates," later in this chapter.

SELF-CHECK

1. You can use any combination of snapping systems (Object or Grid) while you work. True or false?
2. When we open the Shape Style dialog box, what elements can we change of the drawings that we have done?
   a. Size and Scale
   b. Gradation and Hue
   c. Brightness and Contrast
   d. Fill and Stroke
3. If we have drawn two simple shapes in LayOut that are overlapping, such as rectangle and a polygon, name two tools that we can use to modify these shapes.

Apply Your Knowledge You are commissioned to come up with a concept of an exclusive spa complex in the mountains that would include various small huts, each one dedicated for one particular use. Design in SketchUp Pro one hut and copy it five times and scatter them one so that they are all separate from each other. Then, capture a plan view in an orthographic projection and create a scene of it.

Then, using your template in LayOut, insert a viewport that includes that scene. Then, choose a scale that will fit that viewport nicely in your page and then draw a collection of thick arrows that would indicate the different paths that we can take to visit the different buildings.

15.8 CUSTOMIZING LAYOUT WITH TEMPLATES AND SCRAPBOOKS

GOOGLE SKETCHUP PRO IN ACTION

Learn how to build your own Templates and Scrapbooks.

Every time you need to put together a drawing set, all you have to do is open your template (which already includes your logo, title block, layout, and text styles) and insert your model.

Need to add your firm's custom symbols? Just open one of your scrapbooks and drag the symbols onto your pages. Having a collection of your own templates and scrapbooks means you don't have to start over again on your project at the last minute.

15.8.1 Creating Your Own Templates

Most of the design presentations that you (or your firm) put together probably look very similar since they are part of your brand identity. If the presentation documents you make are all variations of a few themes, why not build your own templates and use them every time you need to start a new project?

You can configure LayOut so that your templates appear in the Getting Started dialog box, making it easier to build consistent presentations, quicker.

PATHWAYS TO...
TURN A LAYOUT FILE INTO A TEMPLATE

1. **Build a LayOut file that includes all the elements you want.** These elements may include a title block, a logo, a page number, the company's contact information and a cover page. Before you move to Step 2, make sure that you're viewing the page that you want to use as the thumbnail preview in the template list.
2. **Choose File ⇨ Save as Template.** The Save as Template dialog box opens.
3. **Type a name for your template and then choose a location for your new template.** In the list at the bottom of the dialog box, click the folder (they're all folders) in which you want to include the template you're adding.
4. **Click OK (Save on a Mac).** The next time the Getting Started dialog box appears, your new template will be in it.

15.8.2 Putting Together Your Own Scrapbooks

Most experienced LayOut users make their own scrapbooks of scale figures, cars, trees, drafting symbols, typography and anything they need to use again and again.

Like templates, *scrapbooks* are just LayOut files that have been saved in a special folder on your system. When you open the program, it checks that folder and displays the files it finds in the Scrapbooks panel.

Follow the steps outlined in the Pathways section below to build your own LayOut scrapbook, as shown in Figure 15-22:

PATHWAYS TO... BUILDING YOUR OWN LAYOUT SCRAPBOOK

1. Build a LayOut file with the elements you want to include in your scrapbook.
2. Choose File ⇨ Save as Scrapbook.
3. Type a name for your scrapbook. The Scrapbook Folder list in the Save as Scrapbook dialog box shows the location of the folder on your system where your new scrapbook will be saved. If you prefer to use another folder, you can add one using the Folder panel of the Preferences dialog box.
4. Click OK (Save on a Mac). The next time you restart LayOut, your scrapbook appears at the top of the Scrapbooks panel.

A few notes about making your own scrapbooks:

- **A good size is 6 × 6 inches.** You can choose any paper size for the file you plan to save as a scrapbook, but smaller sheets work better. The scrapbooks that come with LayOut are six inches square.
- **Scrapbooks can have multiple pages.** Nearly all of the default scrapbooks in LayOut have multiple pages. The first page in your document becomes the cover page for the scrapbook; all subsequent pages appear below it in the list. Pay attention to your page names, which appear in the Scrapbooks panel.
- **Use locked layers.** Anything you put on a locked layer can't be dragged out of the scrapbook. Take a look at the People scrapbook that comes with LayOut—the word *People* and the information next to it are on a locked layer. Notice how you can't drag them into your drawing?
- **You can put model viewports into scrapbooks.** Open the Arrows ⇨ 3D ⇨ Curved scrapbook. Drag one of the arrows onto your page. Now double-click it and it becomes a model. Any elements can be put into a scrapbook.

Figure 15-22

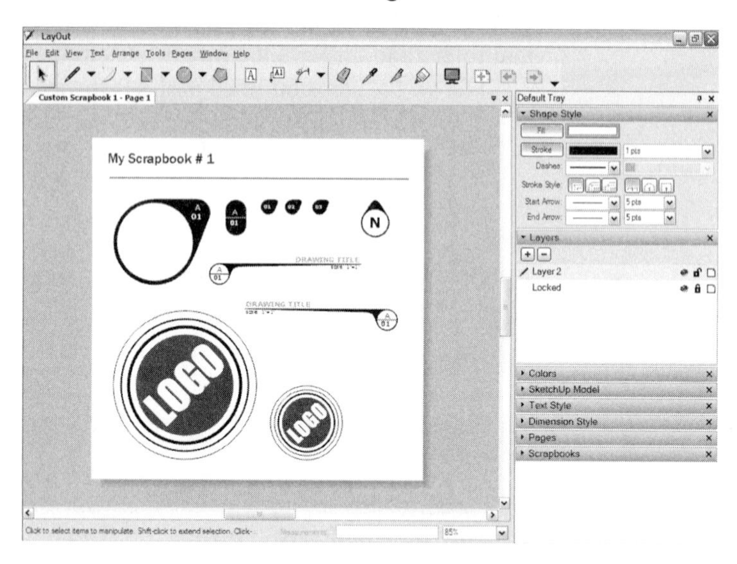

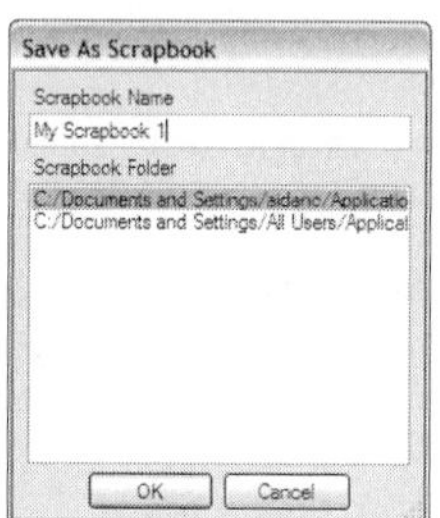

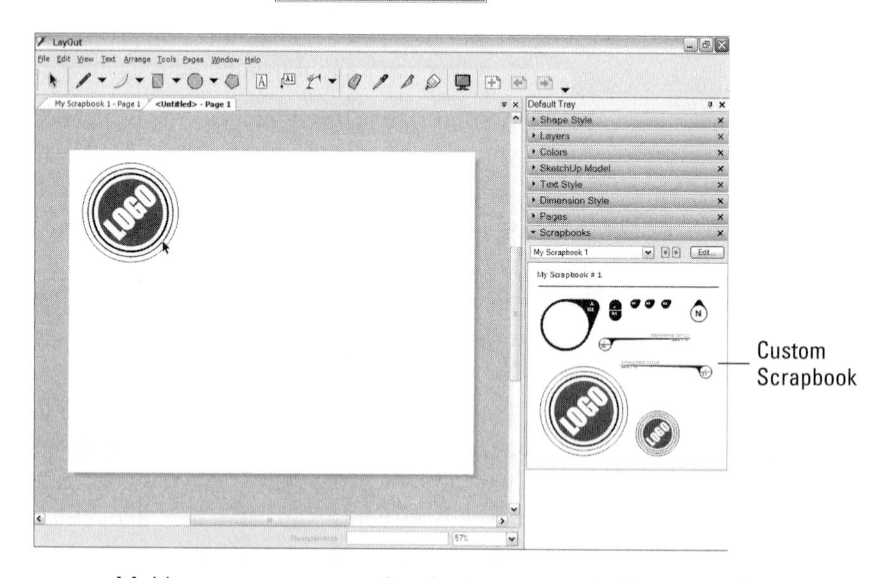

Making your own scrapbooks is easy and ultra-rewarding.

SELF-CHECK

1. You can configure LayOut so that your templates appear in the Getting Started dialog box. True or false?
2. You can create your own Scrapbook that contains the elements that you use most often in your presentations. True or false?
3. What size of scrapbooks is recommended that you work on?
   - **a.** 8 ½ × 11 inches
   - **b.** 9 × 13 inches
   - **c.** 6 × 6 inches
   - **d.** 12 × 12 inches

Apply Your Knowledge Using the exercise you did on the previous section, open the Scrapbook tab and find some tree shapes that you would like to add around the different huts. Also, search on the different sub folders and include a variety of graphics (under the different TB options) such as a scale, a north sign, and a drawing title for your floor plan.

SUMMARY

Section 1

- LayOut is its own program and has its own menus, tools, dialog boxes, and Drawing Window.
- In both the Windows and Mac versions of LayOut, you can have more than one document open at a time.
- You can find most of LayOut's features and functions in its nine dialog boxes.
- The Instructor panel works just like it does in SketchUp Pro; it shows information on whichever tool you happen to be using.

Section 2

- Generally, templates are the quickest way to get started with a new LayOut project.
- LayOut and SketchUp Pro are separate software programs, so you need to launch them individually.
- The Getting Started dialog box shows a list of available ready-made LayOut files, or templates, on the left, with thumbnail previews of each template on the right.
- LayOut can insert raster images, which means TIFFs, JPEGs, GIFs, BMPs, and PNGs are all graphics file formats that save pictures as a myriad of pixels.

Section 3

- Anyone with Adobe Acrobat Reader software (which is free and is already loaded on most computers, but do not confuse it with Adobe Acrobat Pro) can look at a PDF document you create; all you have to do is share it with the person that is expecting it.
- You can export the pages of your file as individual raster images in either JPEG or PNG format.
- It is difficult to find a piece of professional computer-aided drawing (CAD) software that can't read the DWG and DXF formats, which are the industry standard for exchanging CAD files with

people who use apps like AutoCAD and Vectorworks.

- LayOut was designed to help you skip the PowerPoint step by letting you display your presentation in a full-screen view.
- To switch to full-screen in LayOut, you can move back and forth between pages with the arrow keys on your computer, and you can even double-click SketchUp Pro model views to orbit them around.

Section 4

- To know which is the active layer, look for the little red pencil icon; anything you insert, draw, or paste is assigned to the active layer.
- You can use layers to make certain elements appear on more than one page in your document.
- LayOut has two kinds of layers:
  - Unshared
  - Shared

Section 5

- A SketchUp Pro model view that lives in your LayOut document is a *viewport*.
- You can have multiple viewports that show the same model.
- When you insert a SketchUp Pro model into a LayOut document, it shows up in a new viewport.
- If the scenes you created in SketchUp Pro aren't visible in the SketchUp Pro Model panel, you probably forgot to save your model before you switched applications.
- Every time you insert or edit a SketchUp Pro model view, LayOut renders an image of your model to display in the viewport.
- You control how your models look by choosing which method LayOut uses to render each viewport:
  - Raster
  - Vector
  - Hybrid

Section 6

- With the addition of *angular* dimensions in SketchUp Pro 8, LayOut became a full-fledged tool for creating scaled, annotated drawings from your models.
- If you have trouble getting your dimension tool to "see" any of the points in a model viewport, chances are you don't have Object Snaps turned on.
- Use the Dimension Style panel to change the format of text strings and their level of precision.
- Copying a dimension's style and applying it to your other dimensions is relatively simple, by using the Style tool to transfer formatting from one dimension to the other.

Section 7

- LayOut includes a full slate of drawing tools that you can use to create logos, title bars, north arrows, graphic scales, and the drawings you create are *vectors*.
- The LayOut Line tool works just like the Line tool in SketchUp Pro.
- You can use any combination of snapping systems (Object or Grid) while you work and to save time, you can assign a keyboard shortcut that toggles each system on and off.

Section 8

- Every time you need to put together a drawing set, all you have to do is open your template (which already includes your logo, title block, layout, and text styles) and insert your model.
- You can configure LayOut so that your templates appear in the Getting Started dialog box, making it easier to build consistent presentations, quicker.
- Like templates, *scrapbooks* are just LayOut files that have been saved in a special folder on your system.

ASSESS YOUR UNDERSTANDING

SUMMARY QUESTIONS

1. When you insert a SketchUp Pro model or an image in your LayOut document, LayOut creates a file reference that keeps track of where it came from. True or false?
2. What is the first thing you should do to change the overall color scheme or another graphic entity of a LayOut template?
   - **a.** Unlock all the layers.
   - **b.** Open the Shape Styles dialog box.
   - **c.** Insert a SketchUp Pro file.
   - **d.** Click the New tab.
3. Many of the templates that come with LayOut have multiple layers, and some of these layers are locked by default so that you can't accidentally move things around. True or false?
4. When your design changes in SketchUp Pro, the views you've placed in LayOut do not get updated and we have to erase them and re-insert them. True or false?
5. What are the two kinds of layers that we can be have in LayOut?
6. You should hide the layers you aren't using to avoid accidentally moving the wrong things around, or even deleting them. True or false?
7. Let's imagine that you have plan view of your model in your viewport in LayOut but you need to change the drawing scale of it. List the steps that are required to do so, in LayOut.
8. Which of the following must you do to include a view of a SketchUp Pro model in LayOut?
   - **a.** Export the view from SketchUp Pro as an image file.
   - **b.** Place the view in a graphic layout program.
   - **c.** Export the view every time you change the SketchUp Pro file.
   - **d.** Simply insert the SketchUp Pro file.
9. The key to working with SketchUp Pro models you've inserted into LayOut is to right-click them; this opens a context menu full of useful options. True or false?

10. Explain the steps necessary to assign a specific drawing scale to an elevation or a plan view in LayOut.
11. What do we do if we want to edit a clipping mask or a dimension in LayOut?
12. When choosing the option of a full-screen view, LayOut can transition from scene to scene automatically. True or false?

APPLY: WHAT WOULD YOU DO?

1. Name and describe three of LayOut's menus.
2. Why would you use a template to begin your presentation? Name three things that you should know about using templates.
3. What is the difference between the two different kinds of layers in LayOut?
4. You want to insert an image into your LayOut document. How do you do this?
5. You want to export your LayOut document as a PDF file. How do you do this?

BE A LAYOUT DESIGNER

Drawing from Scratch

In LayOut, draw a rectangle, a triangle, a square, and an ellipse. Then move them so that they overlap, and combine them into a single shape using the Split and Join tools.

Create a Template

Create a simple LayOut file with a title block. Then take this LayOut file and turn it into a template.

KEY TERMS

Clipping masks
Shared layers
Snap settings
Stacking order
Unshared layers

CHAPTER 16

TROUBLESHOOTING AND USING ADDITIONAL RESOURCES

Fixing Problems and Enhancing Your Model

Do You Already Know?

- Why faces, colors, and edges aren't behaving the way you expect them to?
- The reasons your computer slows down?
- How to recover from accidental Eraser problems?
- About free and for-pay plug-ins?
- About all of the great SketchUp resources available to you?

For the answers to these questions, go to **www.wiley.com/go/chopra/googlesketchup2e**

What You Will Find Out	What You Will Be Able To Do
16.1 Identifying and troubleshooting common problems.	Understand 10 common modeling problems and how to resolve them.
16.2 What additional tools are available to enhance your modeling experience.	Acquire additional plug-ins and extensions and which ones are free and are available for a small fee.
16.3 What resources are available to help you.	Find free and for-a-fee resources available on the market today.

INTRODUCTION

We've all had an experience where we are working in a program and something goes wrong. It doesn't work the way it is supposed to, or it gives an indecipherable error message, or it just completely crashes. In this chapter, you will evaluate what steps to take when this happens to you as you are working in SketchUp. Not only will you assess what to do when SketchUp is slow or crashes, but you will also assess what to do when faces, colors, and edges are not working the way they should.

However, this chapter is not only about troubleshooting. This chapter is also about plugins that might enhance your SketchUp experience, as well as resources that can help you take your SketchUp skills to the next level. Consequently, in this chapter, you will assess plugins that have specific purposes, and you will also assess various resources available to you.

16.1 TEN SKETCHUP TRAPS AND THEIR SOLUTIONS

GOOGLE SKETCHUP IN ACTION

Learn about 10 common modeling problems and how to resolve them.

The bad news is that every single new SketchUp user runs into certain problems, usually in his or her first couple of hours with the software. The good news is that such predictability means that you can anticipate a great deal of the traps you'll encounter. Here are some of those traps, as well as methods that will help you make sense of what's going on so that you can get on with your life as quickly as possible.

16.1.1 SketchUp Won't Create a Face Where You Want It To

Say that you've dutifully traced all around the boundary of where you'd like SketchUp to create a face, but nothing's happening. To remedy the problem, try checking whether your edges aren't all on the same plane or whether one edge is part of a separate group or component.

To check whether you have a component problem, try hiding groups or components and checking the edges to make sure that they're all in the group or component you think they're in. See Chapter 5 for details.

However, most of the times when SketchUp won't create a face where you think it should, an edge isn't on the *plane* you think it's on. To check whether your edges are coplanar, draw an edge that cuts diagonally across the area where you want a face to appear. If a face appears now, all your edges are not on the same plane. To fix the problem, you have to figure out which edge is the culprit. Try doing this by using the method illustrated in Color Plate 20 and described below:

1. **In the Styles dialog box, change your edge color from All Same to By Axis.** See Chapter 9 for details. Doing this step tells SketchUp to

draw the edges in your model the color of the axis to which they're parallel; edges parallel to the red axis will be red, and so on.

2. **Look carefully at the edges you were hoping would define a face.** Are all the edges the color they're supposed to be? If they're not all supposed to be parallel to the drawing axes, this technique doesn't do much good. But if they are, and one (or more) of them is black (instead of red or green or blue), that edge (or edges) is your problem. Fix it and switch back to All Same when you're done.

16.1.2 Your Faces Are Two Different Colors

In SketchUp, faces have two sides: a front and a back. By default, these two sides are different colors. When you do certain things like use Push/Pull or Follow Me on a face, sometimes the faces on the resulting geometry are "inside out." If it bothers you to have a two-tone model, just right-click the faces you want to flip over and choose Reverse Faces from the context menu. If you have lots of them, you can select them all and then choose Reverse Faces to flip them all at once.

16.1.3 Edges on a Face Won't Sink In

This tends to happen when you're trying to draw a rectangle (or another geometric figure) on a face with one of SketchUp's shape-drawing tools. Ordinarily, the Rectangle tool creates a new face on top of any face you use it on; after that, you can use Push/Pull to create a hole, if you want. If your shape's edges look thick instead of thin, they're not cutting through the face they're drawn on. When that happens, try these approaches:

- **Retrace one of the edges.** This simple method works often.
- **Select Hidden Geometry on the View menu.** Here, you're checking to make sure that the face you just drew isn't crossing any hidden or smoothed edges; if it is, the face you thought was flat might not be.
- **Make sure that the face you drew on isn't part of a group or component.** If it is, undo a few steps and then redraw your shape while you're editing the group or component.

16.1.4 SketchUp Crashed, and You Lost Your Model

Unfortunately, SketchUp crashes happen sometimes. The good news is that SketchUp automatically saves a copy of your file every five minutes. The file that SketchUp autosaves is actually a *separate* file, which it calls *AutoSave_your filename.skp*. So if your file ever gets corrupted in a crash, there's an intact one, ready for you to find and continue working on. The problem is that most people don't even know it's there. So where is it?

- If you've ever saved your file, it's in the same folder as the original.
- If you've never saved your file, it's in your My Documents folder—unless you're on a Mac, in which case it's here: *User folder/Library/Application Support/Google SketchUp 8/SketchUp/Autosave*.

Keep in mind that generally, SketchUp cleans up after itself by deleting the autosaved file when you close your model, and nothing untoward happens.

To minimize the amount of work you lose when software (or hardware) goes south, you should always do two things: save often (compulsively, even), and save numbered copies as you're working. Consider using Save As to create a new copy of your work every half-hour or so and possibly back up your work to an external piece of hardware entirely, onto a network or a cloud storage site.

16.1.5 SketchUp Is Slow

The bigger your model gets, the worse your performance gets, too. What makes a model big? In a nutshell, faces. You should do everything in your power to keep your model as small as you can. Here are some tips for doing that:

- **Reduce the number of sides on your extruded circles and arcs.** See Chapter 6 for instructions on how to do this.
- **Use 2D people and trees instead of 3D ones.** 3D plants and people have *hundreds* of faces each. Consider using 2D ones instead, especially if your model won't be seen much from overhead.

Some models are just big, and you can't do much about it. Here are some tricks for working with very large SketchUp models:

- **Make liberal use of the Outliner and layers.** Explained in detail in Chapter 7, these SketchUp features were specifically designed to let you organize your model into manageable chunks. Hide everything you're not working on at the moment—doing so gives your computer a fighting chance.
- **Use substitution for large numbers of complex components.** For example, insert sticks as placeholders for large sets of 3D trees, cars, and other big components. See the tips for replacing components in Chapter 5 for details.
- **Turn off shadows and switch to a simple style.** It takes a lot of computer power to display shadows, edge effects, and textures in real time on your monitor. When you're working, turn off these elements.
- **Use scenes to navigate between views.** Scenes aren't just for presenting your model—they're also great for working with it. Creating scenes for

the different views you commonly use, and with different combinations of hidden geometry, means that you don't have to orbit, pan, and zoom around your gigantic model. Better yet, deselect Enable Scene Transitions (in the Animation panel of the Model Info dialog box) to speed things up even more.

16.1.6 You Can't Get a Good View of the Inside of Your Model

It's not always easy to work on the inside of something in SketchUp. You can do these things to make it easier, though:

- **Cut into it with sections:** SketchUp's Sections feature lets you cut away parts of your model—temporarily, of course—so that you can get a better view of what's inside. Read Chapter 10 for details on sections.
- **Widen your field of view:** Field of view is basically the amount of your model you can see on the screen at one time. A wider field of view is like having better peripheral vision. You can read all about it in Chapter 10.

16.1.7 A Face Flashes When You Orbit

If you have two faces in the same spot—maybe one is in a separate group or component—you see an effect called **Z-fighting**. What you're witnessing is SketchUp trying to decide which face to display by switching back and forth between them. It's not a good solution, but certainly a logical one, at least for a piece of software. The only way to get rid of Z-fighting is to delete or hide one of the faces.

Z-fighting
The effect resulting from SketchUp trying to decide which face to display by switching back and forth between two faces.

16.1.8 You Can't Move a Component the Way You Want

Some components are set up to automatically *glue* to faces when you insert them into your model. A glued component instance isn't actually glued in one place. Instead, it's glued to the plane of the face you originally placed (or created) it on. For example, if you place a sofa component on the floor of your living room, you can only move it around on that plane, not up and down. This behavior comes in handy when you're dealing with things like furniture. It allows you to use the Move tool to rearrange things without having to worry about accidentally picking them up.

If you can't move your component the way you want to, right-click it and check to see whether Unglue is an option. If it is, choose it. Now you can move your component around however you want. If you can't unglue it, then you would have to click something that is actually glued down to obtain that menu option.

16.1.9 Every Time You Use the Eraser, You Accidentally Delete Something

It is so easy to delete things accidentally with the Eraser tool. Worse yet, you usually don't notice what's missing until it's too late. Here are some tips for erasing more accurately:

- **Orbit around.** Try to make sure that nothing is behind whatever it is you're erasing; use SketchUp's navigation tools to get a view of your model that puts you out of danger.
- **Switch to Wireframe mode.** Choose View ⇨ Edge Style ⇨ Back Edges when you're going to be using the Eraser heavily. That way, you won't have any faces to obstruct your view, and you'll be less likely to erase the wrong edges.
- **Double-check.** Get into the habit of giving your model a quick once-over with the Orbit tool after you do a lot of erasing, just to make sure that you didn't get rid of anything important.

16.1.10 All Your Edges and Faces Are on Different Layers

Using Layers in SketchUp is a dangerous business. Chapter 7 has tips you should follow when using layers, but here's the short version: You should always build everything on Layer0, and only put whole groups or components on other layers if you really need to.

IN THE REAL WORLD

Robert O'Halloran is an interior designer who works in Ireland and came across Google SketchUp when looking at different options for his 3D drawings. He believes that using AutoCAD for this purpose was very time consuming and difficult to master. According to him, within days of using SketchUp he quickly found his way around all the tool bars and has never looked back.

He is currently studying for his BA Honors Degree in interior architecture and uses SketchUp at all possible points throughout his projects. He says that he generally starts working with SketchUp from the initial design stage right through completion of the project. If he feels the need to include a slideshow presentation, he likes to use Microsoft PowerPoint rather than the Layout option as it is more commonly used and has great features. He also utilizes the V-RAY plug-in when working on photo-realistic renderings and Windows Movie Maker or Quicktime when creating animations from his SketchUp files.

If you used layers and things are now messed up, here's what you can do to recover:

1. **Make sure that everything is visible.** Select Hidden Geometry on the View menu; then in the Layers dialog box make all your layers visible. Just make sure that you can see everything in your model.
2. **Choose Edit ⇨ Select All to select everything.**
3. **In the Entity Info dialog box, move everything to Layer0.**
4. **In the Layers dialog box, delete your other layers, telling SketchUp to move anything remaining on them to Layer0.**
5. **Create new layers, and follow the rules in Chapter 7.**

SELF-CHECK

1. What do we do to check if the edges of a plane are coplanar?
2. What do we do to avoid using a lot of computer power when we are designing? Check all that apply:
   a. Use scenes to navigate between views.
   b. Avoid using the Outliner.
   c. Turn off shadows, edge effects, and textures.
   d. Choose 3D components over 2D components.
3. When SketchUp switches back and forth between coplanar, overlapping faces, it is called Z-fighting. True or false?
4. What does it mean when an edge on a face won't sink in?

Apply Your Knowledge Work on a new document in SketchUp of a simple one story house and save it in your computer on a folder where you will find it easily. Then add a bunch of 3D components to the model, such as three cars parked in front, three trees, three persons walking around the house, and some garden paraphernalia. Then apply materials or color to your scenes and choose a viewing style and turn on your shadows. After you have done this, create some scenes in an orthographic projection and in perspective. Then, compare how much quicker it is to navigate through your model by using those scenes, compared to using the Pan, Zoom, and Orbit tools constantly.

As a second test, pretend that you crashed your computer by closing your file without saving it and find the backup file that was automatically generated in your folder, with the extension .skb. Open that backup file by opening first your SketchUp software and then go to File ⇨ Open. How much work did you actually lose?

As a third test, replace all of your 3D components with similar 2D components and check now how much faster you can Pan, Zoom, and Orbit around the model.

16.2 TEN PLUGINS, EXTENSIONS, AND RESOURCES WORTH GETTING

GOOGLE SKETCHUP IN ACTION

Learn about how to acquire additional plug-ins and extensions and which ones are free and are available for a small fee.

The great thing about SketchUp's price is how much room it frees up in your budget for add-ons. This section is a list of ten such add-ons, along with a little bit of information about them, and where you can go to find them. These add-ons have been split into four categories, just to make things clearer: components, Ruby scripts, renderers, and hardware.

16.2.1 Ruby Scripts

Ruby scripts
Mini-programs (scripts) written in a computer programming language called Ruby.

What's a **Ruby script**? Basically, Google provides a way for people to make their own plugins for SketchUp. These plugins are just mini-programs (scripts) written in a computer programming language called Ruby. The best thing about Ruby scripts (Rubies, for short) is that you don't have to know anything about Ruby, or programming in general, to use ones that other people have created.

To install Rubies, you just drop them into a special folder on your computer:

- **Windows:** C:/Program Files/Google/Google SketchUp 8/Plugins.
- **Mac:** Hard Drive/Library/Application Support/Google SketchUp 8/SketchUp/Plugins.

The next time you launch SketchUp, Rubies you put in that folder become available for you to use. How you use them depends on what they do. They might show up on one of the toolbar menus or on your right-click context menu. The more complex ones even come with their own toolbars. Most Rubies also come with a set of instructions that tells you how to use them.

Luckily for those of you who aren't programmers, plenty of people develop and (in some cases) sell Rubies that anyone can use. Here are some of the most useful:

- **Smustard.com:** Smustard.com is a Web site run by a few of these smart folks. You can choose from dozens of helpful Rubies that add functionality to SketchUp. Most are free, but the ones that aren't are well worth the money. (www.smustard.com)
- **Ruby Library Depot:** The Ruby Library Depot is a huge collection of Rubies from helpful software developers around the world. Everything is well-organized and thoroughly vetted. Highly recommended. (http://modelisation.nancy.archi.fr/rld/)
- **SketchUcation:** This combination how-to Web site, blog, and discussion forum is also a terrific source of Rubies. Developers post their plugins in forum discussion threads, where SketchUp über-modelers from around the galaxy gather to test them and provide feedback. The upside is that

you can see instructions and examples for most Rubies. The downside is that some of the threads are 20 pages long. Typing **sketchucation plugins index** into your favorite search engine leads you to the right place on SketchUcation's extensive Web site.

IN THE REAL WORLD

James Douglas Smith is the owner of James Douglas Smith Architects Inc. and one of the principals at VRA Architects in Crown Point. He enjoys working with Google SketchUp because it is an easy to learn program, as well as intuitive. Even after not working with it for awhile, its easy to pick back up, unlike AutoCad. According to him, SketchUp is a really powerful tool and very economical, which gives small practitioners an edge to compete with larger firms in producing extremely professional models and drawings.

He says that once the 3D model is detailed, textured, and populated with components, he creates scenes and sections that depict the design intent and then renders them. His clients love the realism that SketchUp provides and can accurately understand how their needs are being met. He goes on saying that his clients especially enjoy the real time designing they can do together on a laptop while meeting.

His working process starts with the plans in AutoCAD and then he uses the basic plans to complete the model design in great detail with SketchUp. The next step is to produce 2D elevations and sections using *Real Section* and *Zorro*, which are then imported into AutoCAD. He also imports various perspective views of the model into the AutoCAD files with notes and dimensions that better describe the building that he is designing. The presentation's cover sheet will always have a rendered view and he also presents to his clients full color rendered views that are created with *Podium*, *I-Render*, and *Shaderlight*.

Now that you know where to go to find them, which Rubies should you use? Table 16-1 lists some of the most common.

16.2.2 Renderers

One thing SketchUp does not do is create photorealistic renderings. Its styles are great for making your models look hand-drawn, but none of them can make your work look like a photograph. Most SketchUp users are okay with that, but for those who aren't, you can find some nice solutions out there.

Of course, SketchUp Pro's 3D export formats make it possible to render SKP files with just about any of the dozens of powerful renderers on the

Table 16-1 SketchUp Ruby Scripts

Ruby Name	*What It Does*	*Author; Source*	*Free Or Pay*
CAD Cleanup Scripts	If you routinely import 2D CAD drawings to use as a starting point for SketchUp models, you need these scripts. Without going into any detail, here's a list of the CAD cleanup scripts to hunt for on Smustard.com: StrayLines, CloseOpens, MakeFaces, IntersectOverlaps, Flatten, and DeleteShortLines.	Todd, Burch; Smustard.com	Free
Drop	Takes a bunch of groups or components that you've selected and drops them down in the blue direction until each hits a surface. If you're trying to populate a sloping hillside with trees, it's much easier to make a big, flat grid of them in the air and then Drop them to meet the ground.	Octavian Chis; Smustard.com	Free
FredoScale	Adds the ability to taper, twist, bend, and stretch 3D objects in a way you'd never be able to with plain ol' SketchUp.	Fredo6; Ruby Library Depot	Free
Make Fur	Lets you "grow" fur—or grass—on faces in your models. More fun, and more useful, than you'd think.	tak2hata; SketchUcation	Free
Layers Management Ruby Script	Allows you to add a layer that's only visible in your current scene. Great for design iterations where each iteration is on a separate layer that is linked to a unique scene.	Joe Zeh; search the Web for *layers management ruby script*	Free
JointPushPull (recommend JointPushPull for v8)	Wouldn't it be great if you could push/pull curved surfaces to give them thickness? With this plugin, you can. Even better, it comes with great documentation when you download it.	Fredo6; Ruby Library Depot	Free
PresentationBundle	This package of five Rubies helps you use scenes to create better presentations. You can customize the transition time between individual scenes, for instance, and even create really elaborate fly-by animations.	Rick Wilson; Smustard.com	Pay
RoundCorner	Turns sharp edges and corners into smooth rounds and fillets. Shaves literally hours off the time it takes to model almost any manufactured product.	Fredo6; Ruby Library Depot	Free

Ruby Name	*What It Does*	*Author; Source*	*Free Or Pay*
Shape Bender	Lets you bend (and stretch) a 3D shape along a path you designate. Everything from spiral ramps to fancy retail signage just got a heck of a lot easier.	Chris Fullmer; SketchUcation	Free
Simplify Contours	Sometimes *contours* (topography lines) you import from a CAD file are very complex. Running this script on them reduces their complexity, which makes the terrain you use them to generate easier to work with.	Google; search the Web for simplify *contours ruby*	Free
Slicer 4.3	Takes a model and cuts it into even, parallel slices. Indispensable for turning 3D terrain into easy-to-make site models.	TIG; Ruby Library Depot	Free
Soap Skin & Bubble	Make a loop of connected edges, activate this Ruby, and watch in amazement as it automatically "stretches" a surface over them—think circus tent. You can even apply pressure to inflate your new tensile structure right before your eyes.	Josef Leibinger; search the Web for *ruby soapskin bubble*	Free
SubdivideAndSmooth	If you're interested in *organic modeling*—stuff that isn't boxy or otherwise angular—look at this one. Basically, it lets you create smooth, blobby (I mean that in a good way) forms based on *proxies* (boxes, sort of) that you manipulate. It's much cooler and more versatile than it sounds.	Dale Martens; Smustard.com	Pay
SketchyPhysics	An extension that lets you apply real-world physical *constraints* (such as gravity) to your SketchUp models. You can, for instance, set up a row of dominoes, tip one over, and watch them all fall down. Or build a machine with wheels, pistons, and other mechanical parts and then set it in motion.	DarthGak; search the Web for *sketchyphysics*	Free
Tools On (recommend the latest ToolsOnSurface16c)	Surface Use these drawing tools—Line, Offset, Circle, Rectangle, and Freehand—to draw directly on non-flat surfaces like terrain.	Fredo6; Ruby Library Depot	Free
Weld	Takes edges you've selected and welds them together to make a single edge that you can select with a single click. This is *super* handy when you use Follow Me.	Rick Wilson; Smustard.com	Free
Windowizer	Helps you create storefront windows much more quickly than you can manually. Wicked powerful.	Rick Wilson; Smustard.com	Free (donation suggested)

market, but the ones described in the following list have three important things in common: They work with the free version of SketchUp, they were developed with SketchUp in mind, and you don't have to be a computer expert to figure them out. Here's the list:

- **SU Podium:** If you're using a Windows computer, you should probably check out Podium first. It's a plugin that lets you create photorealistic views right inside SketchUp. The newer version has the program available for Mac and PC's and it's definitely worth checking out. (www.suplugins.com)
- **IDX Renditioner:** This software evolved from another renderer, TurboSketch. Like SU Podium, IDX Renditioner is a plugin available for Windows and Mac computers, and you can try a free version that creates smaller 640-×-480-pixels images. It is recommended to try both SU Podium and IDX Renditioner, and seeing which you prefer. (www.idx-design.com)
- **Artlantis R:** If you're really serious about making images that look like photographs, take a good look at Artlantis R. Instead of running as a plugin inside SketchUp, it's a fully functional, separate piece of software that works with lots of other 3D modeling programs. The results it produces are out of this world. (www.artlantis.com)

Again, lots of rendering applications are out there, and depending on what you want to do, different ones might work better than others. If photo rendering is your thing, try plugging these names into your favorite search engine: Cheetah3D, Kerkythea, LightUp, Maxwell Render, Shaderlight, Thea Render, Twilight Render, V-Ray, and Vue.

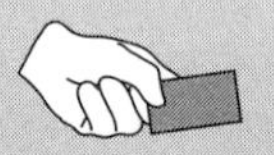

CAREER CONNECTION

For Lateo Soletic, founder of www.sketchupartists.org and owner of www.concepto-illustrations.com, choosing SketchUp was simple, "I used other packages before but after SketchUp I never looked back." He likes to walk his clients over the entire model by capturing perspective scenes that later are included in a presentation folder. His clients like this because that helps them visualize their dream.

Thinking about the features that can be used, Lateo comments that "if a person has an eye for things and some artistic ability to make a right presentation, then SketchUp and a bit of Photoshop are more than enough for a perfect presentation. All of my projects are already sold at this stage."

Sometimes he would experiment with other rendering packages once the presentation is over, but that would be done for his own personal gain. In his own words, "Sometimes I feel that adding too much detail in the renderings can distract the client from the concept itself. I want to avoid comments like 'I don't like the fabric on that sofa.'"

16.2.3 Hardware

All you really need to use SketchUp is a computer with a decent video card, a keyboard, and a mouse. On the other hand, having specialized hardware can come in handy—especially if you find yourself using SketchUp all the time:

- **A better video card:** This isn't really an option if you're on a laptop, but if you use a desktop machine, upgrading your computer's video card (or *graphics card*) is the best way to improve your SketchUp experience. Your *video card* is a piece of hardware that handles the elements that appear onscreen. Not all cards are made to be used with 3D modeling programs, even those that are made specifically for video games. Look for something that supports *OpenGL 1.5* or higher and has a lot of video memory; the more, the better. Check out cards made by NVIDIA and ATI.

SpaceNavigator
A 3D navigation tool that looks a little like an enormous button that is connected to the computer via a USB cable. It enables the user to orbit, pan, and zoom with subtle movements of his or her hand.

- **SpaceNavigator from 3Dconnexion:** Using a scroll-wheel mouse to fly around in three-dimensional space works well for most people, but many SketchUp power users swear by dedicated 3D navigation tools like the SpaceNavigator. It looks a little like an enormous button that sits on your desk, connected to your computer via a USB cable. You use it with whichever hand you aren't using for your mouse; it's an add-on (and not a replacement) for any of the other peripherals in your system. Basically, the SpaceNavigator enables you orbit, pan, and zoom with subtle movements of your hand; it really is a much more natural way to interact with a 3D model. You'll find a bit of a learning curve, but that's nothing for serious SketchUp users. Anything that makes software easier and more fun to use is worth the time it takes to master it.

IN THE REAL WORLD

Interior designer Grace Hwang from Gensler in San Francisco mentions that she uses Google SketchUp in conjunction with Adobe Photoshop and InDesign when creating presentation documents for her clients. There are a couple of steps that she follows to help the modeling process run smoothly, especially when working with more complex models. According to her, any model that is downloaded from the SketchUp 3D warehouse is opened into a blank SketchUp file, where the layers are then condensed to one before being imported into her models. In addition, sometimes when the model gets too big and the purge and fix options do not work, it helps to copy and paste the entire model into a new SketchUp file and work from there. She also says that working without the textures can make the entire process a lot faster.

SELF-CHECK

1. Give the name of one of the rendering programs available for Google SketchUp.
2. Describe two examples of what two specific Rubies can do for us.
3. SketchUp creates photorealistic renderings. True or false?
4. What is the name of the mini-programs that are created by other people to add functionality to SketchUp?

Apply Your Knowledge Chances are that at some point a potential client will ask you to prepare a photorealistic rendition of one or more of your designs and will you need to be ready when that moment arrives. Use the model of the house that you worked on the last exercise and render it using one of the following trial versions:

- Su Podium trial, http://www.suplugins.com/free-evaluation.php and download Artlantis trial, http://activate.imsisoft.com/gennedfreetrial.aspx?productpage=RenProPC

16.3 TEN WAYS TO DISCOVER EVEN MORE

GOOGLE SKETCHUP IN ACTION

Learn about free and for a fee resources available on the market today.

In addition to this book, there are many great resources out there that will help you learn even more about SketchUp. Consequently, the first part of this section is devoted to free resources, all of which are available online to anyone who wants them. The second part of the section describes some other resources that are available for purchase.

16.3.1 Free Resources

Everything in this section requires that you have an Internet connection, so make sure that your computer's online before you try any of these.

- **SketchUp Training resources:** Google publishes really first-rate materials for SketchUp right on its Web site (http://sketchup.google.com/training):
  - *Video Tutorials:* When SketchUp first launched in 2000, it became known for its excellent video tutorials. I can't recommend them highly enough; there's nothing like *seeing* SketchUp in action.
  - *Self-Paced Tutorials:* These are SketchUp files that use scenes to teach different aspects of the program in a "follow along with me" style. If this is how you like to figure things out, have a look.

- **Online Help Center:** Google maintains extensive help centers (Web sites, basically) for all its products. These include hundreds of articles in question-and-answer format, created specifically to help new users along. The SketchUp one is terrific; choose Help ⇨ Help Center from the SketchUp bar menu.
- **SketchUp Help Forum:** The SketchUp Help Forum is an online discussion forum where SketchUp and LayOut users from all over the world can get help, ask questions, and show off their work. (http://www.google.com/support/forum/p/sketchup?hl=en).
- **SketchUcation:** Home of the SketchUp Community Forums, this is easily the largest and most active group of SketchUp users in the world. You find discussions, tutorials, plugins, news, and other useful information about SketchUp and LayOut at SketchUcation. (http://forums.sketchucation.com/)
- **Go-2-School Videos:** The free videos on this site are first-rate, and you can buy DVDs, too. This is discussed in next part of this chapter. (http://www.youtube.com/4sketchupgo2school)
- **The Official Google SketchUp Blog:** Visit the SketchUp blog regularly for news, case studies, tips and tricks, modeler profiles, plugins, and other updates. (http://sketchupdate.blogspot.com).

16.3.2 Resources You Can Purchase

These resources cost a bit of money, but they're worth every penny:

- **Bonnie Roskes' books:** Bonnie Roskes' *The SketchUp Book* was the first such product available, and now she has two new titles. If you think you'd like to get another, bigger book about SketchUp (written with architects and other design pros in mind), check out these books at http://www.3dvinci.net/
- **School DVD:** School's videos are mentioned in the previous section, but School's designers have also produced the world's first SketchUp educational/training DVD, which you can order from the company's Web site (www.go-2-school.com). The production quality on this video is outstanding, and the DVD does an amazing job of teaching SketchUp for both Windows and the Mac.
- **Dennis Fukai's books:** These books are hard to describe. Dennis Fukai has written four of them; each is fully illustrated in SketchUp, and each teaches a different subject. If you want to discover more about using SketchUp in building construction or more about construction itself, or if you just want to be completely inspired by what you can do with SketchUp, have a look at these books. Search for Fukai's name on Amazon (www.amazon.com) or go to his company's Web site (www.insitebuilders.com).

- **SketchUp Pro training:** If you think you might benefit from being able to spend a few hours with a real-live trainer and a handful of other SketchUp students, Google's SketchUp training might be for you. Its trainers travel to different cities, giving training seminars that you can sign up to attend. Check out the following Web site for more information (www.sketchup.google.com/training).

IN THE REAL WORLD

Claudia Kopietz is a Brazilian interior designer who studied at the British Institute of Interior Design in London. She tried learning how to use 3D Max to create her designs but due to its complexity she decided that it was faster and easier to just use SketchUp. According to her, with SketchUp you can just export the file into the program and start working on the 3D model, which saves you not only time but also gives you enough material to show to your clients.

She starts with AutoCAD and then she exports her drawings to SketchUp, where she builds her 3D models. She believes SketchUp is a great program to use on projects that require a quick visualization and especially with clients who do not have a good spatial sense.

SELF-CHECK

1. SketchUp does not have an online help center. True or false?
2. SketchUp and Google offer many free resources for SketchUp users. True or false?
3. SketchUp offers many video tutorials that enable you to see the program in action. True or false?
4. Professional SketchUp trainers tour the country and offer training sessions for a fee. True or false?

Apply Your Knowledge Go to any of the free resources available for SketchUp users, such as http://forums.sketchucation.com/ and explore under Resources or Gallery some topics that you did not know about or simply admire some of the work done by other SketchUp users. Hopefully this will inspire you to become a regular visitor in these forums and contribute as well too.

SUMMARY

Section 1

- To check whether you have a component problem, try hiding groups or components and checking the edges to make sure that they're all in the group or component you think they're in.
- SketchUp automatically saves a copy of your file every five minutes and it is actually a *separate* file, which it calls *AutoSave_your filename.skp*.
- When SketchUp is slow, you should do everything in your power to keep your model as small as you can by:
  - Reducing the number of sides on your extruded circles and arcs.
  - Using 2D people and trees instead of 3D ones.

Section 2

- To install Rubies, you just drop them into a special folder on your computer.
- All you really need to use SketchUp is a computer with a decent video card, a keyboard, and a mouse.

Section 3

- There are many free and for a fee Google SketchUp resources available to you to help you along your modeling journey.

ASSESS YOUR UNDERSTANDING

SUMMARY QUESTIONS

1. What do we do to flip a side that is inside out in SketchUp?
2. Every how often does SketchUp automatically save a copy of your file?
3. 2D plants and people have hundreds of faces each and we should use the 3D equivalent models instead.
4. The SketchUp Help Forum or the Forum of Sketchucation is for anybody in the world to use, we can ask questions, get help or show off our work too. True or false?
5. Name the name of a Web site that sells ready-to-use SketchUp models.
6. SketchUp files that use scenes to teach different aspects of the program in a "follow along with me" style are:
   - **a.** Video tutorials
   - **b.** Self-paced tutorials
   - **c.** School podcasts
   - **d.** In the books by Bonnie Roskes
7. In order to work effectively, Rubies need to be dropped in a specific folder in our computer and many of them come with a set of instructions that tell us how to use them. True or false?
8. If you want to make images that look like photographs, consider using (check all that apply):
   - **a.** SU Podium
   - **b.** Form Fonts
   - **c.** Turbo Sketch
   - **d.** Atlantis R

APPLY: WHAT WOULD YOU DO?

1. You're working on a complex model of a building when SketchUp crashes. Where can you find a copy of your model, and how much work you have lost?
2. If you find yourself using SketchUp all the time, what additional hardware might you want to purchase to become more efficient and productive?
3. You are good at using SketchUp, but you want to get better. What free resources are available to you to help you take your skills to the next level?

BE AN ARCHITECT

Rubies

Download and install the Presentation Bundle of Ruby scripts, which is mentioned earlier in this chapter, http://www.smustard.com/script/PresentationBundle (you will have to pay a bit of money for this Ruby but it is worth it). Use it to create a presentation using one of the last house models you have built for this chapter. Before you start using it be sure to create enough scenes around your house so that the script will work better for you. Experiment with this Ruby, creating either a fly-by animation or a presentation, varying the transition time in between individual scenes.

KEY TERMS

Ruby scripts
SpaceNavigator
Z-fighting

GLOSSARY

3DS	One of the few standard 3D file formats.
Active cut	The section plane that is actually cutting through your model; others section planes are considered "inactive."
Anti-aliasing	A process that fills in the gaps around pixels with similar-colored pixels so that things look smooth.
Artifacts	Smudges and other image degradation introduced by file compression; common in JPEGs.
As-built	A drawing of an existing building.
Clipping masks	Used in LayOut to hide the parts of images that you don't want to see.
Components	Edges and faces grouped together into objects that display certain useful properties.
Coplanar	On the same plane.
Curves-based modelers	Modeling programs that use curves to define lines and surfaces.
DAE (Collada)	A new 3D file format that is gaining acceptance with a large number of gaming and animation programs.
Dormer	A structure above a roof surface that serves to make attic space more usable.
Drawing axes	Three colored lines visible in the SketchUp modeling window that enable users to work in three-dimensional space.
DWG	AutoCAD's native file format and the best one to use for transferring information to that program and other pieces of CAD software.
DXF	Document Exchange Format, a type of DWG that was developed by Autodesk to be the file format that other pieces of software use to transfer data into AutoCAD.
Eave	A part of a roof that overhangs the building.
Edges	In SketchUp, straight lines.
EPIX	A file format that lets you open 2D views of your SketchUp Pro model in Piranesi.
EPS	Encapsulated Postscript, a file format that some people still use to transfer vector information.
Extensions	The line overruns that you can choose to display in the Styles dialog box.

Exterior model	A model of the outside of a building, which does not include interior walls, rooms, or furniture.
Faces	In SketchUp, infinitely thin surfaces.
Fascia	The trim around the edge of a roof's eaves where gutters are sometimes attached.
FBX	A 3D file format used primarily by people in the entertainment industry who use Maya, 3ds Max, or Autodesk VIZ.
Field of view	The amount of your model you're able to see in your modeling window at one time.
Flat roof	A roof that appears flat but is sloped very slightly.
Folds	Edges and faces that are created in place of a single face. It usually occurs when we move an edge or a point into a direction that cannot hold the shape as a single face.
Gable	The pointed section of wall that sits under the peak of a pitched roof.
Gabled roof	A roof with two planes that slope away from a central ridge.
Geo-located objects	Objects that never move and have a fixed position, such as buildings and monuments.
Geo-location	1) The act of telling SketchUp the precise coordinates where the model is situated in the world. 2) The process where SketchUp sets the user's snapshot's latitude and longitude to match Google Earth, and it orients the snapshot in the right cardinal direction.
Geometry	Edges and faces in SketchUp models.
Group	Collection of edges and faces that are grouped together and act like a mini-model within the main model.
Guides	Temporary lines that users can create to work more accurately in SketchUp.
Hip roof	A roof where the sides and ends all slope together.
Inference engine	Objects such as colored shapes, dotted lines, yellow tags, and other similar objects that appear as the user moves his or her cursor around the SketchUp modeling window.
Interior model	A model of the inside of a building, which must take into account interior wall thicknesses, floor heights, ceilings, and furnishings.
Isometric view	A kind of three-dimensional view of a model.
Iteration	The process of doing multiple versions of the same thing.
JPEG	A compressed file type for digital images.
KMZ	The Google Earth 3D file format.
Landing	A platform somewhere around the middle of a set of stairs.
Lathed form	3D form created by spinning a 2D shape around a central axis.

Latitude	Geographic location measured by the angular distance north or south from the earth's equator measured through 90 degrees.
Layers	A collection of geometry (including groups and components) that can be made visible or invisible all at once.
Lightness	The file size of a model, or the number of faces and textures used to build it.
Linear inferences	Helper lines that allow a user to work more accurately.
Lossy	Type of compression that occurs with JPEGs. JPEGs compress file size by degrading image quality.
Mapping	Painting surfaces with pictures using 3D software.
Materials	The colors and textures in SketchUp models.
Meshes	Surfaces made out of triangles.
Modifier key	A button on the keyboard that users can push to take a different action than what they are currently doing.
Move/Scale/Rotate/Shear/ Distort Texture mode	Mode to use when manipulating textures in SketchUp; also called Fixed Pin Mode.
Nest	To embed an object in a separate group or component.
Nonphotorealistic rendering (NPR)	Technology that makes objects look hand-drawn or otherwise not like a photograph. SketchUp offers nonphotorealistic rendering.
Nosing	A bump at the leading edge of a tread on a stair.
OBJ	A 3D file format that can be used to send data to Maya.
Orbit	Ability to look at a SketchUp model from every angle; this is accomplished with the Orbit tool.
Ortographic projection	A common way for three-dimensional objects to be drawn so that they can be built.
Outliner	A dialog box that lists all of the groups and components in a model.
Pan	Sliding the model view around the modeling window; this is accomplished by using the Pan tool.
Parapet	The extension of a building's walls that go up a few feet past the roof.
PDF	Portable Document Format, a file format that can be read by almost anyone and that is great for sending information to vector-illustration programs.
Pitch	The angle of a roof surface.
Pitched roof	A roof that isn't flat.
Plan	A top-down, two-dimensional, nonperspectival view of an object or space. Also referred to as a planimetric view.
Point inferences	Small colored shapes that appear when a SketchUp user moves the cursor over specific parts of their model.

Polygonal modelers	Modeling programs that use straight lines and flat surfaces to define everything; within these modelers, even things that look curvy and aren't actually curvy.
Rake	The part of a gabled roof that overhangs the gable.
Raster	Term describing images that are composed of pixels.
Resolution	An image's pixel density.
Rise	The total vertical distance a staircase climbs.
Riser	The part of a step that connects each thread in the vertical direction.
Rubber banding	Drawing edge segments, automatically starting each new one at the end of the previous one.
Ruby scripts	Mini-programs (scripts) written in a computer programming language called Ruby.
Run	The total horizontal distance a staircase takes up.
Scenes	Saved views of a model.
Section	A from-the-side, two-dimensional, nonperspectival view of an object or space. Also referred to as a sectional view.
Section lines	Lines that occur where section planes create section cuts in a model.
Shared layers	Layers that let you automatically place elements on more than one page.
Shearing	Action that keeps the top and bottom edges of an image parallel while making the image lean to the left or right.
Shed roof	A roof that slopes from one side to the other.
Snap settings	Help you line up elements on your page with a grid or with other elements.
Soffit	The underside of an overhanging eave.
Solid models	Models that are not hollow but are dense throughout.
SpaceNavigator	A 3D navigation tool that looks a little like an enormous button that is connected to the computer via a USB cable. It enables the user to orbit, pan, and zoom with subtle movements of his or her hand.
Stacking order	The arrangement of all elements on the same layer in a document, determining which elements appear to be in front of which.
Stretch Texture mode	Mode to use to edit a texture by stretching it to fit the face it is painted on. Also known as Free Pin Mode.
Stringer	A diagonal piece of structure that supports all the steps in a staircase.
Styles	A collection of settings that determines how the geometry appears in a given SketchUp model.
Surface models	Models that are hollow.

Texture-mapping Painting with textures.

Tread An individual step, or the part of a staircase that you step on.

Unshared layers Layers containing elements that exist only on one page.

User interface The visual elements in a software program that the user uses to interact with the software, such as dialog boxes and buttons.

Valley The place where the bottoms of two roof slopes come together.

Vector Term describing images that consist of instructions written in computer code.

Vertices The endpoints of edges.

VRML An older 3D file format that is still used by some programs.

Watermark A graphic element that can be applied either behind or in front of a model to produce certain effects—such as a faint orthographic plan view or a logo.

XSI A 3D file format for people who use Softimage.

Z-fighting The effect resulting from SketchUp trying to decide which face to display by switching back and forth between two faces.

Zoom Getting closer or further away from the model; this is accomplished by using the Zoom tool.

INDEX

E

H

I

J

K

L

P

S

T

U

V

W